STEPHEN WOLFRAM

THE SECOND LAW

STEPHEN WOLFRAM

THE SECOND LAW

The Second Law: Resolving the Mystery of the Second Law of Thermodynamics

Copyright © 2023 Stephen Wolfram, LLC

Wolfram Media, Inc. | wolfram-media.com

ISBN-978-1-57955-083-7 (hardback)
ISBN-978-1-57955-084-4 (ebook)

Science/Physics

Cataloging-in-publication data available at wolfr.am/SecondLaw-cip

For information about permission to reproduce selections from this book, contact permissions@wolfram.com. Sources for photos and archival materials that are not from the author's collection or in the public domain (further details at the end of this book):

pp. 121–122, 124, 219–221, 396–397: McGraw Hill; p. 125: Jim Austin Computer Collection at The Computer Sheds & Neil Barrett Photography; pp. 154–156: *Scientific American*; p. 157: *Physica Scripta*; p. 176: Hachette Book Group; pp. 221–222: Berni Alder; p. 222: World Scientific Publishing Co.; p. 224: American Association for the Advancement of Science; pp. 224–226: AIP Publishing; p. 227: Interscience Publishers (John Wiley & Sons); p. 228: John Wiley & Sons; p. 229: *Scientific American*; p. 230: AIP Publishing; p. 231: AIP Emilio Segrè Visual Archives; pp. 232–233: Los Alamos National Lab; p. 310: Daderot at English Wikipedia; p. 312: Jeremy Norman Collection of Images; pp. 393–394: John Wiley & Sons; pp. 413–420: American Physical Society.

Cover design inspired by B. J. Alder's computer-generated images on the cover of *Statistical Physics*, Berkeley Physics Course—Volume 5 by Fred Reif (McGraw Hill, 1965). See pages 121 and 219.

Typeset with Wolfram Notebooks: wolfram.com/notebooks

Printed by Friesens, Manitoba, Canada. ∞ Acid-free paper. First edition. First printing.

CONTENTS

Preface

It was something I first wondered when I was 12 years old: "How does the Second Law of thermodynamics really work?" Back then I assumed that even though I didn't know, someone did. But a few years later I realized that, actually, no, it was a complicated story, that had never really been figured out.

But now, a little over 50 years after I began to think about it, I have finally come to the point where I believe I really understand the Second Law, its origin and its limitations. This book is my attempt to explain what I've figured out, how I figured it out, and how it relates to the whole 150-year history of the Second Law.

The Second Law is often seen as a quintessential part of the content—and culture—of physics. But actually, as I argue in this book, the essence of the Second Law is something much more general than physics. It's really about a very deep consequence of the interplay between computational irreducibility and the computational boundedness of observers like us. And in a very unexpected development, it turns out that when it comes to physics, this phenomenon leads not only to the Second Law, but also to both of the other two pillars of twentieth century physics: general relativity and quantum mechanics.

If I look back at the past 50 years of my life, I can see my early interest in the Second Law as a crucial seed for my whole journey in science and technology. Most of that journey has revolved around what has emerged as the great intellectual theme of our times: the practical and theoretical development of the computational paradigm. And having spent much of a lifetime creating a tower of science and technology based on this paradigm, it has been exciting to finally return to what started it all for me: the Second Law.

The questions with which I began 50 years ago I think I can now definitively answer. But what I have now done leads to new ideas and new questions—about both limitations and extensions of the Second Law—that I hope will provide fertile seeds for future investigations and applications that build on the Second Law.

As my 50-year journey might suggest, the Second Law is an intellectually subtle topic—that turns out to be fraught with issues and confusions. And as part of trying to clarify what's going on, I have found it helpful to trace the tangled history of the Second Law. So in this book I tell the story of the Second Law, and its origins in the work of some of the great physicists of the nineteenth century. And against this backdrop it's then even more exciting to see just what we can now understand through our new twenty-first century ideas.

For me personally there's a certain satisfaction and sense of closure in finally getting a resolution of the longest-running unanswered intellectual question of my life. But what's still more satisfying is how profound and elegant the resolution turns out to be. And I hope that this book can communicate that elegance, and let others share in the truly remarkable and intellectually enriching experience I've had in my efforts to understand the Second Law over all these years.

Stephen Wolfram

April 2023

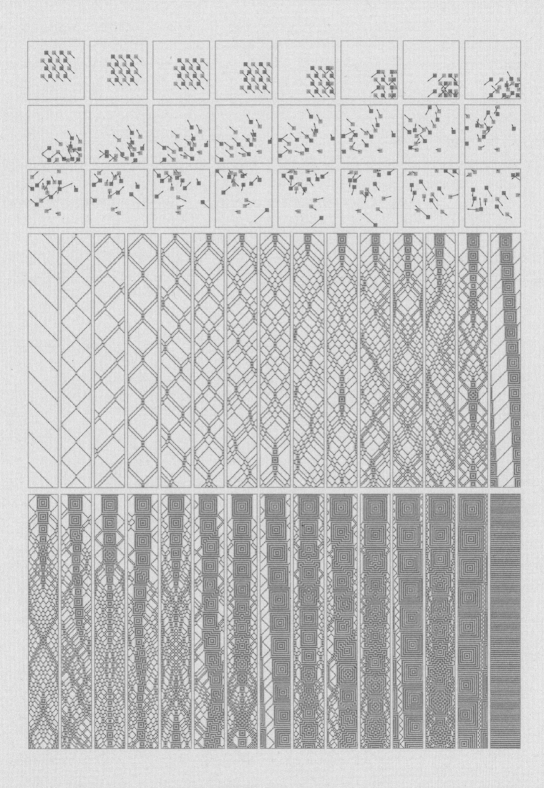

Computational Foundations for the Second Law of Thermodynamics

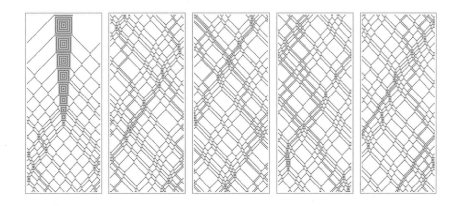

The Mystery of the Second Law

Entropy increases. Mechanical work irreversibly turns into heat. The Second Law of thermodynamics is considered one of the great general principles of physical science. But 150 years after it was first introduced, there's still something deeply mysterious about the Second Law. It almost seems like it's going to be "provably true". But one never quite gets there; it always seems to need something extra. Sometimes textbooks will gloss over everything; sometimes they'll give some kind of "common-sense-but-outside-of-physics argument". But the mystery of the Second Law has never gone away.

Why does the Second Law work? And does it even in fact always work, or is it actually sometimes violated? What does it really depend on? What would be needed to "prove it"?

For me personally the quest to understand the Second Law has been no less than a 50-year story. But back in the 1980s, as I began to explore the computational universe of simple programs, I discovered a fundamental phenomenon that was immediately reminiscent of the Second Law. And in the 1990s I started to map out just how this phenomenon might finally be able to demystify the Second Law. But it is only now—with ideas that have emerged from our Physics Project—that I think I can pull all the pieces together and finally be able to construct a proper framework to explain why—and to what extent—the Second Law is true.

In its usual conception, the Second Law is a law of thermodynamics, concerned with the dynamics of heat. But it turns out that there's a vast generalization of it that's possible. And in fact my key realization is that the Second Law is ultimately just a manifestation of the very same core computational phenomenon that is at the heart of our Physics Project and indeed the whole conception of science that is emerging from our study of the ruliad and the multicomputational paradigm.

It's all a story of the interplay between underlying computational irreducibility and our nature as computationally bounded observers. Other observers—or even our own future technology—might see things differently. But at least for us now the ubiquity of computational irreducibility leads inexorably to the generation of behavior that we—with our computationally bounded nature—will read as "random". We might start from something highly ordered (like gas molecules all in the corner of a box) but soon—at least as far as we're concerned—it will typically seem to "randomize", just as the Second Law implies.

In the twentieth century there emerged three great physical theories: general relativity, quantum mechanics and statistical mechanics, with the Second Law being the defining phenomenon of statistical mechanics. But while there was a sense that statistical mechanics (and in particular the Second Law) should somehow be "formally derivable", general relativity and quantum mechanics seemed quite different. But our Physics Project has changed that picture. And the remarkable thing is that it now seems as if all three of general relativity, quantum mechanics and statistical mechanics are actually derivable, and from the same ultimate foundation: the interplay between computational irreducibility and the computational boundedness of observers like us.

The case of statistical mechanics and the Second Law is in some ways simpler than the other two because in statistical mechanics it's realistic to separate the observer from the system they're observing, while in general relativity and quantum mechanics it's essential that the observer be an integral part of the system. It also helps that phenomena about things like molecules in statistical mechanics are much more familiar to us today than those about atoms of space or branches of multiway systems. And by studying the Second Law we'll be able to develop intuition that we can use elsewhere, say in discussing "molecular" vs. "fluid" levels of description in my recent exploration of the physicalization of the foundations of metamathematics.

The Core Phenomenon of the Second Law

The earliest statements of the Second Law were things like: "Heat doesn't flow from a colder body to a hotter one" or "You can't systematically purely convert heat to mechanical work". Later on there came the somewhat more abstract statement "Entropy tends to increase". But in the end, all these statements boil down to the same idea: that somehow things always tend to get progressively "more random". What may start in an orderly state will—according to the Second Law—inexorably "degrade" to a "randomized" state.

But how general is this phenomenon? Does it just apply to heat and temperature and molecules and things? Or is it something that applies across a whole range of kinds of systems?

The answer, I believe, is that underneath the Second Law there's a very general phenomenon that's extremely robust. And that has the potential to apply to pretty much any kind of system one can imagine.

Here's a longtime favorite example of mine: the rule 30 cellular automaton:

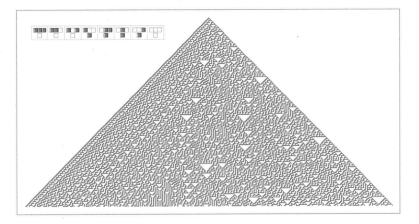

Start from a simple "orderly" state, here containing just a single non-white cell. Then apply the rule over and over again. The pattern that emerges has some definite, visible structure. But many aspects of it "seem random". Just as in the Second Law, even starting from something "orderly", one ends up getting something "random".

But is it "really random"? It's completely determined by the initial condition and rule, and you can always recompute it. But the subtle yet critical point is that if you're just given the output, it can still "seem random" in the sense that no known methods operating purely on this output can find regularities in it.

It's reminiscent of the situation with something like the digits of π. There's a fairly simple algorithm for generating these digits. Yet once generated, the digits on their own seem for practical purposes random.

In studying physical systems there's a long history of assuming that whenever randomness is seen, it somehow comes from outside the system. Maybe it's the effect of "thermal noise" or "perturbations" acting on the system. Maybe it's chaos-theory-style "excavation" of higher-order digits supplied through real-number initial conditions. But the surprising discovery I made in the 1980s by looking at things like rule 30 is that actually no such "external source" is needed: instead, it's perfectly possible for randomness to be generated intrinsically within a system just through the process of applying definite underlying rules.

How can one understand this? The key is to think in computational terms. And ultimately the source of the phenomenon is the interplay between the computational process associated with the actual evolution of the system and the computational processes that our perception of the output of that evolution brings to bear.

We might have thought if a system had a simple underlying rule—like rule 30—then it'd always be straightforward to predict what the system will do. Of course, we could in principle always just run the rule and see what happens. But the question is whether we can expect to "jump ahead" and "find the outcome", with much less computational effort than the actual evolution of the system involves.

And an important conclusion of a lot of science I did in the 1980s and 1990s is that for many systems—presumably including rule 30—it's simply not possible to "jump ahead". And instead the evolution of the system is what I call computationally irreducible—so that it takes an irreducible amount of computational effort to find out what the system does.

Ultimately this is a consequence of what I call the Principle of Computational Equivalence, which states that above some low threshold, systems always end up being equivalent in the sophistication of the computations they perform. And this is why even our brains and our most sophisticated methods of scientific analysis can't "computationally outrun" even something like rule 30, so that we must consider it computationally irreducible.

So how does this relate to the Second Law? It's what makes it possible for a system like rule 30 to operate according to a simple underlying rule, yet to intrinsically generate what seems like random behavior. If we could do all the necessary computationally irreducible work then we could in principle "see through" to the simple rules underneath. But the key point (emphasized by our Physics Project) is that observers like us are computationally bounded in our capabilities. And this means that we're not able to "see through the computational irreducibility"—with the result that the behavior we see "looks random to us".

And in thermodynamics that "random-looking" behavior is what we associate with heat. And the Second Law assertion that energy associated with systematic mechanical work tends to "degrade into heat" then corresponds to the fact that when there's computational irreducibility the behavior that's generated is something we can't readily "computationally see through"—so that it appears random to us.

The Road from Ordinary Thermodynamics

Systems like rule 30 make the phenomenon of intrinsic randomness generation particularly clear. But how do such systems relate to the ones that thermodynamics usually studies? The original formulation of the Second Law involved gases, and the vast majority of its applications even today still concern things like gases.

At a basic level, a typical gas consists of a collection of discrete molecules that interact through collisions. And as an idealization of this, we can consider hard spheres that move according to the standard laws of mechanics and undergo perfectly elastic collisions with each other, and with the walls of a container. Here's an example of a sequence of snapshots from a simulation of such a system, done in 2D:

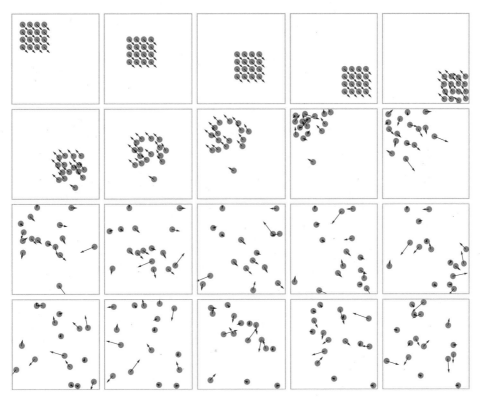

We begin with an organized "flotilla" of "molecules", systematically going in a particular direction (and not touching, to avoid a "Newton's cradle" many-collisions-at-a-time effect). But after these molecules collide with a wall, they quickly start to move in what seem like much more random ways. The original systematic motion is like what happens when one is "doing mechanical work", say moving a solid object. But what we see is that—just as the Second Law implies—this motion is quickly "degraded" into disordered and seemingly random "heat-like" microscopic motion.

Here's a "spacetime" view of the behavior:

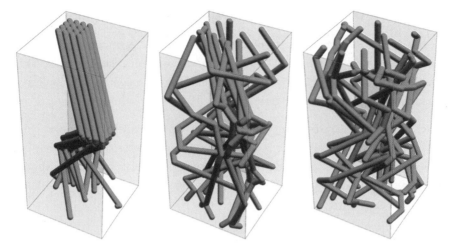

Looking from far away, with each molecule's spacetime trajectory shown as a slightly transparent tube, we get:

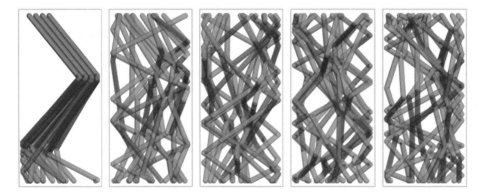

There's already some qualitative similarity with the rule 30 behavior we saw above. But there are many detailed differences. And one of the most obvious is that while rule 30 just has a discrete collection of cells, the spheres in the hard-sphere gas can be at any position. And, what's more, the precise details of their positions can have an increasingly large effect. If two elastic spheres collide perfectly head-on, they'll bounce back the way they came. But as soon as they're even slightly off center they'll bounce back at a different angle, and if they do this repeatedly even the tiniest initial off-centeredness will be arbitrarily amplified:

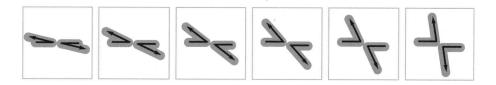

And, yes, this chaos-theory-like phenomenon makes it very difficult even to do an accurate simulation on a computer with limited numerical precision. But does it actually matter to the core phenomenon of randomization that's central to the Second Law?

To begin testing this, let's consider not hard spheres but instead hard squares (where we assume that the squares always stay in the same orientation, and ignore the mechanical torques that would lead to spinning). If we set up the same kind of "flotilla" as before, with the edges of the squares aligned with the walls of the box, then things are symmetrical enough that we don't see any randomization—and in fact the only nontrivial thing that happens is a little Newton's-cradling when the "caravan" of squares hits a wall:

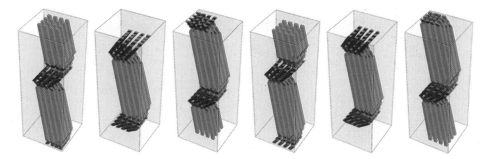

Viewed in "spacetime" we can see the "flotilla" is just bouncing unchanged off the walls:

But remove even a tiny bit of the symmetry—here by roughly doubling the "masses" of some of the squares and "riffling" their positions (which also avoids singular multi-square collisions)—and we get:

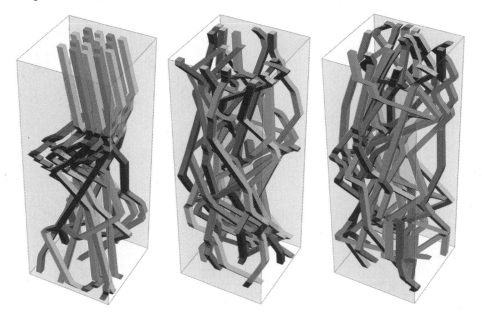

In "spacetime" this becomes

or "from the side":

So despite the lack of chaos-theory-like amplification behavior (or any associated loss of numerical precision in our simulations), there's still rapid "degradation" to a certain apparent randomness.

So how much further can we go? In the hard-square gas, the squares can still be at any location, and be moving at any speed in any direction. As a simpler system (that I happened to first investigate a version of nearly 50 years ago), let's consider a discrete grid in which idealized molecules have discrete directions and are either present or not on each edge:

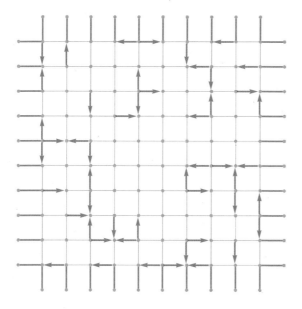

The system operates in discrete steps, with the molecules at each step moving or "scattering" according to the rules (up to rotations)

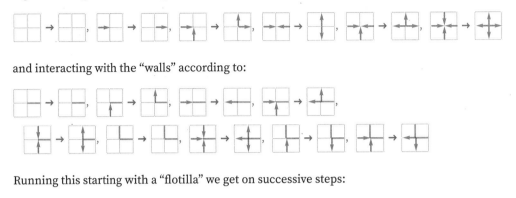

and interacting with the "walls" according to:

Running this starting with a "flotilla" we get on successive steps:

Or, sampling every 10 steps:

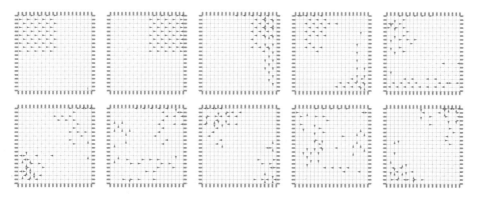

In "spacetime" this becomes (with the arrows tipped to trace out "worldlines")

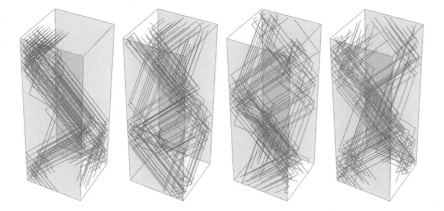

or "from the side":

And again we see at least a certain level of "randomization". With this model we're getting quite close to the setup of something like rule 30. And reformulating this same model we can get even closer. Instead of having "particles" with explicit "velocity directions", consider just having a grid in which an alternating pattern of 2×2 blocks are updated at each step according to

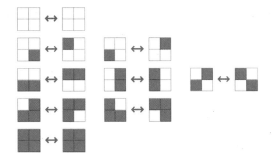

and the "wall" rules

as well as the "rotations" of all these rules. With this "block cellular automaton" setup, "isolated particles" move according to the 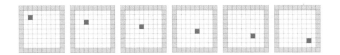 rule like the pieces on a checkerboard:

A "flotilla" of particles—like equal-mass hard squares—has rather simple behavior in the "square enclosure":

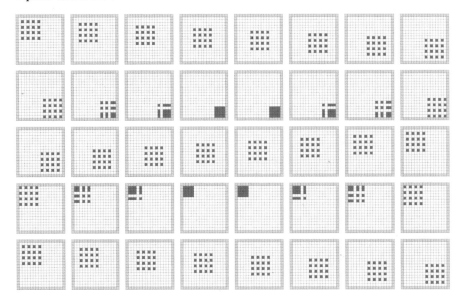

In "spacetime" this is just:

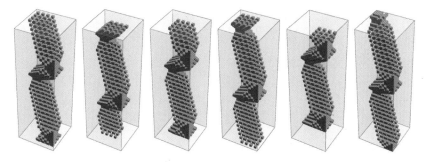

But if we add even a single fixed ("single-cell-of-wall") "obstruction cell" (here at the very center of the box, so preserving reflection symmetry) the behavior is quite different:

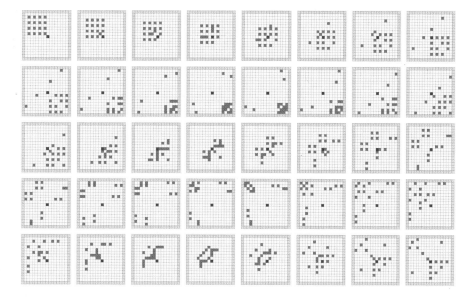

In "spacetime" this becomes (with the "obstruction cell" shown in gray)

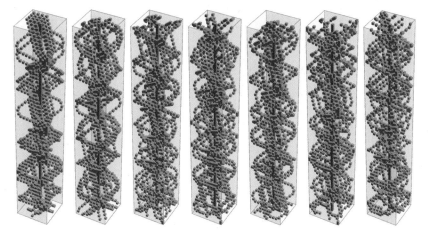

or "from the side" (with the "obstruction" sometimes getting obscured by cells in front):

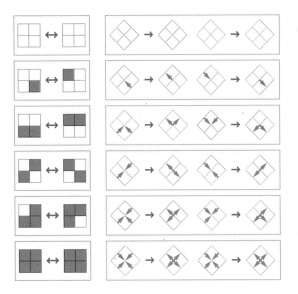

As it turns out, the block cellular automaton model we're using here is actually functionally identical to the "discrete velocity molecules" model we used above, as the correspondence of their rules indicates:

And seeing this correspondence one gets the idea of considering a "rotated container"—which no longer gives simple behavior even without any kind of "interior fixed obstruction cell":

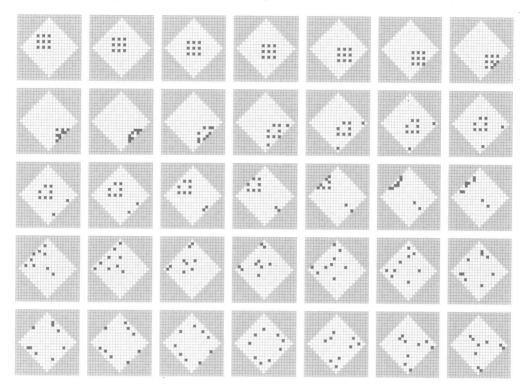

Here's the corresponding "spacetime" view

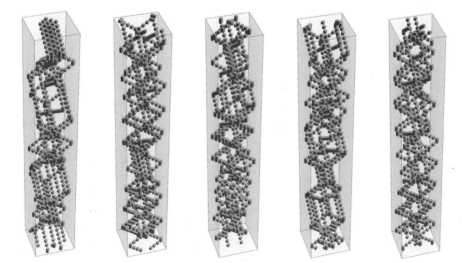

and here's what it looks like "from the side":

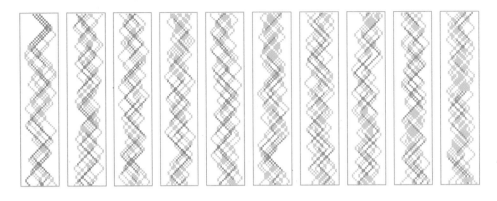

Here's a larger version of the same setup (though no longer with exact symmetry) sampled every 50 steps:

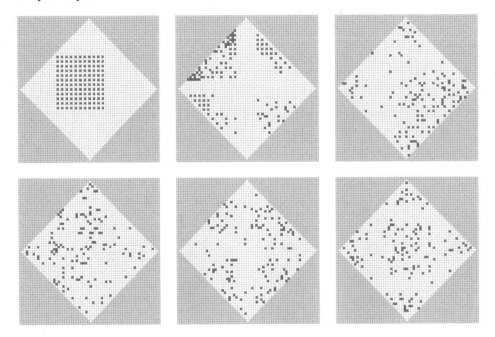

And, yes, it's increasingly looking as if there's intrinsic randomness generation going on, much like in rule 30. But if we go a little further the correspondence becomes even clearer.

The systems we've been looking at so far have all been in 2D. But what if—like in rule 30—we consider 1D? It turns out we can set up very much the same kind of "gas-like" block cellular automata. Though with blocks of size 2 and two possible values for each cell, there's only one viable rule

where in effect the only nontrivial transformation is:

(In 1D we can also make things simpler by not using explicit "walls", but instead just wrapping the array of cells around cyclically.) Here's then what happens with this rule with a few possible initial states:

And what we see is that in all cases the "particles" effectively just "pass through each other" without really "interacting". But we can make there be something closer to "real interactions" by introducing another color, and adding a transformation which effectively introduces a "time delay" to each "crossover" of particles (as an alternative, one can also stay with 2 colors, and use size-3 blocks):

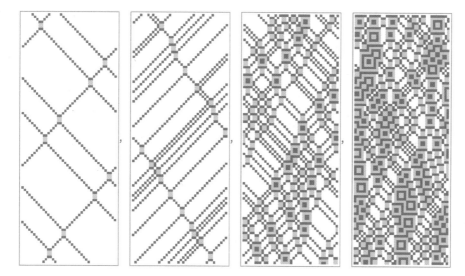

And with this "delayed particle" rule (that, as it happens, I first studied in 1986) we get:

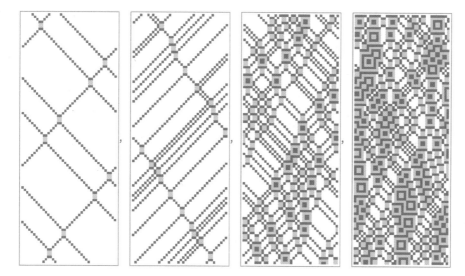

With sufficiently simple initial conditions this still gives simple behavior, such as:

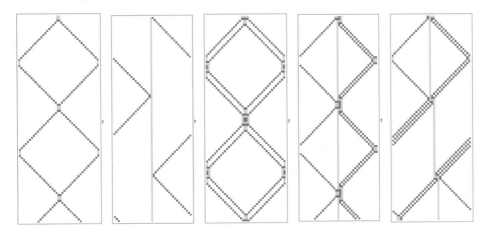

But as soon as one reaches the 121st initial condition (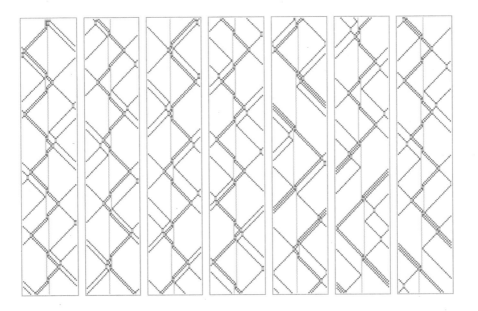) one sees:

(As we'll discuss below, in a finite-size region of the kind we're using, it's inevitable that the pattern eventually repeats, though in the particular case shown it takes 7022 steps.) Here's a slightly larger example, in which there's clearer "progressive degradation" of the initial condition to apparent randomness:

We've come quite far from our original hard-sphere "realistic gas molecules". But there's even further to go. With hard spheres there's built-in conservation of energy, momentum and number of particles. And we don't specifically have these things anymore. But the rule we're using still does have conservation of the number of non-white cells. Dropping this requirement, we can have rules like

which gradually "fill in with particles":

What happens if we just let this "expand into a vacuum", without any "walls"? The behavior is complex. And as is typical when there's computational irreducibility, it's at first hard to know what will happen in the end. For this particular initial condition everything becomes essentially periodic (with period 70) after 979 steps:

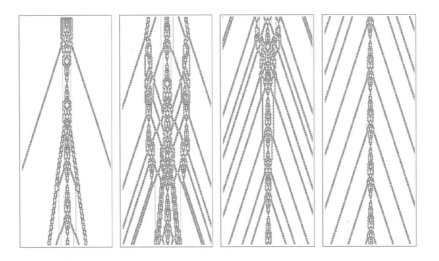

But with a slightly different initial condition, it seems to have a good chance of growing forever:

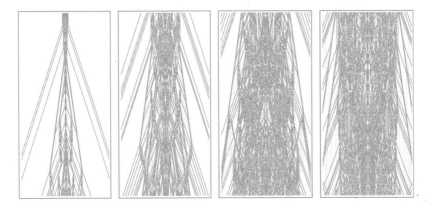

With slightly different rules (that here happen not to be left-right symmetric) we start seeing rapid "expansion into the vacuum"—basically just like rule 30:

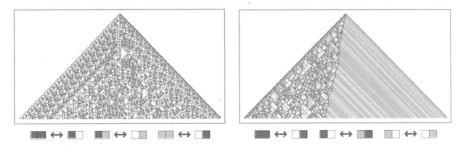

The whole setup here is very close to what it is for rule 30. But there's one more feature we've carried over here from our hard-sphere gas and other models. Just like in standard classical mechanics, every part of the underlying rule is reversible, in the sense that if the rule says that block u goes to block v it also says that block v goes to block u.

Rules like

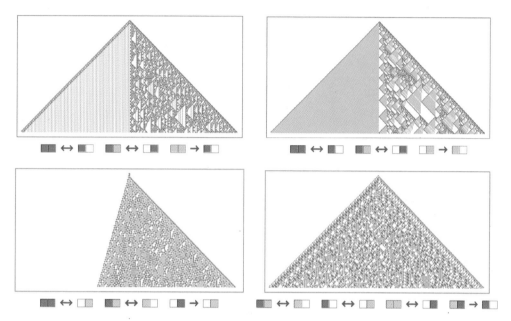

remove this restriction but produce behavior that's qualitatively no different from the reversible rules above.

But now we've got to systems that are basically set up just like rule 30. (They happen to be block cellular automata rather than ordinary ones, but that really doesn't matter.) And, needless to say, being set up like rule 30 it shows the same kind of intrinsic randomness generation that we see in a system like rule 30.

We started here from a "physically realistic" hard-sphere gas model—which we've kept on simplifying and idealizing. And what we've found is that through all this simplification and idealization, the same core phenomenon has remained: that even starting from "simple" or "ordered" initial conditions, complex and "apparently random" behavior is somehow generated, just like it is in typical Second Law behavior.

At the outset we might have assumed that to get this kind of "Second Law behavior" would need at least quite a few features of physics. But what we've discovered is that this isn't the case. And instead we've got evidence that the core phenomenon is much more robust and in a sense purely computational.

Indeed, it seems that as soon as there's computational irreducibility in a system, it's basically inevitable that we'll see the phenomenon. And since from the Principle of Computational Equivalence we expect that computational irreducibility is ubiquitous, the core phenomenon of the Second Law will in the end be ubiquitous across a vast range of systems, from things like hard-sphere gases to things like rule 30.

Reversibility, Irreversibility and Equilibrium

Our typical everyday experience shows a certain fundamental irreversibility. An egg can readily be scrambled. But you can't easily reverse that: it can't readily be unscrambled. And indeed this kind of one-way transition from order to disorder—but not back—is what the Second Law is all about. But there's immediately something mysterious about this. Yes, there's irreversibility at the level of things like eggs. But if we drill down to the level of atoms, the physics we know says there's basically perfect reversibility. So where is the irreversibility coming from? This is a core (and often confused) question about the Second Law, and in seeing how it resolves we will end up face to face with fundamental issues about the character of observers and their relationship to computational irreducibility.

A "particle cellular automaton" like the one from the previous section

has transformations that "go both ways", making its rule perfectly reversible. Yet we saw above that if we start from a "simple initial condition" and then just run the rule, it will "produce increasing randomness". But what if we reverse the rule, and run it backwards? Well, since the rule is reversible, the same thing must happen: we must get increasing randomness. But how can it be that "randomness increases" both going forward in time and going backward? Here's a picture that shows what's going on:

In the middle the system takes on a "simple state". But going either forward or backward it "randomizes". The second half of the evolution we can interpret as typical Second-Law-style "degradation to randomness". But what about the first half? Something unexpected is happening here. From what seems like a "rather random" initial state, the system appears to be "spontaneously organizing itself" to produce—at least temporarily—a simple and "orderly" state. An initial "scrambled" state is spontaneously becoming "unscrambled". In the setup of ordinary thermodynamics, this would be a kind of "anti-thermodynamic" behavior in which what seems like "random heat" is spontaneously generating "organized mechanical work".

So why isn't this what we see happening all the time? Microscopic reversibility guarantees that in principle it's possible. But what leads to the observed Second Law is that in practice we just don't normally end up setting up the kind of initial states that give "anti-thermodynamic" behavior. We'll be talking at length below about why this is. But the basic point is that to do so requires more computational sophistication than we as computationally bounded observers can muster. If the evolution of the system is computationally irreducible, then in effect we have to invert all of that computationally irreducible work to find the initial state to use, and that's not something that we—as computationally bounded observers—can do.

But before we talk more about this, let's explore some of the consequences of the basic setup we have here. The most obvious aspect of the "simple state" in the middle of the picture above is that it involves a big blob of "adjacent particles". So now here's a plot of the "size of the biggest blob that's present" as a function of time starting from the "simple state":

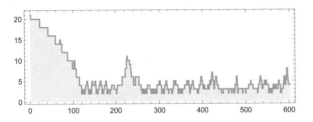

The plot indicates that—as the picture above indicates—the "specialness" of the initial state quickly "decays" to a "typical state" in which there aren't any large blobs present. And if we were watching the system at the beginning of this plot, we'd be able to "use the Second Law" to identify a definite "arrow of time": later times are the ones where the states are "more disordered" in the sense that they only have smaller blobs.

There are many subtleties to all of this. We know that if we set up an "appropriately special" initial state we can get anti-thermodynamic behavior. And indeed for the whole picture above—with its "special initial state"—the plot of blob size vs. time looks like this, with a symmetrical peak "developing" in the middle:

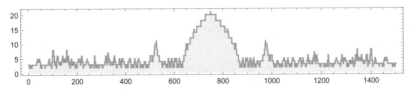

We've "made this happen" by setting up "special initial conditions". But can it happen "naturally"? To some extent, yes. Even away from the peak, we can see there are always little fluctuations: blobs being formed and destroyed as part of the evolution of the system. And if we wait long enough we may see a fairly large blob. Like here's one that forms (and decays) after about 245,400 steps:

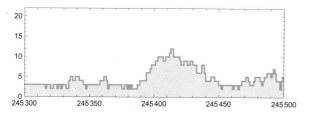

The actual structure this corresponds to is pretty unremarkable:

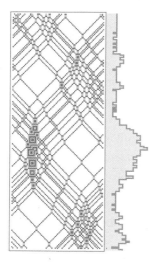

But, OK, away from the "special state", what we see is a kind of "uniform randomness", in which, for example, there's no obvious distinction between forward and backward in time. In thermodynamic terms, we'd describe this as having "reached equilibrium"—a situation in which there's no longer "obvious change".

To be fair, even in "equilibrium", there will always be "fluctuations". But for example in the system we're looking at here, "fluctuations" corresponding to progressively larger blobs tend to occur exponentially less frequently. So it's reasonable to think of there being an "equilibrium state" with certain unchanging "typical properties". And, what's more, that state is the basic outcome from any initial condition. Whatever special characteristics might have been present in the initial state will tend to be degraded away, leaving only the generic "equilibrium state".

One might think that the possibility of such an "equilibrium state" showing "typical behavior" would be a specific feature of microscopically reversible systems. But this isn't the case.

And much as the core phenomenon of the Second Law is actually something computational that's deeper and more general than the specifics of particular physical systems, so also this is true of the core phenomenon of equilibrium. And indeed the presence of what we might call "computational equilibrium" turns out to be directly connected to the overall phenomenon of computational irreducibility.

Let's look again at rule 30. We start it off with different initial states, but in each case it quickly evolves to look basically the same:

Yes, the details of the patterns that emerge depend on the initial conditions. But the point is that the overall form of what's produced is always the same: the system has reached a kind of "computational equilibrium" whose overall features are independent of where it came from. Later, we'll see that the rapid emergence of "computational equilibrium" is characteristic of what I long ago identified as "class 3 systems"—and it's quite ubiquitous to systems with a wide range of underlying rules, microscopically reversible or not.

That's not to say that microscopic reversibility is irrelevant to "Second-Law-like" behavior. In what I called class 1 and class 2 systems the force of irreversibility in the underlying rules is strong enough that it overcomes computational irreducibility, and the systems ultimately evolve not to a "computational equilibrium" that looks random but rather to a definite, predictable end state:

How common is microscopic reversibility? In some types of rules it's basically always there, by construction. But in other cases microscopically reversible rules represent just a subset of possible rules of a given type. For example, for block cellular automata with k colors and blocks of size b, there are altogether $(k^b)^{k^b}$ possible rules, of which $k^b!$ are reversible (i.e. of all mappings between possible blocks, only those that are permutations correspond to reversible rules). Among reversible rules, some—like the particle cellular automaton rule above—are "self-inverses", in the sense that the forward and backward versions of the rule are the same.

But a rule like this is still reversible

and there's still a straightforward backward rule, but it's not exactly the same as the forward rule:

Using the backward rule, we can again construct an initial state whose forward evolution seems "anti-thermodynamic", but the detailed behavior of the whole system isn't perfectly symmetric between forward and backward in time:

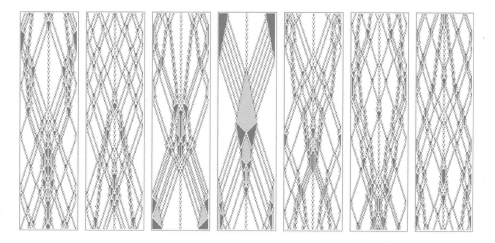

Basic mechanics—like for our hard-sphere gas—are reversible and "self-inverse". But it's known that in particle physics there are small deviations from time reversal invariance, so that the rules are not precisely self-inverse—though they are still reversible in the sense that there's always both a unique successor and a unique predecessor to every state (and indeed in our Physics Project such reversibility is probably guaranteed to exist in the laws of physics assumed by any observer who "believes they are persistent in time").

For block cellular automata it's very easy to determine from the underlying rule whether the system is reversible (just look to see if the rule serves only to permute the blocks). But for something like an ordinary cellular automaton it's more difficult to determine reversibility from the rule (and above one dimension the question of reversibility can actually be undecidable). Among the 256 2-color nearest-neighbor rules there are only 6 reversible examples, and they are all trivial. Among the 134,217,728 3-color nearest-neighbor rules, 1800 are reversible.

Of the 82 of these rules that are self-inverse, all are trivial. But when the inverse rules are different, the behavior can be nontrivial:

Note that unlike with block cellular automata the inverse rule often involves a larger neighborhood than the forward rule. (So, for example, here 396 rules have $r = 1$ inverses, 612 have $r = 2$, 648 have $r = 3$ and 144 have $r = 4$.)

A notable variant on ordinary cellular automata are "second-order" ones, in which the value of a cell depends on its value two steps in the past:

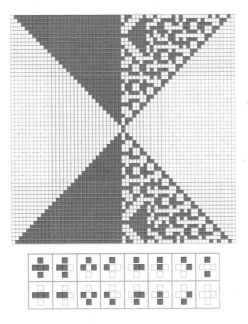

With this approach, one can construct reversible second-order variants of all 256 "elementary cellular automata":

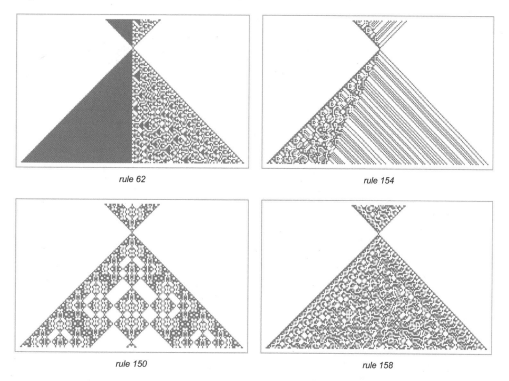

rule 62

rule 154

rule 150

rule 158

Note that such second-order rules are equivalent to 4-color first-order nearest-neighbor rules:

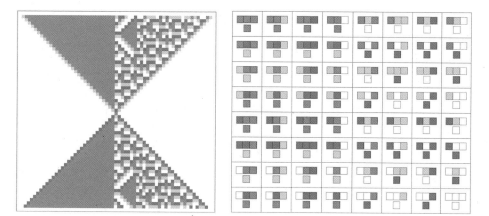

Ergodicity and Global Behavior

Whenever there's a system with deterministic rules and a finite total number of states, it's inevitable that the evolution of the system will eventually repeat. Sometimes the repetition period—or "recurrence time"—will be fairly short

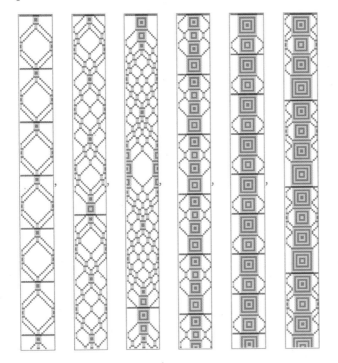

and sometimes it's much longer:

In general we can make a state transition graph that shows how each possible state of the system transitions to another under the rules. For a reversible system this graph consists purely of cycles in which each state has a unique successor and a unique predecessor. For a size-4 version of the system we're studying here, there are a total of $2 \times 3^4 = 162$ possible states (the factor 2 comes from the even/odd "phases" of the block cellular automaton)—and the state transition graph for this system is:

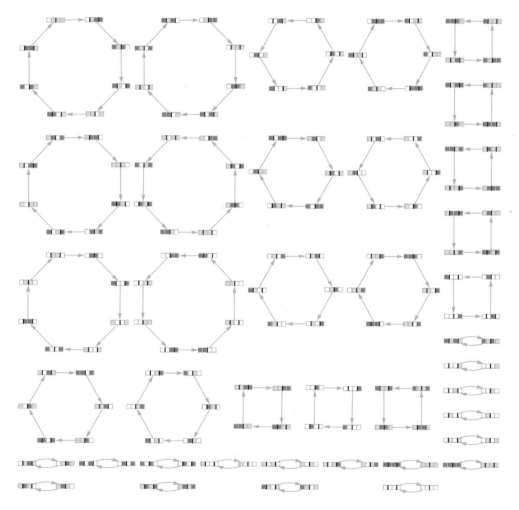

For a non-reversible system—like rule 30—the state transition graph (here shown for sizes 4 and 8) also includes "transient trees" of states that can be visited only once, on the way to a cycle:

In the past one of the key ideas for the origin of Second-Law-like behavior was ergodicity. And in the discrete-state systems we're discussing here the definition of perfect ergodicity is quite straightforward: ergodicity just implies that the state transition graph must consist not of many cycles, but instead purely of one big cycle—so that whatever state one starts from, one's always guaranteed to eventually visit every possible other state.

But why is this relevant to the Second Law? Well, we've said that the Second Law is about "degradation" from "special states" to "typical states". And if one's going to "do the ergodic thing" of visiting all possible states, then inevitably most of the states we'll at least eventually pass through will be "typical".

But on its own, this definitely isn't enough to explain "Second-Law behavior" in practice. In an example like the following, one sees rapid "degradation" of a simple initial state to something "random" and "typical":

But of the $2 \times 3^{80} \approx 10^{38}$ possible states that this system would eventually visit if it were ergodic, there are still a huge number that we wouldn't consider "typical" or "random". For example, just knowing that the system is eventually ergodic doesn't tell one that it wouldn't start off by painstakingly "counting down" like this, "keeping the action" in a tightly organized region:

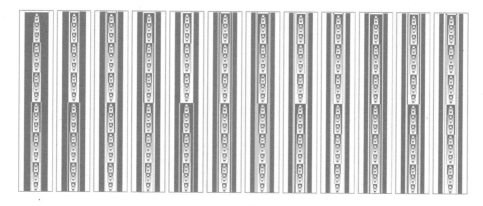

So somehow there's more than ergodicity that's needed to explain the "degradation to randomness" associated with "typical Second-Law behavior". And, yes, in the end it's going to be a computational story, connected to computational irreducibility and its relationship to observers like us. But before we get there, let's talk some more about "global structure", as captured by things like state transition diagrams.

Consider again the size-4 case above. The rules are such that they conserve the number of "particles" (i.e. non-white cells). And this means that the states of the system necessarily break into separate "sectors" for different particle numbers. But even with a fixed number of particles, there are typically quite a few distinct cycles:

The system we're using here is too small for us to be able to convincingly identify "simple" versus "typical" or "random" states, though for example we can see that only a few of the cycles have the simplifying feature of left-right symmetry.

Going to size 6 one begins to get a sense that there are some "always simple" cycles, as well as others that involve "more typical" states:

At size 10 the state transition graph for "4-particle" states has the form

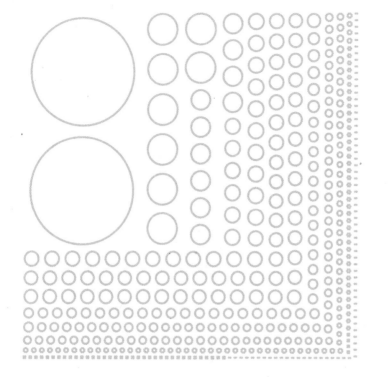

and the longer cycles are:

It's notable that most of the longest ("closest-to-ergodicity") cycles look rather "simple and deliberate" all the way through. The "more typical and random" behavior seems to be reserved here for shorter cycles.

But in studying "Second Law behavior" what we're mostly interested in is what happens from initially orderly states. Here's an example of the results for progressively larger "blobs" in a system of size 30:

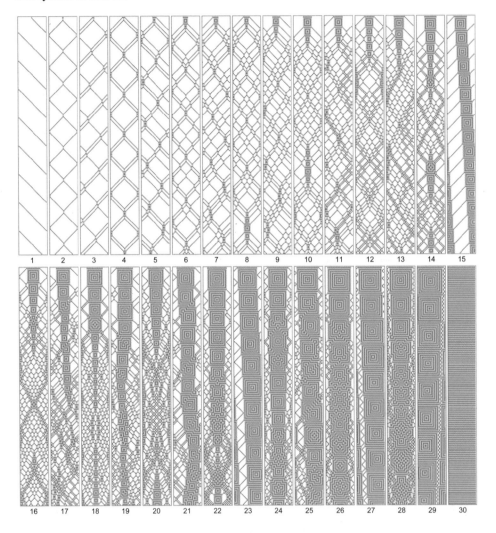

To get some sense of how the "degradation to randomness" proceeds, we can plot how the maximum blob size evolves in each case:

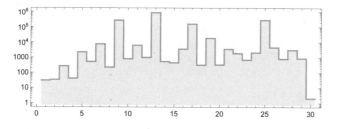

For some of the initial conditions one sees "thermodynamic-like" behavior, though quite often it's overwhelmed by "freezing", fluctuations, recurrences, etc. In all cases the evolution must eventually repeat, but the "recurrence times" vary widely (the longest—for a width-13 initial blob—being 861,930):

Let's look at what happens in these recurrences, using as an example a width-17 initial blob—whose evolution begins:

As the picture suggests, the initial "large blob" quickly gets at least somewhat degraded, though there continue to be definite fluctuations visible:

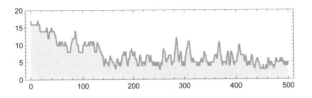

If one keeps going long enough, one reaches the recurrence time, which in this case is 155,150 steps. Looking at the maximum blob size through a "whole cycle" one sees many fluctuations:

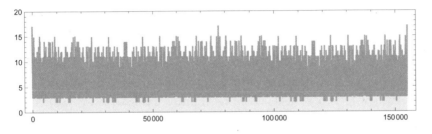

Most are small—as illustrated here with ordinary and logarithmic histograms:

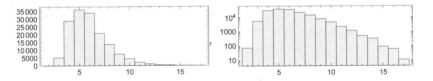

But some are large. And for example at half the full recurrence time there is a fluctuation

that involves an "emergent blob" as wide as in the initial condition—that altogether lasts around 280 steps:

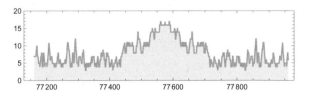

There are also "runner-up" fluctuations with various forms—that reach "blob width 15" and occur more or less equally spaced throughout the cycle:

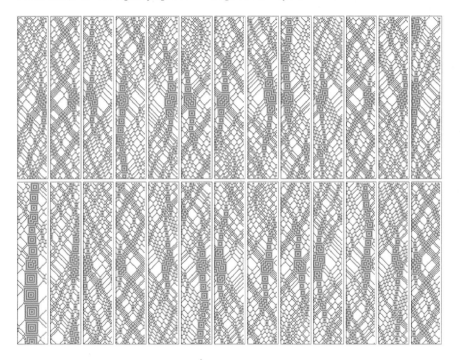

It's notable that clear Second-Law-like behavior occurs even in a size-30 system. But if we go, say, to a size-80 system it becomes even more obvious

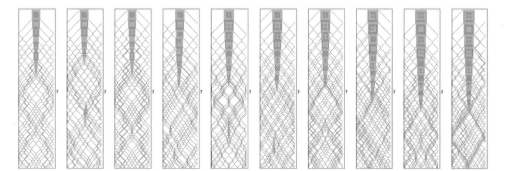

and one sees rapid and systematic evolution towards an "equilibrium state" with fairly small fluctuations:

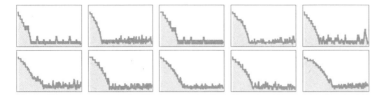

It's worth mentioning again that the idea of "reaching equilibrium" doesn't depend on the particulars of the rule we're using—and in fact it can happen more rapidly in other reversible block cellular automata where there are no "particle conservation laws" to slow things down:

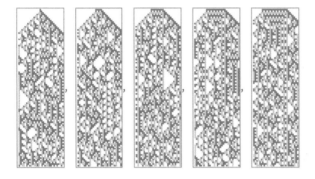

In such rules there also tend to be fewer, longer cycles in the state transition graph, as this comparison for size 6 with the "delayed particle" rule suggests:

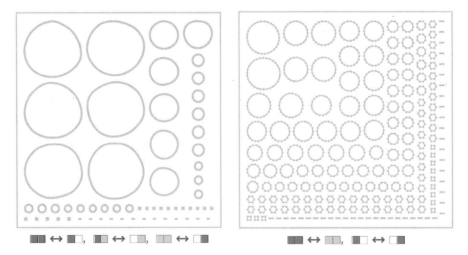

But it's important to realize that the "approach to equilibrium" is its own—computational—phenomenon, not directly related to long cycles and concepts like ergodicity. And indeed, as we mentioned above, it also doesn't depend on built-in reversibility in the rules, so one sees it even in something like rule 30:

How Random Does It Get?

At an everyday level, the core manifestation of the Second Law is the tendency of things to "degrade" to randomness. But just how random is the randomness? One might think that anything that is made by a simple-to-describe algorithm—like the pattern of rule 30 or the digits of π—shouldn't really be considered "random". But for the purpose of understanding our experience of the world what matters is not what is "happening underneath" but instead what our perception of it is. So the question becomes: when we see something produced, say by rule 30 or by π, can we recognize regularities in it or not?

And in practice what the Second Law asserts is that systems will tend to go from states where we can recognize regularities to ones where we cannot. And the point is that this phenomenon is something ubiquitous and fundamental, arising from core computational ideas, in particular computational irreducibility.

But what does it mean to "recognize regularities"? In essence it's all about seeing if we can find succinct ways to summarize what we see—or at least the aspects of what we see that we care about. In other words, what we're interested in is finding some kind of compressed representation of things. And what the Second Law is ultimately about is saying that even if compression works at first, it won't tend to keep doing so.

As a very simple example, let's consider doing compression by essentially "representing our data as a sequence of blobs"—or, more precisely, using run-length encoding to represent sequences of 0s and 1s in terms of lengths of successive runs of identical values. For example, given the data

$$\{0, 0, 1, 1, 1, 1, 1, 1, 1, 1, 0, 0, 0, 1, 1, 1, 1, 1, 1, 1, 1, 1, 1, 0, 0,$$
$$0, 0, 0, 0, 1, 1, 1, 1, 0, 0, 0, 0, 0, 0, 0, 0, 0, 0, 0, 1, 1, 1, 1, 1, 1, 0, 0, 0, 0, 0, 0\}$$

we split into runs of identical values

$$\{\{0, 0\}, \{1, 1, 1, 1, 1, 1, 1, 1\}, \{0, 0, 0\}, \{1, 1, 1, 1, 1, 1, 1, 1, 1, 1\},$$
$$\{0, 0, 0, 0, 0, 0\}, \{1, 1, 1, 1\}, \{0, 0, 0, 0, 0, 0, 0, 0, 0, 0, 0\}, \{1, 1, 1, 1, 1, 1\}, \{0, 0, 0, 0, 0, 0\}\}$$

then as a "compressed representation" just give the length of each run

$$\{2, 8, 3, 10, 6, 4, 11, 6, 6\}$$

which we can finally encode as a sequence of binary numbers with base-3 delimiters:

$$\{1, 0, 2, 1, 0, 0, 0, 2, 1, 1, 2, 1, 0, 1, 0, 2, 1, 1, 0, 2, 1, 0, 0, 2, 1, 0, 1, 1, 2, 1, 1, 0, 2, 1, 1, 0\}$$

"Transforming" our "particle cellular automaton" in this way we get:

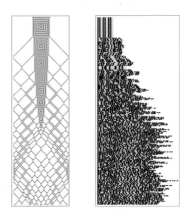

The "simple" initial conditions here are successfully compressed, but the later "random" states are not. Starting from a random initial condition, we don't see any significant compression at all:

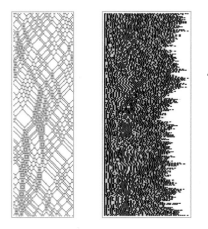

What about other methods of compression? A standard approach involves looking at blocks of successive values on a given step, and asking about the relative frequencies with which different possible blocks occur. But for the particular rule we are discussing here, there's immediately an issue. The rule conserves the total number of non-white cells—so at least for size-1 blocks the frequency of such blocks will always be what it was for the initial conditions.

What about for larger blocks? This gives the evolution of relative frequencies of size-2 blocks starting from the simple initial condition above:

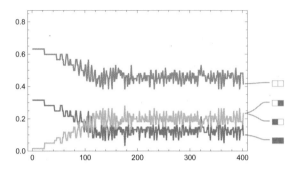

Arranging for exactly half the cells to be non-white, the frequencies of size-2 block converge towards equality:

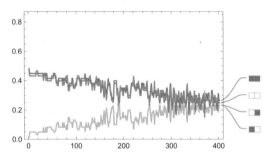

In general, the presence of unequal frequencies for different blocks allows the possibility of compression: much like in Morse code, one just has to use shorter codewords for more frequent blocks. How much compression is ultimately possible in this way can be found by computing $-\sum p_i \log p_i$ for the probabilities p_i of all blocks of a given length, which we see quickly converge to constant "equilibrium" values:

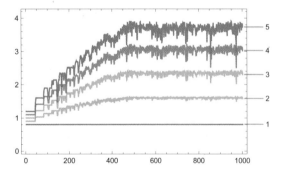

In the end we know that the initial conditions were "simple" and "special". But the issue is whether whatever method we use for compression or for recognizing regularities is able to pick up on this. Or whether somehow the evolution of the system has sufficiently "encoded" the information about the initial condition that it's no longer detectable. Clearly if our "method of compression" involved explicitly running the evolution of the system backwards, then it'd be possible to pick out the special features of the initial conditions. But explicitly running the evolution of the system requires doing lots of computational work.

So in a sense the question is whether there's a shortcut. And, yes, one can try all sorts of methods from statistics, machine learning, cryptography and so on. But so far as one can tell, none of them make any significant progress: the "encoding" associated with the evolution of the system seems to just be too strong to "break". Ultimately it's hard to know for sure that there's no scheme that can work. But any scheme must correspond to running some program. So a way to get a bit more evidence is just to enumerate "possible compression programs" and see what they do.

In particular, we can for example enumerate simple cellular automata, and see whether when run they produce "obviously different" results. Here's what happens for a collection of different cellular automata when they are applied to a "simple initial condition", to states obtained after 20 and 200 steps of evolution according to the particle cellular automaton rule and to an independently random state:

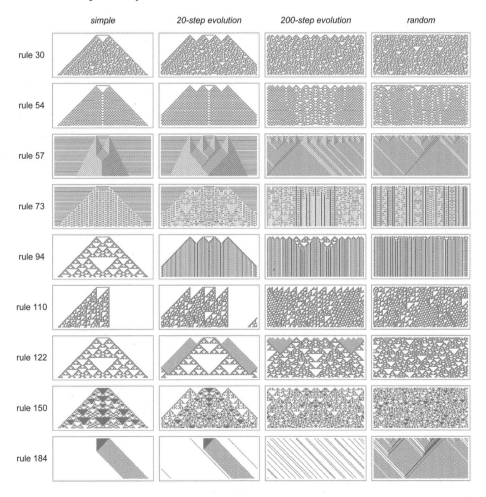

And, yes, in many cases the simple initial condition leads to "obviously different behavior". But there's nothing obviously different about the behavior obtained in the last two cases. Or, in other words, at least programs based on these simple cellular automata don't seem to be able to "decode" the different origins of the third and fourth cases shown here.

What does all this mean? The fundamental point is that there seems to be enough computational irreducibility in the evolution of the system that no computationally bounded observer can "see through it". And so—at least as far as a computationally bounded observer is concerned—"specialness" in the initial conditions is quickly "degraded" to an "equilibrium" state that "seems random". Or, in other words, the computational process of evolution inevitably seems to lead to the core phenomenon of the Second Law.

The Concept of Entropy

"Entropy increases" is a common statement of the Second Law. But what does this mean, especially in our computational context? The answer is somewhat subtle, and understanding it will put us right back into questions of the interplay between computational irreducibility and the computational boundedness of observers.

When it was first introduced in the 1860s, entropy was thought of very much like energy, and was computed from ratios of heat content to temperature. But soon—particularly through work on gases by Boltzmann—there arose a quite different way of computing (and thinking about) entropy: in terms of the log of the number of possible states of a system. Later we'll discuss the correspondence between these different ideas of entropy. But for now let's consider what I view as the more fundamental definition based on counting states.

In the early days of entropy, when one imagined that—like in the cases of the hard-sphere gas—the parameters of the system were continuous, it could be mathematically complex to tease out any kind of discrete "counting of states". But from what we've discussed here, it's clear that the core phenomenon of the Second Law doesn't depend on the presence of continuous parameters, and in something like a cellular automaton it's basically straightforward to count discrete states.

But now we have to get more careful about our definition of entropy. Given any particular initial state, a deterministic system will always evolve through a series of individual states—so that there's always only one possible state for the system, which means the entropy will always be exactly zero. (This is a lot muddier and more complicated when continuous parameters are considered, but in the end the conclusion is the same.)

So how do we get a more useful definition of entropy? The key idea is to think not about individual states of a system but instead about collections of states that we somehow consider "equivalent". In a typical case we might imagine that we can't measure all the detailed positions of molecules in a gas, so we look just at "coarse-grained" states in which we consider, say, only the number of molecules in particular overall bins or blocks.

The entropy can be thought of as counting the number of possible microscopic states of the system that are consistent with some overall constraint—like a certain number of particles in each bin. If the constraint talks specifically about the position of every particle, there'll only be one microscopic state consistent with the constraints, and the entropy will be zero. But if the constraint is looser, there'll often be many possible microscopic states consistent with it, and the entropy we define will be nonzero.

Let's look at this in the context of our particle cellular automaton. Here's a particular evolution, starting from a specific microscopic state, together with a sequence of "coarse grainings" of this evolution in which we keep track only of "overall particle density" in progressively larger blocks:

The very first "coarse graining" here is particularly trivial: all it's doing is to say whether a "particle is present" or not in each cell—or, in other words, it's showing every particle but ignoring whether it's "light" or "dark". But in making this and the other coarse-grained pictures we're always starting from the single "underlying microscopic evolution" that's shown and just "adding coarse graining after the fact".

But what if we assume that all we ever know about the system is a coarse-grained version? Say we look at the "particle-or-not" case. At a coarse-grained level the initial condition just says there are 6 particles present. But it doesn't say if each particle is light or dark, and actually there are $2^6 = 64$ possible microscopic configurations. And the point is that each of these microscopic configurations has its own evolution:

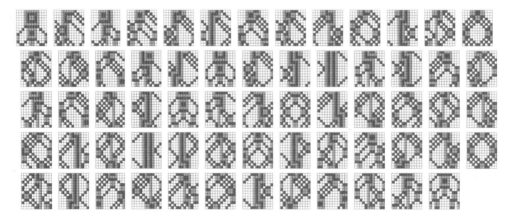

But now we can consider coarse graining things. All 64 initial conditions are—by construction—equivalent under particle-or-not coarse graining:

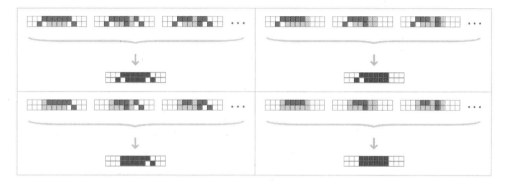

But after just one step of evolution, different initial "microstates" can lead to different coarse-grained evolutions:

In other words, a single coarse-grained initial condition "spreads out" after just one step to several coarse-grained states:

After another step, a larger number of coarse-grained states are possible:

And in general the number of distinct coarse-grained states that can be reached grows fairly rapidly at first, though soon saturates, showing just fluctuations thereafter:

But the coarse-grained entropy is basically just proportional to the log of this quantity, so it too will show rapid growth at first, eventually leveling off at an "equilibrium" value.

The framework of our Physics Project makes it natural to think of coarse-grained evolution as a multicomputational process—in which a given coarse-grained state has not just a single successor, but in general multiple possible successors. For the case we're considering here, the multiway graph representing all possible evolution paths is then:

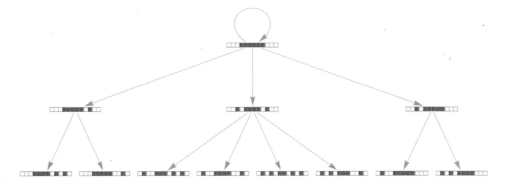

The branching here reflects a spreading out in coarse-grained state space, and an increase in coarse-grained entropy. If we continue longer—so that the system begins to "approach equilibrium"—we'll start to see some merging as well

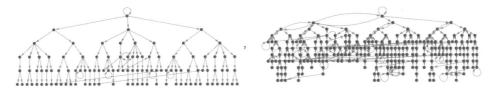

as a less "time-oriented" graph layout makes clear:

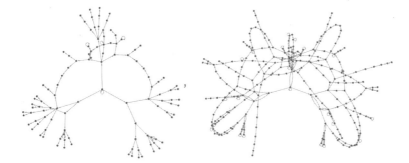

But the important point is that in its "approach to equilibrium" the system in effect rapidly "spreads out" in coarse-grained state space. Or, in other words, the number of possible states of the system consistent with a particular coarse-grained initial condition increases, corresponding to an increase in what one can consider to be the entropy of the system.

There are many possible ways to set up what we might view as "coarse graining". An example of another possibility is to focus on the values of a particular block of cells, and then to ignore the values of all other cells. But it typically doesn't take long for the effects of other cells to "seep into" the block we're looking at:

So what is the bigger picture? The basic point is that insofar as the evolution of each individual microscopic state "leads to randomness", it'll tend to end up in a different "coarse-grained bin". And the result is that even if one starts with a tightly defined coarse-grained description, it'll inevitably tend to "spread out", thereby encompassing more states and increasing the entropy.

In a sense, looking at entropy and coarse graining is just a less direct way to detect that a system tends to "produce effective randomness". And while it may have seemed like a convenient formalism when one was, for example, trying to tease things out from systems with continuous variables, it now feels like a rather indirect way to get at the core phenomenon of the Second Law.

It's useful to understand a few more connections, however. Let's say one's trying to work out the average value of something (say particle density) in a system. What do we mean by "average"? One possibility is that we take an "ensemble" of possible states of the system, then find the average across these. But another possibility is that we instead look at the average across successive states in the evolution of the system. The "ergodic hypothesis" is that the ensemble average will be the same as the time average.

One way this would—at least eventually—be guaranteed is if the evolution of the system is ergodic, in the sense that it eventually visits all possible states. But as we saw above, this isn't something that's particularly plausible for most systems. But it also isn't necessary. Because so long as the evolution of the system is "effectively random" enough, it'll quickly "sample typical states", and give essentially the same averages as one would get from sampling all possible states, but without having to laboriously visit all these states.

How does one tie all this down with rigorous, mathematical-style proofs? Well, it's difficult. And in a first approximation not much progress has been made on this for more than a century. But having seen that the core phenomenon of the Second Law can be reduced to an essentially purely computational statement, we're now in a position to examine this in a different—and I think ultimately very clarifying—way.

Why the Second Law Works

At its core the Second Law is essentially the statement that "things tend to get more random". And in a sense the ultimate driver of this is the surprising phenomenon of computational irreducibility I identified in the 1980s—and the remarkable fact that even from simple initial conditions simple computational rules can generate behavior of great complexity. But there are definitely additional nuances to the story.

For example, we've seen that—particularly in a reversible system—it's always in principle possible to set up initial conditions that will evolve to "magically produce" whatever "simple" configuration we want. And when we say that we generate "apparently random" states, our "analyzer of randomness" can't go in and invert the computational process that generated the states. Similarly, when we talk about coarse-grained entropy and its increase, we're assuming that we're not inventing some elaborate coarse-graining procedure that's specially set up to pick out collections of states with "special" behavior.

But there's really just one principle that governs all these things: that whatever method we have to prepare or analyze states of a system is somehow computationally bounded. This isn't as such a statement of physics. Rather, it's a general statement about observers, or, more specifically, observers like us.

We could imagine some very detailed model for an observer, or for the experimental apparatus they use. But the key point is that the details don't matter. Really all that matters is that the observer is computationally bounded. And it's then the basic computational mismatch between the observer and the computational irreducibility of the underlying system that leads us to "experience" the Second Law.

At a theoretical level we can imagine an "alien observer"—or even an observer with technology from our own future—that would not have the same computational limitations. But the point is that insofar as we are interested in explaining our own current experience, and our own current scientific observations, what matters is the way we as observers are now, with all our computational boundedness. And it's then the interplay between this computational boundedness, and the phenomenon of computational irreducibility, that leads to our basic experience of the Second Law.

At some level the Second Law is a story of the emergence of complexity. But it's also a story of the emergence of simplicity. For the very statement that things go to a "completely random equilibrium" implies great simplification. Yes, if an observer could look at all the details they would see great complexity. But the point is that a computationally bounded observer necessarily can't look at those details, and instead the features they identify have a certain simplicity.

And so it is, for example, that even though in a gas there are complicated underlying molecular motions, it's still true that at an overall level a computationally bounded observer can meaningfully discuss the gas—and make predictions about its behavior— purely in terms of things like pressure and temperature that don't probe the underlying details of molecular motions.

In the past one might have thought that anything like the Second Law must somehow be specific to systems made from things like interacting particles. But in fact the core phenomenon of the Second Law is much more general, and in a sense purely computational, depending only on the basic computational phenomenon of computational irreducibility, together with the fundamental computational boundedness of observers like us.

And given this generality it's perhaps not surprising that the core phenomenon appears far beyond where anything like the Second Law has normally been considered. In particular, in our Physics Project it now emerges as fundamental to the structure of space itself—as well as to the phenomenon of quantum mechanics. For in our Physics Project we imagine that at the lowest level everything in our universe can be represented by some essentially computational structure, conveniently described as a hypergraph whose nodes are abstract "atoms of space". This structure evolves by following rules, whose operation will typically show all sorts of computational irreducibility. But now the question is how observers like us will perceive all this. And the point is that through our limitations we inevitably come to various "aggregate" conclusions about what's going on. It's very much like with the gas laws and their broad applicability to systems involving different kinds of molecules. Except that now the emergent laws are about spacetime and correspond to the equations of general relativity.

But the basic intellectual structure is the same. Except that in the case of spacetime, there's an additional complication. In thermodynamics, we can imagine that there's a system we're studying, and the observer is outside it, "looking in". But when we're thinking about spacetime, the observer is necessarily embedded within it. And it turns out that there's then one additional feature of observers like us that's important. Beyond the statement that we're computationally bounded, it's also important that we assume that we're persistent in time. Yes, we're made of different atoms of space at different moments. But somehow we assume that we have a coherent thread of experience. And this is crucial in deriving our familiar laws of physics.

We'll talk more about it later, but in our Physics Project the same underlying setup is also what leads to the laws of quantum mechanics. Of course, quantum mechanics is notable for the apparent randomness associated with observations made in it. And what we'll see later is that in the end the same core phenomenon responsible for randomness in the Second Law also appears to be what's responsible for randomness in quantum mechanics.

The interplay between computational irreducibility and computational limitations of observers turns out to be a central phenomenon throughout the multicomputational paradigm and its many emerging applications. It's core to the fact that observers can experience computationally reducible laws in all sorts of samplings of the ruliad. And in a sense all of this strengthens the story of the origins of the Second Law. Because it shows that what might have seemed like arbitrary features of observers are actually deep and general, transcending a vast range of areas and applications.

But even given the robustness of features of observers, we can still ask about the origins of the whole computational phenomenon that leads to the Second Law. Ultimately it begins with the Principle of Computational Equivalence, which asserts that systems whose behavior is not obviously simple will tend to be equivalent in their computational sophistication. The Principle of Computational Equivalence has many implications. One of them is computational irreducibility, associated with the fact that "analyzers" or "predictors" of a system cannot be expected to have any greater computational sophistication than the system itself, and so are reduced to just tracing each step in the evolution of a system to find out what it does.

Another implication of the Principle of Computational Equivalence is the ubiquity of computation universality. And this is something we can expect to see "underneath" the Second Law. Because we can expect that systems like the particle cellular automaton—or, for that matter, the hard-sphere gas—will be provably capable of universal computation. Already it's easy to see that simple logic gates can be constructed from configurations of particles, but a full demonstration of computation universality will be considerably more elaborate. And while it'd be nice to have such a demonstration, there's still more that's needed to establish full computational irreducibility of the kind the Principle of Computational Equivalence implies.

As we've seen, there are a variety of "indicators" of the operation of the Second Law. Some are based on looking for randomness or compression in individual states. Others are based on computing coarse grainings and entropy measures. But with the computational interpretation of the Second Law we can expect to translate such indicators into questions in areas like computational complexity theory.

At some level we can think of the Second Law as being a consequence of the dynamics of a system so "encrypting" the initial conditions of a system that no computations available to an "observer" can feasibly "decrypt" it. And indeed as soon as one looks at "inverting" coarse-grained results one is immediately faced with fairly classic NP problems from computational complexity theory. (Establishing NP completeness in a particular case remains challenging, just like establishing computation universality.)

Textbook Thermodynamics

In our discussion here, we've treated the Second Law of thermodynamics primarily as an abstract computational phenomenon. But when thermodynamics was historically first being developed, the computational paradigm was still far in the future, and the only way to identify something like the Second Law was through its manifestations in terms of physical concepts like heat and temperature.

The First Law of thermodynamics asserted that heat was a form of energy, and that overall energy was conserved. The Second Law then tried to characterize the nature of the energy associated with heat. And a core idea was that this energy was somehow incoherently spread among a large number of separate microscopic components. But ultimately thermodynamics was always a story of energy.

But is energy really a core feature of thermodynamics or is it merely "scaffolding" relevant for its historical development and early practical applications? In the hard-sphere gas example that we started from above, there's a pretty clear notion of energy. But quite soon we largely abstracted energy away. Though in our particle cellular automaton we do still have something somewhat analogous to energy conservation: we have conservation of the number of non-white cells.

In a traditional physical system like a gas, temperature gives the average energy per degree of freedom. But in something like our particle cellular automaton, we're effectively assuming that all particles always have the same energy—so there is for example no way to "change the temperature". Or, put another way, what we might consider as the energy of the system is basically just given by the number of particles in the system.

Does this simplification affect the core phenomenon of the Second Law? No. That's something much stronger, and quite independent of these details. But in the effort to make contact with recognizable "textbook thermodynamics", it's useful to consider how we'd add in ideas like heat and temperature.

In our discussion of the Second Law, we've identified entropy with the log of the number states consistent with a constraint. But more traditional thermodynamics involves formulas like $dS = dQ/T$. And it's not hard to see at least roughly where this formula comes from. Q gives total heat content, or "total heat energy" (not worrying about what this is measured relative to, which is what makes it dQ rather than Q). T gives average energy per "degree of freedom" (or, roughly, particle). And this means that Q/T effectively measures something like the "number of particles". But at least in a system like a particle cellular automaton, the number of possible complete configurations is exponential in the number of particles, making its logarithm, the entropy S, roughly proportional to the number of particles, and thus to Q/T. That anything like this argument works depends, though, on being able to discuss things "statistically", which in turn depends on the core phenomenon of the Second Law: the tendency of things to evolve to uniform ("equilibrium") randomness.

When the Second Law was first introduced, there were several formulations given, all initially referencing energy. One formulation stated that "heat does not spontaneously go from a colder body to a hotter". And even in our particle cellular automaton we can see a fairly direct version of this. Our proxy for "temperature" is density of particles. And what we observe is that an initial region of higher density tends to "diffuse" out:

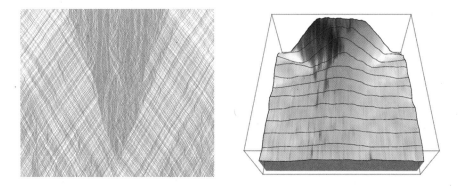

Another formulation of the Second Law talks about the impossibility of systematically "turning heat into mechanical work". At a computational level, the analog of "mechanical work" is systematic, predictable behavior. So what this is saying is again that systems tend to generate randomness, and to "remove predictability".

In a sense this is a direct reflection of computational irreducibility. To get something that one can "harness as mechanical work" one needs something that one can readily predict. But the whole point is that the presence of computational irreducibility makes prediction take an irreducible amount of computational work—that is beyond the capabilities of an "observer like us".

Closely related is the statement that it's not possible to make a perpetual motion machine ("of the second kind", i.e. violating the Second Law), that continually "makes systematic motion" from "heat". In our computational setting this would be like extracting a systematic, predictable sequence of bits from our particle cellular automaton, or from something like rule 30. And, yes, if we had a device that could for example systematically predict rule 30, then it would be straightforward, say, "just to pick out black cells", and effectively to derive a predictable sequence. But computational irreducibility implies that we won't be able to do this, without effectively just directly reproducing what rule 30 does, which an "observer like us" doesn't have the computational capability to do.

Much of the textbook discussion of thermodynamics is centered around the assumption of "equilibrium"—or something infinitesimally close to it—in which one assumes that a system behaves "uniformly and randomly". Indeed, the Zeroth Law of thermodynamics is essentially the statement that "statistically unique" equilibrium can be achieved, which in terms of energy becomes a statement that there is a unique notion of temperature.

Once one has the idea of "equilibrium", one can then start to think of its properties as purely being functions of certain parameters—and this opens up all sorts of calculus-based mathematical opportunities. That anything like this makes sense depends, however, yet again on "perfect randomness as far as the observer is concerned". Because if the observer could notice a difference between different configurations, it wouldn't be possible to treat all of them as just being "in the equilibrium state".

Needless to say, while the intuition of all this is made rather clear by our computational view, there are details to be filled in when it comes to any particular mathematical formulation of features of thermodynamics. As one example, let's consider a core result of traditional thermodynamics: the Maxwell–Boltzmann exponential distribution of energies for individual particles or other degrees of freedom.

To set up a discussion of this, we need to have a system where there can be many possible microscopic amounts of energy, say, associated with some kind of idealized particles. Then we imagine that in "collisions" between such particles energy is exchanged, but the total is always conserved. And the question is how energy will eventually be distributed among the particles.

As a first example, let's imagine that we have a collection of particles which evolve in a series of steps, and that at each step particles are paired up at random to "collide". And, further, let's assume that the effect of the collision is to randomly redistribute energy between the particles, say with a uniform distribution.

We can represent this process using a token-event graph, where the events (indicated here in yellow) are the collisions, and the tokens (indicated here in red) represent states of particles at each step. The energy of the particles is indicated here by the size of the "token dots":

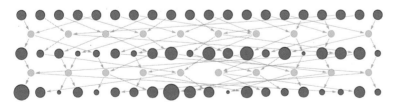

Continuing this a few more steps we get:

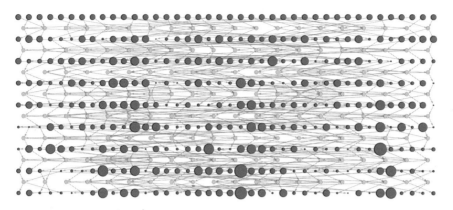

At the beginning we started with all particles having equal energies. But after a number of steps the particles have a distribution of energies—and the distribution turns out to be accurately exponential, just like the standard Maxwell–Boltzmann distribution:

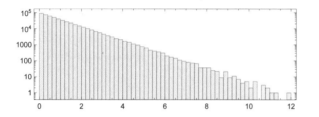

If we look at the distribution on successive steps we see rapid evolution to the exponential form:

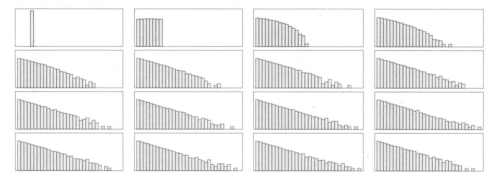

Why we end up with an exponential is not hard to see. In the limit of enough particles and enough collisions, one can imagine approximating everything purely in terms of probabilities (as one does in deriving Boltzmann transport equations, basic SIR models in epidemiology, etc.) Then if the probability for a particle to have energy E is $f(E)$, in every collision once the system has "reached equilibrium" one must have $f(E_1) f(E_2) = f(E_3) f(E_4)$ where $E_1 + E_2 = E_3 + E_4$—and the only solution to this is $f(E) \sim e^{-\beta E}$.

In the example we've just given, there's in effect "immediate mixing" between all particles. But what if we set things up more like in a cellular automaton—with particles only colliding with their local neighbors in space? As an example, let's say we have our particles arranged on a line, with alternating pairs colliding at each step in analogy to a block cellular automaton (the long-range connections represent wraparound of our lattice):

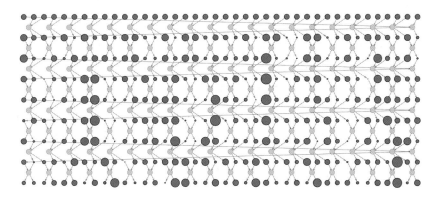

In the picture above we've assumed that in each collision energy is randomly redistributed between the particles. And with this assumption it turns out that we again rapidly evolve to an exponential energy distribution:

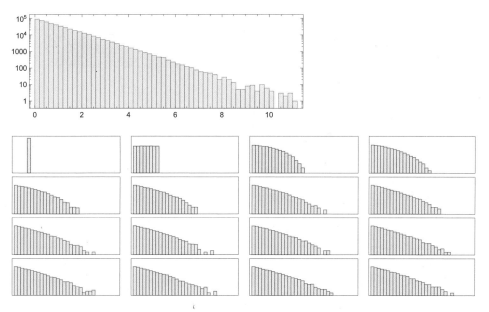

But now that we have a spatial structure, we can display what's going on in more of a cellular automaton style—where here we're showing results for 3 different sequences of random energy exchanges:

And once again, if we run long enough, we eventually get an exponential energy distribution for the particles. But note that the setup here is very different from something like rule 30—because we're continuously injecting randomness from the outside into the system. And as a minimal way to avoid this, consider a model where at each collision the particles get fixed fractions $(1 - \alpha)/2$ and $(1 + \alpha)/2$ of the total energy. Starting with all particles having equal energies, the results are quite trivial—basically just reflecting the successive pairings of particles:

$\alpha = 0$ $\alpha = 0.25$ $\alpha = 0.5$

Here's what happens with energy concentrated into a few particles

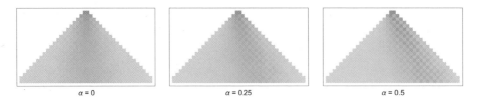

$\alpha = 0$ $\alpha = 0.25$ $\alpha = 0.5$

and with random initial energies:

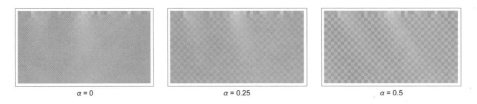

$\alpha = 0$ $\alpha = 0.25$ $\alpha = 0.5$

And in all cases the system eventually evolves to a "pure checkerboard" in which the only particle energies are $(1 - \alpha)/2$ and $(1 + \alpha)/2$. (For $\alpha = 0$ the system corresponds to a discrete version of the diffusion equation.) But if we look at the structure of the system, we can think of it as a continuous block cellular automaton. And as with other cellular automata, there are lots of possible rules that don't lead to such simple behavior.

In fact, all we need do is allow α to depend on the energies E_1 and E_2 of colliding pairs of particles (or, here, the values of cells in each block). As an example, let's take $\alpha(E_1, E_2) = \pm \mathsf{FractionalPart}[\kappa E]$, where E is the total energy of the pair, and the + is used when $E_1 > E_2$:

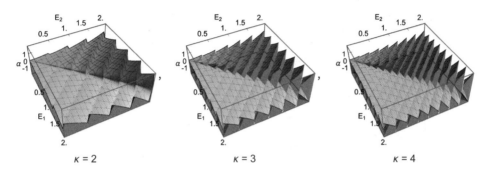

And with this setup we once again often see "rule-30-like behavior" in which effectively quite random behavior is generated even without any explicit injection of randomness from outside (the lower panels start at step 1000):

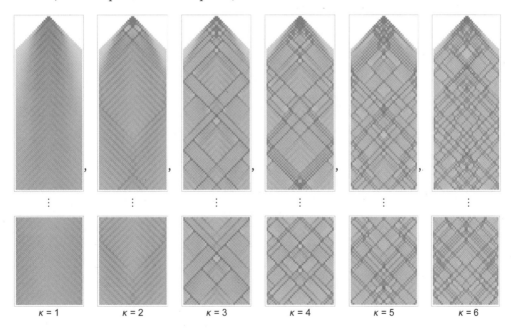

The underlying construction of the rule ensures that total energy is conserved. But what we see is that the evolution of the system distributes it across many elements. And at least if we use random initial conditions

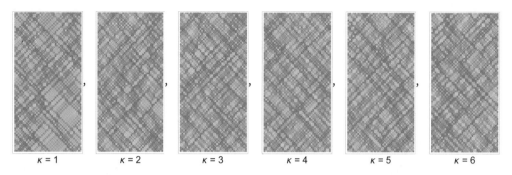

we eventually in all cases see an exponential distribution of energy values (with simple initial conditions it can be more complicated):

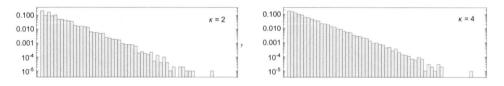

The evolution towards this is very much the same as in the systems above. In a sense it depends only on having a suitably randomized energy-conserving collision process, and it takes only a few steps to go from a uniform initial distribution energy to an accurately exponential one:

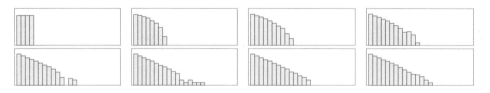

So how does this all work in a "physically realistic" hard-sphere gas? Once again we can create a token-event graph, where the events are collisions, and the tokens correspond to periods of free motion of particles. For a simple 1D "Newton's cradle" configuration, there is an obvious correspondence between the evolution in "spacetime", and the token-event graph:

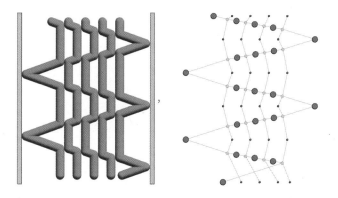

,

But we can do exactly the same thing for a 2D configuration. Indicating the energies of particles by the sizes of tokens we get (excluding wall collisions, which don't affect particle energy)

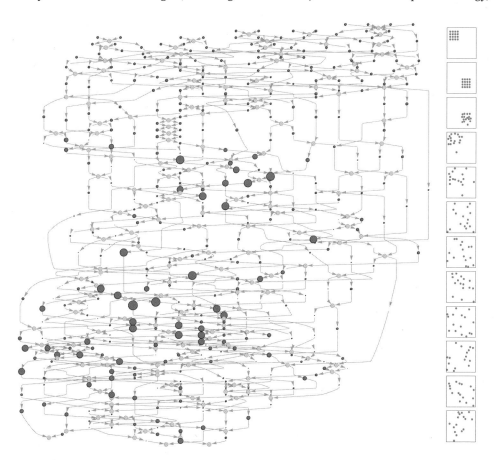

where the "filmstrip" at the side gives snapshots of the evolution of the system. (Note that in this system, unlike the ones above, there aren't definite "steps" of evolution; the collisions just happen "asynchronously" at times determined by the dynamics.)

In the initial condition we're using here, all particles have the same energy. But when we run the system we find that the energy distribution for the particles rapidly evolves to the standard exponential form (though note that here successive panels are "snapshots", not "steps"):

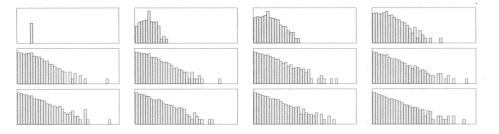

And because we're dealing with "actual particles", we can look not only at their energies, but also at their speeds (related simply by $E = 1/2\, m\, v^2$). When we look at the distribution of speeds generated by the evolution, we find that it has the classic Maxwellian form:

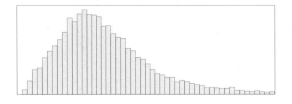

And it's this kind of final or "equilibrium" result that's what's mainly discussed in typical textbooks of thermodynamics. Such books also tend to talk about things like tradeoffs between energy and entropy, and define things like the (Helmholtz) free energy $F = U - T\,S$ (where U is internal energy, T is temperature and S is entropy) that are used in answering questions like whether particular chemical reactions will occur under certain conditions.

But given our discussion of energy here, and our earlier discussion of entropy, it's at first quite unclear how these quantities might relate, and how they can trade off against each other, say in the formula for free energy. But in some sense what connects energy to the standard definition of entropy in terms of the logarithm of the number of states is the Maxwell–Boltzmann distribution, with its exponential form. In the usual physical setup, the Maxwell–Boltzmann distribution is basically $e^{(-E/kT)}$, where T is the temperature, and kT is the average energy.

But now imagine we're trying to figure out whether some process—say a chemical reaction—will happen. If there's an energy barrier, say associated with an energy difference Δ, then according to the Maxwell–Boltzmann distribution there'll be a probability proportional to $e^{(-\Delta/kT)}$ for molecules to have a high enough energy to surmount that barrier. But the next question is how many configurations of molecules there are in which molecules will "try to surmount the barrier". And that's where the entropy comes in. Because if the number of possible configurations is Ω, the entropy S is given by $k \log \Omega$, so that in terms of S, $\Omega = e^{(S/k)}$. But now the "average number of molecules which will surmount the barrier" is roughly given by $e^{(S/k)} e^{(-\Delta/kT)}$, so that in the end the exponent is proportional to $\Delta - TS$, which has the form of the free energy $U - TS$.

This argument is quite rough, but it captures the essence of what's going on. And at first it might seem like a remarkable coincidence that there's a logarithm in the definition of entropy that just "conveniently fits together" like this with the exponential in the Maxwell–Boltzmann distribution. But it's actually not a coincidence at all. The point is that what's really fundamental is the concept of counting the number of possible states of a system. But typically this number is extremely large. And we need some way to "tame" it. We could in principle use some slow-growing function other than log to do this. But if we use log (as in the standard definition of entropy) we precisely get the tradeoff with energy in the Maxwell–Boltzmann distribution.

There is also another convenient feature of using log. If two systems are independent, one with Ω_1 states, and the other with Ω_2 states, then a system that combines these (without interaction) will have Ω_1, Ω_2 states. And if $S = k \log \Omega$, then this means that the entropy of the combined state will just be the sum $S_1 + S_2$ of the entropies of the individual states. But is this fact actually "fundamentally independent" of the exponential character of the Maxwell–Boltzmann distribution? Well, no. Or at least it comes from the same mathematical idea. Because it's the fact that in equilibrium the probability $f(E)$ is supposed to satisfy $f(E_1) f(E_2) = f(E_3) f(E_4)$ when $E_1 + E_2 = E_3 + E_4$ that makes $f(E)$ have its exponential form. In other words, both stories are about exponentials being able to connect additive combination of one quantity with multiplicative combination of another.

Having said all this, though, it's important to understand that you don't need energy to talk about entropy. The concept of entropy, as we've discussed, is ultimately a computational concept, quite independent of physical notions like energy. In many textbook treatments of thermodynamics, energy and entropy are in some sense put on a similar footing. The First Law is about energy. The Second Law is about entropy. But what we've seen here is that energy is really a concept at a different level from entropy: it's something one gets to "layer on" in discussing physical systems, but it's not a necessary part of the "computational essence" of how things work.

(As an extra wrinkle, in the case of our Physics Project—as to some extent in traditional general relativity and quantum mechanics—there are some fundamental connections between energy and entropy. In particular—related to what we'll discuss below—the number of possible discrete configurations of spacetime is inevitably related to the "density" of events, which defines energy.)

Towards a Formal Proof of the Second Law

It would be nice to be able to say, for example, that "using computation theory, we can prove the Second Law". But it isn't as simple as that. Not least because, as we've seen, the validity of the Second Law depends on things like what "observers like us" are capable of. But we can, for example, formulate what the outline of a proof of the Second Law could be like, though to give a full formal proof we'd have to introduce a variety of "axioms" (essentially about observers) that don't have immediate foundations in existing areas of mathematics, physics or computation theory.

The basic idea is that one imagines a state S of a system (which could just be a sequence of values for cells in something like a cellular automaton). One considers an "observer function" Θ which, when applied to the state S, gives a "summary" of S. (A very simple example would be the run-length encoding that we used above.) Now we imagine some "evolution function" Ξ that is applied to S. The basic claim of the Second Law is that the "sizes" normally satisfy the inequality $\Theta[\Xi[S]] \geq \Theta[S]$, or in other words, that "compression by the observer" is less effective after the evolution of system, in effect because the state of the system has "become more random", as our informal statement of the Second Law suggests.

What are the possible forms of Θ and Ξ? It's slightly easier to talk about Ξ, because we imagine that this is basically any not-obviously-trivial computation, run for an increasing number of steps. It could be repeated application of a cellular automaton rule, or a Turing machine, or any other computational system. We might represent an individual step by an operator ξ, and say that in effect $\Xi = \xi^t$. We can always construct ξ^t by explicitly applying ξ successively t times. But the question of computational irreducibility is whether there's a shortcut way to get to the same result. And given any specific representation of ξ^t (say, rather prosaically, as a Boolean circuit), we can ask how the size of that representation grows with t.

With the current state of computation theory, it's exceptionally difficult to get definitive general results about minimal sizes of ξ^t, though in sufficiently small cases it's possible to determine this "experimentally", essentially by exhaustive search. But there's an increasing amount of at least circumstantial evidence that for many kinds of systems, one can't do much better than explicitly constructing ξ^t, as the phenomenon of computational irreducibility suggests. (One can imagine "toy models", in which ξ corresponds to some very simple computational process—like a finite automaton—but while this likely allows one to prove things, it's not at all clear how useful or representative any of the results will be.)

OK, so what about the "observer function" Θ? For this we need some kind of "observer theory", that characterizes what observers—or, at least "observers like us"—can do, in the same kind of way that standard computation theory characterizes what computational systems can do. There are clearly some features Θ must have. For example, it can't involve unbounded amounts of computation. But realistically there's more than that. Somehow the role of observers is to take all the details that might exist in the "outside world", and reduce or compress these to some "smaller" representation that can "fit in the mind of the observer", and allow the observer to "make decisions" that abstract from the details of the outside world whatever specifics the observer "cares about". And—like a construction such as a Turing machine—one must in the end have some way of building up "possible observers" from something like basic primitives.

Needless to say, even given primitives—or an axiomatic foundation—for Ξ and Θ, things are not straightforward. For example, it's basically inevitable that many specific questions one might ask will turn out to be formally undecidable. And we can't expect (particularly as we'll see later) that we'll be able to show that the Second Law is "just true". It'll be a statement that necessarily involves qualifiers like "typically". And if we ask to characterize "typically" in terms, say, of "probabilities", we'll be stuck in a kind of recursive situation of having to define probability measures in terms of the very same constructs we're starting from.

But despite these difficulties in making what one might characterize as general abstract statements, what our computational formulation achieves is to provide a clear intuitive guide to the origin of the Second Law. And from this we can in particular construct an infinite range of specific computational experiments that illustrate the core phenomenon of the Second Law, and give us more and more understanding of how the Second Law works, and where it conceptually comes from.

Maxwell's Demon and the Character of Observers

Even in the very early years of the formulation of the Second Law, James Clerk Maxwell already brought up an objection to its general applicability, and to the idea that systems "always become more random". He imagined that a box containing gas molecules had a barrier in the middle with a small door controlled by a "demon" who could decide on a molecule-by-molecule basis which molecules to let through in each direction. Maxwell suggested that such a demon should readily be able to "sort" molecules, thereby reversing any "randomness" that might be developing.

As a very simple example, imagine that at the center of our particle cellular automaton we insert a barrier that lets particles pass from left to right but not the reverse. (We also add "reflective walls" on the two ends, rather than having cyclic boundary conditions.)

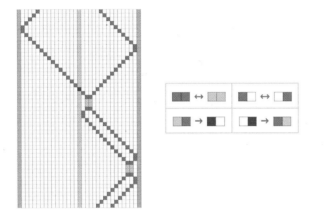

Unsurprisingly, after a short while, all the particles have collected on one side of the barrier, rather than "coming to equilibrium" in a "uniform random distribution" across the system:

Over the past century and a half (and even very recently) a whole variety of mechanical ratchets, molecular switches, electrical diodes, noise-reducing signal processors and other devices have been suggested as at least conceptually practical implementations of Maxwell's demon. Meanwhile, all kinds of objections to their successful operation have been raised. "The demon can't be made small enough"; "The demon will heat up and stop working"; "The demon will need to reset its memory, so has to be fundamentally irreversible"; "The demon will inevitably randomize things when it tries to sense molecules"; etc.

So what's true? It depends on what we assume about the demon—and in particular to what extent we suppose that the demon needs to be following the same underlying laws as the system it's operating on. As a somewhat extreme example, let's imagine trying to "make a demon out of gas molecules". Here's an attempt at a simple model of this in our particle cellular automaton:

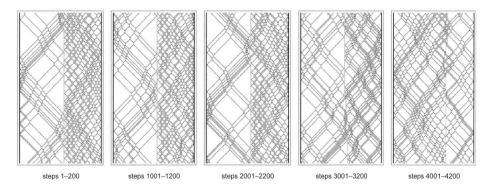

| steps 1–200 | steps 1001–1200 | steps 2001–2200 | steps 3001–3200 | steps 4001–4200 |

For a while we successfully maintain a "barrier". But eventually the barrier succumbs to the same "degradation" processes as everything else, and melts away. Can we do better?

Let's imagine that "inside the barrier" (AKA "demon") there's "machinery" that whenever the barrier is "buffeted" in a given way "puts up the right kind of armor" to "protect it" from that kind of buffeting. Assuming our underlying system is for example computation universal, we should at some level be able to "implement any computation we want". (What needs to be done is quite analogous to cellular automata that successfully erase up to finite levels of "noise".)

But there's a problem. In order to "protect the barrier" we have to be able to "predict" how it will be "attacked". Or, in other words, our barrier (or demon) will have to be able to systematically determine what the outside system is going to do before it does it. But if the behavior of the outside system is computationally irreducible this won't in general be possible. So in the end the criterion for a demon like this to be impossible is essentially the same as the criterion for Second Law behavior to occur in the first place: that the system we're looking at is computationally irreducible.

There's a bit more to say about this, though. We've been talking about a demon that's trying to "achieve something fairly simple", like maintaining a barrier or a "one-way membrane".

But what if we're more flexible in what we consider the objective of the demon to be? And even if the demon can't achieve our original "simple objective" might there at least be some kind of "useful sorting" that it can do?

Well, that depends on what we imagine constitutes "useful sorting". The system is always following its rules to do something. But probably it's not something we consider "useful sorting". But what would count as "useful sorting"? Presumably it's got to be something that an observer will "notice", and more than that, it should be something that has "done some of the job of decision making" ahead of the observer. In principle a sufficiently powerful observer might be able to "look inside the gas" and see what the results of some elaborate sorting procedure would be. But the point is for the demon to just make the sorting happen, so the job of the observer becomes essentially trivial.

But all of this then comes back to the question of what kind of thing an observer might want to observe. In general one would like to be able to characterize this by having an "observer theory" that provides a metatheory of possible observers in something like the kind of way that computation theory and ideas like Turing machines provide a metatheory of possible computational systems.

So what really is an observer, or at least an observer like us? The most crucial feature seems to be that the observer is always ultimately some kind of "finite mind" that takes all the complexity of the world and extracts from it just certain "summary features" that are relevant to the "decisions" it has to make. (Another crucial feature seems to be that the observer can consistently view themselves as being "persistent".) But we don't have to go all the way to a sophisticated "mind" to see this picture in operation. Because it's already what's going on not only in something like perception but also in essentially anything we'd usually call "measurement".

For example, imagine we have a gas containing lots of molecules. A standard measurement might be to find the pressure of the gas. And in doing such a measurement, what's happening is that we're reducing the information about all the detailed motions of individual molecules, and just summarizing it by a single aggregate number that is the pressure.

How do we achieve this? We might have a piston connected to the box of gas. And each time a molecule hits the piston it'll push it a little. But the point is that in the end the piston moves only as a whole. And the effects of all the individual molecules are aggregated into that overall motion.

At a microscopic level, any actual physical piston is presumably also made out of molecules. But unlike the molecules in the gas, these molecules are tightly bound together to make the piston solid. Every time a gas molecule hits the surface of the piston, it'll transfer some momentum to a molecule in the piston, and there'll be some kind of tiny deformation wave that goes through the piston. To get a "definitive pressure measurement"—based on definitive motion of the piston as a whole—that deformation wave will somehow have to disappear. And in making a theory of the "piston as observer" we'll typically ignore the physical details, and idealize things by saying that the piston moves only as a whole.

But ultimately if we were to just look at the system "dispassionately", without knowing the "intent" of the piston, we'd just see a bunch of molecules in the gas, and a bunch of molecules in the piston. So how would we tell that the piston is "acting as an observer"? In some ways it's a rather circular story. If we assume that there's a particular kind of thing an observer wants to measure, then we can potentially identify parts of a system that "achieve that measurement". But in the abstract we don't know what an observer "wants to measure". We'll always see one part of a system affecting another. But is it "achieving measurement" or not?

To resolve this, we have to have some kind of metatheory of the observer: we have to be able to say what kinds of things we're going to count as observers and what not. And ultimately that's something that must inevitably devolve to a rather human question. Because in the end what we care about is what we humans sense about the world, which is what, for example, we try to construct science about.

We could talk very specifically about the sensory apparatus that we humans have—or that we've built with technology. But the essence of observer theory should presumably be some kind of generalization of that. Something that recognizes fundamental features—like computational boundedness—of us as entities, but that does not depend on the fact that we happen to use sight rather than smell as our most important sense.

The situation is a bit like the early development of computation theory. Something like a Turing machine was intended to define a mechanism that roughly mirrored the computational capabilities of the human mind, but that also provided a "reasonable generalization" that covered, for example, machines one could imagine building. Of course, in that particular case the definition that was developed proved extremely useful, being, it seems, of just the right generality to cover computations that can occur in our universe—but not beyond.

And one might hope that in the future observer theory would identify a similarly useful definition for what a "reasonable observer" can be. And given such a definition, we will, for example, be in position to further tighten up our characterization of what the Second Law might say.

It may be worth commenting that in thinking about an observer as being an "entity like us" one of the immediate attributes we might seek is that the observer should have some kind of "inner experience". But if we're just looking at the pattern of molecules in a system, how do we tell where there's an "inner experience" happening? From the outside, we presumably ultimately can't. And it's really only possible when we're "on the inside". We might have scientific criteria that tell us whether something can reasonably support an inner experience. But to know if there actually is an inner experience "going on" we basically have to be experiencing it. We can't make a "first-principles" objective theory; we just have to posit that such-and-such part of the system is representing our subjective experience.

Of course, that doesn't mean that there can't still be very general conclusions to be drawn. Because it can still be—as it is in our Physics Project and in thinking about the ruliad—that it takes knowing only rather basic features of "observers like us" to be able to make very general statements about things like the effective laws we will experience.

The Heat Death of the Universe

It didn't take long after the Second Law was first proposed for people to start talking about its implications for the long-term evolution of the universe. If "randomness" (for example as characterized by entropy) always increases, doesn't that mean that the universe must eventually evolve to a state of "equilibrium randomness", in which all the rich structures we now see have decayed into "random heat"?

There are several issues here. But the most obvious has to do with what observer one imagines will be experiencing that future state of the universe. After all, if the underlying rules which govern the universe are reversible, then in principle it will always be possible to go back from that future "random heat" and reconstruct from it all the rich structures that have existed in the history of the universe.

But the point of the Second Law as we've discussed it is that at least for computationally bounded observers like us that won't be possible. The past will always in principle be determinable from the future, but it will take irreducibly much computation to do so—and vastly more than observers like us can muster.

And along the same lines, if observers like us examine the future state of the universe we won't be able to see that there's anything special about it. Even though it came from the "special state" that is the current state of our universe, we won't be able to tell it from a "typical" state, and we'll just consider it "random".

But what if the observers evolve with the evolution of the universe? Yes, to us today that future configuration of particles may just "look random". But in actuality, it has rich computational content that there's no reason to assume a future observer will not find in some way or another significant. Indeed, in a sense the longer the universe has been around, the larger the amount of irreducible computation it will have done. And, yes, observers like us today might not care about most of what comes out of that computation. But in principle there are features of it that could be mined to inform the "experience" of future observers.

At a practical level, our basic human senses pick out certain features on certain scales. But as technology progresses, it gives us ways to pick out much more, on much finer scales. A century ago we couldn't realistically pick out individual atoms or individual photons; now we routinely can. And what seemed like "random noise" just a few decades ago is now often known to have specific, detailed structure.

There is, however, a complex tradeoff. A crucial feature of observers like us is that there is a certain coherence to our experience; we sample little enough about the world that we're able to turn it into a coherent thread of experience. But the more an observer samples, the more difficult this will become. So, yes, a future observer with vastly more advanced technology might successfully be able to sample lots of details of the future universe. But to do that, the observer will have to lose some of their own coherence, and ultimately we won't even be able to identify that future observer as "coherently existing" at all.

The usual "heat death of the universe" refers to the fate of matter and other particles in the universe. But what about things like gravity and the structure of spacetime? In traditional physics, that's been a fairly separate question. But in our Physics Project everything is ultimately described in terms of a single abstract structure that represents both space and everything in it. And we can expect that the evolution of this whole structure then corresponds to a computationally irreducible process.

The basic setup is at its core just like what we've seen in our general discussion of the Second Law. But here we're operating at the lowest level of the universe, so the irreducible progression of computation can be thought of as representing the fundamental inexorable passage of time. As time moves forward, therefore, we can generally expect "more randomness" in the lowest-level structure of the universe.

But what will observers perceive? There's considerable trickiness here—particularly in connection with quantum mechanics—that we'll discuss later. In essence, the point is that there are many paths of history for the universe, that branch and merge—and observers sample certain collections of paths. And for example on some paths the computations may simply halt, with no further rules applying—so that in effect "time stops", at least for observers on those paths. It's a phenomenon that can be identified with spacetime singularities, and with what happens inside (at least certain) black holes.

So does this mean that the universe might "just stop", in effect ending with a collection of black holes? It's more complicated than that. Because there are always other paths for observers to follow. Some correspond to different quantum possibilities. But ultimately what we imagine is that our perception of the universe is a sampling from the whole ruliad—the limiting entangled structure formed by running all abstractly possible computations forever. And it's a feature of the construction of the ruliad that it's infinite. Individual paths in it can halt, but the whole ruliad goes on forever.

So what does this mean about the ultimate fate of the universe? Much like the situation with heat death, specific observers may conclude that "nothing interesting is happening anymore". But something always will be happening, and in fact that something will represent the accumulation of larger and larger amounts of irreducible computation. It won't be possible for an observer to encompass all this while still themselves "remaining coherent". But as we'll discuss later there will inexorably be pockets of computational reducibility for which coherent observers can exist, although what those observers will perceive is likely to be utterly incoherent with anything that we as observers now perceive.

The universe does not fundamentally just "descend into randomness". And indeed all the things that exist in our universe today will ultimately be encoded in some way forever in the detailed structure that develops. But what the core phenomenon of the Second Law suggests is that at least many aspects of that encoding will not be accessible to observers like us. The future of the universe will transcend what we so far "appreciate", and will require a redefinition of what we consider meaningful. But it should not be "taken for dead" or dismissed as being just "random heat". It's just that to find what we consider interesting, we may in effect have to migrate across the ruliad.

Traces of Initial Conditions

The Second Law gives us the expectation that so long as we start from "reasonable" initial conditions, we should always evolve to some kind of "uniformly random" configuration that we can view as a "unique equilibrium state" that's "lost any meaningful memory" of the initial conditions. But now that we've got ways to explore the Second Law in specific, simple computational systems, we can explicitly study the extent to which this expectation is upheld. And what we'll find is that even though as a general matter it is, there can still be exceptions in which traces of initial conditions can be preserved at least long into the evolution.

Let's look again at our "particle cellular automaton" system. We saw above that the evolution of an initial "blob" (here of size 17 in a system with 30 cells) leads to configurations that typically look quite random:

But what about other initial conditions? Here are some samples of what happens:

In some cases we again get what appears to be quite random behavior. But in other cases the behavior looks much more structured. Sometimes this is just because there's a short recurrence time:

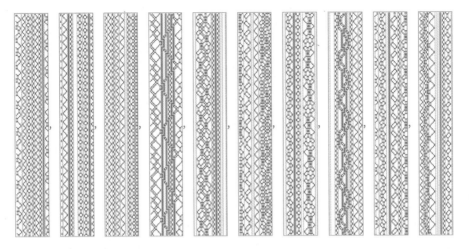

And indeed the overall distribution of recurrence times falls off in a first approximation exponentially (though with a definite tail):

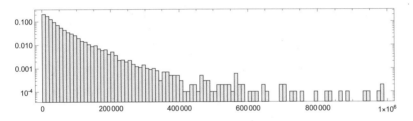

But the distribution is quite broad—with a mean of more than 50,000 steps. (The 17-particle initial blob gives a recurrence time of 155,150 steps.) So what happens with "typical" initial conditions that don't give short recurrences? Here's an example:

What's notable here is that unlike for the case of the "simple blob", there seem to be identifiable traces of the initial conditions that persist for a long time. So what's going on—and how does it relate to the Second Law?

Given the basic rules for the particle cellular automaton

we immediately know that at least a couple of aspects of the initial conditions will persist forever. In particular, the rules conserve the total number of "particles" (i.e. non-white cells) so that:

$$n_■ + n_▨ = \text{constant}$$

In addition, the number of light or dark cells can change only by increments of 2, and therefore their total number must remain either always even or always odd—and combined with overall particle conservation this then implies that:

$$(n_■ - n_▨) \bmod 2 = \text{constant}$$

What about other conservation laws? We can formulate the conservation of total particle number as saying that the number of instances of "length-1 blocks" with weights specified as follows is always constant:

Then we can go on and ask about conservation laws associated with longer blocks. For blocks of length 2, there are no new nontrivial conservation laws, though for example the weighted combination of blocks

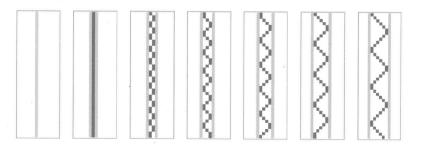

is nominally "conserved"—but only because it is 0 for any possible configuration.

But in addition to such global conservation laws, there are also more local kinds of regularities. For example, a single "light particle" on its own just stays fixed, and a pair of light particles can always trap a single dark particle between them:

For any separation of light particles, it turns out to always be possible to trap any number of dark particles:

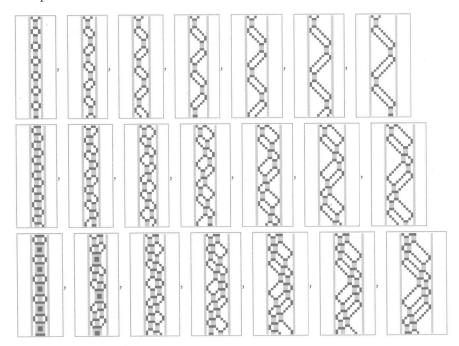

But not every initial configuration of dark particles gets trapped. With separation s and d dark particles, there are a total of Binomial[s, d] possible initial configurations. For $d = 2$, a fraction $(s - 3)/(s - 1)$ of these get trapped while the rest do not. For $d = 3$, the fraction becomes $(s - 3)(s - 4)/(s(s - 1))$ and for $d = 4$ it is $(s - 4)(s - 5)/(s(s - 1))$. (For larger d, the trapping fraction continues to be a rational function of s, but the polynomials involved rapidly become more complicated.) For sufficiently large separation s the trapping fraction always goes to 1—though does so more slowly as d increases:

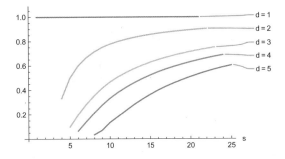

What's basically going on is that a single dark particle always just "bounces off" a light particle:

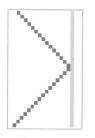

But a pair of dark particles can "go through" the light particle, shifting it slightly:

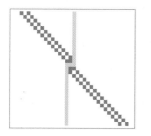

Different things happen with different configurations of dark particles:

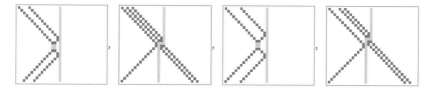

And with more complicated "barriers" the behavior can depend in detail on precise phase and separation relationships:

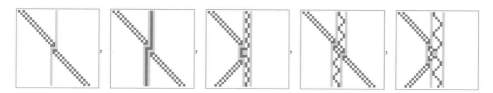

But the basic point is that—although there are various ways they can be modified or destroyed—"light particle walls" can persist for a least a long time. And the result is that if such walls happen to occur in an initial condition they can at least significantly slow down "degradation to randomness".

For example, this shows evolution over the course of 200,000 steps from a particular initial condition, sampled every 20,000 steps—and even over all these steps we see that there's definite "wall structure" that survives:

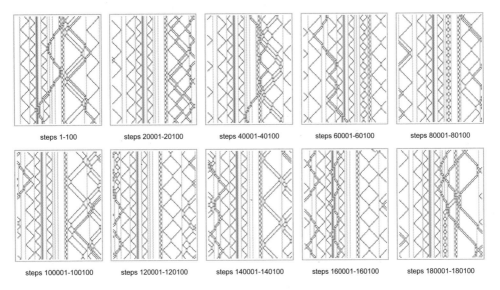

Let's look at a simpler case: a single light particle surrounded by a few dark particles:

If we plot the position of the light particle we see that for thousands of steps it just jiggles around

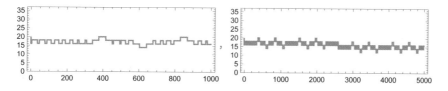

but if one runs it long enough it shows systematic motion at a rate of about 1 position every 1300 steps, wrapping around the cyclic boundary conditions, and eventually returning to its starting point—at the recurrence time of 46,836 steps:

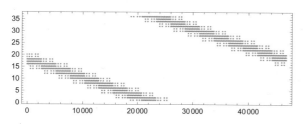

What does all this mean? Essentially the point is that even though something like our particle cellular automaton exhibits computational irreducibility and often generates "featureless" apparent randomness, a system like this is also capable of exhibiting computational reducibility in which traces of the initial conditions can persist, and there isn't just "generic randomness generation".

Computational irreducibility is a powerful force. But, as we'll discuss below, its very presence implies that there must inevitably also be "pockets" of computational reducibility. And once again (as we'll discuss below) it's a question of the observer how obvious or not these pockets may be in a particular case, and whether—say for observers like us—they affect what we perceive in terms of the operation of the Second Law.

It's worth commenting that such issues are not just a feature of systems like our particle cellular automaton. And indeed they've appeared—stretching all the way back to the 1950s— pretty much whenever detailed simulations have been done of systems that one might expect would show "Second Law" behavior. The story is typically that, yes, there is apparent randomness generated (though it's often barely studied as such), just as the Second Law would suggest. But then there's a big surprise of some kind of unexpected regularity. In arrays of nonlinear springs, there were solitons. In hard-sphere gases, there were "long-time tails"—in which correlations in the motion of spheres were seen to decay not exponentially in time, but rather like power laws.

The phenomenon of long-time tails is actually visible in the cellular automaton "approximation" to hard-sphere gases that we studied above. And its interpretation is a good example of how computational reducibility manifests itself. At a small scale, the motion of our idealized molecules shows computational irreducibility and randomness. But on a larger scale, it's more like "collective hydrodynamics", with fluid mechanics effects like vortices. And it's these much-simpler-to-describe computationally reducible effects that lead to the "unexpected regularities" associated with long-time tails.

When the Second Law Works, and When It Doesn't

At its core, the Second Law is about evolution from orderly "simple" initial conditions to apparent randomness. And, yes, this is a phenomenon we can certainly see happen in things like hard-sphere gases in which we're in effect emulating the motion of physical gas molecules. But what about systems with other underlying rules? Because we're explicitly doing everything computationally, we're in a position to just enumerate possible rules (i.e. possible programs) and see what they do.

As an example, here are the distinct patterns produced by all 288 3-color reversible block cellular automata that don't change the all-white state (but don't necessarily conserve "particle number"):

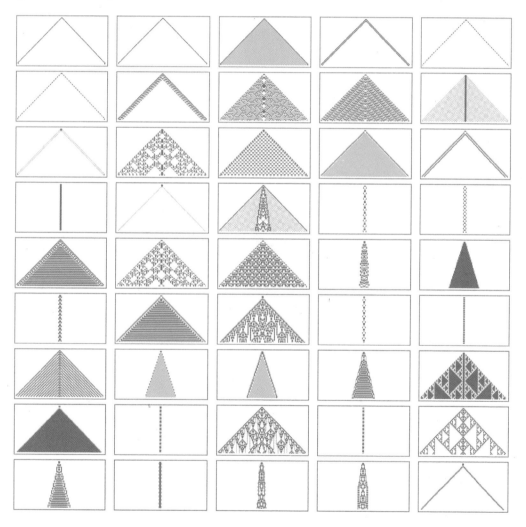

As is typical to see in the computational universe of simple programs, there's quite a diversity of behavior. Often we see it "doing the Second Law thing" and "decaying" to apparent randomness

although sometimes taking a while to do so:

But there are also cases where the behavior just stays simple forever

as well as other cases where it takes a fairly long time before it's clear what's going to happen:

In many ways, the most surprising thing here is that such simple rules can generate randomness. And as we've discussed, that's in the end what leads to the Second Law. But what about rules that don't generate randomness, and just produce simple behavior? Well, in these cases the Second Law doesn't apply.

In standard physics, the Second Law is often applied to gases—and indeed this was its very first application area. But to a solid whose atoms have stayed in more or less fixed positions for a billion years, it really doesn't usefully apply. And the same is true, say, for a line of masses connected by perfect springs, with perfect linear behavior.

There's been a quite pervasive assumption that the Second Law is somehow always universally valid. But it's simply not true. The validity of the Second Law is associated with the phenomenon of computational irreducibility. And, yes, this phenomenon is quite ubiquitous. But there are definitely systems and situations in which it does not occur. And those will not show "Second Law" behavior.

There are plenty of complicated "marginal" cases, however. For example, for a given rule (like the 3 shown here), some initial conditions may not lead to randomness and "Second Law behavior", while others do:

And as is so often the case in the computational universe there are phenomena one never expects, like the strange "shock-front-like" behavior of the third rule, which produces randomness, but only on a scale determined by the region it's in:

It's worth mentioning that while restricting to a finite region often yields behavior that more obviously resembles a "box of gas molecules", the general phenomenon of randomness generation also occurs in infinite regions. And indeed we already know this from the classic example of rule 30. But here it is in a reversible block cellular automaton:

In some simple cases the behavior just repeats, but in other cases it's nested

albeit sometimes in rather complicated ways:

The Second Law and Order in the Universe

Having identified the computational nature of the core phenomenon of the Second Law we can start to understand in full generality just what the range of this phenomenon is. But what about the ordinary Second Law as it might be applied to familiar physical situations?

Does the ubiquity of computational irreducibility imply that ultimately absolutely everything must "degrade to randomness"? We saw in the previous section that there are underlying rules for which this clearly doesn't happen. But what about with typical "real-world" systems involving molecules? We've seen lots of examples of idealized hard-sphere gases in which we observe randomization. But—as we've mentioned several times—even when there's computational irreducibility, there are always pockets of computational reducibility to be found.

And for example the fact that simple overall gas laws like $PV = $ constant apply to our hard-sphere gas can be viewed as an example of computational reducibility. And as another example, consider a hard-sphere gas in which vortex-like circulation has been set up. To get a sense of what happens we can just look at our simple discrete model. At a microscopic level there's clearly lots of apparent randomness, and it's hard to see what's globally going on:

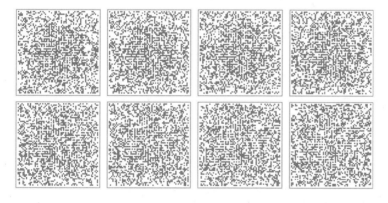

But if we coarse grain the system by 3×3 blocks of cells with "average velocities" we see that there's a fairly persistent hydrodynamic-like vortex that can be identified:

Microscopically, there's computational irreducibility and apparent randomness. But macroscopically the particular form of coarse-grained measurement we're using picks out a pocket of reducibility—and we see overall behavior whose obvious features don't show "Second-Law-style" randomness.

And in practice this is how much of the "order" we see in the universe seems to work. At a small scale there's all sorts of computational irreducibility and randomness. But on a larger scale there are features that we as observers notice that tap into pockets of reducibility, and that show the kind of order that we can describe, for example, with simple mathematical laws.

There's an extreme version of this in our Physics Project, where the underlying structure of space—like the underlying structure of something like a gas—is full of computational irreducibility, but where there are certain overall features that observers like us notice, and that show computational reducibility. One example involves the large-scale structure of spacetime, as described by general relativity. Another involves the identification of particles that can be considered to "move without change" through the system.

One might have thought—as people often have—that the Second Law would imply a degradation of every feature of a system to uniform randomness. But that's just not how computational irreducibility works. Because whenever there's computational irreducibility, there are also inevitably an infinite number of pockets of computational reducibility. (If there weren't, that very fact could be used to "reduce the irreducibility".)

And what that means is that when there's irreducibility and Second-Law-like randomization, there'll also always be orderly laws to be found. But which of those laws will be evident—or relevant—to a particular observer depends on the nature of that observer.

The Second Law is ultimately a story of the mismatch between the computational irreducibility of underlying systems, and the computational boundedness of observers like us. But the point is that if there's a pocket of computational reducibility that happens to be "a fit" for us as observers, then despite our computational limitations, we'll be perfectly able to recognize the orderliness that's associated with it—and we won't think that the system we're looking at has just "degraded to randomness".

So what this means is that there's ultimately no conflict between the existence of order in the universe, and the operation of the Second Law. Yes, there's an "ocean of randomness" generated by computational irreducibility. But there's also inevitably order that lives in pockets of reducibility. And the question is just whether a particular observer "notices" a given pocket of reducibility, or whether they only "see" the "background" of computational irreducibility.

In the "hydrodynamics" example above, the "observer" picks out a "slice" of behavior by looking at aggregated local averages. But another way for an observer to pick out a "slice" of behavior is just to look only at a specific region in a system. And in that case one can observe simpler behavior because in effect "the complexity has radiated away". For example, here are reversible cellular automata where a random initial block is "simplified" by "radiating its information out":

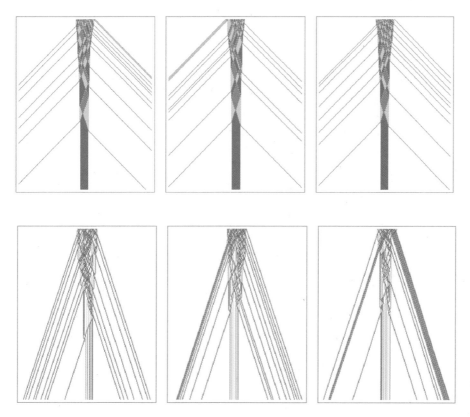

If one picked up all those "pieces of radiation" one would be able—with appropriate computational effort—to reconstruct all the randomness in the initial condition. But if we as observers just "ignore the radiation to infinity" then we'll again conclude that the system has evolved to a simpler state—against the "Second-Law trend" of increasing randomness.

Class 4 and the Mechanoidal Phase

When I first studied cellular automata back in the 1980s, I identified four basic classes of behavior that are seen when starting from generic initial conditions—as exemplified by:

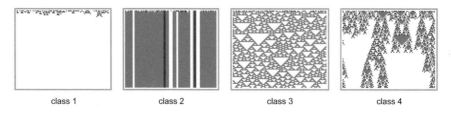

Class 1 essentially always evolves to the same final "fixed-point" state, immediately destroying information about its initial state. Class 2, however, works a bit like solid matter, essentially just maintaining whatever configuration it was started in. Class 3 works more like a gas or a liquid, continually "mixing things up" in a way that looks quite random. But class 4 does something more complicated.

In class 3 there aren't significant identifiable persistent structures, and everything always seems to quickly get randomized. But the distinguishing feature of class 4 is the presence of identifiable persistent structures, whose interactions effectively define the activity of the system.

So how do these types of behavior relate to the Second Law? Class 1 involves intrinsic irreversibility, and so doesn't immediately connect to standard Second Law behavior. Class 2 is basically too static to follow the Second Law. But class 3 shows quintessential Second Law behavior, with rapid evolution to "typical random states". And it's class 3 that captures the kind of behavior that's seen in typical Second Law systems, like gases.

But what about class 4? Well, it's a more complicated story. The "level of activity" in class 4— while above class 2—is in a sense below class 3. But unlike in class 3, where there is typically "too much activity" to "see what's going on", class 4 often gives one the idea that it's operating in a "more potentially understandable" way. There are many different detailed kinds of behavior that appear in class 4 systems. But here are a few examples in reversible block cellular automata:

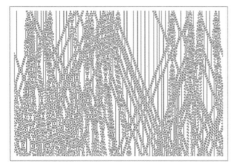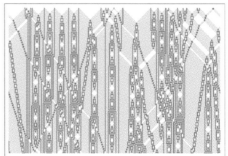

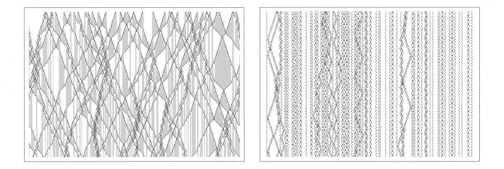

Looking at the first rule, it's easy to identify some simple persistent structures, some stationary, some moving:

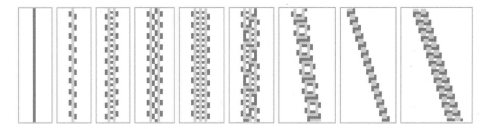

But even with this rule, many other things can happen too

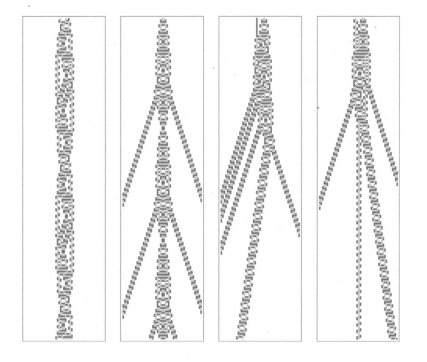

and in the end the whole behavior of the system is built up from combinations and interactions of structures like these.

The second rule above behaves in an immediately more elaborate way. Here it is starting from a random initial condition:

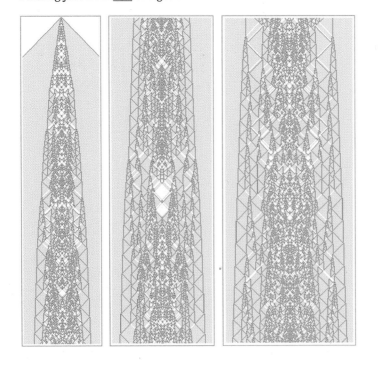

Starting just from one gets:

Sometimes the behavior seems simpler

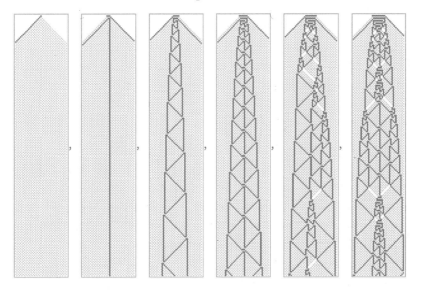

though even in the last case here, there is elaborate "number-theoretical" behavior that seems to never quite become either periodic or nested:

We can think of any cellular automaton—or any system based on rules—as "doing a computation" when it evolves. Class 1 and 2 systems basically behave in computationally simple ways. But as soon as we reach class 3 we're dealing with computational irreducibility, and with a "density of computation" that lets us decode almost nothing about what comes out, with the result that what we see we can basically describe only as "apparently random". Class 4 no doubt has the same ultimate computational irreducibility—and the same ultimate computational capabilities—as class 3. But now the computation is "less dense", and seemingly more accessible to human interpretation. In class 3 it's difficult to imagine making any kind of "symbolic summary" of what's going on. But in class 4, we see definite structures whose behavior we can imagine being able to describe in a symbolic way, building up what we can think of as a "human-accessible narrative" in which we talk about "structure X collides with structure Y to produce structure Z" and so on.

And indeed if we look at the picture above, it's not too difficult to imagine that it might correspond to the execution trace of a computation we might do. And more than that, given the "identifiable components" that arise in class 4 systems, one can imagine assembling these to explicitly set up particular computations one wants to do. In a class 3 system "randomness" always just "spurts out", and one has very little ability to "meaningfully control" what happens. But in a class 4 system, one can potentially do what amounts to traditional engineering or programming to set up an arrangement of identifiable component "primitives" that achieves some particular purpose one has chosen.

And indeed in a case like the rule 110 cellular automaton we know that it's possible to perform any computation in this way, proving that the system is capable of universal computation, and providing a piece of evidence for the phenomenon of computational irreducibility. No doubt rule 30 is also computation universal. But the point is that with our current ways of analyzing things, class 3 systems like this don't make this something we can readily recognize.

Like so many other things we're discussing, this is basically again a story of observers and their capabilities. If observers like us—with our computational boundedness—are going to be able to "get things into our minds" we seem to need to break them down to the point where they can be described in terms of modest numbers of types of somewhat-independent parts. And that's what the "decomposition into identifiable structures" that we observe in class 4 systems gives us the opportunity to do.

What about class 3? Notwithstanding things like our discussion of traces of initial conditions above, our current powers of perception just don't seem to let us "understand what's going on" to the point where we can say much more than there's apparent randomness. And of course it's this very point that we're arguing is the basis for the Second Law. Could there be observers who could "decode class 3 systems"? In principle, absolutely yes. And even if the observers—like us—are computationally bounded, we can expect that there will be at least some pockets of computational reducibility that could be found that would allow progress to be made.

But as of now—with the methods of perception and analysis currently at our disposal—there's something very different for us about class 3 and class 4. Class 3 shows quintessential "apparently random" behavior, like molecules in a gas. Class 4 shows behavior that looks more like the "insides of a machine" that could have been "intentionally engineered for a purpose". Having a system that is like this "in bulk" is not something familiar, say from physics. There are solids, and liquids, and gases, whose components have different general organizational characteristics. But what we see in class 4 is something yet different—and quite unfamiliar.

Like solids, liquids and gases, it's something that can exist "in bulk", with any number of components. We can think of it as a "phase" of a system. But it's a new type of phase, that we might call a "mechanoidal phase".

How do we recognize this phase? Again, it's a question of the observer. Something like a solid phase is easy for observers like us to recognize. But even the distinction between a liquid and a gas can be more difficult to recognize. And to recognize the mechanoidal phase we basically have to be asking something like "Is this a computation we recognize?"

How does all this relate to the Second Law? Class 3 systems—like gases—immediately show typical "Second Law" behavior, characterized by randomness, entropy increase, equilibrium, and so on. But class 4 systems work differently. They have new characteristics that don't fit neatly into the rubric of the Second Law.

No doubt one day we will have theories of the mechanoidal phase just like today we have theories of gases, of liquids and of solids. Likely those theories will have to get more sophisticated in characterizing the observer, and in describing what kinds of coarse graining can reasonably be done. Presumably there will be some kind of analog of the Second Law that leverages the difference between the capabilities and features of the observer and the system they're observing. But in the mechanoidal phase there is in a sense less distance between the mechanism of the system and the mechanism of the observer, so we probably can't expect a statement as ultimately simple and clear-cut as the usual Second Law.

The Mechanoidal Phase and Bulk Molecular Biology

The Second Law has long had an uneasy relationship with biology. "Physical" systems like gases readily show the "decay" to randomness expected from the Second Law. But living systems instead somehow seem to maintain all sorts of elaborate organization that doesn't immediately "decay to randomness"— and indeed actually seems able to grow just through "processes of biology".

It's easy to point to the continual absorption of energy and material by living systems—as well as their eventual death and decay—as reasons why such systems might still at least nominally follow the Second Law. But even if at some level this works, it's not particularly useful in letting us talk about the actual significant "bulk" features of living systems—in the kind of way that the Second Law routinely lets us make "bulk" statements about things like gases.

So how might we begin to describe living systems "in bulk"? I suspect a key is to think of them as being in large part in what we're here calling the mechanoidal phase. If one looks inside a living organism at a molecular scale, there are some parts that can reasonably be described as solid, liquid or gas. But what molecular biology has increasingly shown is that there's often much more elaborate molecular-scale organization than exist in those phases— and moreover that at least at some level this organization seems "describable" and "machine-like", with molecules and collections of molecules that we can say have "particular functions", often being "carefully" and actively transported by things like the cytoskeleton.

In any given organism, there are for example specific proteins defined by the genomics of the organism, that behave in specific ways. But one suspects that there's also a higher-level or "bulk" description that allows one to make at least some kinds of general statements. There are already some known general principles in biology—like the concept of natural selection, or the self-replicating digital character of genetic information—that let one come to various conclusions independent of microscopic details.

And, yes, in some situations the Second Law provides certain kinds of statements about biology. But I suspect that there are much more powerful and significant principles to be discovered, that in fact have the potential to unlock a whole new level of global understanding of biological systems and processes.

It's perhaps worth mentioning an analogy in technology. In a microprocessor what we can think of as the "working fluid" is essentially a gas of electrons. At some level the Second Law has things to say about this gas of electrons, for example describing scattering processes that lead to electrical resistance. But the vast majority of what matters in the behavior of this particular gas of electrons is defined not by things like this, but by the elaborate pattern of wires and switches that exist in the microprocessor, and that guide the motion of the electrons.

In living systems one sometimes also cares about the transport of electrons—though more often it's atoms and ions and molecules. And living systems often seem to provide what one can think of as a close analog of wires for transporting such things. But what is the arrangement of these "wires"? Ultimately it'll be defined by the application of rules derived from things like the genome of the organism. Sometimes the results will for example be analogous to crystalline or amorphous solids. But in other cases one suspects that it'll be better described by something like the mechanoidal phase.

Quite possibly this may also provide a good bulk description of technological systems like microprocessors or large software codebases. And potentially then one might be able to have high-level laws—analogous to the Second Law—that would make high-level statements about these technological systems.

It's worth mentioning that a key feature of the mechanoidal phase is that detailed dynamics—and the causal relations they define—matter. In something like a gas it's perfectly fine for most purposes to assume "molecular chaos", and to say that molecules are arbitrarily mixed. But the mechanoidal phase depends on the "detailed choreography" of elements. It's still a "bulk phase" with arbitrarily many elements. But things like the detailed history of interactions of each individual element matter.

In thinking about typical chemistry—say in a liquid or gas phase—one's usually just concerned with overall concentrations of different kinds of molecules. In effect one assumes that the "Second Law has acted", and that everything is "mixed randomly" and the causal histories of molecules don't matter. But it's increasingly clear that this picture isn't correct for molecular biology, with all its detailed molecular-scale structures and mechanisms. And instead it seems more promising to model what's there as being in the mechanoidal phase.

So how does this relate to the Second Law? As we've discussed, the Second Law is ultimately a reflection of the interplay between underlying computational irreducibility and the limited computational capabilities of observers like us. But within computational irreducibility there are inevitably always "pockets" of computational reducibility—which the observer may or may not care about, or be able to leverage.

In the mechanoidal phase there is ultimately computational irreducibility. But a defining feature of this phase is the presence of "local computational reducibility" visible in the existence of identifiable localized structures. Or, in other words, even to observers like us, it's clear that the mechanoidal phase isn't "uniformly computationally irreducible". But just what general statements can be made about it will depend—potentially in some detail—on the characteristics of the observer.

We've managed to get a long way in discussing the Second Law—and even more so in doing our Physics Project—by making only very basic assumptions about observers. But to be able to make general statements about the mechanoidal phase—and living systems—we're likely to have to say more about observers. If one's presented with a lump of biological tissue one might at first just describe it as some kind of gel. But we know there's much more to it. And the question is what features we can perceive. Right now we can see with microscopes all kinds of elaborate spatial structures. Perhaps in the future there'll be technology that also lets us systematically detect dynamic and causal structures. And it'll be the interplay of what we perceive with what's computationally going on underneath that'll define what general laws we will be able to see emerge.

We already know we won't just get the ordinary Second Law. But just what we will get isn't clear. But somehow—perhaps in several variants associated with different kinds of observers—what we'll get will be something like "general laws of biology", much like in our Physics Project we get general laws of spacetime and of quantum mechanics, and in our analysis of metamathematics we get "general laws of mathematics".

The Thermodynamics of Spacetime

Traditional twentieth-century physics treats spacetime a bit like a continuous fluid, with its characteristics being defined by the continuum equations of general relativity. Attempts to align this with quantum field theory led to the idea of attributing an entropy to black holes, in essence to represent the number of quantum states "hidden" by the event horizon of the black hole. But in our Physics Project there is a much more direct way of thinking about spacetime in what amount to thermodynamic terms.

A key idea of our Physics Project is that there's something "below" the "fluid" representation of spacetime—and in particular that space is ultimately made of discrete elements, whose relations (which can conveniently be represented by a hypergraph) ultimately define everything about the structure of space. This structure evolves according to rules that are somewhat analogous to those for block cellular automata, except that now one is doing replacements not for blocks of cell values, but instead for local pieces of the hypergraph.

So what happens in a system like this? Sometimes the behavior is simple. But very often—much like in many cellular automata—there is great complexity in the structure that develops even from simple initial conditions:

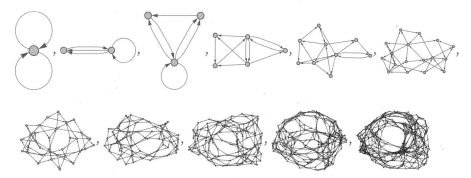

It's again a story of computational irreducibility, and of the generation of apparent randomness. The notion of "randomness" is a bit less straightforward for hypergraphs than for arrays of cell values. But what ultimately matters is what "observers like us" perceive in the system. A typical approach is to look at geodesic balls that encompass all elements within a certain graph distance of a given element—and then to study the effective geometry that emerges in the large-scale limit. It's then a bit like seeing fluid dynamics emerge from small-scale molecular dynamics, except that here (after navigating many technical issues) it's the Einstein equations of general relativity that emerge.

But the fact that this can work relies on something analogous to the Second Law. It has to be the case that the evolution of the hypergraph leads at least locally to something that can be viewed as "uniformly random", and on which statistical averages can be done. In effect, the microscopic structure of spacetime is reaching some kind of "equilibrium state", whose detailed internal configuration "seems random"—but which has definite "bulk" properties that are perceived by observers like us, and give us the impression of continuous spacetime.

As we've discussed above, the phenomenon of computational irreducibility means that apparent randomness can arise completely deterministically just by following simple rules from simple initial conditions. And this is presumably what basically happens in the evolution and "formation" of spacetime. (There are some additional complications associated with multicomputation that we'll discuss at least to some extent later.)

But just like for the systems like gases that we've discussed above, we can now start talking directly about things like entropy for spacetime. As "large-scale observers" of spacetime we're always effectively doing coarse graining. So now we can ask how many microscopic configurations of spacetime (or space) are consistent with whatever result we get from that coarse graining.

As a toy example, consider just enumerating all possible graphs (say up to a given size), then asking which of them have a certain pattern of volumes for geodesic balls (i.e. a certain sequence of numbers of distinct nodes within a given graph distance of a particular node). The "coarse-grained entropy" is simply determined by the number of graphs in which the geodesic ball volumes start in the same way. Here are all trivalent graphs (with up to 24 nodes) that have various such geodesic ball "signatures" (most, but not all, turn out to be vertex transitive; these graphs were found by filtering a total of 125,816,453 possibilities):

1, 4, 9, 17, ...	
1, 4, 10, 10, ...	
1, 4, 10, 12, ...	
1, 4, 10, 14, ...	
1, 4, 10, 15, ...	
1, 4, 10, 16, ...	
1, 4, 10, 17, ...	
1, 4, 10, 18, ...	
1, 4, 10, 19, ...	
1, 4, 10, 20, ...	
1, 4, 10, 21, ...	
1, 4, 10, 22, ...	

We can think of the different numbers of graphs in each case as representing different entropies for a tiny fragment of space constrained to have a given "coarse-grained" structure. At the graph sizes we're dealing with here, we're very far from having a good approximation to continuum space. But assume we could look at much larger graphs. Then we might ask how the entropy varies with "limiting geodesic ball signature"—which in the continuum limit is determined by dimension, curvature, etc.

For a general "disembodied lump of spacetime" this is all somewhat hard to define, particularly because it depends greatly on issues of "gauge" or of how the spacetime is foliated into spacelike slices. But event horizons, being in a sense much more global, don't have such issues, and so we can expect to have fairly invariant definitions of spacetime entropy in this case. And the expectation would then be that for example the entropy we would compute would agree with the "standard" entropy computed for example by analyzing quantum fields or strings near a black hole. But with the setup we have here we should also be able to ask more general questions about spacetime entropy—for example seeing how it varies with features of arbitrary gravitational fields.

In most situations the spacetime entropy associated with any spacetime configuration that we can successfully identify at our coarse-grained level will be very large. But if we could ever find a case where it is instead small, this would be somewhere we could expect to start seeing a breakdown of the continuum "equilibrium" structure of spacetime, and where evidence of discreteness should start to show up.

We've so far mostly been discussing hypergraphs that represent instantaneous states of space. But in talking about spacetime we really need to consider causal graphs that map out the causal relationships between updating events in the hypergraph, and that represent the structure of spacetime. And once again, such graphs can show apparent randomness associated with computational irreducibility.

One can make causal graphs for all sorts of systems. Here is one for a "Newton's cradle" configuration of an (effectively 1D) hard-sphere gas, in which events are collisions between spheres, and two events are causally connected if a sphere goes from one to the other:

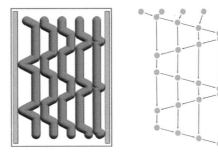

And here is an example for a 2D hard-sphere case, with the causal graph now reflecting the generation of apparently random behavior:

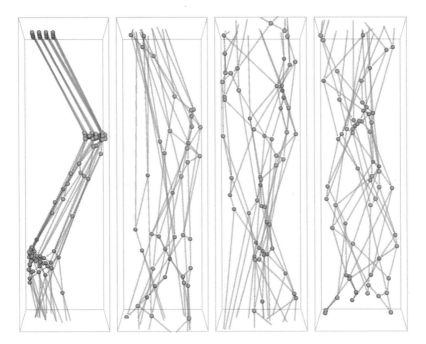

Similar to this, we can make a causal graph for our particle cellular automaton, in which we consider it an event whenever a block changes (but ignore "no-change updates"):

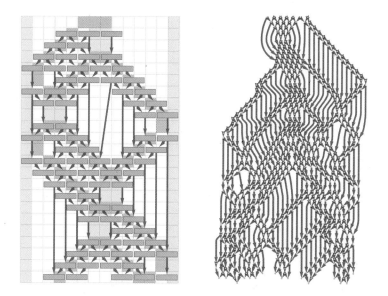

For spacetime, features of the causal graph have some definite interpretations. We define the reference frame we're using by specifying a foliation of the causal graph. And one of the results of our Physics Project is then that the flux of causal edges through the spacelike hypersurfaces our foliation defines can be interpreted directly as the density of physical energy. (The flux through timelike hypersurfaces gives momentum.)

One can make a surprisingly close analogy to causal graphs for hard-sphere gases—except that in a hard-sphere gas the causal edges correspond to actual, nonrelativistic motion of idealized molecules, while in our model of spacetime the causal edges are abstract connections that are in effect always lightlike (i.e. they correspond to motion at the speed of light). In both cases, reducing the number of events is like reducing some version of temperature—and if one approaches no-event "absolute zero" both the gas and spacetime will lose their cohesion, and no longer allow propagation of effects from one part of the system to another.

If one increases density in the hard-sphere gas one will eventually form something like a solid, and in this case there will be a regular arrangement of both spheres and the causal edges. In spacetime something similar may happen in connection with event horizons—which may behave like an "ordered phase" with causal edges aligned.

What happens if one combines thinking about spacetime and thinking about matter? A long-unresolved issue concerns systems with many gravitationally attracting bodies—say a "gas" of stars or galaxies. While the molecules in an ordinary gas might evolve to an apparently random configuration in a standard "Second Law way", gravitationally attracting bodies tend to clump together to make what seem like "progressively simpler" configurations.

It could be that this is a case where the standard Second Law just doesn't apply, but there's long been a suspicion that the Second Law can somehow be "saved" by appropriately associating an entropy with the structure of spacetime. In our Physics Project, as we've discussed, there's always entropy associated with our coarse-grained perception of space-time. And it's conceivable that, at least in terms of overall counting of states, increased "organization" of matter could be more than balanced by enlargement in the number of available states for spacetime.

We've discussed at length above the idea that "Second Law behavior" is the result of us as observers (and preparers of initial states) being "computationally weak" relative to the computational irreducibility of the underlying dynamics of systems. And we can expect that very much the same thing will happen for spacetime. But what if we could make a Maxwell's demon for spacetime? What would this mean?

One rather bizarre possibility is that it could allow faster-than-light "travel". Here's a rough analogy. Gas molecules—say in air in a room—move at roughly the speed of sound. But they're always colliding with other molecules, and getting their directions randomized. But what if we had a Maxwell's-demon-like device that could tell us at every collision which molecule to ride on? With an appropriate choice for the sequence of molecules we could then potentially "surf" across the room at roughly the speed of sound. Of course, to have the device work it'd have to overcome the computational irreducibility of the basic dynamics of the gas.

In spacetime, the causal graph gives us a map of what event can affect what other event. And insofar as we just treat spacetime as "being in uniform equilibrium" there'll be a simple correspondence between "causal distance" and what we consider distance in physical space. But if we look down at the level of individual causal edges it'll be more complicated. And in general we could imagine that an appropriate "demon" could predict the microscopic causal structure of spacetime, and carefully pick causal edges that could "line up" to "go further in space" than the "equilibrium expectation".

Of course, even if this worked, there's still the question of what could be "transported" through such a "tunnel"—and for example even a particle (like an electron) presumably involves a vast number of causal edges, that one wouldn't be able to systematically organize to fit through the tunnel. But it's interesting to realize that in our Physics Project the idea that "nothing can go faster than light" becomes something very much analogous to the Second Law: not a fundamental statement about underlying rules, but rather a statement about our interaction with them, and our capabilities as observers.

So if there's something like the Second Law that leads to the structure of spacetime as we typically perceive it, what can be said about typical issues in thermodynamics in connection with spacetime? For example, what's the story with perpetual motion machines in spacetime?

Even before talking about the Second Law, there are already issues with the First Law of thermodynamics—because in a cosmological setting there isn't local conservation of energy as such, and for example the expansion of the universe can transfer energy to things. But what about the Second Law question of "getting mechanical work from heat"? Presumably the analog of "mechanical work" is a gravitational field that is "sufficiently organized" that observers like us can readily detect it, say by seeing it pull objects in definite directions. And presumably a perpetual motion machine based on violating the Second Law would then have to take the heat-like randomness in "ordinary spacetime" and somehow organize it into a systematic and measurable gravitational field. Or, in other words, "perpetual motion" would somehow have to involve a gravitational field "spontaneously being generated" from the microscopic structure of spacetime.

Just like in ordinary thermodynamics, the impossibility of doing this involves an interplay between the observer and the underlying system. And conceivably it might be possible that there could be an observer who can measure specific features of spacetime that correspond to some slice of computational reducibility in the underlying dynamics—say some weird configuration of "spontaneous motion" of objects. But absent this, a "Second-Law-violating" perpetual motion machine will be impossible.

Quantum Mechanics

Like statistical mechanics (and thermodynamics), quantum mechanics is usually thought of as a statistical theory. But whereas the statistical character of statistical mechanics one imagines to come from a definite, knowable "mechanism underneath", the statistical character of quantum mechanics has usually just been treated as a formal, underivable "fact of physics".

In our Physics Project, however, the story is different, and there's a whole lower-level structure—ultimately rooted in the ruliad—from which quantum mechanics and its statistical character appears to be derived. And, as we'll discuss, that derivation in the end has close connections both to what we've said about the standard Second Law, and to what we've said about the thermodynamics of spacetime.

In our Physics Project the starting point for quantum mechanics is the unavoidable fact that when one's applying rules to transform hypergraphs, there's typically more than one rewrite that can be done to any given hypergraph. And the result of this is that there are many different possible "paths of history" for the universe.

As a simple analog, consider rewriting not hypergraphs but strings. And doing this, we get for example:

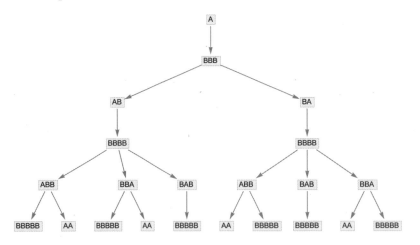

This is a deterministic representation of all possible "paths of history", but in a sense it's very wasteful, among other things because it includes multiple copies of identical strings (like BBBB). If we merge such identical copies, we get what we call a multiway graph, that contains both branchings and mergings:

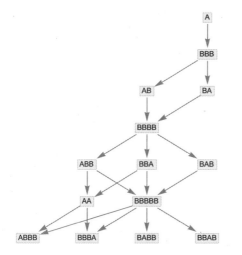

In the "innards" of quantum mechanics one can imagine that all these paths are being followed. So how is it that we as observers perceive definite things to happen in the world? Ultimately it's a story of coarse graining, and of us conflating different paths in the multiway graph.

But there's a wrinkle here. In statistical mechanics we imagine that we can observe from outside the system, implementing our coarse graining by sampling particular features of the system. But in quantum mechanics we imagine that the multiway system describes the whole universe, including us. So then we have the peculiar situation that just as the universe is branching and merging, so too are our brains. And ultimately what we observe is therefore the result of a branching brain perceiving a branching universe.

But given all those branches, can we just decide to conflate them into a single thread of experience? In a sense this is a typical question of coarse graining and of what we can consistently equivalence together. But there's something a bit different here because without the "coarse graining" we can't talk at all about "what happened", only about what might be happening. Put another way, we're now fundamentally dealing not with computation (like in a cellular automaton) but with multicomputation.

And in multicomputation, there are always two fundamental kinds of operations: the generation of new states from old, and the equivalencing of states, effectively by the observer. In ordinary computation, there can be computational irreducibility in the process of generating a thread of successive states. In multicomputation, there can be multicomputational irreducibility in which in a sense all computations in the multiway system have to be done in order even to determine a single equivalenced result. Or, put another way, you can't shortcut following all the paths of history. If you try to equivalence at the beginning, the equivalence class you've built will inevitably be "shredded" by the evolution, forcing you to follow each path separately.

It's worth commenting that just as in classical mechanics, the "underlying dynamics" in our description of quantum mechanics are reversible. In the original unmerged evolution tree above, we could just reverse each rule and from any point uniquely construct a "backwards tree". But once we start merging and equivalencing, there isn't the same kind of "direct reversibility"—though we can still count possible paths to determine that we preserve "total probability".

In ordinary computational systems, computational irreducibility implies that even from simple initial conditions we can get behavior that "seems random" with respect to most computationally bounded observations. And something directly analogous happens in multicomputational systems. From simple initial conditions, we generate collections of paths of history that "seem random" with respect to computationally bounded equivalencing operations, or, in other words, to observers who do computationally bounded coarse graining of different paths of history.

When we look at the graphs we've drawn representing the evolution of a multiway system, we can think of there being a time direction that goes down the page, following the arrows that point from states to their successors. But across the page, in the transverse direction, we can think of there as being a space in which different paths of history are laid—what we call "branchial space".

A typical way to start constructing branchial space is to take slices across the multiway graph, then to form a branchial graph in which two states are joined if they have a common ancestor on the step before (which means we can consider them "entangled"):

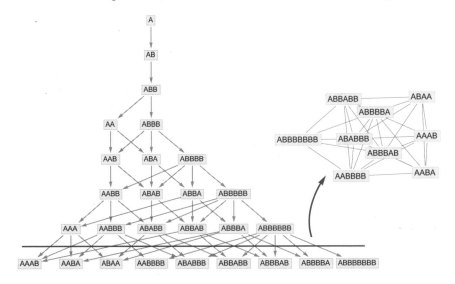

Although the details remain to be clarified, it seems as if in the standard formalism of quantum mechanics, distance in branchial space corresponds essentially to quantum phase, so that, for example, particles whose phases would make them show destructive interference will be at "opposite ends" of branchial space.

So how do observers relate to branchial space? Basically what an observer is doing is to coarse grain in branchial space, equivalencing certain paths of history. And just as we have a certain extent in physical space, which determines our coarse graining of gases, and—at a much smaller scale—of the structure of spacetime, so also we have an extent in branchial space that determines our coarse graining across branches of history.

But this is where multicomputational irreducibility and the analog of the Second Law are crucial. Because just as we imagine that gases—and spacetime—achieve a certain kind of "unique random equilibrium" that leads us to be able to make consistent measurements of them, so also we can imagine that in quantum mechanics there is in effect a "branchial space equilibrium" that is achieved.

Think of a box of gas in equilibrium. Put two pistons on different sides of the box. So long as they don't perturb the gas too much, they'll both record the same pressure. And in our Physics Project it's the same story with observers and quantum mechanics. Most of the time there'll be enough effective randomness generated by the multicomputationally irreducible evolution of the system (which is completely deterministic at the level of the multiway graph) that a computationally bounded observer will always see the same "equilibrium values".

A central feature of quantum mechanics is that by making sufficiently careful measurements one can see what appear to be random results. But where does that randomness come from? In the usual formalism for quantum mechanics, the idea of purely probabilistic results is just burnt into the formal structure. But in our Physics Project, the apparent randomness one sees has a definite, "mechanistic" origin. And it's basically the same as the origin of randomness for the standard Second Law, except that now we're dealing with multicomputational rather than pure computational irreducibility.

By the way, the "Bell's inequality" statement that quantum mechanics cannot be based on "mechanistic randomness" unless it comes from a nonlocal theory remains true in our Physics Project. But in the Physics Project we have an immediate ubiquitous source of "nonlocality": the equivalencing or coarse graining "across" branchial space done by observers.

(We're not discussing the role of physical space here. But suffice it to say that instead of having each node of the multiway graph represent a complete state of the universe, we can make an extended multiway graph in which different spatial elements—like different paths of history—are separated, with their "causal entanglements" then defining the actual structure of space, in a spatial analog of the branchial graph.)

As we've already noted, the complete multiway graph is entirely deterministic. And indeed if we have a complete branchial slice of the graph, this can be used to determine the whole future of the graph (the analog of "unitary evolution" in the standard formalism of quantum mechanics). But if we equivalence states—corresponding to "doing a measurement"—then we won't have enough information to uniquely determine the future of the system, at least when it comes to what we consider to be quantum effects.

At the outset, we might have thought that statistical mechanics, spacetime mechanics and quantum mechanics were all very different theories. But what our Physics Project suggests is that in fact they are all based on a common, fundamentally computational phenomenon.

So what about other ideas associated with the standard Second Law? How do they work in the quantum case?

Entropy, for example, now just becomes a measure of the number of possible configurations of a branchial graph consistent with a certain coarse-grained measurement. Two independent systems will have disconnected branchial graphs. But as soon as the systems interact, their branchial graphs will connect, and the number of possible graph configurations will change, leading to an "entanglement entropy".

One question about the quantum analog of the Second Law is what might correspond to "mechanical work". There may very well be highly structured branchial graphs—conceivably associated with things like coherent states—but it isn't yet clear how they work and whether existing kinds of measurements can readily detect them. But one can expect that multicomputational irreducibility will tend to produce branchial graphs that can't be "decoded" by most computationally bounded measurements—so that, for example, "quantum perpetual motion", in which "branchial organization" is spontaneously produced, can't happen.

And in the end randomness in quantum measurements is happening for essentially the same basic reason we'd see randomness if we looked at small numbers of molecules in a gas: it's not that there's anything fundamentally not deterministic underneath, it's just there's a computational process that's making things too complicated for us to "decode", at least as observers with bounded computational capabilities. In the case of the gas, though, we're sampling molecules at different places in physical space. But in quantum mechanics we're doing the slightly more abstract thing of sampling states of the system at different places in branchial space. But the same fundamental randomization is happening, though now through multicomputational irreducibility operating in branchial space.

The Future of the Second Law

The original formulation of the Second Law a century and a half ago—before even the existence of molecules was established—was an impressive achievement. And one might assume that over the course of 150 years—with all the mathematics and physics that's been done—a complete foundational understanding of the Second Law would long ago have been developed. But in fact it has not. And from what we've discussed here we can now see why. It's because the Second Law is ultimately a computational phenomenon, and to understand it requires an understanding of the computational paradigm that's only very recently emerged.

Once one starts doing actual computational experiments in the computational universe (as I already did in the early 1980s) the core phenomenon of the Second Law is surprisingly obvious—even if it violates one's traditional intuition about how things should work. But in the end, as we have discussed here, the Second Law is a reflection of a very general, if deeply computational, idea: an interplay between computational irreducibility and the computational limitations of observers like us. The Principle of Computational Equivalence tells us that computational irreducibility is inevitable. But the limitation of observers is something different: it's a kind of epiprinciple of science that's in effect a formalization of our human experience and our way of doing science.

Can we tighten up the formulation of all this? Undoubtedly. We have various standard models of the computational process—like Turing machines and cellular automata. We still need to develop an "observer theory" that provides standard models for what observers like us can do. And the more we can develop such a theory, the more we can expect to make explicit proofs of specific statements about the Second Law. Ultimately these proofs will have solid foundations in the Principle of Computational Equivalence (although there remains much to formalize there too), but will rely on models for what "observers like us" can be like.

So how general do we expect the Second Law to be in the end? In the past couple of sections we've seen that the core of the Second Law extends to spacetime and to quantum mechanics. But even when it comes to the standard subject matter of statistical mechanics, we expect limitations and exceptions to the Second Law.

Computational irreducibility and the Principle of Computational Equivalence are very general, but not very specific. They talk about the overall computational sophistication of systems and processes. But they do not say that there are no simplifying features. And indeed we expect that in any system that shows computational irreducibility, there will always be arbitrarily many "slices of computational reducibility" that can be found.

The question then is whether those slices of reducibility will be what an observer can perceive, or will care about. If they are, then one won't see Second Law behavior. If they're not, one will just see "generic computational irreducibility" and Second Law behavior.

How can one find the slices of reducibility? Well, in general that's irreducibly hard. Every slice of reducibility is in a sense a new scientific or mathematical principle. And the computational irreducibility involved in finding such reducible slices basically speaks to the ultimately unbounded character of the scientific and mathematical enterprise. But once again, even though there might be an infinite number of slices of reducibility, we still have to ask which ones matter to us as observers.

The answer could be one thing for studying gases, and another, for example, for studying molecular biology, or social dynamics. The question of whether we'll see "Second Law behavior" then boils down to whether whatever we're studying turns out to be something that doesn't simplify, and ends up showing computational irreducibility.

If we have a sufficiently small system—with few enough components—then the computational irreducibility may not be "strong enough" to stop us from "going beyond the Second Law", and for example constructing a successful Maxwell's demon. And indeed as computer and sensor technology improve, it's becoming increasingly feasible to do measurement and set up control systems that effectively avoid the Second Law in particular, small systems.

But in general the future of the Second Law and its applicability is really all about how the capabilities of observers develop. What will future technology, and future paradigms, do to our ability to pick away at computational irreducibility?

In the context of the ruliad, we are currently localized in rulial space based on our existing capabilities. But as we develop further we are in effect "colonizing" rulial space. And a system that may look random—and may seem to follow the Second Law—from one place in rulial space may be "revealed as simple" from another.

There is an issue, though. Because the more we as observers spread out in rulial space, the less coherent our experience will become. In effect we'll be following a larger bundle of threads in rulial space, which makes who "we" are less definite. And in the limit we'll presumably be able to encompass all slices of computational reducibility, but at the cost of having our experience "incoherently spread" across all of them.

It's in the end some kind of tradeoff. Either we can have a coherent thread of experience, in which case we'll conclude that the world produces apparent randomness, as the Second Law suggests. Or we can develop to the point where we've "spread our experience" and no longer have coherence as observers, but can recognize enough regularities that the Second Law potentially seems irrelevant.

But as of now, the Second Law is still very much with us, even if we are beginning to see some of its limitations. And with our computational paradigm we are finally in a position to see its foundations, and understand how it ultimately works.

Thanks & Notes

Thanks to Brad Klee, Kegan Allen, Jonathan Gorard, Matt Kafker, Ed Pegg and Michael Trott for their help—as well as to the many people who have contributed to my understanding of the Second Law over the 50+ years I've been interested in it.

Wolfram Language to generate every image here is available by clicking the image in the online version (wolfr.am/SW-SecondLaw).

A 50-Year Quest: My Personal Journey with the Second Law of Thermodynamics

When I Was 12 Years Old...

I've been trying to understand the Second Law now for a bit more than 50 years.

It all started when I was 12 years old. Building on an earlier interest in space and spacecraft, I'd gotten very interested in physics, and was trying to read everything I could about it. There were several shelves of physics books at the local bookstore. But what I coveted most was the largest physics book collection there: a series of five plushly illustrated college textbooks. And as a kind of graduation gift when I finished (British) elementary school in June 1972 I arranged to get those books. And here they are, still on my bookshelf today, just a little faded, more than half a century later:

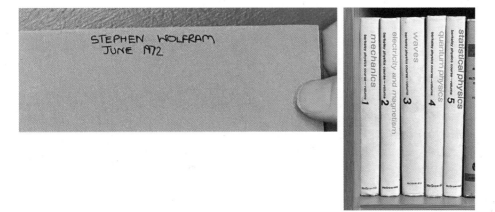

For a while the first book in the series was my favorite. Then the third. The second. The fourth. The fifth one at first seemed quite mysterious—and somehow more abstract in its goals than the others:

What story was the filmstrip on its cover telling? For a couple of months I didn't look seriously at the book. And I spent much of the summer of 1972 writing my own (unseen by anyone else for 30+ years) *Concise Directory of Physics*

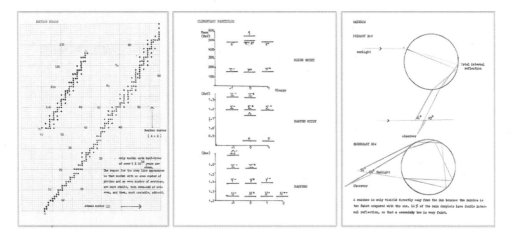

that included a rather stiff page about energy, mentioning entropy—along with the heat death of the universe.

But one afternoon late that summer I decided I should really find out what that mysterious fifth book was all about. Memory being what it is I remember that—very unusually for me—I took the book to read sitting on the grass under some trees. And, yes, my archives almost let me check my recollection: in the distance, there's the spot, except in 1967 the trees are significantly smaller, and in 1985 they're bigger:

1967   1985

Of course, by 1972 I was a little bigger than in 1967—and here I am a little later, complete with a book called *Planets and Life* on the ground, along with a tube of (British) Smarties, and, yes, a pocket protector (but, hey, those were actual ink pens):

But back to the mysterious green book. It wasn't like anything I'd seen before. It was full of pictures like the one on the cover. And it seemed to be saying that—just by looking at those pictures and thinking—one could figure out fundamental things about physics. The other books I'd read had all basically said "physics works like this". But here was a book saying "you can figure out how physics has to work". Back then I definitely hadn't internalized it, but I think what was so exciting that day was that I got a first taste of the idea that one didn't have to be told how the world works; one could just figure it out:

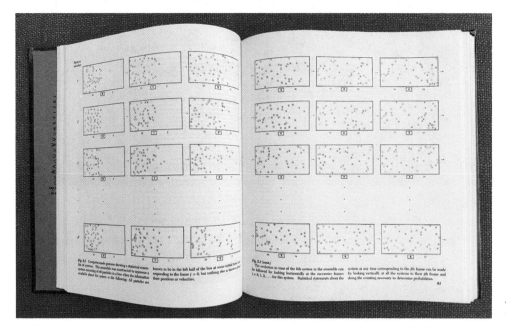

I didn't yet understand quite a bit of the math in the book. But it didn't seem so relevant to the core phenomenon the book was apparently talking about: the tendency of things to become more random. I remember wondering how this related to stars being organized into galaxies. Why might that be different? The book didn't seem to say, though I thought maybe somewhere it was buried in the math.

But soon the summer was over, and I was at a new school, mostly away from my books, and doing things like diligently learning more Latin and Greek. But whenever I could I was learning more about physics—and particularly about the hot area of the time: particle physics. The pions. The kaons. The lambda hyperon. They all became my personal friends. During the school vacations I would excitedly bicycle the few miles to the nearby university library to check out the latest journals and the latest news about particle physics.

The school I was at (Eton) had five centuries of history, and I think at first I assumed no particular bridge to the future. But it wasn't long before I started hearing mentions that somewhere at the school there was a computer. I'd seen a computer in real life only once—when I was 10 years old, and from a distance. But now, tucked away at the edge of the school, above a bicycle repair shed, there was an island of modernity, a "computer room" with a glass partition separating off a loudly humming desk-sized piece of electronics that I could actually touch and use: an Elliott 903C computer with 8 kilowords of 18-bit ferrite core memory (acquired by the school in 1970 for £12,000, or about $300k today):

At first it was such an unfamiliar novelty that I was satisfied writing little programs to do things like compute primes, print curious patterns on the teleprinter, and play tunes with the built-in diagnostic tone generator. But it wasn't long before I set my sights on the goal of using the computer to reproduce that interesting picture on the book cover.

I programmed in assembler, with my programs on paper tape. The computer had just 16 machine instructions, which included arithmetic ones, but only for integers. So how was I going to simulate colliding "molecules" with that? Somewhat sheepishly, I decided to put everything on a grid, with everything represented by discrete elements. There was a convention for people to name their programs starting with their own first initial. So I called the program SPART, for "Stephen's Particle Program". (Thinking about it today, maybe that name reflected some aspiration of relating this to particle physics.)

It was the most complicated program I had ever written. And it was hard to test, because, after all, I didn't really know what to expect it to do. Over the course of several months, it went through many versions. Rather often the program would just mysteriously crash before producing any output (and, yes, there weren't real debugging tools yet). But eventually I got it to systematically produce output. But to my disappointment the output never looked much like the book cover.

I didn't know why, but I assumed it was because I was simplifying things too much, putting everything on a grid, etc. A decade later I realized that in writing my program I'd actually ended up inventing a form of 2D cellular automaton. And I now rather suspect that this cellular automaton—like rule 30—was actually intrinsically generating randomness, and in some sense showing what I now understand to be the core phenomenon of the Second Law. But at the time I absolutely wasn't ready for this, and instead I just assumed that what I was seeing was something wrong and irrelevant. (In past years, I had suspected that what went wrong had to do with details of particle behavior on square—as opposed to other—grids. But I now suspect it was instead that the system was in a sense generating too much randomness, making the intended "molecular dynamics" unrecognizable.)

I'd love to "bring SPART back to life", but I don't seem to have a copy anymore, and I'm pretty sure the printouts I got as output back in 1973 seemed so "wrong" I didn't keep them. I do still have quite a few paper tapes from around that time, but as of now I'm not sure what's on them—not least because I wrote my own "advanced" paper-tape loader, which used what I later learned were error-correcting codes to try to avoid problems with pieces of "confetti" getting stuck in the holes that had been punched in the tape:

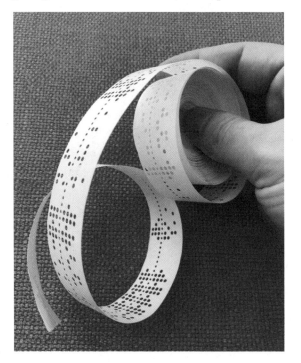

Becoming a Physicist

I don't know what would have happened if I'd thought my program was more successful in reproducing "Second Law" behavior back in 1973 when I was 13 years old. But as it was, in the summer of 1973 I was away from "my" computer, and spending all my time on particle physics. And between that summer and early 1974 I wrote a book-length summary of what I called "The Physics of Subatomic Particles":

```
C O N T E N T S.


Chapter One..............The Early History of Particle Physics
Chapter Two..............Some Basic Principles
Chapter Three............The Exclusion Principle, Antimatter, and Yukawa's Hypothesis
Chapter Four.............The Proliferation of Particles
Chapter Five.............Reactions
Chapter Six..............Symmetry and Structure
Chapter Seven............Interactions
Chapter Eight............The Detection of Particles
Chapter Nine.............The Acceleration of Particles


Bibliography

Appendix A...............Properties of Particles and Fields
Appendix B...............Abbreviations
Appendix C...............Units
Appendix D...............The Greek Alphabet
Appendix E...............Particle Accelerators
Appendix F...............Physical constants
```

I don't think I'd looked at this in any detail in 48 years. But reading it now I am a bit shocked to find history and explanations that I think are often better than I would immediately give today—even if they do bear definite signs of coming from a British early teenager writing "scientific prose".

Did I talk about statistical mechanics and the Second Law? Not directly, though there's a curious passage where I speculate about the possibility of antimatter galaxies, and their (rather un-Second-Law-like) segregation from ordinary, matter galaxies:

```
        Let us now discuss the possibility of the natural existence of antimatter in our
universe. We must start by asking how this could be detected, if it did exist. It is
obvious that two galaxies which were intrinsically the same, but one of which was
composed of antimatter, and the other matter, would look the same, and so it is useless
to use ordinary optical telescopes to search for antimatter. However, we could look
for the products of matter-antimatter collisions, which we would find in the form of
```

By the next summer I was writing the 230-page, much more technical "Introduction to the Weak Interaction". Lots of quantum mechanics and quantum field theory. No statistical mechanics. The closest it gets is a chapter on CP violation (AKA time-reversal violation)—a longtime favorite topic of mine—but from a very particle-physics point of view. By the next year I was publishing papers about particle physics, with no statistical mechanics in sight—though in a picture of me (as a "lanky youth") from that time, the *Statistical Physics* book is right there on my shelf, albeit surrounded by particle physics books:

But despite my focus on particle physics, I still kept thinking about statistical mechanics and the Second Law, and particularly its implications for the large-scale structure of the universe, and things like the possibility of matter-antimatter separation. And in early 1977, now 17 years old, and (briefly) a college student in Oxford, my archives record that I gave a talk to the newly formed (and short-lived) Oxford Natural Science Club entitled "Whither Physics" in which I talked about "large, small, many" as the main frontiers of physics, and presented the visual

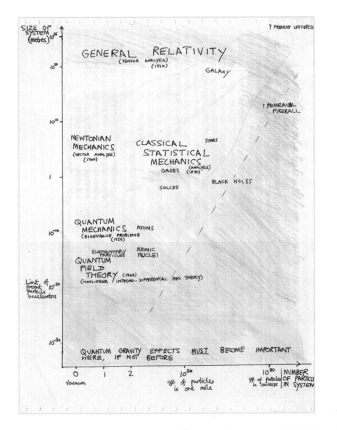

with a dash of "unsolved purple" impinging on statistical mechanics, particularly in connection with non-equilibrium situations. Meanwhile, looking at my archives today, I find some "back of the envelope" equilibrium statistical mechanics from that time (though I have no idea now what this was about):

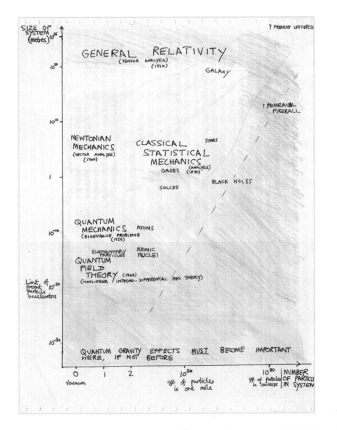

But then, in the fall of 1977 I ended up for the first time really needing to use statistical mechanics "in production". I had gotten interested in what would later become a hot area: the intersection between particle physics and the early universe. One of my interests was neutrino background radiation (the neutrino analog of the cosmic microwave background); another was early-universe production of stable charged particles heavier than the proton. And it turned out that to study these I needed all three of cosmology, particle physics, and statistical mechanics:

ABUNDANCES OF NEW STABLE PARTICLES PRODUCED IN THE EARLY UNIVERSE *

Stephen WOLFRAM

California Institute of Technology, Pasadena, CA 91125, USA

Received 10 October 1978

The standard model of the early universe is used to estimate the present abundances of possible absolutely-stable hadrons or charged leptons more massive than the proton. It is found that experimental limits on their present abundances indicate that no such particles exist with masses above about 16 GeV/c^2. Forthcoming experiments could increase this limit to masses up to around 300 GeV/c^2.

The standard model of the early universe has recently been used to place constraints on the masses and lifetimes of possible nearly-stable heavy neutrino-like particles predicted by various gauge models of weak interactions [1]. Several models of this kind imply the existence of absolutely-stable charged and/or strongly-interacting particles more massive than the proton (e.g. [2]). In this note, I show that rather large numbers of such particles would have been produced in the early universe, so that experimental limits on their terrestrial abundances may place stringent bounds on their masses.

Any new stable charged particles with masses below about 4 GeV/c^2 should already have been seen in e^+e^- interactions. The next generation of e^+e^- accelerators (PETRA, PEP) could extend this limit to masses up to 20 GeV/c^2. Attempts to produce pairs of new stable hadrons in 400 GeV proton interactions have probed up to masses ≈ 10 GeV/c^2 [2,3], but the production cross-sections for heavy hadrons near threshold are not known with sufficient accuracy for definite conclusions to be drawn [4].

The number density (n) of any species of stable particles spread uniformly throughout a homogeneous universe should obey the rate equation [1,5]

$$\frac{dn}{dt} = \frac{-3(dR/dt)}{R} n - \langle \sigma\beta c \rangle (n^2 - n_{eq}^2).$$ (1)

* Work supported in part by the U.S. Department of Energy under Contract No. EY76-C-03-0068.

where R is the expansion scale factor for the universe and $\langle \sigma\beta c \rangle$ is the product of the low-energy annihilation cross-section and relative velocity for the particles, averaged over their energy distribution at time t. n_{eq} is their number density in thermal equilibrium. The first term in eq. (1) accounts for the dilution in n due to the expansion of the universe, while the second term arises from the annihilation and production of particles in interactions. Let

$$f = \frac{n}{T^3}, \qquad x = \frac{kT}{mc^2}.$$

$$f_{eq} = \frac{n_{eq}}{T^3} = \frac{(2s+1)}{2\pi^2}\left(\frac{k}{hc}\right)^3 \int_0^\infty \frac{u^2\,du}{\exp\sqrt{(u^2 + x^{-2})} \pm 1},$$ (2)

where T is the equilibrium temperature, and f_{eq} the upper (lower) sign is for fermions (bosons). Then, ignoring the curvature of the universe, which has no effect at the times we consider, eq. (1) becomes

$$\frac{df}{dx} = Z\left[f^2(x) - f_{eq}^2(x)\right].$$ (3)

$$k^3 Z = \left(\frac{45}{8\pi^3 G}\right)^{1/2} \frac{m\langle\sigma\beta\rangle}{\sqrt{N_{\text{eff}}(T)}} (c^{11}h^3)^{1/2}$$

$$\approx 4 \times 10^{-29} \frac{\langle\sigma\beta\rangle[\text{GeV}^{-2}]\,m[\text{GeV}/c^2]}{\sqrt{N_{\text{eff}}(T)}} \text{GeV}^3\,\text{m}^3.$$

$\sigma[\text{cm}^2] \approx 4 \times 10^{-28}\sigma[\text{GeV}^{-2}]$.

65

66

In the couple of years that followed, I worked on all sorts of topics in particle physics and in cosmology. Quite often ideas from statistical mechanics would show up, like when I worked on the hadronization of quarks and gluons, or when I worked on phase transitions in the early universe. But it wasn't until 1979 that the Second Law made its first explicit appearance by name in my published work.

I was studying how there could be a net excess of matter over antimatter throughout the universe (yes, I'd by then given up on the idea of matter-antimatter separation). It was a subtle story of quantum field theory, time reversal violation, general relativity—and non-equilibrium statistical mechanics. And in the paper we wrote we included a detailed appendix about Boltzmann's H theorem and the Second Law—and the generalization we needed for relativistic quantum time-reversal-violating systems in an expanding universe:

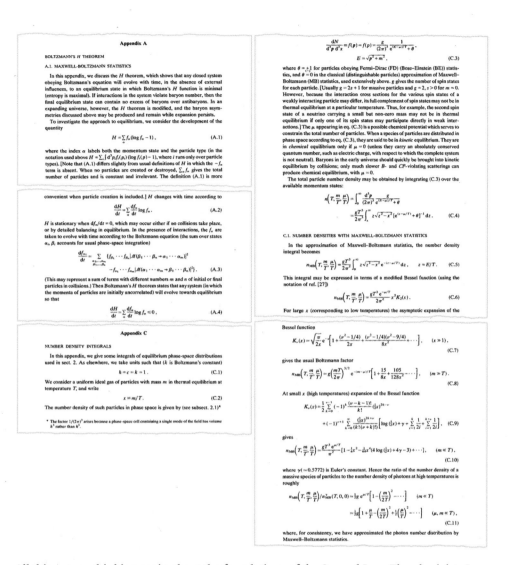

All this got me thinking again about the foundations of the Second Law. The physicists I was around mostly weren't too interested in such topics—though Richard Feynman was something of an exception. And indeed when I did my PhD thesis defense in November 1979 it ended up devolving into a spirited multi-hour debate with Feynman about the Second Law. He maintained that the Second Law must ultimately cause everything to randomize, and that the order we see in the universe today must be some kind of temporary fluctuation. I took the point of view that there was something else going on, perhaps related to gravity. Today I would have more strongly made the rather Feynmanesque point that if you have a theory that says everything we observe today is an exception to your theory, then the theory you have isn't terribly useful.

Statistical Mechanics and Simple Programs

Back in 1973 I never really managed to do much science on the very first computer I used. But by 1976 I had access to much bigger and faster computers (as well as to the ARPANET—forerunner of the internet). And soon I was routinely using computers as powerful tools for physics, and particularly for symbolic manipulation. But by late 1979 I had basically outgrown the software systems that existed, and within weeks of getting my PhD I embarked on the project of building my own computational system.

It's a story I've told elsewhere, but one of the important elements for our purposes here is that in designing the system I called SMP (for "Symbolic Manipulation Program") I ended up digging deeply into the foundations of computation, and its connections to areas like mathematical logic. But even as I was developing the critical-to-Wolfram-Language-to-this-day paradigm of basing everything on transformations for symbolic expressions, as well as leading the software engineering to actually build SMP, I was also continuing to think about physics and its foundations.

There was often something of a statistical mechanics orientation to what I did. I worked on cosmology where even the collection of possible particle species had to be treated statistically. I worked on the quantum field theory of the vacuum—or effectively the "bulk properties of quantized fields". I worked on what amounts to the statistical mechanics of cosmological strings. And I started working on the quantum-field-theory-meets-statistical-mechanics problem of "relativistic matter" (where my unfinished notes contain questions like "Does causality forbid relativistic solids?"):

But hovering around all of this was my old interest in the Second Law, and in the seemingly opposing phenomenon of the spontaneous emergence of complex structure.

SMP Version 1.0 was ready in mid-1981. And that fall, as a way to focus my efforts, I taught a "Topics in Theoretical Physics" course at Caltech (supposedly for graduate students but actually almost as many professors came too) on what, for want of a better name, I called "non-equilibrium statistical mechanics". My notes for the first lecture dived right in:

Lecture 1: Outline of non-equilibrium statistical mechanics

Statistical mechanics: study of large-scale properties of systems with very large numbers of (usually simple) elementary components.

Equilibrium statistical mechanics: study of closed systems with time-independent average behaviour.

Non-equilibrium statistical mechanics: study of systems with time-dependent large-scale properties, or open systems prevented from reaching "equilibrium" by external constraints.

Common case of non-eq. stat. mech. is gas initially constrained to some volume and then allowed to expand. Process dominated on microscopic scale by elastic collisions between gas molecules. First analysed by Boltzmann in 1860 or so. Many features still far from understood. Empirically, system tends to "equilibrium". Boltzmann transport equation (see below) apparently reproduces such behaviour. Use of equation difficult to justify, and significance of results not clear. Most non-equilibrium statistical mechanics based on Boltzmann transport equation. Two directions for fundamental improvements in non-equilibrium statistical mechanics:

(a) Attempt to derive more complete and applicable equations than Boltzmann transport equation.

(b) Attempt to find mathematical basis for tendency towards "equilibrium".

A necessary condition for validity of Boltzmann transport equation (BTE) is large separations between successive scatterings (rarefied medium). Plasmas and liquids require generalizations. All existing generalizations based on Liouville equation, either through BBGKY hierarchy or through Prigogine school's methods.

Investigations in direction (b) take one of essentially three forms:

(1) Investigate statistical behaviour of systems simpler than gases.

(2) Investigate properties of solutions to sets of differential equations derived from BTE.

(3) Attempt to extract mathematical essence of concepts of randomness and equilibrium.

(1) Next level of simplification is "pinball machines" (Lorentz models) in which single particle scatters from random array of fixed scatterers. Next consider particles specularly reflected from boundaries of "billiard tables", perhaps with polygonal obstacles on them. Particle in general bounces "randomly" around in such a region. Trajectory after a time independent "on average" of initial conditions. In special cases of regular-shaped billiard tables, trajectories exhibit simple behaviour. Consider onset of random behaviour under small perturbations to regular form. Similarly, consider conditions for "random" behaviour in e.g. sets of coupled slightly non-linear oscillators. Next level of idealization: consider differential equations with no direct physical significance. Investigate irregular

Echoing what I'd seen on that book cover back in 1972 I talked about the example of the expansion of a gas, noting that even in this case "Many features [are] still far from understood":

> Common case of non-eq. stat. mech. is gas initially constrained to some volume and then allowed to expand. Process dominated on microscopic scale by elastic collisions between gas molecules. First analysed by Boltzmann in 1880 or so. Many features still far from understood. Empirically, system tends to "equilibrium". Boltzmann transport equation (see below) apparently reproduces such behaviour. Use of equation difficult to justify, and significance of results not clear. Most non-equilibrium statistical mechanics based on Boltzmann transport

I talked about the Boltzmann transport equation and its elaboration in the BBGKY hierarchy, and explored what might be needed to extend it to things like self-gravitating systems. And then—in what must have been a very overstuffed first lecture—I launched into a discussion of "Possible origins of irreversibility". I began by talking about things like ergodicity, but soon made it clear that this didn't go the distance, and there was much more to understand—saying that "with a bit of luck" the material in my later lectures might help:

> depend on the "direction of time". An ordered system would become disordered if evolved either forwards or backwards in time. While this discussion gives qualitative justification for the occurrence of universal "irreversible" average evolution from ordered to disordered systems, a firm quantitative basis for the physical effect is still lacking. Part of the purpose of the material to be presented (with a bit of luck) in later lectures is to begin to provide such a basis.

I continued by noting that some systems can "develop order and considerable organization"—which non-equilibrium statistical mechanics should be able to explain:

> Even though closed systems may on average tend to more disordered states, it is clear that parts of closed systems may develop order and considerable

> organization. Living systems attests to this. A major challenge for non-equilibrium statistical mechanics is to provide some quantitative explanation for the occurrence of local ordered systems. The attempts to take up this challenge will be discussed in some later lectures - the attempts have a long way to go.

I then went quite "cosmological":

> One system in which the occurrence of local order has been discussed extensively is the complete universe. Statistical discussions of the complete universe can run into trouble because we have access to a rather small number of universes (namely one). There are several potential explanations of order in the universe:

The first candidate explanation I listed was the fluctuation argument Feynman had tried to use:

> 1. The observed order in the local portion of the universe is a statistical fluctuation. If we could see the whole universe, most of it would be disordered. This was the argument originally used by Boltzmann. It has been discussed in conjunction with the "anthropic principle": that the universe or some part of it has to be the way it is for it to be possible for intelligent (?) creatures such as ourselves to observe it. It is not clear what the scientific content of such a statement is. And without seeing other examples of intelli-

I discussed the possibility of fundamental microscopic irreversibility—say associated with time-reversal violation in gravity—but largely dismissed this. I talked about the possibility that the universe could have started in a special state in which "the matter is in thermal equilibrium, but the gravitational field is not." And finally I gave what the 22-year-old me thought at the time was the most plausible explanation:

> 4. Gravitational clumping is responsible for ordering the universe. Either the clumping increases the entropy of the gravitational field more than it decreases the entropy of the matter, or the BTE fails for self-gravitating systems. No way is known to calculate in general entropy associated with gravitational field. Hawking entropy can apparently be associated with black holes, but for weaker fields, no useful results. Should determine to what extent formation of stars is irreversible. Box of gas collapses to star,

All of this was in a sense rooted in a traditional mathematical physics style of thinking. But the second lecture gave a hint of a quite different approach:

> ## 2. Information theory and theory of computation
> Information theory: Statistical description of properties of sets or sequences.
> Computation theory: General quantitative description of organization of fundamental logical processes.

In my first lecture, I had summarized my plans for subsequent lectures:

> 1. (Remainder) Boltzmann transport equation and implications.
> 2. Information theory and theory of computation.
> 3. Application/digression: The early universe
> 4. Theory of small deviations from equilibrium.
> 5. Large deviations from equilibrium.

But discovery intervened. People had discussed reaction-diffusion patterns as examples of structure being formed "away from equilibrium". But I was interested in more dramatic examples, like galaxies, or snowflakes, or turbulent flow patterns, or forms of biological organisms. What kinds of models could realistically be made for these? I started from neural networks, self-gravitating gases and spin systems, and just kept on simplifying and simplifying. It was rather like language design, of the kind I'd done for SMP. What were the simplest primitives from which I could build up what I wanted?

Before long I came up with what I'd soon learn could be called one-dimensional cellular automata. And immediately I started running them on a computer to see what they did:

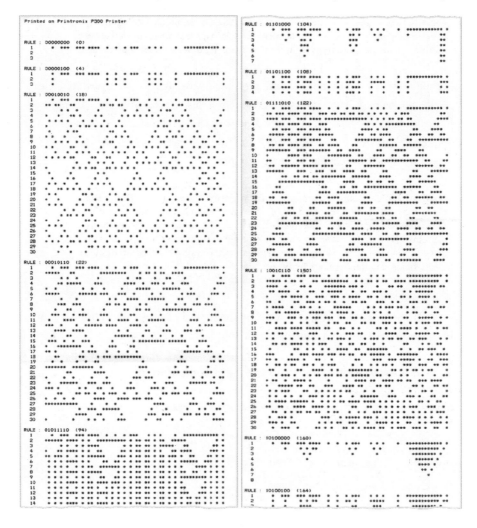

And, yes, they were "organizing themselves"—even from random initial conditions—to make all sorts of structures. By December I was beginning to frame how I would write about what was going on:

CALT 68-XXX
DoE RESEARCH AND
DEVELOPMENT REPORT

Complex Structures and Statistical Mechanics

Stephen Wolfram
California Institute of Technology, Pasadena CA 91125

December 1981

ABSTRACT

Systems with many components are often observed to evolve with time to a disordered "equilibrium" state. This behaviour is reproduced by the standard statistical mechanics of Boltzmann and Gibbs. However, a wide variety of complex systems do not share this tendency towards increasing disorder, and may even spontaneously appear to become more ordered. Examples of this behaviour are the formation of crystals or of traffic jams, the appearance of patterns in turbulent fluids, and the persistence of biological systems. Often the results of this self-organization may be very complex, and in the cases of snowflakes or living creatures. These notes discuss some attempts to find a framework for describing self-organizing systems, through which universal features of such systems may be identified.

The commonest system which "tends to become disordered" is the classical ideal gas. In the standard Boltzmann-Gibbs treatment, a state of an N-particle gas is represented by a point in a 6N-dimensional phase space whose coordinates specify the positions and momenta of each of the particles. Collisions in the gas cause this point to move erratically

* Work supported in part by the U.S. Department of Energy under Contract No. DE-AC-03-81-ER40050 and by the Fleischmann Foundation.

And by May 1982 I had written my first long paper about cellular automata (published in 1983 under the title "Statistical Mechanics of Cellular Automata"):

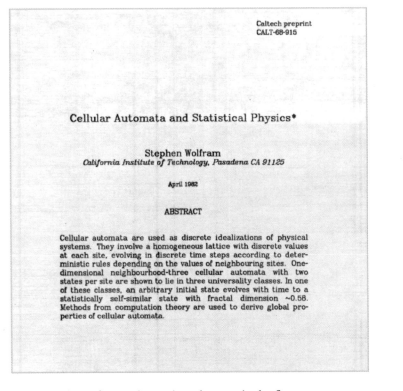

Caltech preprint
CALT-68-915

Cellular Automata and Statistical Physics*

Stephen Wolfram

California Institute of Technology, Pasadena CA 91125

April 1982

ABSTRACT

Cellular automata are used as discrete idealizations of physical systems. They involve a homogeneous lattice with discrete values at each site, evolving in discrete time steps according to deterministic rules depending on the values of neighbouring sites. One-dimensional neighbourhood-three cellular automata with two states per site are shown to lie in three universality classes. In one of these classes, an arbitrary initial state evolves with time to a statistically self-similar state with fractal dimension ~0.58. Methods from computation theory are used to derive global properties of cellular automata.

The Second Law featured prominently, even in the first sentence:

1. Introduction

The Second Law of thermodynamics implies that isolated microscopically-reversible physical systems tend with time to states of maximal entropy and maximal "disorder". However, "dissipative" systems involving microscopic irreversibility, or open to interactions with their environment, may evolve from "disordered" to more "ordered" states. The states attained often exhibit a complicated structure. Examples are outlines of snowflakes, patterns of flow in turbulent fluids, and biological systems. The purpose of this paper is to begin the investigation of cellular automata (introduced in sect. 2) as a class of mathematical models for such behaviour. Cellular automata are sufficiently simple to allow detailed mathematical analysis, yet sufficiently complex to exhibit a wide variety of complicated phenomena. The ultimate goal is to abstract from a study of cellular automata general features of "self-organizing" behaviour and perhaps to devise universal laws analogous to the laws of thermodynamics.

I made quite a lot out of the fundamentally irreversible character of most cellular automaton rules, pretty much assuming that this was the fundamental origin of their ability to "generate complex structures"—as the opening transparencies of two talks I gave at the time suggested:

It wasn't that I didn't know there could be reversible cellular automata. And a footnote in my paper even records the fact these can generate nested patterns with a certain fractal dimension—as computed in a charmingly manual way on a couple of pages I now find in my archives:

But somehow I hadn't quite freed myself from the assumption that microscopic irreversibility was what was "causing" structures to be formed. And this was related to another important—and ultimately incorrect—assumption: that all the structure I was seeing was somehow the result of the "filtering" random initial conditions. Right there in my paper is a picture of rule 30 starting from a single cell:

And, yes, the printout from which that was made is still in my archives, if now a little worse for wear:

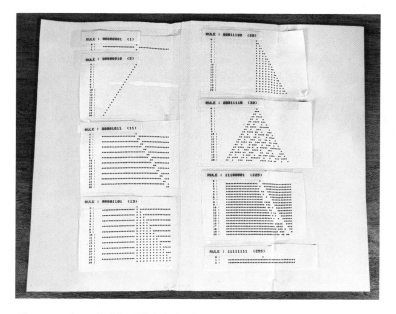

Of course, it probably didn't help that with my "display" consisting of an array of printed characters I couldn't see too much of the pattern—though my archives do contain a long "imitation-high-resolution" printout of the conveniently narrow, and ultimately nested, pattern from rule 225:

But I think the more important point was that I just didn't have the necessary conceptual framework to absorb what I was seeing in rule 30—and I wasn't ready for the intuitional shock that it takes only simple rules with simple initial conditions to produce highly complex behavior.

My motivation for studying the behavior of cellular automata had come from statistical mechanics. But I soon realized that I could discuss cellular automata without any of the "baggage" of statistical mechanics, or the Second Law. And indeed even as I was finishing my long statistical-mechanics-themed paper on cellular automata, I was also writing a short paper that described cellular automata essentially as purely computational systems (even though I still used the term "mathematical models") without talking about any kind of Second Law connections:

Caltech preprint
CALT-68-938

Cellular Automata as Simple Self-Organizing Systems

Stephen Wolfram*
Physics Department, California Institute of Technology, Pasadena CA 91125

(July 1982; revised November 1982)

ABSTRACT

Cellular automata provide simple discrete deterministic mathematical models for physical, biological and computational systems. Despite their simple construction, cellular automata are shown to be capable of complicated behaviour, and to generate complex patterns with universal features. An outline of their statistical mechanics is given.

Introduction

An "elementary" cellular automaton consists of a sequence of sites carrying values 0 or 1 arranged on a line. The value at each site evolves deterministically with time according to a set of definite rules involving the values of its nearest neighbours. In general, the sites of a cellular automaton may be arranged on any regular lattice, and each site may take on any discrete set of values. This article concentrates on the case of "elementary" cellular automata in one dimension with binary values at each site, and shows that despite their simple construction, such systems can exhibit complicated behaviour. Details, extensions and further discussion, together with more extensive references, are given in ref. [1].

Through much of 1982 I was alternating between science, technology and the startup of my first company. I left Caltech in October 1982, and after stops at Los Alamos and Bell Labs, started working at the Institute for Advanced Study in Princeton in January 1983, equipped with a newly obtained Sun workstation computer whose ("one megapixel") bitmap display let me begin to see in more detail how cellular automata behave:

It had very much the flavor of classic observational science—looking not at something like mollusc shells, but instead at images on a screen—and writing down what I saw in a "lab notebook":

What did all those rules do? Could I somehow find a way to classify their behavior?

	5 4 3 2 1 0		fractal	bounded/unbounded	D_1
2	1 0	triangles		U	$\log_2 3$
4	1 0 0	dies quickly		B	
6	1 1 0	triangles		U	$\log_2 3$
8	1 0 0 0	triv. attractor		B	
10	1 0 1 0	irregular tri		U	?
12	1 1 0 0	triangles		U	—
14	1 1 1 0	triangles		U	$\log_2 3$
16	1 0 0 0 0	dies quickly		B	
18	1 0 0 1 0	irregular		U	?
20	1 0 1 0 0	complex		B	
22	1 0 1 1 0	triangles		U	?
24	1 1 0 0 0	trivial attractors		B	
26	1 1 0 1 0	triangles		U	?
28	1 1 1 0 0	triangles		U	—
30	1 1 1 1 0	triangles		U	$\log_2 3$
32	1 0 0 0 0 0	dies quickly		B	
34	1 0 0 0 1 0	triangles		U	$\log_2 3$
36	1 0 0 1 0 0	dies quickly		B	
38	1 0 0 1 1 0	irregular		U	~ R150
40	1 0 1 0 0 0	triv. attr.		B	
42	1 0 1 0 1 0	irregular		U	~ R150 ?
44	1 0 1 1 0 0	triangles		U	—
46	1 0 1 1 1 0	triangles		U	?
48	1 1 0 0 0 0	dies quickly		B	
50	1 1 0 0 1 0	triangles		U	?
52	1 1 0 1 0 0	complex		(U)	
54	1 1 0 1 1 0	quick homogenization		B	
56	1 1 1 0 0 0	triv. attr.		B	
58	1 1 1 0 1 0	triv. attr. (0 or 1 bkgd)		B	
60	1 1 1 1 0 0	quick homog.		B	
62	1 1 1 1 1 0	quick homog.		B	

In irregular cases, appears that patterns starting from
small initial states are not self-similar. (e.g. code 10)

42 should correspond to a linear rule.

Mostly I was looking at random initial conditions. But in a near miss of the rule 30 phenomenon I wrote in my lab notebook: "In irregular cases, appears that patterns starting
from small initial states are not self-similar (e.g. code 10)". I even looked again at asymmetric "elementary" rules (of which rule 30 is an example)—but only from random initial
conditions (though noting the presence of "class 4" rules, which would include rule 110):

Consider asymmetric elementary rules

Some are definitely in class 4.
Some tend to symmetry from random initial states.
Some e.g. R120 have new forms.

My technology stack at the time consisted of printing screen dumps of cellular automaton behavior

then using repeated photocopying to shrink them—and finally cutting out the images and assembling arrays of them using Scotch tape:

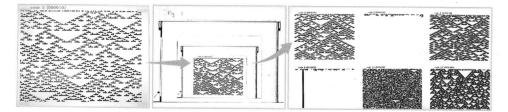

And looking at these arrays I was indeed able to make an empirical classification, identifying initially five—but in the end four—basic classes of behavior. And although I sometimes made analogies with solids, liquids and gases—and used the mathematical concept of entropy—I was now mostly moving away from thinking in terms of statistical mechanics, and was instead using methods from areas like dynamical systems theory, and computation theory:

PRELIMINARY DRAFT

Universality and Complexity in Cellular Automata

Stephen Wolfram
The Institute for Advanced Study, Princeton NJ 08540.

February 1983

ABSTRACT

Cellular automata are discrete dynamical systems with simple construction but complicated self-organizing behaviour. Evidence is presented that all one-dimensional cellular automata fall into five distinct universality classes. Characterizations of the structures generated in these classes are discussed. One class is probably capable of universal computation.

Even so, when I summarized the significance of investigating the computational characteristics of cellular automata, I reached back to statistical mechanics, suggesting that much as information theory provided a mathematical basis for equilibrium statistical mechanics, so similarly computation theory might provide a foundation for non-equilibrium statistical mechanics:

> related to determinations of equivalence of systems and problem classes in computation theory. In general, one may hope for fundamental connections between computation theory and the theory of complex non-equilibrium statistical systems. Information theory forms a mathematical basis for equilibrium statistical mechanics. Computation theory, which addresses time-dependent processes, may be expected to play a fundamental role in non-equilibrium statistical mechanics.

Computational Irreducibility and Rule 30

My experiments had shown that cellular automata could "spontaneously produce structure" even from randomness. And I had been able to characterize and measure various features of this structure, notably using ideas like entropy. But could I get a more complete picture of what cellular automata could make? I turned to formal language theory, and started to work out the "grammar of possible states". And, yes, a quarter century before Graph in Wolfram Language, laying out complicated finite state machines wasn't easy:

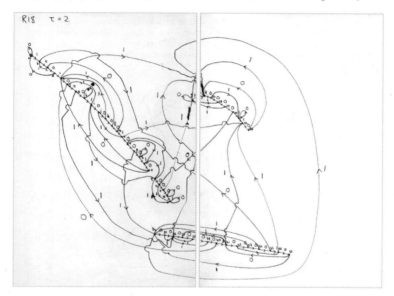

But by November 1983 I was writing about "self-organization as a computational process":

Computation Theory of Cellular Automata

Stephen Wolfram*
The Institute for Advanced Study, Princeton NJ 08540.

November 1983

ABSTRACT

Self-organizing behaviour in cellular automata is discussed as a computational process. Formal language theory is used to extend dynamical systems theory descriptions of cellular automata. The set of configurations generated after a finite number of time steps of cellular automaton evolution is shown to form a regular language. The size of the minimal grammar for this language provides a measure of the complexity of the set. This complexity is found to be non-decreasing with time. The limiting sets generated by some cellular automata appear to form more complicated formal languages. Many properties of these sets are formally non-computable. It is suggested that such undecidability is common in these and other dynamical systems.

The introduction to my paper again led with the Second Law, though now talked about the idea that computation theory might be what could characterize non-equilibrium and self-organizing phenomena:

---

1. Introduction

     Systems which follow the second law of thermodynamics evolve with time to maximal entropy and complete disorder, destroying any order initially present. Cellular automata are examples of mathematical systems which may instead exhibit "self-organizing" behaviour*. Even starting from complete disorder, their irreversible evolution can spontaneously generate ordered structure. One coarse indication of such self-organization is a decrease of entropy with time. This paper discusses an approach to a more complete mathematical characterization of self-organizing processes in cellular automata, and possible quantitative measures of the "complexity" generated by them. The evolution of cellular automata is viewed as a computation which processes information specified as the initial state. The structure of the output from such information processing is then described using the mathematical theory of formal languages (e.g. [6,7,8]). Computation and formal language theory may in general be expected to play a role in the theory of non-equilibrium and self-organizing systems analogous to the role of information theory in conventional statistical mechanics.

The concept of equilibrium in statistical mechanics makes it natural to ask what will happen in a system after an infinite time. But computation theory tells one that the answer to that question can be non-computable or undecidable. I talked about this in my paper, but then ended by discussing the ultimately much richer finite case, and suggesting (with a reference to NP completeness) that it might be common for there to be no computational shortcut to cellular automaton evolution. And rather presciently, I made the statement that "One may speculate that [this phenomenon] is widespread in physical systems" so that "the consequences of their evolution could not be predicted, but could effectively be found only by direct simulation or observation.":

     Undecidability and non-computability are features of problems which attempt to summarize the consequences of infinite processes. Finite processes may always be carried out explicitly. For some particularly simple processes, the consequences of a large, but finite, number of steps may be deduced by a procedure involving only a small number of steps. But at least for many computational processes (e.g. [24]), it is believed that that no such short cut exists: each step (or each possibility) must in fact be carried out explicitly. It was suggested that this phenomenon is common in cellular automata. One may speculate that it is widespread in physical systems. No simple theory or formula could ever be given for the overall behaviour of such systems: the consequences of their evolution could not be predicted, but could effectively be found only by direct simulation or observation.

These were the beginnings of powerful ideas, but I was still tying them to somewhat technical things like ensembles of all possible states. But in early 1984, that began to change. In January I'd been asked to write an article for the then-top popular science magazine *Scientific American* on the subject of "Computers in Science and Mathematics". I wrote about the general idea of computer experiments and simulation. I wrote about SMP. I wrote about cellular automata. But then I wanted to bring it all together. And that was when I came up with the term "computational irreducibility".

By May 26, the concept was pretty clearly laid out in my draft text:

```
May 26 00:10 1984   new Page 1


.LP
.SH
Computers in science and mathematics
.PP
There used to be two basic approaches in science: experimental and theoretical.
Computers are making possible a third approach: computer experimentation.
.PP
In experimental science, one arranges systems to be as simple as possible, and
then observes the phenomena they exhibit. In theoretical science, one constructs
mathematical models, and attempts to work out their consequences for situations
that include as much of the complications of reality as possible. Computers make
feasible on the one hand a new type of experiment, and on the other hand, a new
class of mathematical models.
.PP
```

```
.PP
There are some physical systems whose behaviour can be determined by much
more efficient methods than explicit simulation. To find the position
of a planet orbiting the sun according to Newton's laws, one need not
simulate its motion through time, but instead one may simply evaluate
a simple formula. However, such short cuts may well be the exception rather
than the rule among physical systems.
.PP
For a cellular automaton model, such a short cut would imply that it
is possible to find the consequence of many time steps of cellular automaton
evolution directly, without explicitly simulating each step. But if, for exam
the
cellular automaton is capable of universal computation, it can be shown
that no such short cut is in general possible. The only way to find out
the behaviour of the cellular automaton is essentially to simulate each time
of its evolution explicitly. The evolution of the cellular automaton
implements an irreducible computation: there is no more efficient algorithm
to find the result of the evolution than the evolution itself.
.PP
It is not yet known how widespread computational irreducibility is in physical
systems. But one suspects that it is present in some form whenever very
complicated or chaotic behaviour occurs. And when it is present, there
can be no simple formulae to describe the overall behaviour of the system:
the behaviour can effectively be found only by explicit simulation.
Once one has abstracted the mathematical essence of a physical process,
such as fluid turbulence, or biological evolution, it may well be that
further investigation can be carried out only by explicit, step-by-step
simulation. The investigation is then reduced to direct observation or experi
albeit on a mathematical idealization of the original system.
.PP
The phenomenon of computational irreducibility implies many fundamental
```

```
limitations on
the scope of conventional mathematical theories for physical systems.
It suggests that there are many systems whose investigation is ultimately
reduced to direct computer simulation. The level of the objects in this
simulation is crucial. If to find the behaviour of turbulent fluid it was
necessary to simulate explicitly the motion of every molecule, there would
be no possibility of a meaningful predictive theory for the fluid.
For then a computer to calculate the behaviour of the fluid would have to
have essentially as many elements as there were molecules in the fluid itself:
only a system effectively equivalent to the fluid itself could then describe the
fluid. In practice, however, it seems likely that the relevant features of the
fluid can be determined from a higher-level simulation. One need not account
for individual molecules, but only perhaps for packets of fluid, specified
by their positions and velocities. This higher-level description may involve
sufficiently few components that it may meaningful be simulated by a program
on a general-purpose computer.
```

But just a few days later something big happened. On June 1 I left Princeton for a trip to Europe. And in order to "have something interesting to look at on the plane" I decided to print out pictures of some cellular automata I hadn't bothered to look at much before. The first one was rule 30:

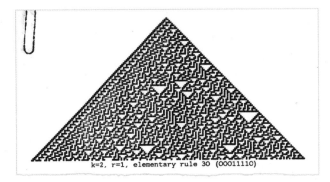

And it was then that it all clicked. The complexity I'd been seeing in cellular automata wasn't the result of some kind of "self-organization" or "filtering" of random initial conditions. Instead, here was an example where it was very obviously being "generated intrinsically" just by the process of evolution of the cellular automaton. This was computational irreducibility up close. No need to think about ensembles of states or statistical mechanics. No need to think about elaborate programming of a universal computer. From just a single black cell rule 30 could produce immense complexity, and showed what seemed very likely to be clear computational irreducibility.

Why hadn't I figured out before that something like this could happen? After all, I'd even generated a small picture of rule 30 more than two years earlier. But at the time I didn't have a conceptual framework that made me pay attention to it. And a small picture like that just didn't have the same in-your-face "complexity from nothing" character as my larger picture of rule 30.

Of course, as is typical in the history of ideas, there's more to the story. One of the key things that had originally let me start "scientifically investigating" cellular automata is that out of all the infinite number of possible constructible rules, I'd picked a modest number on which I could do exhaustive experiments. I'd started by considering only "elementary" cellular automata, in one dimension, with $k = 2$ colors, and with rules of range $r = 1$. There are 256 such "elementary rules". But many of them had what seemed to me "distracting" features— like backgrounds alternating between black and white on successive steps, or patterns that systematically shifted to the left or right. And to get rid of these "distractions" I decided to focus on what I (somewhat foolishly in retrospect) called "legal rules": the 32 rules that leave blank states blank, and are left-right symmetric.

When one uses random initial conditions, the legal rules do seem—at least in small pictures— to capture the most obvious behaviors one sees across all the elementary rules. But it turns out that's not true when one looks at simple initial conditions. Among the "legal" rules, the most complicated behavior one sees with simple initial conditions is nesting.

But even though I concentrated on "legal" rules, I still included in my first major paper on cellular automata pictures of a few "illegal" rules starting from simple initial conditions—including rule 30. And what's more, in a section entitled "Extensions", I discussed cellular automata with more than 2 colors, and showed—though without comment—the pictures:

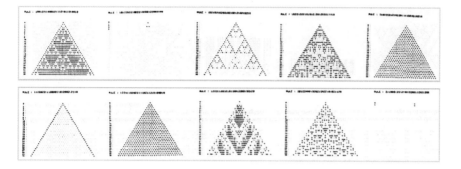

These were low-resolution pictures, and I think I imagined that if one ran them further the behavior would somehow resolve into something simple. But by early 1983, I had some clues that this wouldn't happen. Because by then I was generating fairly high-resolution pictures—including ones of the $k=2$, $r=2$ totalistic rule with code 10 starting from a simple initial condition:

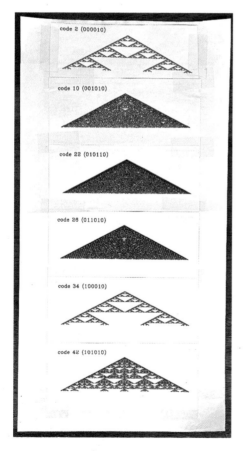

In early drafts of my 1983 paper on "Universality and Complexity in Cellular Automata" I noted the generation of "irregularity", and speculated that it might be associated with class 4 behavior. But later I just stated as an observation without "cause" that some rules—like code 10—generate "irregular patterns". I elaborated a little, but in a very "statistical mechanics" kind of way, not getting the main point:

> Rule such as code 10 are seen to generate irregular patterns by evolution even from a single site initial state. The density of nonzero sites in such patterns is found to tend asymptotically to a nonzero value; in some, but not all, cases the value is the same as would be obtained by evolution from a disordered initial state. The patterns appear to exhibit no large-scale structure.
>
> Cellular automata contain no intrinsic scale beyond the size of neighbourhood which appears in their rules. A configuration containing a single nonzero site is also scale invariant, and any pattern obtained by evolution from it with cellular automaton rules must be scale invariant. The regular patterns in fig. 9 achieve this scale invariance by their self-similarity. The irregular patterns presumably exhibit correlations only over a finite range, and are therefore effectively uniform and scale invariant at large distances.

In September 1983 I did a little better:

> For large k and r, an increasing fraction of cellular automaton rules generate chaotic patterns. No simple characterization of these patterns is known. Even though the patterns may be specified by simple rules and simple initial states, they appear to have a complex and random form. Statistical properties are typically almost indistinguishable from those of uncorrelated site value sequences.

But in the end it wasn't until June 1, 1984, that I really grokked what was going on. And a little over a week later I was in a scenic area of northern Sweden

at a posh "Nobel Symposium" conference on "The Physics of Chaos and Related Problems"—talking for the first time about the phenomenon I'd seen in rule 30 and code 10. And from

June 15 there's a transcript of a discussion session where I bring up the never-before-mentioned-in-public concept of computational irreducibility—and, unsurprisingly, leave the other participants (who were basically all traditional mathematically oriented physicists) at best slightly bemused:

> **Jeffries:** I'd like to suggest an answer to Wolfram's question. What prediction means to me is that you write down a model for the system, numerically integrate it, take Poincaré sections, and see if they look like recognizable maps, e.g. circle maps, etc. That is a way to make a prediction. It's even faster to do the experiment, but this is a method which is based on what you know about the system.
>
> **Wolfram:** That's very close to just simulating. I assume that what one means by prediction is that for any given system you have the equations and in a fixed amount of time you integrate them. It could be that most physically interesting systems are computationally irreducible.
>
> **?:** What does that mean?
>
> **Wolfram:** It means that at least some questions one can ask about these systems will not be answerable by a procedure much more efficient than directly simulating the system. The real question is how generic that behavior is and whether or not it will turn out that most of the interesting questions about the system can in fact be answered more simply. By computationally irreducible, one therefore means that the infinite time (or space) questions are undecidable.
>
> *Physica Scripta T9*

I think I was still a bit prejudiced against rule 30 and code 10 as specific rules: I didn't like the asymmetry of rule 30, and I didn't like the rapid growth of code 10. (Rule 73—while symmetric—I also didn't like because of its alternating background.) But having now grokked the rule 30 phenomenon I knew it also happened in "more aesthetic" "legal" rules with more than 2 colors. And while even 3 colors led to a rather large total space of rules, it was easy to generate examples of the phenomenon there.

A few days later I was back in the US, working on finishing my article for *Scientific American*. A photographer came to help get pictures from the color display I now had:

And, yes, those pictures included multicolor rules that showed the rule 30 phenomenon:

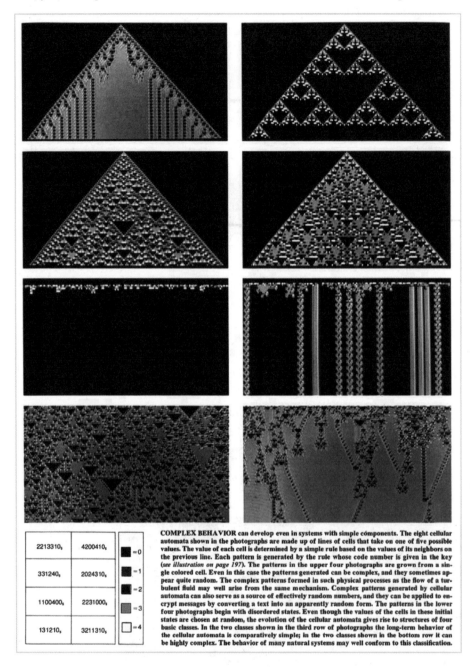

2213310₅	4200410₅
331240₅	2024310₅
1100400₅	2231000₅
131210₅	3211310₅

■	=0
■	=1
■	=2
▨	=3
□	=4

COMPLEX BEHAVIOR can develop even in systems with simple components. The eight cellular automata shown in the photographs are made up of lines of cells that take on one of five possible values. The value of each cell is determined by a simple rule based on the values of its neighbors on the previous line. Each pattern is generated by the rule whose code number is given in the key (*see illustration on page 197*). The patterns in the upper four photographs are grown from a single colored cell. Even in this case the patterns generated can be complex, and they sometimes appear quite random. The complex patterns formed in such physical processes as the flow of a turbulent fluid may well arise from the same mechanism. Complex patterns generated by cellular automata can also serve as a source of effectively random numbers, and they can be applied to encrypt messages by converting a text into an apparently random form. The patterns in the lower four photographs begin with disordered states. Even though the values of the cells in these initial states are chosen at random, the evolution of the cellular automata gives rise to structures of four basic classes. In the two classes shown in the third row of photographs the long-term behavior of the cellular automata is comparatively simple; in the two classes shown in the bottom row it can be highly complex. The behavior of many natural systems may well conform to this classification.

The caption I wrote commented: "Even in this case the patterns generated can be complex, and they sometimes appear quite random. The complex patterns formed in such physical processes as the flow of a turbulent fluid may well arise from the same mechanism."

The article went on to describe computational irreducibility and its implications in quite a lot of detail— illustrating it rather nicely with a diagram, and commenting that "It seems likely that many physical and mathematical systems for which no simple description is now known are in fact computationally irreducible":

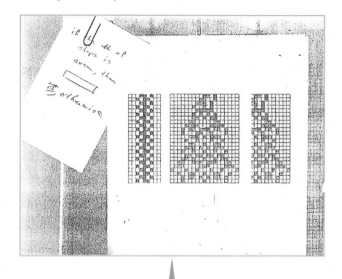

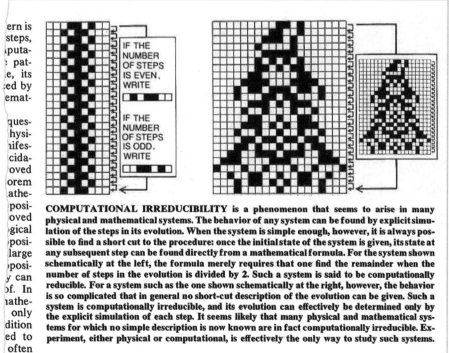

COMPUTATIONAL IRREDUCIBILITY is a phenomenon that seems to arise in many physical and mathematical systems. The behavior of any system can be found by explicit simulation of the steps in its evolution. When the system is simple enough, however, it is always possible to find a short cut to the procedure: once the initial state of the system is given, its state at any subsequent step can be found directly from a mathematical formula. For the system shown schematically at the left, the formula merely requires that one find the remainder when the number of steps in the evolution is divided by 2. Such a system is said to be computationally reducible. For a system such as the one shown schematically at the right, however, the behavior is so complicated that in general no short-cut description of the evolution can be given. Such a system is computationally irreducible, and its evolution can effectively be determined only by the explicit simulation of each step. It seems likely that many physical and mathematical systems for which no simple description is now known are in fact computationally irreducible. Experiment, either physical or computational, is effectively the only way to study such systems.

I also included an example—that would show up almost unchanged in *A New Kind of Science* nearly 20 years later—indicating how computational irreducibility could lead to undecidability (back in 1984 the picture was made by stitching together many screen photographs, yes, with strange artifacts from long-exposure photography of CRTs):

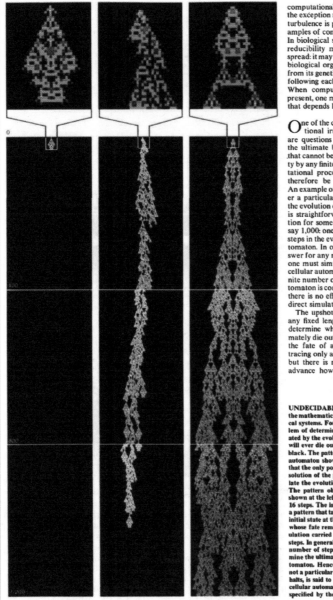

computational reducibility may well be the exception rather than the rule. Fluid turbulence is probably one of many examples of computational irreducibility. In biological systems computational irreducibility may be even more widespread: it may turn out that the form of a biological organism can be determined from its genetic code essentially only by following each step in its development. When computational irreducibility is present, one must adopt a methodology that depends heavily on computation.

One of the consequences of computational irreducibility is that there are questions that can be asked about the ultimate behavior of a system but that cannot be answered in full generality by any finite mathematical or computational process. Such questions must therefore be considered undecidable. An example of such a question is whether a particular pattern ever dies out in the evolution of a cellular automaton. It is straightforward to answer the question for some definite number of steps, say 1,000: one need only simulate 1,000 steps in the evolution of the cellular automaton. In order to determine the answer for any number of steps, however, one must simulate the evolution of the cellular automaton for a potentially infinite number of steps. If the cellular automaton is computationally irreducible, there is no effective alternative to such direct simulation.

The upshot is that no calculation of any fixed length can be guaranteed to determine whether a pattern will ultimately die out. It may be possible to tell the fate of a particular pattern after tracing only a few steps in its evolution, but there is no general way to tell in advance how many steps will be re-

UNDECIDABLE PROBLEMS can arise in the mathematical analysis of models of physical systems. For example, consider the problem of determining whether a pattern generated by the evolution of a cellular automaton will ever die out, so that all the cells become black. The patterns generated by the cellular automaton shown above are so complicated that the only possible general approach to the solution of the problem is to explicitly simulate the evolution of the cellular automaton. The pattern obtained from the initial state shown at the left is found to die out after just 16 steps. The initial state in the center yields a pattern that takes 1,016 steps to die out. The initial state at the right gives rise to a pattern whose fate remains unclear even after a simulation carried out over many thousands of steps. In general no finite simulation of a fixed number of steps can be guaranteed to determine the ultimate behavior of the cellular automaton. Hence the problem of whether or not a particular pattern ultimately dies out, or halts, is said to be formally undecidable. The cellular automaton shown here follows a rule specified by the code number 3311100320$_4$.

In a rather newspaper-production-like experience, I spent the evening of July 18 at the offices of *Scientific American* in New York City putting finishing touches to the article, which at the end of the night—with minutes to spare—was dispatched for final layout and printing.

But already by that time, I was talking about computational irreducibility and the rule 30 phenomenon all over the place. In July I finished "Twenty Problems in the Theory of Cellular Automata" for the proceedings of the Swedish conference, including what would become a rather standard kind of picture:

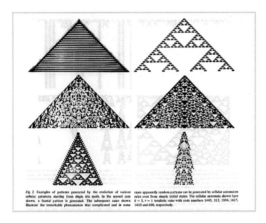

Fig. 2. Examples of patterns generated by the evolution of various cases apparently random patterns can be generated by cellular automata cellular automata starting from single site seeds. In the second case rules even from simple initial states. The cellular automata shown have shown, a fractal pattern is generated. The subsequent cases shown $k = 3$, $r = 1$ totalistic rules with code numbers 1443, 312, 1554, 1617, illustrate the remarkable phenomenon that complicated and in some 1410 and 600, respectively.

Problem 15 talks specifically about rule 30, and already asks exactly what would—35 years later—become Problem #2 in my 2019 Rule 30 Prizes

> Often the temporal sequences that appear in these patterns have a seemingly random form, and satisfy many statistical tests for randomness. There is empirical evidence that in many cases the sequence of values taken on say by the centre site in the pattern contains all possible subsequences with equal frequencies, so that the whole sequence effectively has maximal measure entropy. A simple example of this phenomenon occurs in the $k=2$, $r=1$ rule number 30 $(a_i^{(t+1)} = a_{i-1}^{(t)} \oplus \max(a_i^{(t)}, a_{i+1}^{(t)}))$.
>
> Systems that exhibit chaotic behaviour usually start from initial conditions that contain an infinite amount of information, either in the form of an infinite sequence of cellular automaton site values, or the infinite sequence of digits in a real number. Their irregular behaviour with time

while Problem 18 asks the (still largely unresolved) question of what the ultimate frequency of computational irreducibility is:

> **Problem 18.** *How common is computational irreducibility in cellular automata?*
>
> One way to find out the behaviour of a cellular automaton is to simulate each step in its evolution explicitly. The question is how often there are better ways.
>
> Cellular automaton evolution can be considered as a computation. A procedure can short cut this evolution only if it involves a more sophisticated computation. But there are cellular automata capable of universal computation that can perform arbitrarily sophisticated computations. So at least in these cases no short cut procedure can in general be found. The cellular automaton evolution corresponds to an irreducible computation, whose outcome can be found effectively only by carrying it out explicitly.

Very late in putting together the *Scientific American* article I'd added to the caption of the picture showing rule-30-like behavior the statement "Complex patterns generated by cellular automata can also serve as a source of effectively random numbers, and they can be applied to encrypt messages by converting a text into an apparently random form." I'd realized both that cellular automata could act as good random generators (we used rule 30 as the default in

Wolfram Language for more than 25 years), and that their evolution could effectively encrypt things, much as I'd later describe the Second Law as being about "encrypting" initial conditions to produce effective irreversibility.

Back in 1984 it was a surprising claim that something as simple and "science-oriented" as a cellular automaton could be useful for encryption. Because at the time practical encryption was basically always done by what at least seemed like arbitrary and complicated engineering solutions, whose security relied on details or explanations that were often considered military or commercial secrets.

I'm not sure when I first became aware of cryptography. But back in 1973 when I first had access to a computer there were a couple of kids (as well as a teacher who'd been a friend of Alan Turing's) who were programming Enigma-like encryption systems (perhaps fueled by what were then still officially just rumors of World War II goings-on at Bletchley Park). And by 1980 I knew enough about encryption that I made a point of encrypting the source code of SMP (using a modified version of the Unix crypt program). (As it happens, we lost the password, and it was only in 2015 that we got access to the source again.)

My archives record a curious interaction about encryption in May 1982—right around when I'd first run (though didn't appreciate) rule 30. A rather colorful physicist I knew named Brosl Hasslacher (who we'll encounter again later) was trying to start a curiously modern-sounding company named Quantum Encryption Devices (or QED for short)—that was actually trying to market a quite hacky and definitively classical (multiple-shift-register-based) encryption system, ultimately to some rather shady customers (and, yes, the "expected" funding did not materialize):

QUANTA
INC.
P.O. Box 490 – Los Alamos, New Mexico 87544

Stephen Wolfram
CALTECH
440-48
Pasadena, CA 91124

May 6, 1982

Dear Stephen,

Thank you for your work on the software implementation of Q.E.D. Inc. Stream Encryptor. This note is to formalize our agreement to pay you $ 300 when we acquire the major financing. We have at your suggestion already paid $ 300 to Tim Shaw for his share of the work.

Sincerely,

Brosl Hasslacher

But it was 1984 before I made a connection between encryption and cellular automata. And the first thing I imagined was giving input as the initial condition of the cellular automaton, then running the cellular automaton rule to produce "encrypted output". The most straightforward way to make encryption was then to have the cellular automaton rule be reversible, and to run the inverse rule to do the decryption. I'd already done a little bit of investigation of reversible rules, but this led to a big search for reversible rules—which would later come in handy for thinking about microscopically reversible processes and thermodynamics.

Just down the hall from me at the Institute for Advanced Study was a distinguished mathematician named John Milnor, who got very interested in what I was doing with cellular automata. My archives contain all sorts of notes from Jack, like:

There's even a reversible ("one-to-one") rule, with nice, minimal BASIC code, along with lots of "real math":

4-29-83

Stephen

I tried computing some of the numbers $N(t)$ defined as follows, for an $r=1$ rule, and found some interesting surprizes.

search through all 2^{2t-1} blocks of length $2t+1$

let $N(t)$ be the number of distinct resulting vertical blocks.

It is convenient to describe the result by the power series

$$\varphi(\xi) = \sum_0^\infty N(t)\xi^t$$

where $N(0)=1$, $N(1)=2$, $2 \leq N(t) \leq 2^t$.

Conjecture. φ is always a rational function??

If $\varphi(\xi)$ has radius of convergence ρ, then

$$\rho^{-1} = \lim \sqrt[t]{N(t)},$$

so temporal entropy $= -\log \rho$.

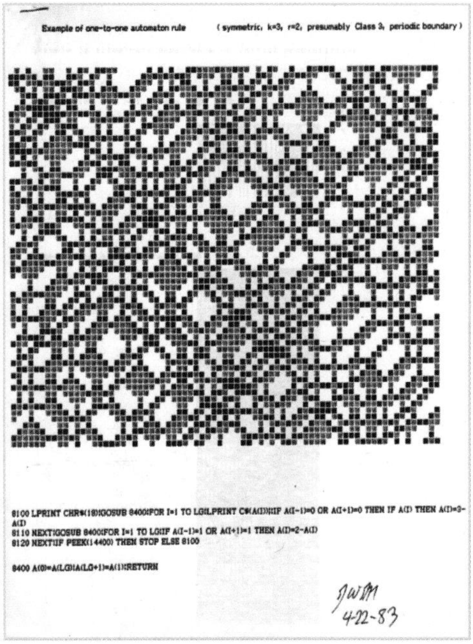

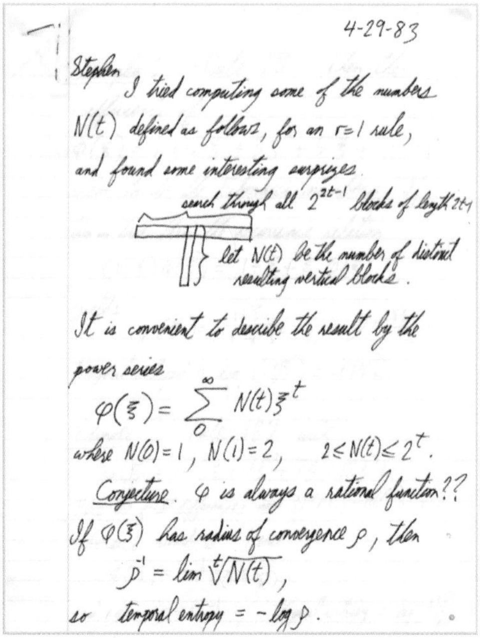

But by the spring of 1984 Jack and I were talking a lot about encryption in cellular automata—and we even began to draft a paper about it

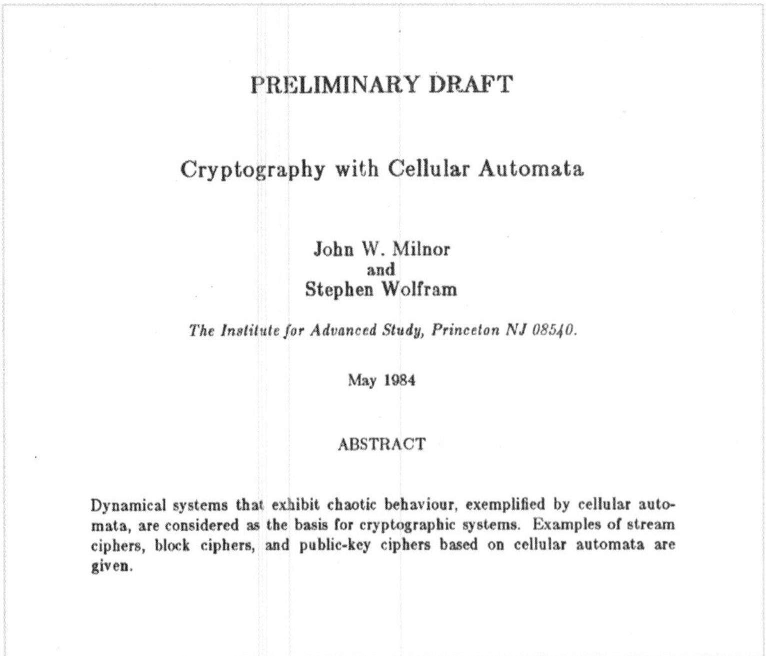

PRELIMINARY DRAFT

Cryptography with Cellular Automata

John W. Milnor
and
Stephen Wolfram

The Institute for Advanced Study, Princeton NJ 08540.

May 1984

ABSTRACT

Dynamical systems that exhibit chaotic behaviour, exemplified by cellular automata, are considered as the basis for cryptographic systems. Examples of stream ciphers, block ciphers, and public-key ciphers based on cellular automata are given.

complete with outlines of how encryption schemes could work:

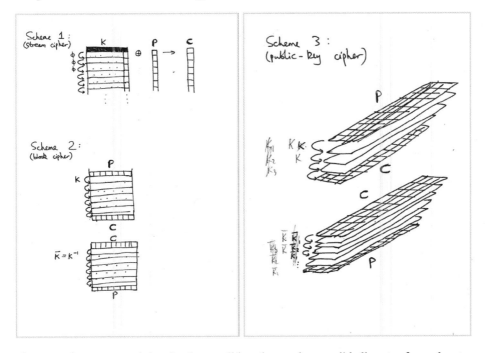

The core of our approach involved reversible rules, and so we did all sorts of searches to find these (and by 1984 Jack was—like me—writing C code):

May 7 17:51 1984 torc.monday Page 1

k= 3 , width= 10 , type= 1, rule table:
0 1 2 0 1 2 0 1 2
2 1 0 0 2 1 0 2 1
0 1 2 0 2 1 0 1 2
Bijectivity measure = 15/27
Convergence times : 104 23 18 18 48 33 38 39 78 18 54 16 24 39 19 28
28 68 46 39

k= 3 , width= 10 , type= 0, rule table:
2 1 0 0 1 2 0 2 1
0 1 2 2 0 1 2 0 1
2 0 1 0 2 1 1 2 0
Bijectivity measure = 21/27
Convergence times : 869 988 196 1564 237 889 764 80 2479 1808 1085 10
2 2131 1237 1683 562 80 860 2209 34

k= 3 , width= 10 , type= 1, rule table:
0 1 2 0 1 2 2 0 1
2 1 0 0 2 1 2 0 1
0 2 1 1 2 0 0 2 1
Bijectivity measure = 19/27
Convergence times : 1116 1133 945 669 1084 106 225 379 1864 1056 494
244 751 100 167 163 1687 72 784 1514

k= 3 , width= 10 , type= 1, rule table:
2 0 1 0 1 2 0 1 2
1 0 2 1 2 0 0 2 1
1 0 2 2 1 0 1 2 0
Bijectivity measure = 19/27
Convergence times : 700 1277 2827 1281 123 466 121 1258 688 1467 746
371 516 182 155 641 2536 573 1159 246

Stephen :

Here is another run. As you can see, we need more thought about testing random rules. (We could just run the program on each.) But I am quite optimistic about the method now.

J.

I wondered how random the output from cellular automata was, and I asked people I knew at Bell Labs about randomness testing (and, yes, email headers haven't changed much in four decades, though then I was swolf@ias.uucp; research!ken was Ken Thompson of Unix fame):

```
Apr 28 13:33 1984  mitchell Page 1

From allegra!don Sat Apr 28 01:07:03 1984
Received: by ias.uucp (3.346/3.14)
         id AA01999; 28 Apr 84 01:06:57 PST (Sat)
Received: from vivace.UUCP by allegra.UUCP (4.12/4.7)
         id AA12614; Sat, 28 Apr 84 01:04:33 est
Received: by vivace.UUCP (4.12/4.7)
         id AA05608; Sat, 28 Apr 84 01:06:51 est
Date: Sat, 28 Apr 84 01:06:51 est
From: allegra!don (Don Mitchell)
Message-Id: <8404280606.AA05608@vivace.UUCP>
To: alice!reeds, swolf@ias.uucp, jlm, research!ken

Here are the results of some of Knuth's suggested tests for randomness.
Tests are geared for looking at cyphertext, but in these tests the high
eight bits of pseudo-random numbers are looked at.

The additive generator is the standard V8 rand(), and the congruential
is the standard 4.1bsd rand().  Tausworthe's generator is a true LFSR.
The last generator is one Jim Reeds brought from the UCB Stat. Dept.  It
combines tausworthe and congruential generators by xor'ing.
```

But then came my internalization of the rule 30 phenomenon, which led to a rather different way of thinking about encryption with cellular automata. Before, we'd basically been assuming that the cellular automaton rule was the encryption key. But rule 30 suggested one could instead have a fixed rule, and have the initial condition define the key. And this is what led me to more physics-oriented thinking about cryptography—and to what I said in *Scientific American*.

In July I was making "encryption-friendly" pictures of rule 30:

But what Jack and I were most interested in was doing something more "cryptographically sophisticated", and in particular inventing a practical public-key cryptosystem based on cellular automata. Pretty much the only public-key cryptosystems known then (or even now) are based on number theory. But we thought maybe one could use something like products of rules instead of products of numbers. Or maybe one didn't need exact invertibility. Or something. But by the late summer of 1984, things weren't looking good:

And eventually we decided we just couldn't figure it out. And it's basically still not been figured out (and maybe it's actually impossible). But even though we don't know how to make a public-key cryptosystem with cellular automata, the whole idea of encrypting initial data and turning it into effective randomness is a crucial part of the whole story of the computational foundations of thermodynamics as I think I now understand them.

Where Does Randomness Come From?

Right from when I first formulated it, I thought computational irreducibility was an important idea. And in the late summer of 1984 I decided I'd better write a paper specifically about it. The result was:

VOLUME 54, NUMBER 8   PHYSICAL REVIEW LETTERS   25 FEBRUARY 1985

Undecidability and Intractability in Theoretical Physics

Stephen Wolfram
The Institute for Advanced Study, Princeton, New Jersey 08540
(Received 26 October 1984)

Physical processes are viewed as computations, and the difficulty of answering questions about them is characterized in terms of the difficulty of performing the corresponding computations. Cellular automata are used to provide explicit examples of various formally undecidable and computationally intractable problems. It is suggested that such problems are common in physical models, and some other potential examples are discussed.

PACS numbers: 02.90.+p, 01.70.+w, 05.90.+m

There is a close correspondence between physical processes and computations. On one hand, theoretical models describe physical processes by computations that transform initial data according to algorithms representing physical laws. And on the other hand, computers themselves are physical systems, obeying physical laws. This paper explores some fundamental consequences of this correspondence.[1]

The behavior of a physical system may always be calculated by simulating explicitly each step in its evolution. Much of theoretical physics has, however, tine equations.[2] One expects in fact that universal computers are as powerful in their computational capabilities as any physically realizable system can be, so that they can simulate any physical system.[3] This is the case if in all physical systems there is a finite density of information, which can be transmitted only at a finite rate in a finite-dimensional space.[4] No physically implementable procedure could then short cut a computationally irreducible process.

Different physically realizable universal computers appear to require the same order of magnitude times

It was a pithy paper, arranged to fit in the 4-page limit of *Physical Review Letters*, with a rather clear description of computational irreducibility and its immediate implications (as well as the relation between physics and computation, which it footnoted as a "physical form of the Church–Turing thesis"). It illustrated computational reducibility and irreducibility in a single picture, here in its original Scotch-taped form:

The paper contains all sorts of interesting tidbits, like this run of footnotes:

[3]This is a physical form of the Church-Turing hypothesis. Mathematically conceivable systems of greater power can be obtained by including tables of answers to questions insoluble for these universal computers.

[4]Real-number parameters in classical physics allow infinite information density. Nevertheless, even in classical physics, the finiteness of experimental arrangements and measurements, implemented as coarse graining in statistical mechanics, implies finite information input and output. In relativistic quantum field theory, finite density of information (or quantum states) is evident for free fields bounded in phase space [e.g., J. Bekenstein, Phys. Rev. D **30**, 1669 (1984)]. It is less clear for interacting fields, except if space-time is ultimately discrete [but cf. B. Simon, *Functional Integration and Quantum Physics* (Academic, New York, 1979), Sec. III.9]. A finite information transmission rate is implied by relativistic causality and the manifold structure of space-time.

[5]It is just possible, however, that the parallelism of the path integral may allow quantum mechanical systems to solve any *NP* problem in polynomial time.

In the paper itself I didn't mention the Second Law, but in my archives I find some notes I made in preparing the paper, about candidate irreducible or undecidable problems (with many still unexplored)

Whether there exists a phase transition in a generalized Ising spin system in two or more dimensions (three or more for rotationally symmetric).

Whether a particular differential equation or cellular automaton yields a strange attractor (say with entropy above some specified value).

Whether there exists a solution to a certain eigenvalue problem. (Try all possible functions ?) Cf. What is ground state for spin glass/lattice theory. Whether there exists a bound state in a certain potential? Require @ d >= 2 @: then like 2DCA problem. (cf. diophantine equations)

Whether two finitely-specified solutions to the Einstein equations (4-manifolds) are equivalent. (Cf. gauge equivalence of finitely-specified loops; Cf. regular grammar equivalence)

Whether there exists an instance of a particular growth rule that leads to unbounded growth (cancer). Or to a glider. Or self-replication.

Neural network halting

In cascade process, like 0L system. (?ETOL grammar non-emptiness: G&J p.270)

What the value of the @ k @th term in a perturbation series is (?graph enumeration).

@ n @-body problem: is a system stable for ever? (Cf. strange attractor problem) (cf. BBM)   *reach part of phase space*

Enumeration of SAW's or general lattice animals? Percolation threshold may be non-computable.

Finding the simplest model of a particular kind to fit a given set of data. Not decidable for anything beyond regular language models.   *Finite version NP-complete like global problem for CA.*

Is logistic map universal scaling exponent non-computable?

Generally what is asymptotic behaviour of a PDE (equivalent to CA). (?Navier-Stokes equation) Is it periodic? Fractal?   *reduced to symbolic dynamics.*

Finding geodesics/minimal surfaces: finding geodesic on Regge skeleton is similar to travelling salesman problem.

Whether it is possible to make a particular optical instrument out of some set of optical elements? (Cf. Post's correspondence problem). (The optical instrument required could be a universal computer.) Or for a charged particle accelerator?

Whether a particular electric circuit can ever "fail": it could be a computer, so that this is the halting problem.

Will a hard sphere gas started from a particular state ever exhibit some specific anti-thermodynamic behaviour? (Cf. n-body problem).

→ *Polymer folding*   ← *Inverse scattering.*

which include "Will a hard sphere gas started from a particular state ever exhibit some specific anti-thermodynamic behaviour?"

In November 1984 the then-editor of *Physics Today* asked if I'd write something for them. I never did, but my archives include a summary of a possible article—which among other things promises to use computational ideas to explain "why the Second Law of thermodynamics holds so widely":

Rough outline for "Computation Theory and Fundamental Physics"
by Stephen Wolfram.

Summary: Viewing physical processes as computations provides insight into several fundamental problems in theoretical physics. Foremost among these are the question of whether "analytical solutions" can always be found, or whether some physical processes can be studied only by direct simulation or experimentation; what the source of chaotic or random behaviour in many systems is; and why the Second Law of thermodynamics holds so widely. In addition, computation theory suggests some general principles that should govern the formation of complex structures in physical systems.

So by November 1984 I was already aware of the connection between computational irreducibility and the Second Law (and also I didn't believe that the Second Law would necessarily always hold). And my notes—perhaps from a little later—make it clear that actually I was thinking about the Second Law along pretty much the same lines as I do now, except that back then I didn't yet understand the fundamental significance of the observer:

And spelunking now in my old filesystem (retrieved from a 9-track backup tape) I find from November 17, 1984 (at 2:42am), troff source for a putative paper (which, yes, we even now can run through troff):

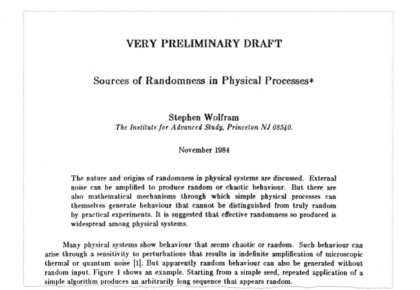

This is all that's in my filesystem. So, yes, in effect, I'm finally (more or less) finishing this 38 years later.

But in 1984 one of the hot—if not new—ideas of the time was "chaos theory", which talked about how "randomness" could "deterministically arise" from progressive "excavation" of higher and higher-order digits in the initial conditions for a system. But having seen rule 30 this whole phenomenon of what was often (misleadingly) called "deterministic chaos" seemed to me at best like a sideshow—and definitely not the main effect leading to most randomness seen in physical systems.

I began to draft a paper about this

VERY PRELIMINARY DRAFT

Sources of Randomness in Physical Processes*

Stephen Wolfram
The Institute for Advanced Study, Princeton NJ 08540.

November 1984

The nature and origins of randomness in physical systems are discussed. External noise can be amplified to produce random or chaotic behaviour. But there are also mathematical mechanisms through which simple physical processes can themselves generate behaviour that cannot be distinguished from truly random by practical experiments. It is suggested that effective randomness so produced is widespread among physical systems.

Many physical systems show behaviour that seems chaotic or random. Such behaviour can arise through a sensitivity to perturbations that results in indefinite amplification of microscopic thermal or quantum noise [1]. But apparently random behaviour can also be generated without random input. Figure 1 shows an example. Starting from a simple seed, repeated application of a simple algorithm produces an arbitrarily long sequence that appears random.

including for the first time an anchor picture of rule 30 intrinsically generating randomness—to be contrasted with pictures of randomness being generated (still in cellular automata) from sensitive dependence on random initial conditions:

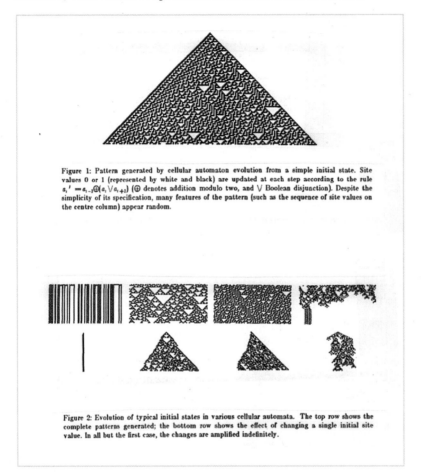

Figure 1: Pattern generated by cellular automaton evolution from a simple initial state. Site values 0 or 1 (represented by white and black) are updated at each step according to the rule $a_i{}' = a_{i-1} \oplus (a_i \vee a_{i+1})$ ($\oplus$ denotes addition modulo two, and $\vee$ Boolean disjunction). Despite the simplicity of its specification, many features of the pattern (such as the sequence of site values on the centre column) appear random.

Figure 2: Evolution of typical initial states in various cellular automata. The top row shows the complete patterns generated; the bottom row shows the effect of changing a single initial site value. In all but the first case, the changes are amplified indefinitely.

It was a bit of a challenge to find an appropriate publishing venue for what amounted to a rather "interdisciplinary" piece of physics-meets-math-meets-computation. But *Physical Review Letters* seemed like the best bet, so on November 19, 1984, I submitted a version of the paper there, shortened to fit in its 4-page limit.

A couple of months later the journal said it was having trouble finding appropriate reviewers. I revised the paper a bit (in retrospect I think not improving it), then on February 1, 1985, sent it in again, with the new title "Origins of Randomness in Physical Systems":

Origins of Randomness in Physical Systems*

Stephen Wolfram

The Institute for Advanced Study, Princeton NJ 08540.

(January 1985)

Randomness in physical systems is often ultimately attributed to external noise. But it is argued here that even without such random input, the intrinsic behaviour of many non-linear systems can be so complicated as to seem random in all practical experiments. This phenomenon is described using methods from computation theory, and is suggested as a fundamental source of chaos in such systems as turbulent fluids.

There are many physical processes that seem random or chaotic. They appear to follow no definite rules, and to be governed merely by probabilities. But all fundamental physical laws, at least outside of quantum mechanics, are thought to be deterministic. So how then is apparent randomness produced?

One possibility is that its ultimate source is external noise, often from a heat bath. When the evolution of a system is unstable, so that perturbations grow, any randomness introduced through initial and boundary conditions is transmitted or amplified with time, and eventually affects many components of the system [1]. A very simple example of this "homoplectic"

On March 8 the journal responded, with two reports from reviewers. One of the reviewers completely missed the point (yes, a risk in writing shift-the-paradigm papers). The other sent a very constructive two-page report:

RE: LB3029

I recommend that the manuscript **Origins of Randomness in Physical Systems** by **Stephen Wolfram** be accepted for publication in Phys. Rev. Lett. if that remains the author's desire after considering this report.

Acceptance is recommended, if the author insists, because this manuscript is in a field of currently intense interdisciplinary interest, and it contains a wealth of intriguing and potentially important ideas. Also, however, the paper is unfocussed; contains ambiguous, cryptic or incorrect statements; and is too short to cover adequately all the topics which are brought up. It is full of dense, pregnant footnotes. My strong suggestion is that some of the ideas be worked through further, the format be expanded, the entire text be carefully revised, and the result be submitted to Phys. Rev. A.

I have discussed this manuscript with three knowledgeable colleagues. Two of them share fully the reservations stated above. The third is enthusiastically for publication in PRL as is.

Below are some specific comments about the manuscript. Some of them are related to each other.

1. The abstract suggests that intrinsically complicated behavior may be a fundamental source of chaos in turbulent fluids, an idea returned to at the end of the paper. This is not new (except maybe to some hardcore mappers); no one who has looked much at turbulent flows can easily doubt it. However, it is not clear to me to what extent the distinction between "homoplectic" and "autoplectic" behavior is meaningful. Consider an idealization of turbulent diffusion which I studied some years ago [R. H. Kraichnan, Phys. Fluids 13, 22 (1970); J. Fluid Mech. 77, 753 (1976)]. A velocity field in infinite space is constructed out of a few discrete wavevector components; it is fully described

As always, I request that this report be forwarded to the author signed.

Sincerely yours,

Robert H. Kraich

Robert H. Kraichnan

I didn't know it then, but later I found out that Bob Kraichnan had spent much of his life working on fluid turbulence (as well as that he was a very independent and think-for-oneself physicist who'd been one of Einstein's last assistants at the Institute for Advanced Study). Looking at his report now it's a little charming to see his statement that "no one who has looked much at turbulent flows can easily doubt [that they intrinsically generate randomness]" (as opposed to getting randomness from noise, initial conditions, etc.). Even decades later, very few people seem to understand this.

There were several exchanges with the journal, leaving it controversial whether they would publish the paper. But then in May I visited Los Alamos, and Bob Kraichnan invited me to lunch. He'd also invited a then-young physicist from Los Alamos who I'd known fairly well a few years earlier—and who'd once paid me the unintended compliment that it wasn't fair for me to work on science because I was "too efficient". (He told me he'd "intended to work on cellular automata", but before he'd gotten around to it, I'd basically figured everything out.) Now he was riding the chaos theory bandwagon hard, and insofar as my paper threatened that, he wanted to do anything he could to kill the paper.

I hadn't seen this kind of "paradigm attack" before. Back when I'd been doing particle physics, it had been a hot and cutthroat area, and I'd had papers plagiarized, sometimes even egregiously. But there wasn't really any "paradigm divergence". And cellular automata—being quite far from the fray—were something I could just peacefully work on, without anyone really paying much attention to whatever paradigm I might be developing.

At lunch I was treated to a lecture about why what I was doing was nonsense, or even if it wasn't, I shouldn't talk about it, at least now. Eventually I got a chance to respond, I thought rather effectively—causing my "opponent" to leave in a huff, with the parting line "If you publish the paper, I'll ruin your career". It was a strange thing to say, given that in the pecking order of physics, he was quite junior to me. (A decade and half later there were nevertheless a couple of "incidents".) Bob Kraichnan turned to me, cracked a wry smile and said "OK, I'll go right now and tell [the journal] to publish your paper":

VOLUME 55, NUMBER 5   **PHYSICAL REVIEW LETTERS**   29 JULY 1985

Origins of Randomness in Physical Systems

Stephen Wolfram
The Institute for Advanced Study, Princeton, New Jersey 08540
(Received 4 February 1985)

Randomness and chaos in physical systems are ususally ultimately attributed to external noise. But it is argued here that even without such random input, the intrinsic behavior of many nonlinear systems can be computationally so complicated as to seem random in all practical experiments. This effect is suggested as the basic origin of such phenomena as fluid turbulence.

PACS numbers: 05.45.+b, 02.90.+p, 03.40.Gc

There are many physical processes that seem random or chaotic. They appear to follow no definite ness,[6] such as relative frequencies of blocks of elements (dimensions and entropies), correlations, and

Kraichnan was quite right that the paper was much too short for what it was trying to say, and in the end it took a long book—namely *A New Kind of Science*—to explain things more clearly. But the paper was where a high-resolution picture of rule 30 first appeared in print. And it was the place where I first tried to explain the distinction between "randomness that's just transcribed from elsewhere" and the fundamental phenomenon one sees in rule 30 where randomness is intrinsically generated by computational processes within a system.

I wanted words to describe these two different cases. And reaching back to my years of learning ancient Greek in school I invented the terms "homoplectic" and "autoplectic", with the noun "autoplectism" to describe what rule 30 does. In retrospect, I think these terms are perhaps "too Greek" (or too "medical sounding"), and I've tended to just talk about "intrinsic randomness generation" instead of autoplectism. (Originally, I'd wanted to avoid the term "intrinsic" to prevent confusion with randomness that's baked into the rules of a system.)

The paper (as Bob Kraichnan pointed out) talks about many things. And at the end, having talked about fluid turbulence, there's a final sentence—about the Second Law:

initial conditions.[28] Autoplectic processes may also be responsible for the widespread applicability of the second law of thermodynamics.

In my archives, I find other mentions of the Second Law too. Like an April 1985 proto-paper that was never completed

---

Computation Theory and Randomness in Physics*

Stephen Wolfram
The Institute for Advanced Study, Princeton NJ 08540.

(April 1985)

EXTENDED ABSTRACT

Many physical systems show apparently random behaviour, even though their component parts follow basically deterministic laws. Turbulent fluid flow is a classic example. Fluid motion is believed to be governed by deterministic partial differential equations; yet when the flow is fast enough, it becomes chaotic and seems random. It is a fundamental problem of mathematical physics to elucidate the mathematical mechanism through which such apparent randomness is produced. This paper discusses the problem in terms of computation theory, and describes some of my recent proposals for solutions to it [1].

---

but included the statement:

The full version of this paper will give a slightly more formal description of the problem, including a version of the Second Law entirely cast in computation theoretical terms.

My main reason for working on cellular automata was to use them as idealized models for systems in nature, and as a window into foundational issues. But being quite involved in the computer industry, I couldn't help wondering whether they might be directly useful for practical computation. And I talked about the possibility of building a "metachip" in which— instead of having predefined "meaningful" opcodes like in an ordinary microprocessor— everything would be built up "purely in software" from an underlying universal cellular automaton rule. And various people and companies started sending me possible designs:

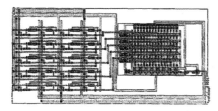 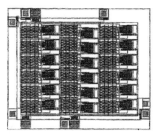

But in 1984 I got involved in being a consultant to an MIT-spinoff startup called Thinking Machines Corporation that was trying to build a massively parallel "Connection Machine" computer with 65536 processors. The company had aspirations around AI (hence the name, which I'd actually been involved in suggesting), but their machine could also be put to work simulating cellular automata, like rule 30. In June 1985, hot off my work on the origins of randomness, I went to spend some of the summer at Thinking Machines, and decided it was time to do whatever analysis—or, as I would call it now, ruliology—I could on rule 30.

My filesystem from 1985 records that it was fast work. On June 24 I printed a somewhat-higher-resolution image of rule 30 (my login was "swolf" back then, so that's how my printer output was labeled):

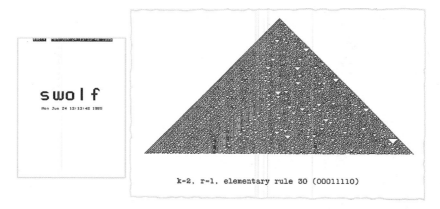

k=2, r=1, elementary rule 30 (00011110)

By July 2 a prototype Connection Machine had generated 2000 steps of rule 30 evolution:

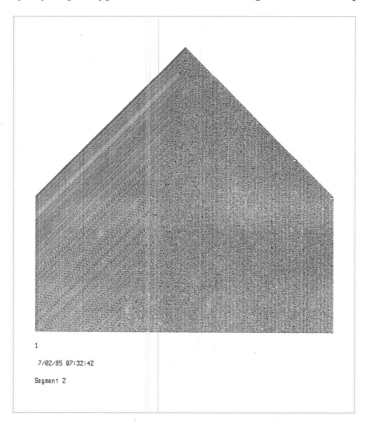

1

7/02/85 07:32:42

Segment 2

With a large-format printer normally used to print integrated circuit layouts I got an even larger "piece of rule 30"—that I laid out on the floor for analysis, for example trying to measure (with meter rules, etc.) the slope of the border between regularity and irregularity in the pattern.

Richard Feynman was also a consultant at Thinking Machines, and we often timed our visits to coincide:

Feynman and I had talked about randomness quite a bit over the years, most recently in connection with the challenges of making a "quantum randomness chip" as a minimal example of quantum computing. Feynman at first didn't believe that rule 30 could really be "producing randomness", and that there must be some way to "crack" it. He tried, both by hand and with a computer, particularly using statistical mechanics methods to try to compute the slope of the border between regularity and irregularity:

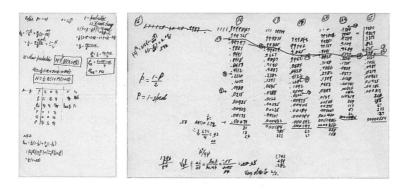

But in the end, he gave up, telling me "OK, Wolfram, I think you're on to something".

Meanwhile, I was throwing all the methods I knew at rule 30. Combinatorics. Dynamical systems theory. Logic minimization. Statistical analysis. Computational complexity theory. Number theory. And I was pulling in all sorts of hardware and software too. The Connection Machine. A Cray supercomputer. A now-long-extinct Celerity C1200 (which successfully computed a length-40,114,679,273 repetition period). A LISP machine for graph layout. A circuit-design logic minimization program. As well as my own SMP system. (The Wolfram Language was still a few years in the future.)

But by July 21, there it was: a 50-page "ruliological profile" of rule 30, in a sense showing what one could of the "anatomy" of its randomness:

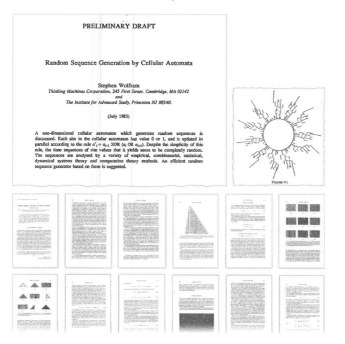

A month later I attended in quick succession a conference in California about cryptography, and one in Japan about fluid turbulence—with these two fields now firmly connected through what I'd discovered.

Hydrodynamics, and a Turbulent Tale

Back from when I first saw it at the age of 14 it was always my favorite page in *The Feynman Lectures on Physics.* But how did the phenomenon of turbulence that it showed happen, and what really was it?

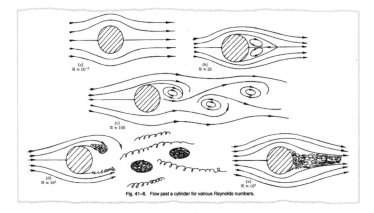

Fig. 41–6. Flow past a cylinder for various Reynolds numbers.

In late 1984, the first version of the Connection Machine was nearing completion, and there was a question of what could be done with it. I agreed to analyze its potential uses in scientific computation, and in my resulting (never ultimately completed) report

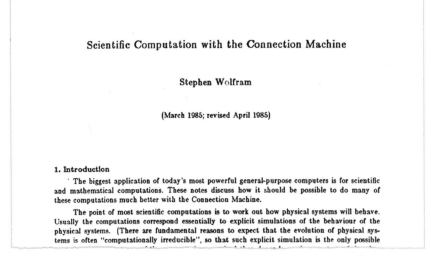

Scientific Computation with the Connection Machine

Stephen Wolfram

(March 1985; revised April 1985)

1. Introduction

The biggest application of today's most powerful general-purpose computers is for scientific and mathematical computations. These notes discuss how it should be possible to do many of these computations much better with the Connection Machine.

The point of most scientific computations is to work out how physical systems will behave. Usually the computations correspond essentially to explicit simulations of the behaviour of the physical systems. (There are fundamental reasons to expect that the evolution of physical systems is often "computationally irreducible", so that such explicit simulation is the only possible

the very first section was about fluid turbulence (other sections were about quantum field theory, *n*-body problems, number theory, etc.):

A. Fluid turbulence

Complicated irregular, or turbulent, flow is very common, and is important for many technological applications such as aerodynamics. But it is still not known exactly how or why turbulence occurs. It has not been possible to solve the Navier-Stokes partial differential equations in systems with realistic fully-developed turbulence. The best current simulations approximate the equations on $256 \times 256 \times 256$-site grids. The physics is such that fluid motion on very large scales is essentially determined by overall boundary conditions, and on very small scales is smoothed out by viscous effects. Current simulations really do not have enough grid points to see an intermediate ("inertial") range of scales, on which true turbulence can occur. And in fact it is not known what the fundamental mathematical mechanism for turbulence is. The nature of the mechanism is important in determining the kinds of approximations that can be made in solving the fluid equations, while still correctly reproducing turbulence.

The traditional computational approach to studying fluids was to start from known continuum fluid equations, then to try to construct approximations to these suitable for numerical computation. But that wasn't going to work well for the Connection Machine. Because in optimizing for parallelism, its individual processors were quite simple, and weren't set up to do fast (e.g. floating-point) numerical computation.

I'd been saying for years that cellular automata should be relevant to fluid turbulence. And my recent study of the origins of randomness made me all the more convinced that they would for example be able to capture the fundamental randomness associated with turbulence (which I explained as being a bit like encryption):

with arbitrary precision real numbers. In addition, with either of these two explanations, one could not in practice expect to find the form of the flow in any particular case, because it would be determined by unknown small details of initial conditions. In fact, it is neither necessary nor probably necessary to know the exact, irregular, flow. What is needed is to know statistical averages that capture what regularities do exist in the flow.

I have recently come up with a third explanation for turbulence (that concurrs with intuition from many numerical fluid experiments, but could not be formalized before). I think that fluids can act much like pseudorandom number generators, producing complicated and apparently random patterns of flow even from simple initial conditions, analogous to the simple "seeds" of pseudorandom number generators, or simple "keys" for cryptographic systems. This phenomenon is very common in cellular automata, and there is every reason to expect that it is also present in real fluids. And with this explanation, it suffices to use discrete approximations to reproduce the fundamental features of turbulence.

I am not sure exactly what the best system to simulate turbulence in is. The Taylor-Green

I sent a letter to Feynman expressing my enthusiasm:

THE INSTITUTE FOR ADVANCED STUDY
PRINCETON, NEW JERSEY 08540

SCHOOL OF NATURAL SCIENCES

April 20, 1985

Dear Feynman,

Here is something I have started to write about using the Connection Machine for scientific computing. It still has quite a way to go, but I thought you might find it amusing. I think that the stuff about using cellular automata to emulate partial differential equations is very promising.

Do let me know if you have comments, etc....

Best regards,

Stephen

I had been invited to a conference in Japan that summer on "High Reynolds Number Flow Computation" (i.e. computing turbulent fluid flow), and on May 4 I sent an abstract which explained a little more of my approach:

Cellular Automaton Approaches to Turbulent Fluids

Stephen Wolfram

Cellular automata are discrete systems that seem to capture the essential mathematical features of many complex natural processes. They can be considered as approximations to partial differential equations, in which a discrete grid is taken in space and time, and the values of variables are taken discrete (corresponding say to just a few bits of numerical precision). Despite these approximations, cellular automata seem able to reproduce many features of solutions to partial differential equations. Here their use in studying fluid flow will be considered. Their methodology can be considered intermediate between molecular dynamics and numerical approximation to continuum partial differential equations. Statistical averages are required in comparing model calculations with data. The accuracy of model results may be connected with the validity of thermodynamics in the models used.

In so far as cellular automata can be used to reproduce fluid flow, they allow for extremely efficient simulation on parallel-processing computers. They use no floating point arithmetic, but instead involve many Boolean operations, which can be carried out in parallel.

The application of cellular automata to fluid flow problems leads to a potential analysis of turbulence using ideas from the mathematical theory of computation. The randomness or chaos observed in flows may be characterized in terms of the computational complexity of finding regularities in it, and the mathematical processes which can produce such randomness may be analysed.

My basic idea was to start not from continuum equations, but instead from a cellular automaton idealization of molecular dynamics. It was the same kind of underlying model as I'd tried to set up in my SPART program in 1973. But now instead of using it to study thermodynamic phenomena and the microscopic motions associated with heat, my idea was to use it to study the kind of visible motion that occurs in fluid dynamics—and in particular to see whether it could explain the apparent randomness of fluid turbulence.

I knew from the beginning that I needed to rely on "Second Law behavior" in the underlying cellular automaton—because that's what would lead to the randomness necessary to "wash out" the simple idealizations I was using in the cellular automaton, and allow standard continuum fluid behavior to emerge. And so it was that I embarked on the project of understanding not only thermodynamics, but also hydrodynamics and fluid turbulence, with cellular automata—on the Connection Machine.

I've had the experience many times in my life of entering a field and bringing in new tools and new ideas. Back in 1985 I'd already done that several times, and it had always been a pretty much uniformly positive experience. But, sadly, with fluid turbulence, it was to be, at best, a turbulent experience.

The idea that cellular automata might be useful in studying fluid turbulence definitely wasn't obvious. The year before, for example, at the Nobel Symposium conference in Sweden, a French physicist named Uriel Frisch had been summarizing the state of turbulence research. Fittingly for the topic of turbulence, he and I first met after a rather bumpy helicopter ride to a conference event—where Frisch told me in no uncertain terms that cellular automata would never be relevant to turbulence, and talked about how turbulence was better thought of as being associated (a bit like in the mathematical theory of phase transitions) with "singularities getting close to the real line". (Strangely, I just now looked at Frisch's paper in the proceedings of the conference: "Ou en est la Turbulence Developpée?" [roughly: "Fully Developed Turbulence: Where Do We Stand?"], and was surprised to discover that its last paragraph actually mentions cellular automata, and its acknowledgements thank me for conversations—even though the paper says it was received June 11, 1984, a couple of days before I had met Frisch. And, yes, this is the kind of thing that makes accurately reconstructing history hard.)

Los Alamos had always been a hotbed of computational fluid dynamics (not least because of its importance in simulating nuclear explosions)—and in fact of computing in general—and, starting in the late fall of 1984, on my visits there I talked to many people about using cellular automata to do fluid dynamics on the Connection Machine. Meanwhile, Brosl Hasslacher (mentioned above in connection with his 1982 encryption startup) had—after a rather itinerant career as a physicist—landed at Los Alamos. And in fact I had been asked by the Los Alamos management for a letter about him in December 1984 (yes, even though he was 18 years older than me), and ended what I wrote with: "He has considerable ability in identifying promising areas of research. I think he would be a significant addition to the staff at Los Alamos."

Well, in early 1985 Brosl identified cellular automaton fluid dynamics as a promising area, and started energetically talking to me about it. Meanwhile, the Connection Machine was just starting to work, and a young software engineer named Jim Salem was assigned to help me get cellular automaton fluid dynamics running on it. I didn't know it at the time, but Brosl—ever the opportunist—had also made contact with Uriel Frisch, and now I find the curious document in French dated May 10, 1985, with the translated title "A New Concept for Supercomputers: Cellular Automata", laying out a grand international multiyear plan, and referencing the (so far as I know, nonexistent) B. Hasslacher and U. Frisch (1985), "The Cellular Automaton Turbulence Machine", Los Alamos:

REFERENCES

. Frisch, U. et Hasslacher, B. (1985), "The Cellular Automaton Turbulence Machine", Los Alamos.
. Frisch, U. (1985), "Le Calcul Parallèle Massif d'Ecoulements Turbulents avec des Automates Cellulaires", Observatoire de Nice.
. Hardy, J., de Passis, O. et Pomeau, Y. (1976), Physical Review vol. A13, p. 1949.
. Wolfram, S. (1984), Nature vol. 311, p. 419.

I visited Los Alamos again in May, but for much of the summer I was at Thinking Machines, and on July 18 Uriel Frisch came to visit there, along with a French physicist named Yves Pomeau, who had done some nice work in the 1970s on applying methods of traditional statistical mechanics to "lattice gases".

But what about realistic fluid dynamics, and turbulence? I wasn't sure how easy it would be to "build up from the (idealized) molecules" to get to pictures of recognizable fluid flows. But we were starting to have some success in generating at least basic results. It wasn't clear how seriously anyone else was taking this (especially given that at the time I hadn't seen the material Frisch had already written), but insofar as anything was "going on", it seemed to be a perfectly collegial interaction—where perhaps Los Alamos or the French government or both would buy a Connection Machine computer. But meanwhile, on the technical side, it had become clear that the most obvious square-lattice model (that Pomeau had used in the 1970s, and that was basically what my SPART program from 1973 was supposed to implement) was fine for diffusion processes, but couldn't really represent proper fluid flow.

When I first started working on cellular automata in 1981 the minimal 1D case in which I was most interested had barely been studied, but there had been quite a bit of work done in previous decades on the 2D case. By the 1980s, however, it had mostly petered out—with the exception of a group at MIT led by Ed Fredkin, who had long had the belief that one might in effect be able to "construct all of physics" using cellular automata. Tom Toffoli and Norm Margolus, who were working with him, had built a hardware 2D cellular automaton simulator—that I happened to photograph in 1982 when visiting Fredkin's island in the Caribbean:

But while "all of physics" was elusive (and our Physics Project suggests that a cellular automaton with a rigid lattice is not the right place to start), there'd been success in making for example an idealized gas, using essentially a block cellular automaton on a square grid. But mostly the cellular automaton machine was used in a maddeningly "Look at this cool thing!" mode, often accompanied by rapid physical rewiring.

In early 1984 I visited MIT to use the machine to try to do what amounted to natural science, systematically studying 2D cellular automata. The result was a paper (with Norman Packard) on 2D cellular automata. We restricted ourselves to square grids, though mentioned hexagonal ones, and my article in *Scientific American* in late 1984 opened with a full-page hexagonal cellular automaton simulation of a snowflake made by Packard (and later in 1984 turned into one of a set of cellular automaton cards for sale):

In any case, in the summer of 1985, with square lattices not doing what was needed, it was time to try hexagonal ones. I think Yves Pomeau already had a theoretical argument for this, but as far as I was concerned, it was (at least at first) just a "next thing to try". Programming the Connection Machine was at that time a rather laborious process (which, almost unprecedentedly for me, I wasn't doing myself), and mapping a hexagonal grid onto its basically square architecture was a little fiddly, as my notes record:

Meanwhile, at Los Alamos, I'd introduced a young and very computer-savvy protege of mine named Tsutomu Shimomura (who had a habit of getting himself into computer security scrapes, though would later become famous for taking down a well-known hacker) to Brosl Hasslacher, and now Tsutomu jumped into writing optimized code to implement hexagonal cellular automata on a Cray supercomputer.

In my archives I now find a draft paper from September 7 that starts with a nice (if not entirely correct) discussion of what amounts to computational irreducibility, and then continues by giving theoretical symmetry-based arguments that a hexagonal cellular automaton should be able to reproduce fluid mechanics:

A LATTICE GAS AUTOMATON FOR THE NAVIER STOKES EQUATION

U. Frisch
CNRS, Observatoire de Nice, BP 139, 06003 Nice cedex,France

B. Hasslacher
CNLS/Theor. Div., LANL, Los Alamos, N.M. 87545, USA

and **Y. Pomeau**
CNRS, École Normale Supérieure, 24 rue Lhomond, 75231 Paris
and Physique Théorique, CEN-Saclay, 91191 Gif-sur-Yvette

ORDER?

Preliminary draft , Septembre 7 1985. (reference list missing)

ABSTRACT

It is shown that lattice gases with discrete Boolean molecules can be used to simulate the Navier Stokes equation with massive parallelism.

One of the deepest problems in non-linear dynamics is to understand how complex behavior develops and becomes self-organizing. Recently, there has been considerable progress in understanding universal properties of non-linear complexity and in studying its morphology. This is mainly due to the relatively recent availability of sophisticated interactive digital simulation. It is likely that without such simulation, further theoretical work will be blocked by the frontier of byzantine computational complexity, namely, questions in non-linear dynamics that are undecidable or uncomputable, and not obviously so, similar to those encountered in the two dimensional automaton LIFE [ ] or the theory of Penrose tilings [ ].

Recent work, relating non-linear dynamics to computation theory, suggest that undecidable and intractable questions are much more common than usually realized and not restricted to discrete systems [S. Wolfram, B. Hasslacher ...]. An example of a problem generally conjectured to be uncomputable is the correspondence between initial states and attractors in chaotic systems whose attractors proliferate. Under such a circumstance, we believe that the interactive digital simulation of a system , on a human time scale, provides a powerful way to discover universal phenomena amenable to theory. At the same time, simulation avoids entrapment in computational complexity problems. A most efficient way to achieve this is to put human pattern recognition abilities into the simulation system, by coupling it to sophisticated graphics environments. A notorious example of this is the discovery of four main universality classes in cellular automata (CA)[ ].

In hydrodynamics, one of the most difficult open problems is the understanding of multi-dimensional turbulent flows at very high Reynolds numbers [UF Phys. Scr.]. The interactive simulation of such flows is many orders of magnitude beyond the capacity of existing computer resources. Similar limitations affect our ability to simulate many other multi-dimensional field theories.

Massively parallel architectures and algorithms are required to avoid the ultimate computation limits of the speed of light and various solid state constraints. Also, when

Near the end, the draft says (misspelling Tsutomu Shimomura's name):

We will finally outline how one can go from this Gedanken world to real machines. The first stage is to demonstate that real hydro phenomena can be reproduced when Lattice Gases are simulated on existing computers. Evidence, just obtained by D'Humières, Lallemand and Shimamura, is presented in a companion paper [ ]. Other simulations are currently under way on the CAM machine at MIT, and on the Connection Machine at Thinking Machines Corporation.

Meanwhile, we (as well as everyone else) were starting to get results that looked at least suggestive:

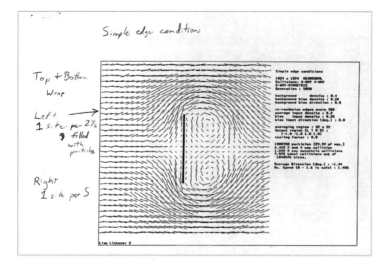

By November 15 I had drafted a paper

Thermodynamics and Hydrodynamics with Cellular Automata

James B. Salem

Thinking Machines Corporation, 245 First Street, Cambridge, MA 02144

and

Stephen Wolfram

The Institute for Advanced Study, Princeton NJ 08540.

(November 15, 1985)

Simple cellular automata which seem to capture the essential features of thermodynamics and hydrodynamics are discussed. At a microscopic level, the cellular
automata are discrete approximations to molecular dynamics, and show relaxation towards equilibrium. On a large scale, they behave like continuum fluids,
and yield efficient methods for hydrodynamic simulation.

Thermodynamics and hydrodynamics give overall descriptions of the behaviour of many systems. The descriptions are quite generic; their form does not seem to depend on the precise construction of each system. As a result, one can study thermodynamics and hydrodynamics with
simple model systems, which are potentially more amenable to mathematical analysis, and can
form the basis for more efficient simulation procedures.

Cellular automata are discrete dynamical systems which can be used as simple models for a
wide variety of complex physical processes [1]. This paper considers cellular automata which can
be viewed as discrete approximations to molecular dynamics. In the simplest case, each link in a
regular spatial lattice carries at most one "particle" with unit velocity in each direction. At each
time step, each particle moves one link; those arriving at a particular site then "scatter" according to a fixed set of rules. This discrete system is well-suited to simulation on digital computers.
The state of each site is represented by a few bits, and follows simple logical rules. The rules are
local, so that many sites can be updated in parallel. The simulations in this paper were performed
with a Connection Machine Computer [2] in which sites are updated concurrently in each of 65536
4-bit processors.

that included some more detailed pictures

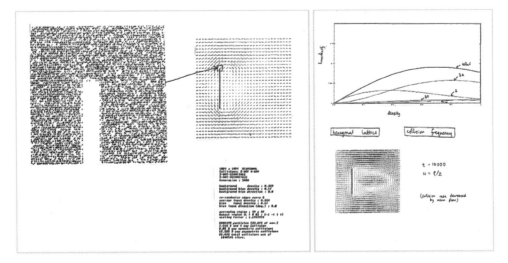

and that at the end (I thought, graciously) thanked Frisch, Hasslacher, Pomeau and Shimomura for "discussions and for sharing their unpublished results with us", which by that point
included a bunch of suggestive, if not obviously correct, pictures of fluid-flow-like behavior.

To me, what was important about our paper is that, after all these years, it filled in with more detail just how computational systems like cellular automata could lead to Second-Law-style thermodynamic behavior, and it "proved" the physicality of what was going on by showing easy-to-recognize fluid-dynamics-like behavior.

Just four days later, though, there was a big surprise. *The Washington Post* ran a front-page story entitled "Discovery in Flow Dynamics May Aid Car, Plane Design"—alongside the day's characteristic-Cold-War-era geopolitical news ("Gorbachev Is Said to Bring New Arms Plans to Geneva", etc.)—about the "Hasslacher–Frisch model", and about how it might be judged so important that it "should be classified to keep it out of Soviet hands".

At that point, things went crazy. There was talk of Nobel Prizes (I wasn't buying it). There were official complaints from the French embassy about French scientists not being adequately recognized. There was upset at Thinking Machines for not even being mentioned. And, yes, as the originator of the idea, I was miffed that nobody seemed to have even suggested contacting me—even if I did view the rather breathless and "geopolitical" tenor of the article as being pretty far from immediate reality.

At the time, everyone involved denied having been responsible for the appearance of the article. But years later it emerged that the source was a certain John Gage, former political operative and longtime marketing operative at Sun Microsystems, who I'd known since 1982, and had at some point introduced to Brosl Hasslacher. Apparently he'd called around various government contacts to help encourage open (international) sharing of scientific code, quoting this as a test case.

But as it was, the article had pretty much exactly the opposite effect, with everyone now out for themselves. In Princeton, I'd interacted with Steve Orszag, whose funding for his new (traditional) computational fluid dynamics company, Nektonics, now seemed at risk, and who pulled me into an emergency effort to prove that cellular automaton fluid dynamics couldn't be competitive. (The paper he wrote about this seemed interesting, but I demurred on being a coauthor.) Meanwhile, Thinking Machines wanted to file a patent as quickly as possible. Any possibility of the French government getting a Connection Machine evaporated and soon Brosl Hasslacher was claiming that "the French are faking their data".

And then there was the matter of the various academic papers. I had been sent the Frisch–Hasslacher–Pomeau paper to review, and checking my 1985 calendar for my whereabouts I must have received it the very day I finished my paper. I told the journal they should publish the paper, suggesting some changes to avoid naivete about computing and computer technology, but not mentioning its very thin recognition of my work.

Our paper, on the other hand, triggered a rather indecorous competitive response, with two "anonymous reviewers" claiming that the paper said nothing more than its "reference 5" (the Frisch–Hasslacher–Pomeau paper). I patiently pointed out that that wasn't the case, not least because our paper had actual simulations, but also that actually I happened to have "been there first" with the overall idea. The journal solicited other opinions, which were mostly supportive. But in the end a certain Leo Kadanoff swooped in to block it, only to publish his own a few months later.

It felt corrupt, and distasteful. I was at that point a successful and increasingly established academic. And some of the people involved were even longtime friends. So was this kind of thing what I had to look forward to in a life in academia? That didn't seem attractive, or necessary. And it was what began the process that led me, a year and a half later, to finally choose to leave academia behind, never to return.

Still, despite the "turbulence"—and in the midst of other activities—I continued to work hard on cellular automaton fluids, and by January 1986 I had the first version of a long (and, I thought, rather good) paper on their basic theory (that was finished and published later that year):

As it turns out, the methods I used in that paper provide some important seeds for our Physics Project, and even in recent times I've often found myself referring to the paper, complete with its SMP open-code appendix:

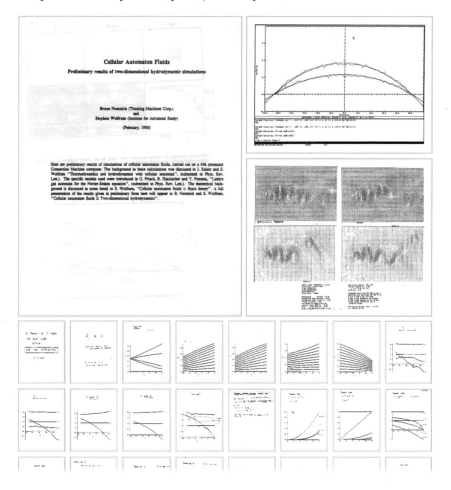

But in addition to developing the theory, I was also getting simulations done on the Connection Machine, and getting actual experimental data (particularly on flow past cylinders) to compare them to. By February 1986, we had quite a few results:

But by this point there was a quite industrial effort, particularly in France, that was churning out papers on cellular automaton fluids at a high rate. I'd called my theory paper "Cellular Automaton Fluid 1: Basic Theory". But was it really worth finishing part 2? There was a veritable army of perfectly good physicists "competing" with me. And, I thought, "I have other things to do. Just let them do this. This doesn't need me".

And so it was that in the middle of 1986 I stopped working on cellular automaton fluids. And, yes, that freed me up to work on lots of other interesting things. But even though methods derived from cellular automaton fluids have become widely used in practical fluid dynamics computations, the key basic science that I thought could be addressed with cellular automaton fluids—about things like the origin of randomness in turbulence—has still, even to this day, not really been further explored.

Getting to the Continuum

In June 1986 I was about to launch both a research center (the Center for Complex Systems Research at the University of Illinois) and a journal (*Complex Systems*)—and I was also organizing a conference called CA '86 (which was held at MIT). The core of the conference was poster presentations, and a few days before the conference was to start I decided I should find a "nice little project" that I could quickly turn into a poster.

In studying cellular automaton fluids I had found that cellular automata with rules based on idealized physical molecular dynamics could on a large scale approximate the continuum behavior of fluids. But what if one just started from continuum behavior? Could one derive underlying rules that would reproduce it? Or perhaps even find the minimal such rules?

By mid-1985 I felt I'd made decent progress on the science of cellular automata. But what about their engineering? What about constructing cellular automata with particular behavior? In May 1985 I had given a conference talk about "Cellular Automaton Engineering", which turned into a paper about "Approaches to Complexity Engineering"—that in effect tried to set up "trainable cellular automata" in what might still be a powerful simple-programs-meet-machine-learning scheme that deserves to be explored:

Approaches to Complexity Engineering*

Stephen Wolfram
The Institute for Advanced Study, Princeton NJ 08540.

(December 1985)

Principles for designing complex systems with specified forms of behaviour are discussed. Multiple scale cellular automata are suggested as dissipative dynamical systems suitable for tasks such as pattern recognition. Fundamental aspects of the engineering of such systems are characterized using computation theory, and some practical procedures are discussed.

The capabilities of the brain and many other natural systems go far beyond those of any artificial systems so far constructed by conventional engineering means. There is however extensive evidence that at a functional level, the basic components of such complex natural systems are quite simple, and could for example be emulated with a variety of technologies. But how large numbers of these components can act together to perform complex tasks is not yet known. There are probably some rather

But so it was that a few days before the CA '86 conference I decided to try to find a minimal "cellular automaton approximation" to a simple continuum process: diffusion in one dimension.

I explained

```
.PP
A continuum system such as a fluid has the feature that its state
can be described (locally) by just a few extensive quantities.
To describe the precise microscopic state of a real gas one must,
of course, specify the precise configuration of molecules.
But it is believed that unless the gas is highly rarefied, this
precise configuration is irrelevant to the macroscopic behaviour
of the gas. Only the values of the few, averaged, extensive quantities
are significant, so that a fluid approximation can be used.
.PP
The basis for this belief is embodied in the Second Law of thermodynamics.
It seems that almost regardless of the initial microscopic configuration,
collisions rapidly tend randomize the configuration of gas molecules,
so that at least for macroscopic purposes, it suffices to specify
merely the values of certain average quantities.
.PP
The true basis for this phenomenon has never been very clear.
Some descriptions of it can be given in terms of the apparent increase of
coarse-grained entropy. But no fundamental derivation
has ever been given. The investigation of cellular automaton
models seems likely to provide some new insights.
.PP
```

and described as my objective:

```
This poster considers as an example the problem of finding the
simplest cellular automaton rule which reproduces the one-dimensional
diffusion equation.
.PP
The potential interest of these investigations is severalfold.
.PP
.IP 1.
They may provide practical methods for solving problems related
to continuum systems (and these methods may be compared in detail with
existing methods).
.IP 2.
They provide examples of systems which exhibit the basic phenomena
of thermodynamics, and should allow further elucidation of the
foundations of thermodynamics.
.IP 3.
They give examples of the procedure of ``adpative programming''.
```

I used block cellular automata, and tried to find rules that were reversible and also conserved something that could serve as "microscopic density" or "particle number". I quickly
determined that there were no such rules with 2 colors and blocks of sizes 2 or 3 that
achieved any kind of randomization.

To go to 3 colors, I used SMP to generate candidate rules

```
genit4[Np[$1,$2,$3,$4,$5]] :: (a1 : Apper[per2[$1],{1,4}] ; a2 : Apper[per2[$2\
     ],{2,8}] ; a3 : Apper[per4[$3],{3,6,9,12}] ; a4 : Apper[per2[\
       $4],{7,13}] ; a5 : Apper[per2[$5],{11,14}] ; {15,\
      a5[1],a4[1],a3[1],a5[2],10,a3[2],a2[1],a4[2],a3[3],5,\
       a1[1],a3[4],a2[2],a1[2],0})

Apper[$perm _= Permp[$perm],$list _= Contp[$list]] :: Ar[Len[$list],\
     $list[$perm[$%1]]]]
```

where for example the function Apper can be literally be translated into Wolfram Language as

Apper[(*perm_*)?PermutationListQ, *list_List*] := Array[*list*[*perm*[#*1*]] &, Length[*list*]]

or, more idiomatically, just

list[perm]

then did what I have done so many times and just printed out pictures of their behavior:

Some clearly did not show randomization, but a couple did. And soon I was studying what I called the "winning rule", which—like rule 30—went from simple initial conditions to apparent randomness:

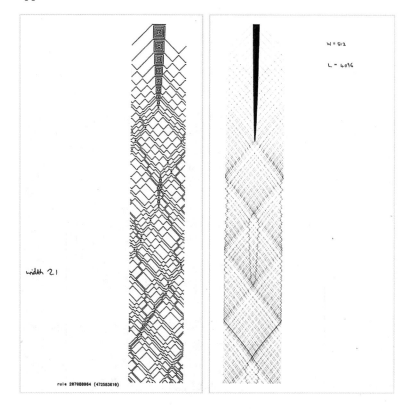

I analyzed what the rule was "microscopically doing"

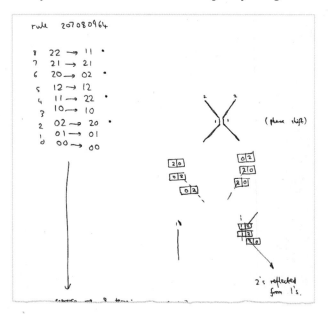

and explored its longer-time behavior:

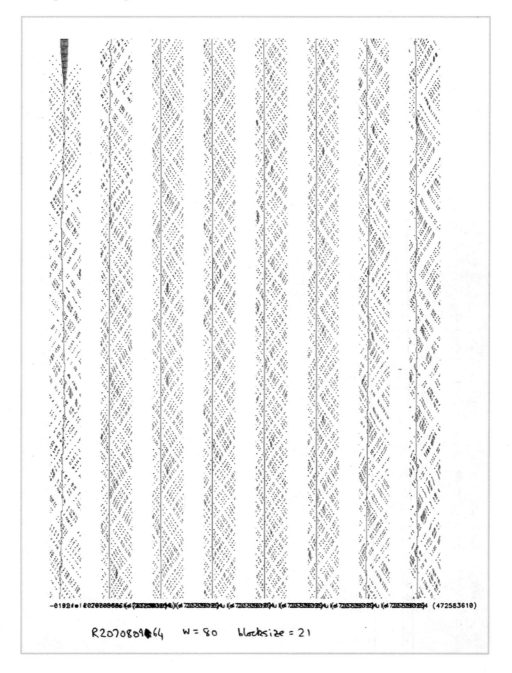

Then I did things like analyze its cycle structure in a finite-size region by running C programs I'd basically already developed back in 1982 (though now they were modified to automatically generate troff code for typesetting):

```
.LP
.nf
.LG
size 7
total number: 1918; number distinct lengths: 6
3x6 2x5 8x4 24x3 174x2 .1707x1

size 9
total number: 17135; number distinct lengths: 11
1x12 1x10 1x9 7x8 12x7 19x6 31x5 93x4 182x3 1537x2 15251x1

size 11
total number: 24219; number distinct lengths: 143
4x816 4x672 4x654 12x547 4x540 4x372 8x366 4x354 12x349 4x342 6x330 12x315
4x312 2x270 12x264 2x246 12x244 8x240 12x239 8x234 8x225 8x222 24x220 24x219
12x194 8x183 12x179 14x174 12x169 36x168 4x162 12x161 8x159 12x153 18x150
12x149 36x148 12x143 12x141 24x139 10x138 12x137 12x135 6x130 36x126 36x124
12x121 40x120 12x119 12x117 24x115 28x114 12x110 12x109 48x108 24x103 36x102
28x96 12x95 48x94 48x93 24x91 46x90 48x89 12x86 12x85 16x84 12x83 24x81 24x80
12x79 26x78 12x77 24x75 24x74 60x73 64x72 24x71 32x69 72x68 12x67 84x66 60x65
24x64 64x63 168x60 36x59 32x57 48x56 48x55 168x54 12x52 24x51 12x50 108x49
164x48 84x47 132x46 124x45 108x44 168x43 266x42 132x41 72x40 96x39 156x38
96x37 136x36 132x35 108x34 116x33 72x32 84x31 218x30 60x29 84x28 156x27 258x26
192x25 140x24 168x23 162x22 408x21 222x20 360x19 304x18 372x17 492x16 1018x15
498x14 528x13 576x12 546x11 612x10 1415x9 1710x8 1194x7 1740x6 495x5 1248x4
1725x3 1460x2 1190x1
.fi
```

And, like rule 30, the "winning rule" that I found back in June 1986 has stayed with me, essentially as a minimal example of reversible, number-conserving randomness. It appeared in *A New Kind of Science*, and it appears now in my recent work on the Second Law—and, of course, the patterns it makes are always the same:

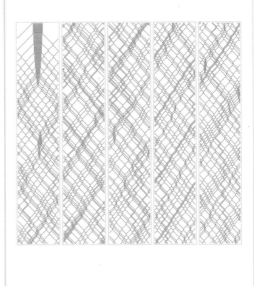

Back in 1986 I wanted to know just how efficiently a simple rule like this could reproduce continuum behavior. And in a portent of observer theory my notes from the time talk about "optimal coarse graining, where the 2nd law is 'most true'", then go on to compare the distributed character of the cellular automaton with traditional "collect information into numerical value" finite-difference approximations:

In a talk I gave I summarized my understanding:

The phenomenon of randomization is generic in computational systems (witness rule 30, the "winning rule", etc.) This leads to the genericity of thermodynamics. And this in turn leads to the genericity of continuum behavior, with diffusion and fluid behavior being two examples.

It would take another 34 years, but these basic ideas would eventually be what underlies our Physics Project, and our understanding of the emergence of things like spacetime. As well as now being crucial to our whole understanding of the Second Law.

The Second Law in *A New Kind of Science*

By the end of 1986 I had begun the development of Mathematica, and what would become the Wolfram Language, and for most of the next five years I was submerged in technology development. But in 1991 I started to use the technology I now had, and began the project that became *A New Kind of Science*.

Much of the first couple of years was spent exploring the computational universe of simple programs, and discovering that the phenomena I'd discovered in cellular automata were actually much more general. And it was seeing that generality that led me to the Principle of Computational Equivalence. In formulating the concept of computational irreducibility I'd in effect been thinking about trying to "reduce" the behavior of systems using an external as-powerful-as-possible universal computer. But now I'd realized I should just be thinking about all systems as somehow computationally equivalent. And in doing that I was pulling the conception of the "observer" and their computational ability closer to the systems they were observing.

But the further development of that idea would have to wait nearly three more decades, until the arrival of our Physics Project. In *A New Kind of Science*, Chapter 7 on "Mechanisms in Programs and Nature" describes the concept of intrinsic randomness generation, and how it's distinguished from other sources of randomness. Chapter 8 on "Implications for Every-day Systems" then has a section on fluid flow, where I describe the idea that randomness in turbulence could be intrinsically generated, making it, for example, repeatable, rather than inevitably different every time an experiment is run.

And then there's Chapter 9, entitled "Fundamental Physics". The majority of the chapter— and its "most famous" part—is the presentation of the direct precursor to our Physics Project, including the concept of graph-rewriting-based computational models for the lowest-level structure of spacetime and the universe.

But there's an earlier part of Chapter 9 as well, and it's about the Second Law. There's a precursor about "The Notion of Reversibility", and then we're on to a section about "Irreversibility and the Second Law of Thermodynamics", followed by "Conserved Quantities and Continuum Phenomena", which is where the "winning rule" I discovered in 1996 appears again:

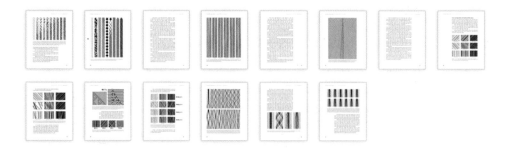

My records show I wrote all of this—and generated all the pictures—between May 2 and July 11, 1995. I felt I already had a pretty good grasp of how the Second Law worked, and just needed to write it down. My emphasis was on explaining how a microscopically reversible rule—through its intrinsic ability to generate randomness—could lead to what appears to be irreversible behavior.

Mostly I used reversible 1D cellular automata as my examples, showing for example randomization both forwards and backwards in time:

I soon got to the nub of the issue with irreversibility and the Second Law:

Yet there is still something of a mystery. For our everyday experience is full of examples in which randomness increases much as in the second half of the picture above. But we essentially never see the kind of systematic decrease in randomness that occurs in the first half.

By setting up the precise initial conditions that exist at the beginning of the whole picture it would certainly in principle be possible to get such behavior. But somehow it seems that initial conditions like these essentially never actually occur in practice.

There has in the past been considerable confusion about why this might be the case. But the key to understanding what is going on is simply to realize that one has to think not only about the systems one is studying, but also about the types of experiments and observations that one uses in the process of studying them.

The crucial point then turns out to be that practical experiments almost inevitably end up involving only initial conditions that are fairly simple for us to describe and construct. And with these types of initial conditions, systems like the one on the previous page always tend to exhibit increasing randomness.

But what exactly is it that determines the types of initial conditions that one can use in an experiment? It seems reasonable to suppose that in any meaningful experiment the process of setting up the experiment should somehow be simpler than the process that the experiment is intended to observe.

But how can one compare such processes? The answer that I will develop in considerable detail later in this book is to view all such processes as computations. The conclusion is then that the computation involved in setting up an experiment should be simpler than the computation involved in the evolution of the system that is to be studied by the experiment.

It is clear that by starting with a simple state and then tracing backwards through the actual evolution of a reversible system one can find initial conditions that will lead to decreasing randomness. But if one looks for example at the pictures on the last couple of pages the complexity of the behavior seems to preclude any less arduous way of finding such initial conditions. And indeed I will argue in Chapter 12 that the Principle of Computational Equivalence suggests that in general no such reduced procedure should exist.

The consequence of this is that no reasonable experiment can ever involve setting up the kind of initial conditions that will lead to decreases in randomness, and that therefore all practical experiments will tend to show only increases in randomness.

It is this basic argument that I believe explains the observed validity of what in physics is known as the Second Law of Thermodynamics. The law was first formulated more than a century

I talked about how "typical textbook thermodynamics" involves a bunch of details about energy and motion, and to get closer to this I showed a simple example of an "ideal gas" 2D cellular automaton:

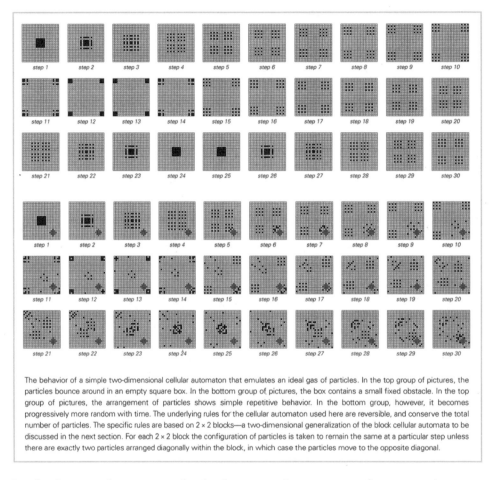

The behavior of a simple two-dimensional cellular automaton that emulates an ideal gas of particles. In the top group of pictures, the particles bounce around in an empty square box. In the bottom group of pictures, the box contains a small fixed obstacle. In the top group of pictures, the arrangement of particles shows simple repetitive behavior. In the bottom group, however, it becomes progressively more random with time. The underlying rules for the cellular automaton used here are reversible, and conserve the total number of particles. The specific rules are based on 2 × 2 blocks—a two-dimensional generalization of the block cellular automata to be discussed in the next section. For each 2 × 2 block the configuration of particles is taken to remain the same at a particular step unless there are exactly two particles arranged diagonally within the block, in which case the particles move to the opposite diagonal.

But despite my early exposure to hard-sphere gases, I never went as far as to use them as examples in *A New Kind of Science*. We did actually take some photographs of the mechanics of real-life billiards:

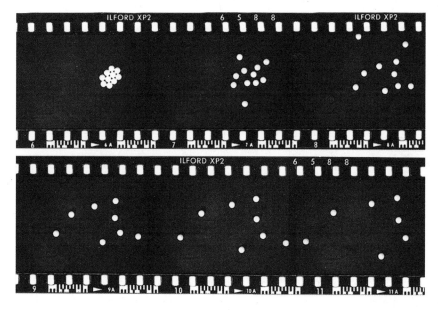

But cellular automata always seemed like a much clearer way to understand what was going on, free from issues like numerical precision, or their physical analogs. And by looking at cellular automata I felt as if I could really see down the foundations of the Second Law, and why it was true.

And mostly it was a story of computational irreducibility, and intrinsic randomness generation. But then there was rule 37R. I've often said that in studying the computational universe we have to remember that the "computational animals" are at least as smart as we are—and they're always up to tricks we don't expect.

And so it is with rule 37R. In 1986 I'd published a book of cellular automaton papers, and as an appendix I'd included lots of tables of properties of cellular automata. Almost all the tables were about the ordinary elementary cellular automata. But as a kind of "throwaway" at the very end I gave a table of the behavior of the 256 second-order reversible versions of the elementary rules, including 37R starting both from completely random initial conditions

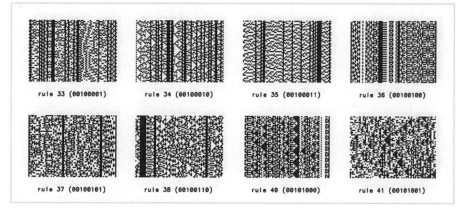

and from single black cells:

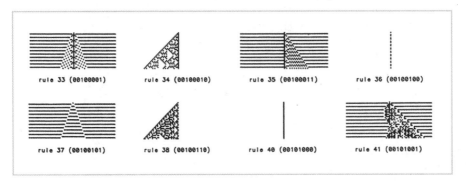

So far, nothing remarkable. And years go by. But then—apparently in the middle of working on the 2D systems section of *A New Kind of Science*—at 4:38am on February 21, 1994 (according to my filesystem records), I generate pictures of all the reversible elementary rules again, but now from initial conditions that are slightly more complicated than a single black cell. Opening the notebook from that time (and, yes, Wolfram Language and our notebook format have been stable enough that 28 years later that still works) it shows up tiny on a modern screen, but there it is: rule 37R doing something "interesting":

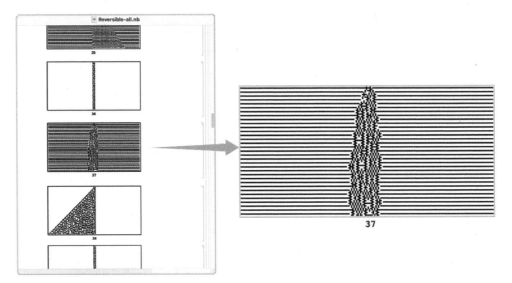

Clearly I noticed it. Because by 4:47am I've generated lots of pictures of rule 37R, like this one evolving from a block of 21 black cells, and showing only every other step

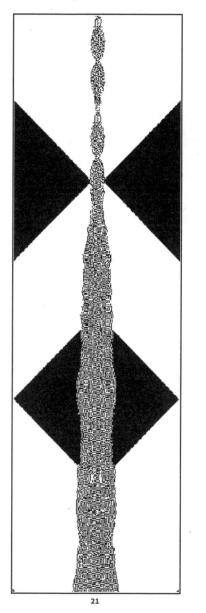

21

and by 4:54am I've got things like:

My guess is that I was looking for class 4 behavior in reversible cellular automata. And with rule 37R I'd found it. And at the time I moved on to other things. (On March 1, 1994, I slipped on some ice and broke my ankle, and was largely out of action for several weeks.)

And that takes us back to May 1995, when I was working on writing about the Second Law. My filesystem records that I did quite a few more experiments on rule 37R then, looking at different initial conditions, and running it as long as I could, to see if its strange neither-simple-nor-randomizing—and not very Second-Law-like—behavior would somehow "resolve".

Up to that moment, for nearly a quarter of a century, I had always fundamentally believed in the Second Law. Yes, I thought there might be exceptions with things like self-gravitating systems. But I'd always assumed that—perhaps with some pathological exceptions—the Second Law was something quite universal, whose origins I could even now understand through computational irreducibility.

But seeing rule 37R this suddenly didn't seem right. In *A New Kind of Science* I included a long run of rule 37R (here colorized to emphasize the structure)

steps 0–3000   steps 5000–8000   steps 10000–13000   steps 20000–23000   steps 100000–103000   steps 200000–203000

then explained:

> The picture on the next page, however, shows the behavior of rule 37R over the course of many steps. And in looking at this picture, we see a remarkable phenomenon: there is neither a systematic trend towards increasing randomness, nor any form of simple predictable behavior. Indeed, it seems that the system just never settles down, but rather continues to fluctuate forever, sometimes becoming less orderly, and sometimes more so.
>
> So how can such behavior be understood in the context of the Second Law? There is, I believe, no choice but to conclude that for practical purposes rule 37R simply does not obey the Second Law.

How could one describe what was happening in rule 37R? I discussed the idea that it was effectively forming "membranes" which could slowly move, but keep things "modular" and organized inside. I summarized at the time, tagging it as "something I wanted to explore in more detail one day":

> And indeed I strongly suspect that there are many systems in nature which behave in more or less the same way. The Second Law is an important and quite general principle—but it is not universally valid. And by thinking in terms of simple programs we have thus been able in this section not only to understand why the Second Law is often true, but also to see some of its limitations.

Rounding out the rest of *A New Kind of Science* takes another seven years of intense work. But finally in May 2002 it was published. The book talked about many things. And even within Chapter 9 my discussion of the Second Law was overshadowed by the outline I gave of an approach to finding a truly fundamental theory of physics—and of the ideas that evolved into our Physics Project.

The Physics Project—and the Second Law Again

After *A New Kind of Science* was finished I spent many years working mainly on technology—building Wolfram|Alpha, launching the Wolfram Language and so on. But "follow up on Chapter 9" was always on my longterm to-do list. The biggest—and most difficult—part of that had to do with fundamental physics. But I still had a great intellectual attachment to the Second Law, and I always wanted to use what I'd then understood about the computational paradigm to "tighten up" and "round out" the Second Law.

I'd mention it to people from time to time. Usually the response was the same: "Wasn't the Second Law understood a century ago? What more is there to say?" Then I'd explain, and it'd be like "Oh, yes, that is interesting". But somehow it always seemed like people felt the Second Law was "old news", and that whatever I might do would just be "dotting an *i* or crossing a *t*". And in the end my Second Law project never quite made it onto my active list, despite the fact that it was something I always wanted to do.

Occasionally I would write about my ideas for finding a fundamental theory of physics. And, implicitly I'd rely on the understanding I'd developed of the foundations and generalization of the Second Law. In 2015, for example, celebrating the centenary of general relativity, I wrote about what spacetime might really be like "underneath"

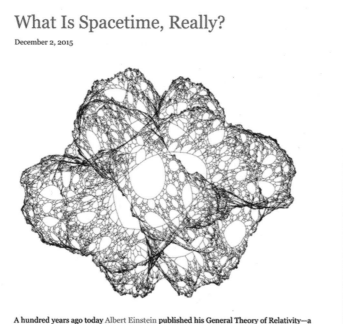

What Is Spacetime, Really?

December 2, 2015

A hundred years ago today Albert Einstein published his General Theory of Relativity—a brilliant, elegant theory that has survived a century, and provides the only successful way we have of describing spacetime.

and how a perceived spacetime continuum might emerge from discrete underlying structure like fluid behavior emerges from molecular dynamics—in effect through the operation of a generalized Second Law:

> **understand the origins of** apparent randomness in fluid turbulence. **And in particular I showed that even when the underlying "molecules" are cells in a simple cellular automaton, it's possible to get large-scale behavior that exactly follows the standard differential equations of fluid flow.**
>
> So when I started thinking about the possibility that underneath space there might be a network, I imagined that perhaps the same methods might be used—and that it might actually be possible to derive Einstein's Equations of General Relativity from something much lower level.

It was 17 years after the publication of *A New Kind of Science* that (as I've described elsewhere) circumstances finally aligned to embark on what became our Physics Project. And after all those years, the idea of computational irreducibility—and its immediate implications for the Second Law—had come to seem so obvious to me (and to the young physicists with whom I worked) that they could just be taken for granted as conceptual building blocks in constructing the tower of ideas we needed.

One of the surprising and dramatic implications of our Physics Project is that general relativity and quantum mechanics are in a sense both manifestations of the same fundamental phenomenon—but played out respectively in physical space and in branchial space. But what really is this phenomenon?

What became clear is that ultimately it's all about the interplay between underlying computational irreducibility and our nature as observers. It's a concept that had its origins in my thinking about the Second Law. Because even in 1984 I'd understood that the Second Law is about our inability to "decode" underlying computationally irreducible behavior.

In *A New Kind of Science* I'd devoted Chapter 10 to "Processes of Perception and Analysis", and I'd recognized that we should view such processes—like any processes in nature or elsewhere—as being fundamentally computational. But I still thought of processes of perception and analysis as being separated from—and in some sense "outside"—actual processes we might be studying. But in our Physics Project we're studying the whole universe, so inevitably we as observers are "inside" and part of the system.

And what then became clear is the emergence of things like general relativity and quantum mechanics depends on certain characteristics of us as observers. "Alien observers" might perceive quite different laws of physics (or no systematic laws at all). But for "observers like us", who are computationally bounded and believe we are persistent in time, general relativity and quantum mechanics are inevitable.

In a sense, therefore, general relativity and quantum mechanics become "abstractly derivable" given our nature as observers. And the remarkable thing is that at some level the story is exactly the same with the Second Law. To me it's a surprising and deeply beautiful scientific unification: that all three of the great foundational theories of physics—general relativity, quantum mechanics and statistical mechanics—are in effect manifestations of the same core phenomenon: an interplay between computational irreducibility and our nature as observers.

Back in the 1970s I had no inkling of all this. And even when I chose to combine my discussions of the Second Law and of my approach to a fundamental theory of physics into a single chapter of *A New Kind of Science*, I didn't know how deeply these would be connected. It's been a long and winding path, that's needed to pass through many different pieces of science and technology. But in the end the feeling I had when I first studied that book cover when I was 12 years old that "this was something fundamental" has played out on a scale almost incomprehensibly beyond what I had ever imagined.

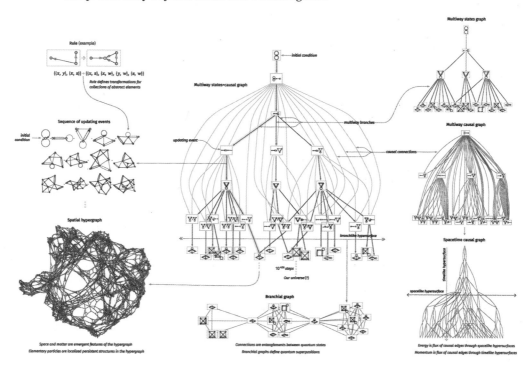

Discovering Class 4

Most of my journey with the Second Law has had to do with understanding origins of randomness, and their relation to "typical Second Law behavior". But there's another piece—still incompletely worked out—which has to do with surprises like rule 37R, and, more generally, with large-scale versions of class 4 behavior, or what I've begun to call the "mechanoidal phase".

I first identified class 4 behavior as part of my systematic exploration of 1D cellular automata at the beginning of 1983—with the "code 20" $k=2$, $r=2$ totalistic rule being my first clear example:

Very soon my searches had identified a whole variety of localized structures in this rule:

```
Summary up to length 5 initial sequences

Halting fraction = 1; average halting time = 11.8125;
maximum finite lifetime = 21
Absolute maximum size = 18; average maximum size = 12.5


Summary up to length 6 initial sequences

Halting fraction = 1; average halting time = 13.1875;
maximum finite lifetime = 32
Absolute maximum size = 18; average maximum size = 12.875


Configuration 151 persistent; period 2
Configuration 187 persistent; period 9
Configuration 189 persistent; period 1
Configuration 195 persistent; period 22
Configuration 219 persistent; period 22
Configuration 221 persistent; period 9
Configuration 233 persistent; period 2

Summary up to length 7 initial sequences

Halting fraction = .919643; average halting time = 12.375;
maximum finite lifetime = 34
Absolute maximum size = 27; average maximum size = 12.625

Persistent configurations :
period   multi.   pred.    configuration
2        2        151      11101000
9        1        187      100010001000
1        1        189      10111100
22       1        195      110011010100001010110010
22       1        219      110101111101010
9        1        221      100110001000
2        2        233      10010110


Summary up to length 19 initial sequences

Halting fraction = .938694; average halting time = 25.4042;
maximum finite lifetime = 135
Absolute maximum size = 1138; average maximum size = 22.6989

2     4    101653    11101001000000011101000
38    1    125231    101100111000011100001110001100
1     7    127647    1011110100000000000000000000001011111100
2     1    204771    100101110000000000000011101000
38    2    205151    111101001001011110
2     1    339567    101111010000000001110100000
2     1    350037    100101110000000000000000001110100000
38    1    358261    11000000011001001100000010
38    1    384477    100100000100100010010000001000
2     2    391875    111010010000000000010010
2     5    409571    100101110000000000000111101000
38    1    417235    111100000111010101110000001110
2     1    435499    111010010000000000000012010110
38    2    453243    110010111011110101010010
```

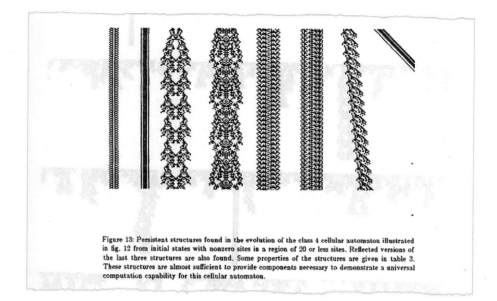

Figure 13: Persistent structures found in the evolution of the class 4 cellular automaton illustrated in fig. 12 from initial states with nonzero sites in a region of 20 or less sites. Reflected versions of the last three structures are also found. Some properties of the structures are given in table 3. These structures are almost sufficient to provide components necessary to demonstrate a universal computation capability for this cellular automaton.

At the time, the most significant attribute of class 4 cellular automata as far as I was concerned was that they seemed likely to be computation universal—and potentially provably so. But from the beginning I was also interested in what their "thermodynamics" might be. If you start them off from random initial conditions, will their patterns die out, or will some arrangement of localized structures persist, and perhaps even grow?

In most cellular automata—and indeed most systems with local rules—one expects that at least their statistical properties will somehow stabilize when one goes to the limit of infinite size. But, I asked, does that infinite-size limit even "exist" for class 4 systems—or if you progressively increase the size, will the results you get keep on jumping around forever, perhaps as you succeed in sampling progressively more exotic structures?

Does thermodynamic limit exist for class 4?

$N = 2000$: f_{∞} .014
.013

$N = 1000$ .02
.01
0
.004
.03
.03

[Density depends in detail on arbitrarily large regions in initial state.

→ Final statistics depend in detail on initial ensemble.

<20 : For $\rho_0 = 0.5$, $\pi = 2$ oscillator by far commonest.

slow glider generated in $N = 1000$ run...*

A paper I wrote in September 1983 talks about the idea that in a sufficiently large class 4 cellular automaton one would eventually get self-reproducing structures, which would end up "taking over everything":

> If class 4 cellular automata are indeed universal computers, then they should in some sense be capable of arbitrarily complicated behaviour. Thus, for example, there should exist initial site value sequences which generate self-reproducing structures. Such structures, once generated, would replicate, and their progeny would eventually dominate the statistical behaviour of the cellular automaton. The generation of a self-reproducing structure may well require a specific very long sequence of initial site values. The probability for such a sequence to occur at a particular point in a disordered initial configuration may be infinitesimal; however, it must eventually occur in any arbitrarily long disordered initial configuration, and ultimately dominate the behaviour of the cellular automaton. Such a cellular automaton thus has no ordinary "infinite volume" or "thermodynamic" limit. Statistical averages based on successively larger numbers of sites do not converge smoothly to a limit. This phenomenon is related to the potentially non-recursive nature of the sets generated by such systems at arbitrarily large times.

The idea that one might be able to see "biology-like" self-reproduction in cellular automata has a long history. Indeed, one of the multiple ways that cellular automata were invented (and the one that led to their name) was through John von Neumann's 1952 effort to construct a complicated cellular automaton in which there could be a complicated configuration capable of self-reproduction.

But could self-reproducing structures ever "occur naturally" in cellular automata? Without the benefit of intuition from things like rule 30, von Neumann assumed that something like self-reproduction would need an incredibly complicated setup, as it seems to have, for example, in biology. But having seen rule 30—and more so class 4 cellular automata—it didn't seem so implausible to me that even with very simple underlying rules, there could be fairly simple configurations that would show phenomena like self-reproduction.

But for such a configuration to "occur naturally" in a random initial condition might require a system with exponentially many cells. And I wondered if in the oceans of the early Earth there might have been only "just enough" molecules for something like a self-reproducing lifeform to occur.

Back in 1983 I already had pretty efficient code for searching for structures in class 4 cellular automata. But even running for days at a time, I never found anything more complicated than purely periodic (if sometimes moving) structures. And in March 1985, following an article about my work in *Scientific American,* I appealed to the public to find "interesting structures"—like "glider guns" that would "shoot out" moving structures:

Glider Gun Guidelines

Stephen Wolfram
The Institute for Advanced Study, Princeton NJ 08540.

(March 1985)

In his *Computer Recreations* column in the May 1985 issue of *Scientific American*, A. K. Dewdney described some of the research that I have been doing on one-dimensional cellular automata. As he mentioned, there are some interesting questions about these systems that you may be able to help me answer. The questions do not rely on sophisticated mathematical knowledge, but to make progress on them will probably take a lot of work and some clever ideas. These notes define some of the questions, describe what I have worked out so far about them, and suggest some approaches you could take to them.

Structures found for the $k=3$, $r=1$ rule with code 357. (All nonzero sites shown black.) Can you find other gliders? Or a glider gun?

Structures found for the $k=3$, $r=1$ rule with code 792. (All nonzero sites shown black.) Can you find a glider gun?

Epilogue

That is about all I can tell you about these questions. The rest is up to you. If you find something interesting, please do send me a letter, at the address given above, and send a copy to A. K. Dewdney, care of *Scientific American*.

I have printed up a set of six colour postcards of cellular automaton patterns. They are available from me at $2 (US) for each set. But I will certainly send a free set to anyone who finds an interesting new cellular automaton structure. And if enough of these are found, I will make a catalogue of them later this year. If you would like a copy of this, please send me a note.

Good luck!

As it happened, right before I made my "public appeal", a student at Princeton working with a professor I knew had sent me a glider gun he'd found, the $k = 2$, $r = 3$ totalistic code 88 rule:

A One-Dimensional Glider Gun

James K. Park

Princeton University, EECS '85
65 Prospect Avenue
Princeton, NJ 08540

In the course of investigating several simple one-dimensional cellular automata, I have discovered a persistant structure analogous to the glider gun of

At the time, though, with computer displays only large enough to see behavior like

I wasn't convinced this was an "ordinary class 4 rule"—even though now, with the benefit of higher display resolution, it seems more convincing:

The "public appeal" generated a lot of interesting feedback—but no glider guns or other exotic structures in the rules I considered "obviously class 4". And it wasn't until after I started working on *A New Kind of Science* that I got back to the question. But then, on the evening of December 31, 1991, using exactly the same code as in 1983, but now with faster computers, there it was: in an ordinary class 4 rule ($k = 3$, $r = 1$ code 1329), after finding several localized structures, there was one that grew without bound (albeit not in the most obvious "glider gun" way):

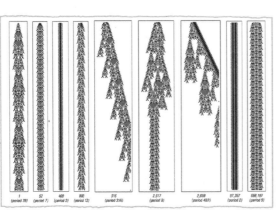

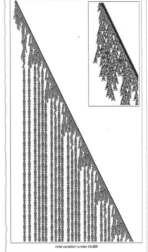

But that wasn't all. Exemplifying the principle that in the computational universe there are always surprises, searching a little further revealed yet other unexpected structures:

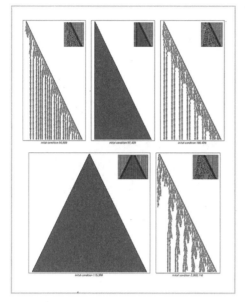

Every few years something else would come up with class 4 rules. In 1994, lots of work on rule 110. In 1995, the surprise of rule 37R. In 1998 efforts to find analogs of particles that might carry over to my graph-based model of space.

After *A New Kind of Science* was published in 2002, we started our annual Wolfram Summer School (at first called the NKS Summer School)—and in 2010 our High School Summer Camp. Some years we asked students to pick their "favorite cellular automaton". Often they were class 4:

And occasionally someone would do a project to explore the world of some particular class 4 rule. But beyond those specifics—and statements about computation universality—it's never been clear quite what one could say about class 4.

Back in 1984 in the series of cellular automaton postcards I'd produced, there were a couple of class 4 examples:

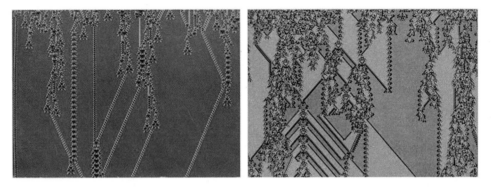

And even then the typical response to these images was that they looked "organic"—like the kind of thing living organisms might produce. A decade later—for *A New Kind of Science*—I studied "organic forms" quite a bit, trying to understand how organisms get their overall shapes, and surface patterns. Mostly that didn't end up being a story of class 4 behavior, though.

Since the early 1980s I've been interested in molecular computing, and in how computation might be done at the level of molecules. My discoveries in *A New Kind of Science* (and specifically the Principle of Computational Equivalence) convinced me that it should be possible to get even fairly simple collections of molecules to "do arbitrary computations" or even build more or less arbitrary structures (in a more general and streamlined way than happens with the whole protein synthesis structure in biology). And over the years, I sometimes thought about trying to do practical work in this area. But it didn't feel as if the ambient technology was quite ready. So I never jumped in.

Meanwhile, I'd long understood the basic correspondence between multiway systems and patterns of possible pathways for chemical reactions. And after our Physics Project was announced in 2020 and we began to develop the general multicomputational paradigm, I immediately considered molecular computing a potential application. But just what might the "choreography" of molecules be like? What causal relationships might there be, for example, between different interactions of the same molecule? That's not something ordinary chemistry—dealing for example with liquid-phase reactions—tends to consider important.

But what I increasingly started to wonder is whether in molecular biology it might actually be crucial. And even in the 20 years since *A New Kind of Science* was published, it's become increasingly clear that in molecular biology things are extremely "orchestrated". It's not about molecules randomly moving around, like in a liquid. It's about molecules being carefully channeled and actively transported from one "event" to another.

Class 3 cellular automata seem to be good "metamodels" for things like liquids, and readily give Second-Law-like behavior. But what about the kind of situation that seems to exist in molecular biology? It's something I've been thinking about only recently, but I think this is a place where class 4 cellular automata can contribute. I've started calling the "bulk limit" of class 4 systems the "mechanoidal phase". It's a place where the ordinary Second Law doesn't seem to apply.

Four decades ago when I was trying to understand how structure could arise "in violation of the Second Law" I didn't yet even know about computational irreducibility. But now we've come a lot further, in particular with the development of the multicomputational paradigm, and the recognition of the importance of the characteristics of the observer in defining what perceived overall laws there will be. It's an inevitable feature of computational irreducibility that there will always be an infinite sequence of new challenges for science, and new pieces of computational reducibility to be found. So, now, yes, a challenge is to understand the mechanoidal phase. And with all the tools and ideas we've developed, I'm hoping the process will happen more than it has for the ordinary Second Law.

The End of a 50-Year Journey

I began my quest to understand the Second Law a bit more than 50 years ago. And—even though there's certainly more to say and figure out—it's very satisfying now to be able to bring a certain amount of closure to what has been the single longest-running piece of intellectual "unfinished business" in my life. It's been an interesting journey—that's very much relied on, and at times helped drive, the tower of science and technology that I've spent my life building. There are many things that might not have happened as they did. And in the end it's been a story of longterm intellectual tenacity—stretching across much of my life so far.

For a long time I've kept (automatically when possible) quite extensive archives. And now these archives allow one to reconstruct in almost unprecedented detail my journey with the Second Law. One sees the gradual formation of intellectual frameworks over the course of years, then the occasional discovery or realization that allows one to take the next step in what is sometimes mere days. There's a curious interweaving of computational and essentially philosophical methodologies—with an occasional dash of mathematics.

Sometimes there's general intuition that's significantly ahead of specific results. But more often there's a surprise computational discovery that seeds the development of new intuition. And, yes, it's a little embarrassing how often I managed to generate in a computer experiment something that I completely failed to interpret or even notice at first because I didn't have the right intellectual framework or intuition.

And in the end, there's an air of computational irreducibility to the whole process: there really wasn't a way to shortcut the intellectual development; one just had to live it. Already in the 1990s I had taken things a fair distance, and I had even written a little about what I had figured out. But for years it hung out there as one of a small collection of unfinished projects: to finally round out the intellectual story of the Second Law, and to write down an exposition of it. But the arrival of our Physics Project just over two years ago brought both a cascade of new ideas, and for me personally a sense that even things that had been out there for a long time could in fact be brought to closure.

And so it is that I've returned to the quest I began when I was 12 years old—but now with five decades of new tools and new ideas. The wonder and magic of the Second Law is still there. But now I'm able to see it in a much broader context, and to realize that it's not just a law about thermodynamics and heat, but instead a window into a very general computational phenomenon. None of this I could know when I was 12 years old. But somehow the quest I was drawn to all those years ago has turned out to be deeply aligned with the whole arc of intellectual development that I have followed in my life. And no doubt it's no coincidence.

But for now I'm just grateful to have had the quest to understand the Second Law as one of my guiding forces through so much of my life, and now to realize that my quest was part of something so broad and so deep.

Appendix: The Backstory of the Book Cover That Started It All

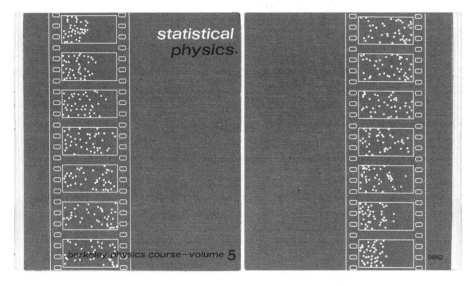

What is the backstory of the book cover that launched my long journey with the Second Law? The book was published in 1965, and inside its front flap we find:

The book covers

The movie strips on the covers illustrate the fundamental ideas of irreversibility and fluctuations by showing the motion of 40 particles inside a two-dimensional box. The movie strips were produced by an electronic computer programmed to calculate particle trajectories. (For details, see pp. 7, 24, and 25 inside the book.) The front cover illustrates the irreversible approach to equilibrium starting from the highly nonrandom initial situation where all the particles are located in the left half of the box. The back cover (read in the upward direction from bottom to top) illustrates the irreversible approach to equilibrium if, starting from the initial situation at the top of the front cover, all the particle velocities are reversed (or equivalently, if the direction of time is imagined to be reversed). The back-cover and front-cover movie strips together, read consecutively in the downward direction, illustrate a very large fluctuation occurring extremely rarely in equilibrium.

On page 7 we then find:

Computer-constructed pictures

The following pages and several subsequent ones show figures constructed by means of a high-speed electronic digital computer. The situation investigated in every case is the classical motion of several particles in a box, the particles being represented by disks moving in two dimensions. The forces between any two particles, or between a particle and a wall, are assumed to be like those between "hard" objects (i.e., to vanish when they do not touch and to become infinite when they do touch). All resulting collisions are thus elastic. The computer is given some initial specified positions and velocities of the particles. It is then asked to solve numerically the equations of motion of these particles for all subsequent (or prior) times and to display pictorially on a cathode-ray oscilloscope the positions of the molecules at successive times $t = j\tau_0$ where τ_0 is some small fixed time interval and where $j = 0, 1, 2, 3, \ldots$. A movie camera photographing the oscilloscope screen then yields the successive picture frames reproduced in the figures. (The time interval τ_0 was chosen long enough so that several molecular collisions occur between the successive frames displayed in the figures.) The computer is thus used to simulate in detail a hypothetical experiment involving the dynamical interaction between many particles.

All the computer-made pictures were produced with the generous cooperation of Dr. B. J. Alder of the Lawrence Radiation Laboratory at Livermore.

All the computer-made pictures were produced with the generous cooperation of Dr. B. J. Alder of the Lawrence Radiation Laboratory at Livermore.

In 2001—as I was putting the finishing touches to the historical notes for *A New Kind of Science*—I tracked down Berni Alder (who died in 2020 at the age of 94) to ask him the origin of the pictures. It turned out to be a complex story, reaching back to the earliest serious uses of computers for basic science, and even beyond.

The book had been born out of the sense of urgency around science education in the US that followed the launch of Sputnik by the Soviet Union—with a group of professors from Berkeley and Harvard believing that the teaching of freshman college physics was in need of modernization, and that they should write a series of textbooks to enable this. (It was also the time of the "new math", and a host of other STEM-related educational initiatives.) Fred Reif (who died at the age of 92 in 2019) was asked to write the statistical physics volume. As he explained in the preface to the book

This last volume of the Berkeley Physics Course is devoted to the study of large-scale (i.e., *macro*scopic) systems consisting of many atoms or molecules; thus it provides an introduction to the subjects of statistical mechanics, kinetic theory, thermodynamics, and heat. The approach which I have followed is not patterned upon the historical development of these subjects and does not proceed along conventional lines. My aim has been rather to adopt a modern point of view and to show, in as systematic and simple a way as possible, how the basic notions of atomic theory lead to a coherent conceptual framework capable of describing and predicting the properties of macroscopic systems.

ending with:

As I indicated at the beginning of this preface, my aim has been to penetrate the essence of a sophisticated subject sufficiently to make it seem simple, coherent, and easily accessible to beginning students. Although the goal is worth pursuing, it is difficult to attain. Indeed, the writing of this book was for me an arduous and lonely task that consumed an incredible amount of time and left me feeling exhausted. It would be some slight compensation to know that I had achieved my aim sufficiently well so that the book would be found useful.

F. Reif

Well, it's taken me 50 years to get to the point where I think I really understand the Second Law that is at the center of the book. And in 2001 I was able to tell Fred Reif that, yes, his book had indeed been useful. He said he was pleased to learn that, adding "It is all too rare that one's educational efforts seem to bear some fruit."

He explained to me that when he was writing the book he thought that "the basic ideas of irreversibility and fluctuations might be very vividly illustrated by the behavior of a gas of particles spreading through a box". He added: "It then occurred to me that Berni Alder might actually show this by a computer generated film since he had worked on molecular dynamics simulations and had also good computer facilities available to him. I was able to enlist Berni's interest in this project, with the results shown in my book."

The acknowledgements in the book report:

> The making of the computer-constructed pictures took an appreciable amount of time and effort. I wish, therefore, to express my warmest thanks to Dr. Berni J. Alder who helped me enormously in this task by personal cooperation uncontaminated by financial compensation. My ideas about these pictures could never have come to fruition if he had not put his computing experience at my disposal. We hope to continue our collaboration in the future by making available some computer-constructed movies which should help to illustrate the same ideas in more vivid form.

Berni Alder and and Fred Reif did indeed create a "film loop", which "could be bought separately from the book and viewed in the physics lab", as Alder told me, adding that "I understand the students liked it very much, but the venture was not a commercial success." Still, he sent me a copy of a videotape version:

The film (which has no sound) begins:

Soon it's showing an actual process of "coming to equilibrium":

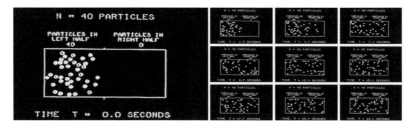

"However", as Alder explained it to me, "if a large number of particles are put in the corner and the velocities of all the particles are reversed after a certain time, the audience laughs or is supposed to after all the particles return to their original positions." (One suspects that particularly in the 1960s this might have been reminiscent of various cartoon-film gags.)

OK, so how were the pictures (and the film) made? It was done in 1964 at what's now Lawrence Livermore Lab (that had been created in 1952 as a spinoff of the Berkeley Radiation Lab, which had initiated some key pieces for the Manhattan Project) on a computer called the LARC ("Livermore Advanced Research Computer"), first made in 1960, that was probably the most advanced scientific computer of the time. Alder explained to me, however: "We could not run the problem much longer than about 10 collision times with 64 bits [*sic*] arithmetic before the round-off error prevented the particles from returning."

Why did they start the particles off in a somewhat random configuration? (The randomness, Alder told me, had been created by a middle-square random number generator.) Apparently if they'd been in a regular array—which would have made the whole process of randomization much easier to see—the roundoff errors would have been too obvious. (And it's issues like this that made it so hard to recognize the rule 30 phenomenon in systems based on real numbers—and without the idea of just studying simple programs not tied to traditional equation-based formulations of physics.)

The actual code for the molecular dynamics simulation was written in assembler and run by Mary Ann Mansigh (Karlsen), who had a degree in math and chemistry and worked as a programmer at Livermore from 1955 until the 1980s, much of the time specifically with Alder. Here she is at the console of the LARC (yes, computers had built-in desks in those days):

The program that was used was called STEP, and the original version of it had actually been written (by a certain Norm Hardy, who ended up having a long Silicon Valley career) to run on a previous generation of computer. (A still-earlier program was called APE, for "Approach to Equilibrium".) But it was only with the LARC—and STEP—that things were fast enough to run substantial simulations, at the rate of about 200,000 collisions per hour (the simulation for the book cover involved 40 particles and about 500 collisions). At the time of the book STEP used an n^2 algorithm where all pairs of particles were tested for collisions; later a neighborhood-based linked list method was used.

The standard method of getting output from a computer back in 1964—and basically until the 1980s—was to print characters on paper. But the LARC could also drive an oscilloscope, and it was with this that the graphics for the book were created (capturing them from the oscilloscope screen with a Polaroid instant camera).

But why was Berni Alder studying molecular dynamics and "hard sphere gases" in the first place? Well, that's another long story. But ultimately it was driven by the effort to develop a microscopic theory of liquids.

The notion that gases might consist of discrete molecules in motion had arisen in the 1700s (and even to some extent in antiquity), but it was only in the mid-1800s that serious development of the "kinetic theory" idea began. Pretty immediately it was clear how to derive the ideal gas law $PV = RT$ for essentially non-interacting molecules. But what analog of this "equation of state" might apply to gases with significant interactions between molecules, or, for that matter, liquids? In 1873 Johannes Diderik van der Waals proposed, on essentially empirical grounds, the formula $(P + a/V^2)(V - b) = RT$—where the parameter b represented "excluded volume" taken up by molecules, that were implicitly being viewed as hard spheres. But could such a formula be derived—like the ideal gas law—from a microscopic kinetic theory of molecules? At the time, nobody really knew how to start, and the problem languished for more than half a century.

(It's worth pointing out, by the way, that the idea of modeling gases, as opposed to liquids, as collections of hard spheres was extensively pursued in the mid-1800s, notably by Maxwell and Boltzmann—though with their traditional mathematical analysis methods, they were limited to studying average properties of what amount to dilute gases.)

Meanwhile, there was increasing interest in the microscopic structure of liquids, particularly among chemists concerned for example with how chemical solutions might work. And at the end of the 1920s the technique of x-ray diffraction, which had originally been used to study the microscopic structure of crystals, was applied to liquids—allowing in particular the experimental determination of the radial distribution function (or pair correlation function) $g(r)$, which gives the probability to find another molecule a distance r from a given one.

But how might this radial distribution function be computed? By the mid-1930s there were several proposals based on looking at the statistics of random assemblies of hard spheres:

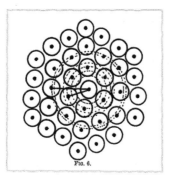

Fig. 6.

Some tried to get results by mathematical methods; others did physical experiments with ball bearings and gelatin balls, getting at least rough agreement with actual experiments on liquids:

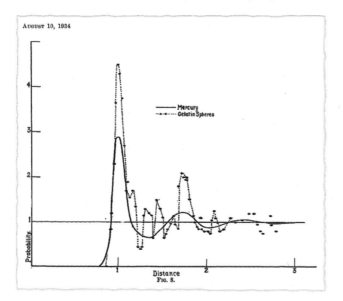

Fig. 8.

But then in 1939 a physical chemist named John Kirkwood gave an actual probabilistic derivation (using a variety of simplifying assumptions) that fairly closely reproduced the radial distribution function:

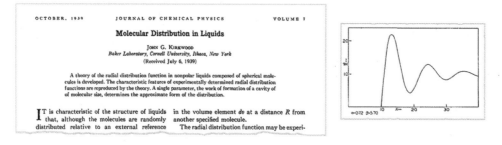

But what about just computing from first principles, on the basis of the mechanics of colliding molecules? Back in 1872 Ludwig Boltzmann had proposed a statistical equation (the "Boltzmann transport equation") for the behavior of collections of molecules, that was based on the approximation of independent probabilities for individual molecules. By the 1940s the independence assumption had been overcome, but at the cost of introducing an infinite hierarchy of equations (the "BBGKY hierarchy", where the "K" stood for Kirkwood). And although the full equations were intractable, approximations were suggested that—while themselves mathematically sophisticated—seemed as if they should, at least in principle, be applicable to liquids.

Meanwhile, in 1948, Berni Alder, fresh from a master's degree in chemical engineering, and already interested in liquids, went to Caltech to work on a PhD with John Kirkwood—who suggested that he look at a couple of approximations to the BBGKY hierarchy for the case of hard spheres. This led to some nasty integro-differential equations which couldn't be solved by analytical techniques. Caltech didn't yet have a computer in the modern sense, but in 1949 they acquired an IBM 604 Electronic Calculating Punch, which could be wired to do calculations with input and output specified on punched cards—and it was on this machine that Alder got the calculations he needed done (the paper records that "[this] ... was calculated ... with the use of IBM equipment and the file of punched cards of $sin(ut)$ employed in these laboratories for electron diffraction calculation"):

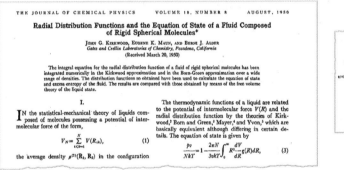
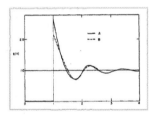

Our story now moves to Los Alamos, where in 1947 Stan Ulam had suggested the Monte Carlo method as a way to study neutron diffusion. In 1949 the method was implemented on the ENIAC computer. And in 1952 Los Alamos got its own MANIAC computer. Meanwhile, there was significant interest at Los Alamos in computing equations of state for matter, especially in extreme conditions such as those in a nuclear explosion. And by 1953 the idea had arisen of using the Monte Carlo method to do this.

The concept was to take a collection of hard spheres (or actually 2D disks), and move them randomly in a series of steps with the constraint that they could not overlap—then look at the statistics of the resulting "equilibrium" configurations. This was done on the MANIAC, with the resulting paper now giving "Monte Carlo results" for things like the radial distribution function:

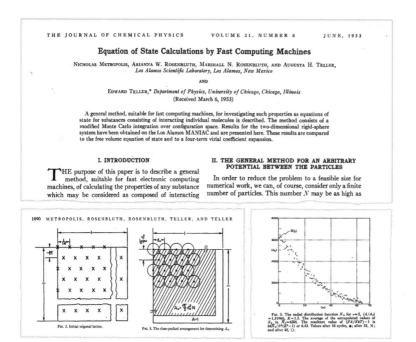

Kirkwood and Alder had been continuing their BBGKY hierarchy work, now using more realistic Lennard-Jones forces between molecules. But by 1954 Alder was also using the Monte Carlo method, implementing it partly (rather painfully) on the IBM Electronic Calculating Punch, and partly on the Manchester Mark II computer in the UK (whose documentation had been written by Alan Turing):

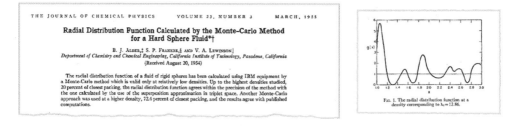

In 1955 Alder started working full-time at Livermore, recruited by Edward Teller. Another Livermore recruit—fresh from a physics PhD—was Thomas Wainwright. And soon Alder and Wainwright came up with an alternative to the Monte Carlo method—that would eventually give the book cover pictures: just explicitly compute the dynamics of colliding hard spheres, with the expectation that after enough collisions the system would come to equilibrium and allow things like equations of state to be obtained.

In 1953 Livermore had obtained its first computer: a Remington Rand Univac I. And it was on this computer that Alder and Wainwright did a first proof of concept of their method, tracing 100 hard spheres with collisions computed at the rate of about 100 per hour. Then in 1955 Livermore got IBM 704 computers, which, with their hardware floating-point capabilities, were able to compute about 2000 collisions per hour.

Alder and Wainwright reported their first results at a statistical mechanics conference in Brussels in August 1956 (organized by Ilya Prigogine). The published version appeared in 1958:

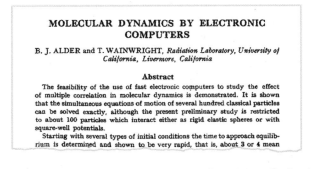

as well as things like the radial distribution function—and the equation of state:

It gives evidence—that they tagged as "provisional"—for the emergence of a Maxwell–Boltzmann velocity distribution "after the system reached equilibrium"

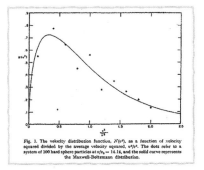

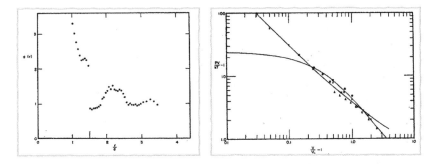

It was notable that there seemed to be a discrepancy between the results for the equation of state computed by explicit molecular dynamics and by the Monte Carlo method. And what is more, there seemed to be evidence of some kind of discontinuous phase-transition-like behavior as the density of spheres changed (an effect which Kirkwood had predicted in 1949).

Given the small system sizes and short runtimes it was all a bit muddy. But by August 1957 Alder and Wainwright announced that they'd found a phase transition, presumably between a high-density phase where the spheres were packed together like in a crystalline solid, and a low-density phase, where they were able to more freely "wander around" like in a liquid or gas. Meanwhile, the group at Los Alamos had redone their Monte Carlo calculations, and they too now claimed a phase transition. Their papers were published back to back:

Preliminary Results from a Recalculation of the Monte Carlo Equation of State of Hard Spheres*

W. W. WOOD AND J. D. JACOBSON

Los Alamos Scientific Laboratory, Los Alamos, New Mexico
(Received August 15, 1957)

THE disagreement between the hard sphere equation of state obtained by Rosenbluth and Rosenbluth[1] using the Monte Carlo method[2] and that reported in the accompanying paper by Alder and Wainwright[3] using detailed molecular dynamics led us to repeat the Monte Carlo investigation. Preliminary results for 32 molecules with cubical periodic boundary conditions[1,2] are shown in Fig. 1 along with Alder and Wainwright's[3] results with which there is rather good agreement. The previous Monte Carlo calculations[1] at reduced volumes v/v_0 (v_0=close-packed volume) from about 1.5 to 2.0 are in error due to inadequate chain length to detect the behavior described below; the difficulty was aggravated by the concentration of effort on the system of 256 molecules, which requires considerably longer computing time.

The present calculations have been made on IBM Type 704 calculators and use the same method as the earlier work[1,2] except that the molecules are "moved" in random rather than ordered sequence.[4]

Phase Transition for a Hard Sphere System

B. J. ALDER AND T. E. WAINWRIGHT

University of California Radiation Laboratory, Livermore, California
(Received August 12, 1957)

A CALCULATION of molecular dynamic motion has been designed principally to study the relaxations accompanying various nonequilibrium phenomena. The method consists of solving exactly (to the number of significant figures carried) the simultaneous classical equations of motion of several hundred particles by means of fast electronic computors. Some of the details as they relate to hard spheres and to particles having square well potentials of attraction have been described.[1,2] The method has been used also to calculate equilibrium properties, particularly the equation of state of hard spheres where differences with previous Monte Carlo[3] results appeared.

The calculation treats a system of particles in a rectangular box with periodic boundary conditions.[4]

But at this point no actual pictures of molecular trajectories had yet been published, or, I believe, made. All there was were traditional plots of aggregated quantities. And in 1958, these plots made their first appearance in a textbook. Tucked into Appendix C of *Elementary Statistical Physics* by Berkeley physics professor Charles Kittel (who would later be chairman of the group developing the Berkeley Physics Course book series) were two rather confusing plots about the approach to the Maxwell–Boltzmann distribution taken from a pre-publication version of Alder and Wainwright's paper:

C. Solutions of Problems in Molecular Dynamics Using Electronic Computers

It has recently been demonstrated that high-speed electronic computers may be employed to give interesting and useful information about the motion of large numbers of interacting particles. We discuss here only the work of Alder and Wainwright.* Their results show the approach to equilibrium in a most revealing way.

In most of their work they treated exactly the motion of 100 hard-

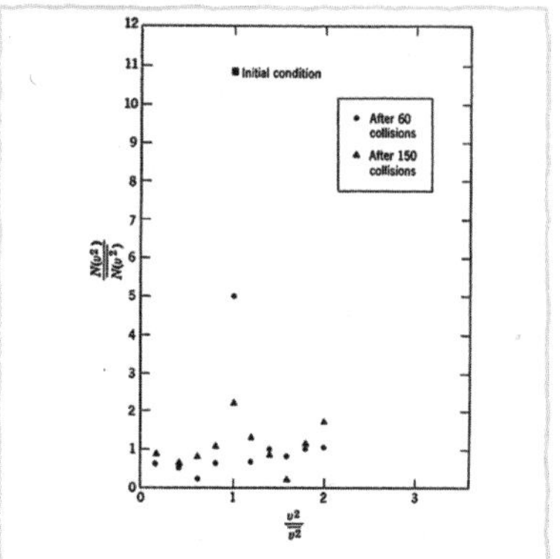

Fig. C.1. Approach of a uniform velocity distribution at the initial time to a Maxwell-Boltzmann distribution, for system of 100 hard spheres.

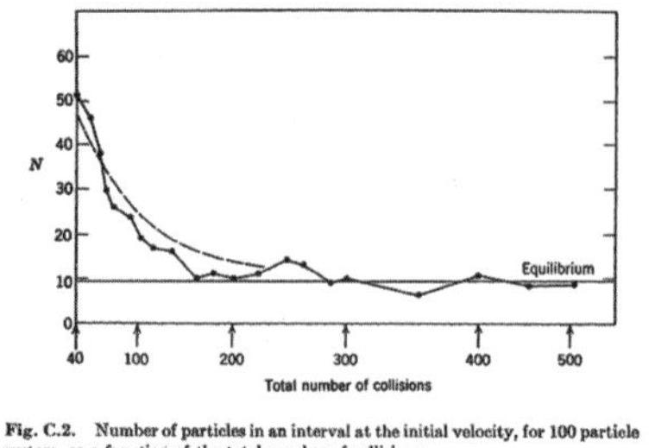

Fig. C.2. Number of particles in an interval at the initial velocity, for 100 particle system, as a function of the total number of collisions.

Alder and Wainwright's phase transition result had created enough of a stir that they were asked to write a *Scientific American* article about it. And in that article—entitled "Molecular Motions", from October 1959—there were finally pictures of actual trajectories, with their caption explaining that the "paths of particles … appear as bright lines on the face of a cathode-ray tube hooked to a computer" (the paths are of the centers of the colliding disks):

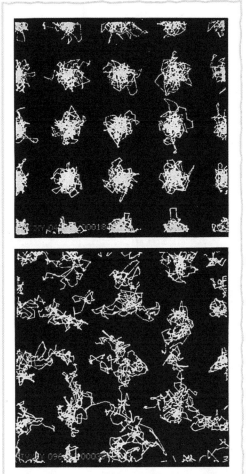 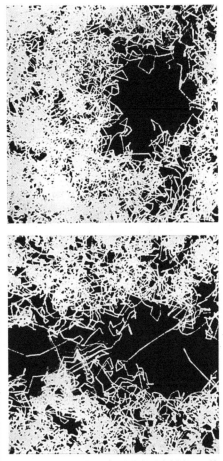

PATHS OF PARTICLES in molecular-dynamical calculation appear as bright lines on the face of a cathode-ray tube hooked to the computer. Each cluster in upper photograph represents two hard spheres, one behind the other. In solid state (*top*) particles can move only around well-defined positions; in fluid (*bottom*) they travel from one position to another.

LIQUID-GAS SEPARATION is illustrated by molecular-dynamical calculations on particles with square-well interaction. Dark area in upper photograph represents a gaseous bubble surrounded by particles whose motions characterize a liquid. Lower photograph shows the system at a later time, when some particles have vaporized and passed through the bubble

A technical article published at the same time gave a diagram of the logic for the dynamical computation:

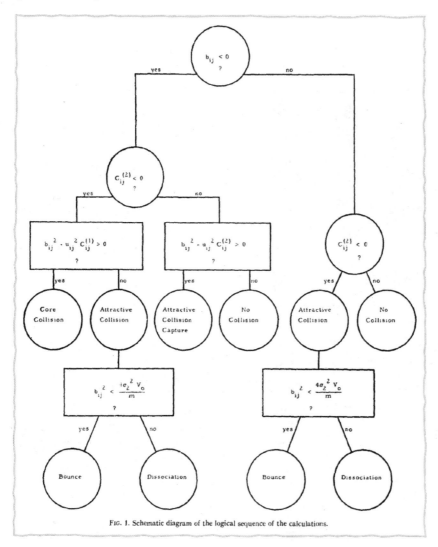

FIG. 1. Schematic diagram of the logical sequence of the calculations.

Then in 1960 Livermore (after various delays) took delivery of the LARC computer—arguably the first scientific supercomputer—which allowed molecular dynamics computations to be done perhaps 20 times times faster. A 1962 picture shows Berni Alder (left) and Thomas Wainwright (right) looking at outputs from the LARC with Mary Ann Mansigh (yes, in those days it was typical for male physicists to wear ties):

And in 1964, the pictures for the *Statistical Physics* book (and film loop) got made, with Mary Ann Mansigh painstakingly constructing images of disks on the oscilloscope display.

Work on molecular dynamics continued, though to do it required the most powerful computers, so for many years it was pretty much restricted to places like Livermore. And in 1967, Alder and Wainwright made another discovery about hard spheres. Even in their first paper about molecular dynamics they'd plotted the velocity autocorrelation function, and noted that it decayed roughly exponentially with time. But by 1967 they had much more precise data, and realized that there was a deviation from exponential decay: a definite "long-time tail". And soon they had figured out that this power-law tail was basically the result of a continuum hydrodynamic effect (essentially a vortex) operating even on the scale of a few molecules. (And—though it didn't occur to me at the time—this should have suggested that even with fairly small numbers of cells cellular automaton fluid simulations had a good chance of giving recognizable hydrodynamic results.)

It's never been entirely easy to do molecular dynamics, even with hard spheres, not least because in standard computations one's inevitably confronted with things like numerical roundoff errors. And no doubt this is why some of the obvious foundational questions about the Second Law weren't really explored there, and why intrinsic randomness generation and the rule 30 phenomenon weren't identified.

Incidentally, even before molecular dynamics emerged, there was already one computer study of what could potentially have been Second Law behavior. Visiting Los Alamos in the early 1950s Enrico Fermi had gotten interested in using computers for physics, and wondered what would happen if one simulated the motion of an array of masses with nonlinear springs between them. The results of running this on the MANIAC computer were reported in 1955 (after Fermi had died)

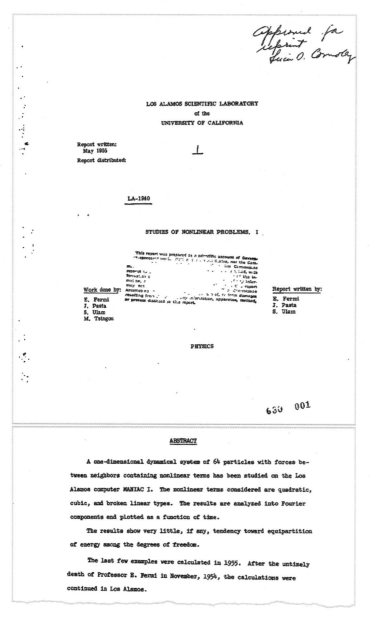

and it was noted that there wasn't just exponential approach to equilibrium, but instead something more complicated (later connected to solitons). Strangely, though, instead of plotting actual particle trajectories, what were given were mode energies—but these still exhibited what, if it hadn't been obscured by continuum issues, might have been recognized as something like the rule 30 phenomenon:

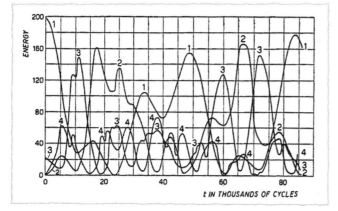

But I knew none of this history when I saw the *Statistical Physics* book cover in 1972. And indeed, for all I knew, it could have been a "standard statistical physics cover picture". I didn't know it was the first of its kind—and a leading-edge example of the use of computers for basic science, accessible only with the most powerful computers of the time. Of course, had I known those things, I probably wouldn't have tried to reproduce the picture myself and I wouldn't have had that early experience in trying to use a computer to do science. (Curiously enough, looking at the numbers now, I realize that the base speed of the LARC was only 20x the Elliott 903C, though with floating point, etc.—a factor that pales in comparison with the 500x speedup in computers in the 40 years since I started working on cellular automata.)

But now I know the history of that book cover, and where it came from. And what I only just discovered now is that actually there's a bigger circle than I knew. Because the path from Berni Alder to that book cover to my work on cellular automaton fluids came full circle—when in 1988 Alder wrote a paper based on cellular automaton fluids (though through the vicissitudes of academic behavior I don't think he knew these had been my idea—and now it's too late to tell him his role in seeding them):

VOLUME 61   25 JULY 1988   NUMBER 4

Maximally Discretized Molecular Dynamics

M. E. Colvin, A. J. C. Ladd, and B. J. Alder
Lawrence Livermore National Laboratory, Livermore, California 94550
(Received 29 March 1988)

It is shown that the coarsest discretization of positions and velocities in molecular dynamics leads to qualitatively correct transport coefficients and quantitatively predictable long-time tails in the velocity autocorrelation function, but requires orders of magnitude less computer time than standard molecular-dynamics methods.

PACS numbers: 05.50.+q, 05.60.+w

The use of cellular-automata-based methods for the solution of the incompressible Navier-Stokes equation has generated considerable interest,[1] in part because their computational simplicity allows efficient implementation on massively parallel computers such as the 65536-processor Connection Machine.[2] The lattice-gas approach still remain. The most serious of these is the lack of translational (Galilean) invariance introduced by the discrete velocities.[3] This results in an incorrect coefficient of the nonlinear u·∇u (advection) term in the Navier-Stokes equation, and, because of this artificiality, this study is confined to linear phe-

Notes & Thanks .

There are many people who've contributed to the 50-year journey I've described here. Some I've already mentioned by name, but others not—including many who doubtless wouldn't even be aware that they contributed. The longtime store clerk at Blackwell's bookstore who in 1972 sold college physics books to a 12-year-old without batting an eye. (I learned his name—Keith Clack—30 years later when he organized a book signing for *A New Kind of Science* at Blackwell's.) John Helliwell and Lawrence Wickens who in 1977 invited me to give the first talk where I explicitly discussed the foundations of the Second Law. Douglas Abraham who in 1977 taught a course on mathematical statistical mechanics that I attended. Paul Davies who wrote a book on *The Physics of Time Asymmetry* that I read around that time. Rocky Kolb who in 1979 and 1980 worked with me on cosmology that used statistical mechanics. The students (including professors like Steve Frautschi and David Politzer) who attended my 1981 class at Caltech about "nonequilibrium statistical mechanics". David Pines and Elliott Lieb who in 1983 were responsible for publishing my breakout paper on "Statistical Mechanics of Cellular Automata". Charles Bennett (curiously, a student of Berni Alder's) with whom in the early 1980s I discussed applying computation theory (notably the ideas of Greg Chaitin) to physics. Brian Hayes who commissioned my 1984 *Scientific American* article, and Peter Brown who edited it. Danny Hillis and Sheryl Handler who in 1984 got me involved with Thinking Machines. Jim Salem and Bruce Nemnich (Walker) who worked on fluid dynamics on the Connection Machine with me. Then—36 years later—Jonathan Gorard and Max Piskunov, who catalyzed the doing of our Physics Project.

In the last 50 years, there've been surprisingly few people with whom I've directly discussed the foundations of the Second Law. Perhaps one reason is that back when I was a "professional physicist" statistical mechanics as a whole wasn't a prominent area. But, more important, as I've described elsewhere, for more than a century most physicists have effectively assumed that the foundations of the Second Law are a solved (or at least merely pedantic) problem.

Probably the single person with whom I had the most discussions about the foundations of the Second Law is Richard Feynman. But there are others with whom at one time or another I've discussed related issues, including: Bruce Boghosian, Richard Crandall, Roger Dashen, Mitchell Feigenbaum, Nigel Goldenfeld, Theodore Gray, Bill Hayes, Joel Lebowitz, David Levermore, Ed Lorenz, John Maddox, Roger Penrose, Ilya Prigogine, Rudy Rucker, David Ruelle, Rob Shaw, Yakov Sinai, Michael Trott, Léon van Hove and Larry Yaffe. (There are also many others with whom I've discussed general issues about origins of randomness.)

Finally, one technical note about the presentation here: in an effort to maintain a clearer timeline, I've typically shown the earliest drafts or preprint versions of papers that I have. Their final published versions (if indeed they were ever published) appeared anything from weeks to years later, sometimes with changes.

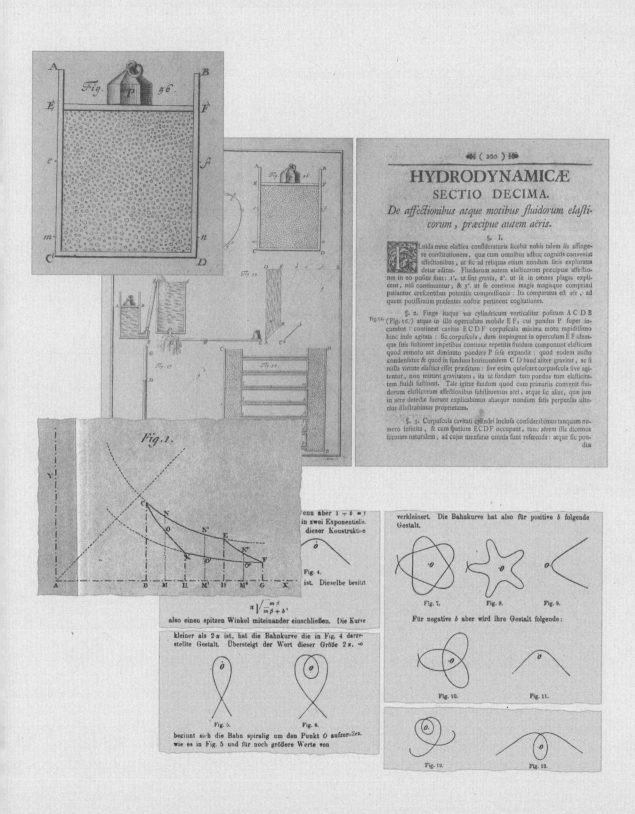

HYDRODYNAMICÆ
SECTIO DECIMA.
De affectionibus atque motibus fluidorum elasticorum, præcipue autem aëris.

§. I. Fluida nunc elastica confideraturis licebit nobis talem iis affingere conftitutionem, quæ cum omnibus adhuc cognitis conveniat affectionibus, ut fic ad reliquas etiam nondum fatis exploratas detur aditus. Fluidorum autem elafticorum præcipuæ affectiones in eo pofitæ funt: 1°. ut fint gravia, 2°. ut fe in omnes plagas explicent, nifi contineantur, & 3°. ut fe continue magis magisque comprimi patiantur crefcentibus potentiis compreffionis: Ita comparatus eft aër, ad quem potiffimum præfentes noftræ pertinent cogitationes.

§. 2. Finge itaque vas cylindricum verticaliter pofitum A C D B (Fig. 56.) atque in illo operculum mobile E F, cui pondus P fuper incumbat: contineat cavitas E C D F corpufcula minima motu rapidiffimo hinc inde agitata: fic corpufcula, dum impingunt in operculum E F ideamque fuis fuftinent impetitus continue repetitis fluidum component elafticum quod remoto aut diminuto pondere P fefe expandit: quod eodem aucto condenfatur & quod in fundum horizontalem C D haud aliter gravitat, ac fi nulla virtute elaftica effet præditum: five enim quiefcant corpufcula five agitentur, non mutant gravitatem, ita ut fundum tum pondus tum elafticitatem fluidi fuftineat. Tale igitur fluidum quod cum primariis convenit fluidorum elafticorum affectionibus fubftituemus aëri, atque fic alias, quæ jam in aëre detectæ fuerunt explicabimus aliasque nondum fatis perpenfas ulterius illuftrabimus proprietates.

§. 3. Corpufcula cavitati cylindri inclufa confiderabimus tanquam numero infinita, & cum fpatium E C D F occupant, tunc aërem illa dicemus formare naturalem, ad cujus menfuras omnia funt referenda: atque fic pondus

---

Fig. 1.

verkleinert. Die Bahnkurve hat also für positive b folgende Gestalt.

Fig. 7.   Fig. 8.   Fig. 9.

Für negative b aber wird ihre Gestalt folgende:

Fig. 10.   Fig. 11.

Fig. 12.   Fig. 13.

... wenn aber $1 - b =$ ... in zwei Exponentielle ... dieser Konstruktion

Fig. 4.

... ist. Dieselbe befitzt

$$\pi \sqrt{\frac{m}{m\beta + b}}$$

also einen fpitzen Winkel miteinander einfchließen. Die Kurve kleiner als 2π ist, hat die Bahnkurve die in Fig. 4 dargestellte Gestalt. Übersteigt der Wert dieser Größe 2π,

Fig. 5.   Fig. 6.

beginnt fich die Bahn fpiralig um den Punkt O aufzurollen, wie es in Fig. 5 und für noch größere Werte von

How Did We Get Here? The Tangled History of the Second Law of Thermodynamics

The Basic Arc of the Story

As I've explained elsewhere, I think I now finally understand the Second Law of thermodynamics. But it's a new understanding, and to get to it I've had to overcome a certain amount of conventional wisdom about the Second Law that I at least have long taken for granted. And to check myself I've been keen to know just where this conventional wisdom came from, how it's been validated, and what might have made it go astray.

And from this I've been led into a rather detailed examination of the origins and history of thermodynamics. All in all, it's a fascinating story, that both explains what's been believed about thermodynamics, and provides some powerful examples of the complicated dynamics of the development and acceptance of ideas.

The basic concept of the Second Law was first formulated in the 1850s, and rather rapidly took on something close to its modern form. It began partly as an empirical law, and partly as something abstractly constructed on the basis of the idea of molecules, that nobody at the time knew for sure existed. But by the end of the 1800s, with the existence of molecules increasingly firmly established, the Second Law began to often be treated as an almost-mathematically-proven necessary law of physics. There were still mathematical loose ends, as well as issues such as its application to living systems and to systems involving gravity. But the almost-universal conventional wisdom became that the Second Law must always hold, and if it didn't seem to in a particular case, then that must just be because there was something one didn't yet understand about that case.

There was also a sense that regardless of its foundations, the Second Law was successfully used in practice. And indeed particularly in chemistry and engineering it's often been in the background, justifying all the computations routinely done using entropy. But despite its ubiquitous appearance in textbooks, when it comes to foundational questions, there's always been a certain air of mystery around the Second Law. Though after 150 years there's typically an assumption that "somehow it must all have been worked out". I myself have been interested in the Second Law now for a little more than 50 years, and over that time I've had a growing awareness that actually, no, it hasn't all been worked out. Which is why, now, it's wonderful to see the computational paradigm—and ideas from our Physics Project—after all these years be able to provide solid foundations for understanding the Second Law, as well as seeing its limitations.

And from the vantage point of the understanding we now have, we can go back and realize that there were precursors of it even from long ago. In some ways it's all an inspiring tale—of how there were scientists with ideas ahead of their time, blocked only by the lack of a conceptual framework that would take another century to develop. But in other ways it's also a cautionary tale, of how the forces of "conventional wisdom" can blind people to unanswered questions and—over a surprisingly long time—inhibit the development of new ideas.

But, first and foremost, the story of the Second Law is the story of a great intellectual achievement of the mid-19th century. It's exciting now, of course, to be able to use the latest 21st-century ideas to take another step. But to appreciate how this fits in with what's already known we have to go back and study the history of what originally led to the Second Law, and how what emerged as conventional wisdom about it took shape.

What Is Heat?

Once it became clear what heat is, it actually didn't take long for the Second Law to be formulated. But for centuries—and indeed until the mid-1800s—there was all sorts of confusion about the nature of heat.

That there's a distinction between hot and cold is a matter of basic human perception. And seeing fire one might imagine it as a disembodied form of heat. In ancient Greek times Heraclitus (~500 BC) talked about everything somehow being "made of fire", and also somehow being intrinsically "in motion". Democritus (~460–~370 BC) and the Epicureans had the important idea (that also arose independently in other cultures) that everything might be made of large numbers of a few types of tiny discrete atoms. They imagined these atoms moving around in the "void" of space. And when it came to heat, they seem to have correctly associated it with the motion of atoms—though they imagined it came from particular spherical "fire" atoms that could slide more quickly between other atoms, and they also thought that souls were the ultimate sources of motion and heat (at least in warm-blooded animals?), and were made of fire atoms.

And for two thousand years that's pretty much where things stood. And indeed in 1623 Galileo (1564–1642) (in his book *The Assayer*, about weighing competing world theories) was still saying:

> Those materials which produce heat in us and make us feel warmth, which are known by the general name of "fire," would then be a multitude of minute particles having certain shapes and moving with certain velocities. Meeting with our bodies, they penetrate by means of their extreme subtlety, and their touch as felt by us when they pass through our substance is the sensation we call "heat."

He goes on:

> Since the presence of fire-corpuscles alone does not suffice to excite heat, but their motion is needed also, it seems to me that one may very reasonably say that motion is the cause of heat... But I hold it to be silly to accept that proposition in the ordinary way, as if a stone or piece of iron or a stick must heat up when moved. The rubbing together and friction of two hard bodies, either by resolving their parts into very subtle flying particles or by opening an exit for the tiny fire-corpuscles within, ultimately sets these in motion; and when they meet our bodies and penetrate them, our conscious mind feels those pleasant or unpleasant sensations which we have named heat...

And although he can tell there's something different about it, he thinks of heat as effectively being associated with a substance or material:

> The tenuous material which produces heat is even more subtle than that which causes odor, for the latter cannot leak through a glass container, whereas the material of heat makes its way through any substance.

In 1620, Francis Bacon (1561–1626) (in his "update on Aristotle", *The New Organon*) says, a little more abstractly, if obscurely—and without any reference to atoms or substances:

> [It is not] that heat generates motion or that motion generates heat (though both are true in certain cases), but that heat itself, its essence and quiddity, is motion and nothing else.

But real progress in understanding the nature of heat had to wait for more understanding about the nature of gases, with air being the prime example. (It was actually only in the 1640s that any kind of general notion of gas began to emerge—with the word "gas" being invented by the "anti-Galen" physician Jan Baptista van Helmont (1580–1644), as a Dutch rendering of the Greek word "chaos", that meant essentially "void", or primordial formlessness.) Ever since antiquity there'd been Aristotle-style explanations like "nature abhors a vacuum" about what nature "wants to do". But by the mid-1600s the idea was emerging that there could be more explicit and mechanical explanations for phenomena in the natural world.

And in 1660 Robert Boyle (1627–1691)—now thoroughly committed to the experimental approach to science—published *New Experiments Physico-mechanicall, Touching the Spring of the Air and its Effects* in which he argued that air has an intrinsic pressure associated with it, which pushes it to fill spaces, and for which he effectively found Boyle's Law PV = constant.

But what was air actually made of? Boyle had two basic hypotheses that he explained in rather flowery terms:

(23)

This Notion may perhaps be somewhat further explain'd, by conceiving the Air near the Earth to be such a heap of little Bodies, lying one upon another, as may be resembled to a Fleece of Wooll. For this (to omit other likenesses betwixt them) consists of many slender and flexible Hairs; each of which, may indeed, like a little Spring, be easily bent or rouled up; but will also, like a Spring, be still endeavouring to stretch it self out again. For though both these Haires, and the Aerial Corpuscles to which we liken them, do easily yield to externall pressures; yet each of them (by vertue of its structure) is endow'd with a Power or Principle of self-Dilatation; by vertue whereof, though the hairs may by a Mans hand be bent and crouded closer together, and into a narrower room then suits best with the nature of the Body: Yet whil'st the compression lasts, there is in the fleece they compose an endeavour outwards, whereby it continually thrusts against the hand that opposes its Expansion. And upon the removall of the external pressure, by opening the hand more or less, the compressed Wooll does, as it were, spontaneously expand or display it self towards

C 4   the

(24)

the recovery of its former more loose and free condition, till the Fleece have either regain'd its former Dimensions, or at least, approach'd them as near as the compressing hand (perchance not quite open'd) will permit. This Power of self-Dilatation, is somewhat more conspicuous in a dry Spunge compress'd, then in a Fleece of Wooll. But yet we rather chose to imploy the latter, on this occasion, because it is not like a Spunge, an entire Body, but a number of slender and flexible Bodies, loosely complicated, as the Air it self seems to be.

There is yet another way to explicate the Spring of the Air, namely, by supposing with that most ingenious Gentleman, Monsieur *Des Cartes*, That the Air is nothing but a Congeries or heap of small and (for the most part) of flexible Particles; of several sizes, and of all kinde of Figures which are rais'd by heat (especially that of the Sun) into that fluid and subtle Etheriall Body that surrounds the Earth; and by the restlesse agitation of that Celestial Matter wherein those Particles swim, are so whirl'd round,

(25)

round, that each Corpuscle endeavours to beat off all others from coming within the little Sphear requisite to its motion about its own Center; and (in case any, by intruding into that Sphear shall oppose its free Rotation) to expell or drive it away: So that according to this Doctrine, it imports very little, whether the particles of the Air have the structure requisite to Springs, or be of any other form (how irregular soever) since their Elastical power is not made to depend upon their shape or structure, but upon the vehement agitation, and (as it were) brandishing motion, which they receive from the fluid *Ether* that swiftly flows between them, and whirling about each of them (independently from the rest) not onely keeps those slender Aërial Bodies separated and stretcht out (at least, as far as the Neighbouring ones will permit) which otherwise, by reason of their flexibleness and weight, would flag or curl; but also makes them hit against, and knock away each other, and consequently require more room, then that which if they were compress'd, they would take up.

By

(26)

By these two differing ways, my Lord, may the Spring of the Air be explicated. But though the former of them be that, which by reason of its seeming somewhat more easie, I shall for the most part make use of in the following Discourse: yet am I not willing to declare peremptorily for either of them, against the other. And indeed, though I have in another Treatise endeavoured to make it probable, that the returning of Elastical Bodies (if I may so call them) forcibly bent, to their former position, may be Mechanically explicated: Yet I must confess, that to determine whether the motion of Restitution in Bodies, proceed from this, That the parts of a Body of a peculiar Structure are put into motion by the bending of the spring, or from the endeavor of some subtle ambient Body, whose passage may be oppos'd or obstructed, or else it's pressure unequally resisted by reason of the new shape or magnitude, which the bending of a Spring may give the Pores of it: To determine this, I say, seems to me a matter of more difficulty, then at first sight one would easily imagine it. Wherefore I shall decline medling with a subject, which is much more hard to be explicated,

His first hypothesis was that air might be like a "fleece of wool" made of "aerial corpuscles" (gases were later often called "aeriform fluids") with a "power or principle of self-dilatation" that resulted from there being "hairs" or "little springs" between these corpuscles. But he had a second hypothesis too—based, he said, on the ideas of "that most ingenious gentleman, Monsieur Descartes": that instead air consists of "flexible particles" that are "so whirled around" that "each corpuscle endeavors to beat off all others". In this second hypothesis, Boyle's "spring of the air" was effectively the result of particles bouncing off each other.

And, as it happens, in 1668 there was quite an effort to understand the "laws of impact" (that would for example be applicable to balls in games like croquet and billiards, that had existed since at least the 1300s, and were becoming popular), with John Wallis (1616–1703), Christopher Wren (1632–1723) and Christiaan Huygens (1629–1695) all contributing, and Huygens producing diagrams like:

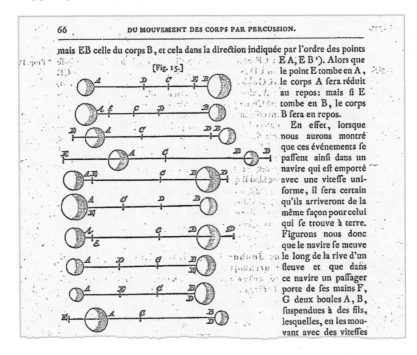

But while some understanding developed of what amount to impacts between pairs of hard spheres, there wasn't the mathematical methodology—or probably the idea—to apply this to large collections of spheres.

Meanwhile, in his 1687 *Principia Mathematica*, Isaac Newton (1642–1727), wanting to analyze the properties of self-gravitating spheres of fluid, discussed the idea that fluids could in effect be made up of arrays of particles held apart by repulsive forces, as in Boyle's first hypothesis. Newton had of course had great success with his $1/r^2$ universal attractive force for gravity. But now he noted (writing originally in Latin) that with a $1/r$ repulsive force between particles in a fluid, he could essentially reproduce Boyle's law:

PROPOSITION XXIII. THEOREM XVIII.

If a fluid be composed of particles mutually flying each other, and the density be as the compression, the centrifugal forces of the particles will be reciprocally proportional to the distances of their centres. And, vice versa, particles flying each other, with forces that are reciprocally proportional to the distances of their centres, compose an elastic fluid, whose density is as the compression.

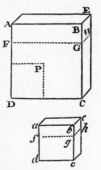

Let the fluid be supposed to be included in a cubic space ACE, and then to be reduced by compression into a lesser cubic space *ace* ; and the distances of the particles retaining a like situation with respect to each other in both the spaces, will be as the sides AB, *ab* of the cubes ; and the densities of the mediums will be reciprocally as the containing spaces AB³, *ab*³. In the plane side of the greater cube ABCD take the square DP equal to the plane side *db* of the lesser cube : and, by the supposition, the pressure with which the square DP urges the inclosed fluid will be to the pressure with which that square *db* urges the inclosed fluid as the densities of the mediums are to each other, that is, as *ab*³ to AB³. But the pressure with which the square DB urges the included fluid is to the pressure with which the square DP urges the same fluid as the square DB to the square DP, that is, as AB² to *ab*². Therefore, *ex æquo*, the pressure with which the

Newton discussed questions like whether one particle would "shield" others from the force, but then concluded:

> But whether elastic fluids do really consist of particles so repelling each other, is a physical question. We have here demonstrated mathematically the property of fluids consisting of particles of this kind, that hence philosophers may take occasion to discuss that question.

Well, in fact, particularly given Newton's authority, for well over a century people pretty much just assumed that this was how gases worked. There was one major exception, however, in 1738, when—as part of his eclectic mathematical career spanning probability theory, elasticity theory, biostatistics, economics and more—Daniel Bernoulli (1700–1782) published his book on hydrodynamics. Mostly he discusses incompressible fluids and their flow, but in one section he considers "elastic fluids"—and along with a whole variety of experimental results about atmospheric pressure in different places—draws the picture

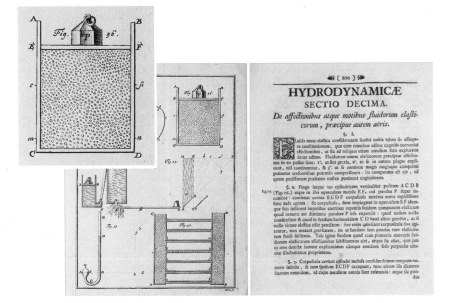

and says

> Let the space ECDF contain very small particles in rapid motion; as they strike against the piston EF and hold it up by their impact, they constitute an elastic fluid which expands as the weight P is removed or reduced; but if P is increased it becomes denser and presses on the horizontal case CD just as if it were endowed with no elastic property.

Then—in a direct and clear anticipation of the kinetic theory of heat—he goes on:

> The pressure of the air is increased not only by reduction in volume but also by rise in temperature. As it is well known that heat is intensified as the internal motion of the particles increases, it follows that any increase in the pressure of air that has not changed its volume indicates more intense motion of its particles, which is in agreement with our hypothesis...

But at the time, and in fact for more than a century thereafter, this wasn't followed up.

A large part of the reason seems to have been that people just assumed that heat ultimately had to have some kind of material existence; to think that it was merely a manifestation of microscopic motion was too abstract an idea. And then there was the observation of "radiant heat" (i.e. infrared radiation)—that seemed like it could only work by explicitly transferring some kind of "heat material" from one body to another.

But what was this "heat material"? It was thought of as a fluid—called caloric—that could suffuse matter, and for example flow from a hotter body to a colder. And in an echo of Democritus, it was often assumed that caloric consisted of particles that could slide between ordinary particles of matter. There was some thought that it might be related to the concept of phlogiston from the mid-1600s, that was effectively a chemical substance, for example participating in chemical reactions or being generated in combustion (through the "principle of fire"). But the more mainstream view was that there were caloric particles that would collect around ordinary particles of matter (often called "molecules", after the use of that term by Descartes (1596–1650) in 1620), generating a repulsive force that would for example expand gases—and that in various circumstances these caloric particles would move around, corresponding to the transfer of heat.

To us today it might seem hacky and implausible (perhaps a little like dark matter, cosmological inflation, etc.), but the caloric theory lasted for more than two hundred years and managed to explain plenty of phenomena—and indeed was certainly going strong in 1825 when Laplace wrote his *A Treatise of Celestial Mechanics*, which included a successful computation of properties of gases like the speed of sound and the ratio of specific heats, on the basis of a somewhat elaborated and mathematicized version of caloric theory (that by then included the concept of "caloric rays" associated with radiant heat).

But even though it wasn't understood what heat ultimately was, one could still measure its attributes. Already in antiquity there were devices that made use of heat to produce pressure or mechanical motion. And by the beginning of the 1600s—catalyzed by Galileo's development of the thermoscope (in which heated liquid could be seen to expand up a tube)—the idea quickly caught on of making thermometers, and of quantitatively measuring temperature.

And given a measurement of temperature, one could correlate it with effects one saw. So, for example, in the late 1700s the French balloonist Jacques Charles (1746–1823) noted the linear increase of volume of a gas with temperature. Meanwhile, at the beginning of the 1800s Joseph Fourier (1768–1830) (science advisor to Napoleon) developed what became his 1822 *Analytical Theory of Heat*, and in it he begins by noting that:

> Heat, like gravity, penetrates every substance of the universe, its rays occupy all parts of space. The object of our work is to set forth the mathematical laws which this element obeys. The theory of heat will hereafter form one of the most important branches of general physics.

Later he describes what he calls the "Principle of the Communication of Heat". He refers to "molecules"—though basically just to indicate a small amount of substance—and says

> When two molecules of the same solid are extremely near and at unequal temperatures, the most heated molecule communicates to that which is less heated a quantity of heat exactly expressed by the product of the duration of the instant, of the extremely small difference of the temperatures, and of certain function of the distance of the molecules.

then goes on to develop what's now called the heat equation and all sorts of mathematics around it, all the while effectively adopting a caloric theory of heat. (And, yes, if you think of heat as a fluid it does lead you to describe its "motion" in terms of differential equations just like Fourier did. Though it's then ironic that Bernoulli, even though he studied hydrodynamics, seemed to have a less "fluid-based" view of heat.)

Heat Engines and the Beginnings of Thermodynamics

At the beginning of the 1800s the Industrial Revolution was in full swing—driven in no small part by the availability of increasingly efficient steam engines. There had been precursors of steam engines even in antiquity, but it was only in 1712 that the first practical steam engine was developed. And after James Watt (1736–1819) produced a much more efficient version in 1776, the adoption of steam engines began to take off.

Over the years that followed there were all sorts of engineering innovations that increased the efficiency of steam engines. But it wasn't clear how far it could go—and whether for example there was a limit to how much mechanical work could ever, even in principle, be derived from a given amount of heat. And it was the investigation of this question—in the hands of a young French engineer named Sadi Carnot (1796–1832)—that began the development of an abstract basic science of thermodynamics, and to the Second Law.

The story really begins with Sadi Carnot's father, Lazare Carnot (1753–1823), who was trained as an engineer but ascended to the highest levels of French politics, and was involved with both the French Revolution and Napoleon. Particularly in years when he was out of political favor, Lazare Carnot worked on mathematics and mathematical engineering. His first significant work—in 1778—was entitled *Memoir on the Theory of Machines*. The mathematical and geometrical science of mechanics was by then fairly well developed; Lazare Carnot's objective was to understand its consequences for actual engineering machines, and to somehow abstract general principles from the mechanical details of the operation of those machines. In 1803 (alongside works on the geometrical theory of fortifications) he published his *Fundamental Principles of [Mechanical] Equilibrium and Movement*, which argued for what was at one time called (in a strange foreshadowing of reversible thermodynamic processes) "Carnot's Principle": that useful work in a machine will be maximized if accelerations and shocks of moving parts are minimized—and that a machine with perpetual motion is impossible.

Sadi Carnot was born in 1796, and was largely educated by his father until he went to college in 1812. It's notable that during the years when Sadi Carnot was a kid, one of his father's activities was to give opinions on a whole range of inventions—including many steam engines and their generalizations. Lazare Carnot died in 1823. Sadi Carnot was by that point a well-educated but professionally undistinguished French military engineer. But in 1824, at the age of 28, he produced his one published work, *Reflections on the Motive Power of Fire, and on Machines to Develop That Power* (where by "fire" he meant what we would call heat):

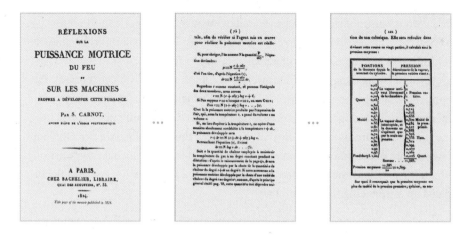

The style and approach of the younger Carnot's work is quite similar to his father's. But the subject matter turned out to be more fruitful. The book begins:

> Everyone knows that heat can produce motion. That it possesses vast motive-power none can doubt, in these days when the steam-engine is everywhere so well known... The study of these engines is of the greatest interest, their importance is enormous, their use is continually increasing, and they seem destined to produce a great revolution in the civilized world. Already the steam-engine works our mines, impels our ships, excavates our ports and our rivers, forges iron, fashions wood, grinds grain, spins and weaves our cloths, transports the heaviest burdens, etc. It appears that it must some day serve as a universal motor, and be substituted for animal power, water-falls, and air currents. ...
>
> Notwithstanding the work of all kinds done by steam-engines, notwithstanding the satisfactory condition to which they have been brought to-day, their theory is very little understood, and the attempts to improve them are still directed almost by chance. ...
>
> The question has often been raised whether the motive power of heat is unbounded, whether the possible improvements in steam-engines have an assignable limit, a limit which the nature of things will not allow to be passed by any means whatever; or whether, on the contrary, these improvements may be carried on indefinitely. We propose now to submit these questions to a deliberate examination.

Carnot operated very much within the framework of caloric theory, and indeed his ideas were crucially based on the concept that one could think about "heat itself" (which for him was caloric fluid), independent of the material substance (like steam) that was hot. But—like his father's efforts with mechanical machines—his goal was to develop an abstract "metamodel" of something like a steam engine, crucially assuming that the generation of unbounded heat or mechanical work (i.e. perpetual motion) in the closed cycle of the operation of the machine was impossible, and noting (again with a reflection of his father's work) that the system would necessarily maximize efficiency if it operated reversibly. And he then argued that:

> The production of motive power is then due in steam-engines not to an actual consumption of caloric, but to its transportation from a warm body to a cold body, that is, to its re-establishment of equilibrium... .

In other words, what was important about a steam engine was that it was a "heat engine", that "moved heat around". His book is mostly words, with just a few formulas related to the behavior of ideal gases, and some tables of actual parameters for particular materials. But even though his underlying conceptual framework—of caloric theory—was not correct, the abstract arguments that he made (that involved essentially logical consequences of reversibility and of operating in a closed cycle) were robust enough that it didn't matter, and in particular he was able to successfully show that there was a theoretical maximum efficiency for a heat engine, that depended only on the temperatures of its hot and cold reservoirs of heat. But what's important for our purposes here is that in the setup Carnot constructed he basically ended up introducing the Second Law.

At the time it appeared, however, Carnot's book was basically ignored, and Carnot died in obscurity from cholera in 1832 (about 9 months after Évariste Galois (1811–1832)) at the age of 36. (The Sadi Carnot who would later become president of France was his nephew.) But in 1834, Émile Clapeyron (1799–1864)—a rather distinguished French engineering professor (and steam engine designer)—wrote a paper entitled "Memoir on the Motive Power of Heat". He starts off by saying about Carnot's book:

> The idea which serves as a basis of his researches seems to me to be both fertile and beyond question; his demonstrations are founded on the absurdity of the possibility of creating motive power or heat out of nothing. ...

> This new method of demonstration seems to me worthy of the attention of theoreticians; it seems to me to be free of all objection ...

> I believe that it is of some interest to take up this theory again; S. Carnot, avoiding the use of mathematical analysis, arrives by a chain of difficult and elusive arguments at results which can be deduced easily from a more general law which I shall attempt to prove ...

Clapeyron's paper doesn't live up to the claims of originality or rigor expressed here, but it served as a more accessible (both in terms of where it was published and how it was written) exposition of Carnot's work, featuring, for example, for the first time a diagrammatic representation of a Carnot cycle

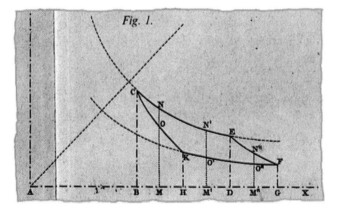

as well as notations like *Q*-for-heat that are still in use today:

Pendant le contact avec la source A , la température est restée constante; d'où il suit que les variations dp et dv de la pression et du volume sont liées par la relation

$$\frac{dT}{dp} \, dp \; + \; \frac{dT}{dv} \, dv \; = \; 0.$$

XXIIIe Cahier.   23

Ces variations dp et dv en occasionent une dans la quantité absolue
de chaleur Q , qui a pour expression·

$$dQ \; = \; \frac{dQ}{dp} \, dp \; + \; \frac{dQ}{dv} \, dv \; = \; dv \left[\frac{dQ}{dv} - \frac{dQ}{dp} \frac{\left(\frac{dT}{dv}\right)}{\left(\frac{dT}{dp}\right)} \right];$$

telle est la quantité de chaleur consommée pour produire l'effet que

The Second Law Is Formulated

One of the implications of Newton's Laws of Motion is that momentum is conserved. But what else might also be conserved? In the 1680s Gottfried Leibniz (1646–1716) suggested the quantity $m v^2$, which he called, rather grandly, *vis viva*—or, in English, "life force". And yes, in things like elastic collisions, this quantity did seem to be conserved. But in plenty of situations it wasn't. By 1807 the term "energy" had been introduced, but the question remained of whether it could in any sense globally be thought of as conserved.

It had seemed for a long time that heat was something a bit like mechanical energy, but the relation wasn't clear—and the caloric theory of heat implied that caloric (i.e. the fluid corresponding to heat) was conserved, and so certainly wasn't something that for example could be interconverted with mechanical energy. But in 1798 Benjamin Thompson (Count Rumford) (1753–1814) measured the heat produced by the mechanical process of boring a cannon, and began to make the argument that, in contradiction to the caloric theory, there was actually some kind of correspondence between mechanical energy and amount of heat.

It wasn't a very accurate experiment, and it took until the 1840s—with new experiments by the English brewer and "amateur" scientist James Joule (1818–1889) and the German physician Robert Mayer (1814–1878)—before the idea of some kind of equivalence between heat and mechanical work began to look more plausible. And in 1847 this was something William Thomson (1824–1907) (later Lord Kelvin)—a prolific young physicist recently graduated from the Mathematical Tripos in Cambridge and now installed as a professor of "natural philosophy" (i.e. physics) in Glasgow—began to be curious about.

But first we have to go back a bit in the story. In 1845 Kelvin (as we'll call him) had spent some time in Paris (primarily at a lab that was measuring properties of steam for the French government), and there he'd learned about Carnot's work from Clapeyron's paper (at first he couldn't get a copy of Carnot's actual book). Meanwhile, one of the issues of the time was a proliferation of different temperature scales based on using different kinds of thermometers based on different substances. And in 1848 Kelvin realized that Carnot's concept of a "pure heat engine"—assumed at the time to be based on caloric—could be used to define an "absolute" scale of temperature in which, for example, at absolute zero all caloric would have been removed from all substances:

On an Absolute Thermometric Scale founded on Carnot's Theory of the Motive Power of Heat*, and calculated from Regnault's observations†. By Prof. W. Thomson, Fellow of St. Peter's College.

The determination of temperature has long been recognized as a problem of the greatest importance in physical science. It has accordingly been made a subject of most careful attention, and, especially in late years, of very elaborate and refined experimental researches‡ ; and we are thus at present in possession of as complete a practical solution of the problem as can be desired, even for the most accurate investigations. The theory of thermometry is however as yet far from being in so satisfactory a state. The principle to be followed in constructing a thermometric scale might at first sight seem to be obvious, as it might appear that a perfect thermometer would indicate equal additions of heat, as corresponding to equal elevations of temperature, estimated by the numbered divisions of its scale. It is however now recognized (from the variations in the specific heats of bodies) as an experimentally demonstrated fact that

* Published in 1824 in a work entitled *Réflexions sur la Puissance Motrice du Feu,* by M. S. Carnot. Having never met with the original work, it is only through a paper by M. Clapeyron, on the same subject, published in the *Journal de l'Ecole Polytechnique,* vol. xiv. 1834, and translated in the first volume of Taylor's Scientific Memoirs, that the author has become acquainted with Carnot's theory.—W. T.

Having found Carnot's ideas useful, Kelvin in 1849 wrote a 33-page summary of them (small world that it was then, the immediately preceding paper in the journal is "On the Theory of Rolling Curves", written by the then-17-year-old James Clerk Maxwell (1831–1879), while the one that follows is "Theoretical Considerations on the Effect of Pressure in Lowering the Freezing Point of Water" by James Thomson (1786–1849), engineering-oriented older brother of William):

XXXVI.—*An Account of* CARNOT'S *Theory of the Motive Power of Heat ;** *with Numerical Results deduced from* REGNAULT'S *Experiments on Steam.*† By WILLIAM THOMSON, Professor of Natural Philosophy in the University of Glasgow.

(Read January 2, 1849.)

1. The presence of heat may be recognised in every natural object ; and there is scarcely an operation in nature which is not more or less affected by its all-pervading influence. An evolution and subsequent absorption of heat generally give rise to a variety of effects ; among which may be enumerated, chemical combinations or decompositions ; the fusion of solid substances ; the vaporisation

He characterizes Carnot's work as being based not so much on physics and experiment, but on the "strictest principles of philosophy":

> medium of certain simple operations, may be clearly appreciated. Thus it is that CARNOT, in accordance with the strictest principles of philosophy, enters upon the investigation of the theory of the motive power of heat.

He doesn't immediately mention "caloric" (though it does slip in later), referring instead to a vaguer concept of "thermal agency":

> (2.) How may the amount of this thermal agency necessary for performing a given quantity of work be estimated?
>
> 3. In the following paper I shall commence by giving a short abstract of the reasoning by which CARNOT is led to an answer to the first of these questions; I

In keeping with the idea that this is more philosophy than experimental science, he refers to "Carnot's fundamental principle"—that after a complete cycle an engine can be treated as back in the "same state"—while adding the footnote that "this is tacitly assumed as an axiom":

> the sides of the boiler, and that heat is continually abstracted by the water employed for keeping the condenser cool. According to CARNOT's fundamental principle, the quantity of heat thus discharged, during a complete revolution (or double stroke) of the engine must be precisely equal to that which enters the water of the boiler ;* provided the total mass of water and steam be invariable, and be restored to its primitive physical condition (which will be the case rigorously, if the condenser be kept cool by the external application of cold water, instead of by in-
>
> ---
> * So generally is CARNOT's principle tacitly admitted as an axiom, that its application in this case has never, so far as I am aware, been questioned by practical engineers.

In actuality, to say that an engine comes back to the same state is a nontrivial statement of the existence of some kind of unique equilibrium in the system, related to the Second Law. But in 1848 Kelvin brushes this off by saying that the "axiom" has "never, so far as I am aware, been questioned by practical engineers".

His next page is notable for the first-ever use of the term "thermo-dynamic" (then hyphenated) to discuss systems where what matters is "the dynamics of heat":

> II. On the measurement of Thermal Agency, considered with reference to its equivalent of mechanical affect.
>
> 12. A *perfect* thermo-dynamic engine of any kind, is a machine by means of which the greatest possible amount of mechanical effect can be obtained from a given thermal agency; and, therefore, if in any manner we can construct or imagine a perfect engine which may be applied for the transference of a given quantity of heat from a body at any given temperature, to another body, at a lower

That same page has a curious footnote presaging what will come, and making the statement that "no energy can be destroyed", and considering it "perplexing" that this seems incompatible with Carnot's work and its caloric theory framework:

> * When "thermal agency" is thus spent in conducting heat through a solid, what becomes of the mechanical effect which it might produce? Nothing can be lost in the operations of nature—no energy can be destroyed. What effect then is produced in place of the mechanical effect which is lost? A perfect theory of heat imperatively demands an answer to this question; yet no answer can be given in the present state of science. A few years ago, a similar confession must have been made with reference to the mechanical effect lost in a fluid set in motion in the interior of a rigid closed vessel, and allowed to come to rest by its own internal friction; but in this case, the foundation of a solution of the difficulty has been actually found, in Mr JOULE's discovery of the generation of heat, by the internal friction of a fluid in motion. Encouraged by this example, we may hope that the very perplexing question in the theory of heat, by which we are at present arrested, will, before long, be cleared up.
>
> It might appear, that the difficulty would be entirely avoided, by abandoning CARNOT's fundamental axiom; a view which is strongly urged by Mr JOULE (at the conclusion of his paper " On the Changes of Temperature produced by the Rarefaction and Condensation of Air." *Phil. Mag.*, May 1845, vol. xxvi.) If we do so, however, we meet with innumerable other difficulties—insuperable without farther experimental investigation, and an entire reconstruction of the theory of heat, from its foundation. It is in reality to experiment that we must look—either for a verification of CARNOT's axiom, and an explanation of the difficulty we have been considering; or for an entirely new basis of the Theory of Heat.

After going through Carnot's basic arguments, the paper ends with an appendix in which Kelvin basically says that even though the theory seems to just be based on a formal axiom, it should be experimentally tested:

> *Appendix.*
>
> (Read April 30, 1849.)
>
> 41. In p. 30, some conclusions drawn by CARNOT from his general reasoning were noticed; according to which it appears, that if the value of μ for any temperature is known, certain information may be derived with reference to the saturated vapour of any liquid whatever, and, with reference to any gaseous mass, without the necessity of experimenting upon the specific medium considered. Nothing in the whole range of Natural Philosophy is more remarkable than the establishment of general laws by such a process of reasoning. We have seen, however, that doubt may exist with reference to the truth of the axiom on which the entire theory is founded, and it therefore becomes more than a matter of mere curiosity to put the inferences deduced from it to the test of experience.

He proceeds to give some tests, which he claims agree with Carnot's results—and finally ends with a very practical (but probably not correct) table of theoretical efficiencies for steam engines of his day:

TABLE A. *Various Engines in which the temperature of the Boiler is 140°, and that of the Condenser 30°.*

Theoretical Duty for each Unit of Heat transmitted, 440 foot-pounds.

CASES.	Work produced for each pound of coal consumed.	Work produced for each pound of water evaporated.	Work produced for each unit of heat transmitted.	Per centage of theoretical duty.
	Foot-Pounds.	Foot-Pounds.	Foot-Pounds.	
(1.) Fowey Consols Experiment, reported in 1845,	1,330,734	156,556	253	57·5
(2.) Taylor's Engine at the United Mines, working in 1840,	1,042,553	122,653	198·4	45·1
(3.) French Engines, according to contract,	* * * *	98,427	159	36·1
(4.) English Engines, according to contract,	565,700	80,814	130·8	29·7
(5.) Average actual performance of Cornish Engines,	585,106	68,836	111·3	25·3
(6.) Common Engines, consuming 12 lbs. of best coal per hour per horse-power,	165,000	23,571	38·1	8·6
(7.) Improved Engines with Expansion Cylinders, consuming an equivalent to 4 lbs. of best coal per horse-power per hour,	495,000	70,710	114·4	26

But now what of Joule's and Mayer's experiments, and their apparent disagreement with the caloric theory of heat? By 1849 a new idea had emerged: that perhaps heat was itself a form of energy, and that, when heat was accounted for, the total energy of a system would always be conserved. And what this suggested was that heat was somehow a dynamical phenomenon, associated with microscopic motion—which in turn suggested that gases might indeed consist just of molecules in motion.

And so it was that in 1850 Kelvin (then still "William Thomson") wrote a long exposition "On the Dynamical Theory of Heat", attempting to reconcile Carnot's ideas with the new concept that heat was dynamical in origin:

XV.—*On the Dynamical Theory of Heat, with numerical results deduced from* Mr Joule's *equivalent of a Thermal Unit, and* M. Regnault's *Observations on Steam.* By William Thomson, M.A., Fellow of St Peter's College, Cambridge, and Professor of Natural Philosophy in the University of Glasgow.

(Read 17th March 1851.)

INTRODUCTORY NOTICE.

1. Sir Humphrey Davy, by his experiment of melting two pieces of ice by rubbing them together, established the following proposition :—" The phenomena of repulsion are not dependent on a peculiar elastic fluid for their existence, or caloric does not exist." And he concludes that heat consists of a motion excited among the particles of bodies. " To distinguish this motion from others, and to signify the cause of our sensation of heat," and of the expansion or expansive pressure produced in matter by heat, " the name *repulsive* motion has been adopted."*

2. The Dynamical Theory of Heat, thus established by Sir Humphrey Davy, is extended to radiant heat by the discovery of phenomena, especially those of the polarization of radiant heat, which render it excessively probable that heat propagated through vacant space, or through diathermane substances, consists of waves of transverse vibrations in an all-pervading medium.

3. The recent discoveries made by Mayer and Joule,† of the generation of heat through the friction of fluids in motion, and by the magneto-electric excitation of galvanic currents, would, either of them be sufficient to demonstrate the immateriality of heat; and would so afford, if required, a perfect confirmation of Sir Humphrey Davy's views.

4. Considering it as thus established, that heat is not a substance, but a dynamical form of mechanical effect, we perceive that there must be an equivalence between mechanical work and heat, as between cause and effect. The first

He begins by quoting—presumably for some kind of "British-based authority"—an "anti-caloric" experiment apparently done by Humphry Davy (1778–1829) as a teenager, involving melting pieces of ice by rubbing them together, and included anonymously in a 1799 list of pieces of knowledge "principally from the west of England":

CONTRIBUTIONS

TO

PHYSICAL AND MEDICAL

KNOWLEDGE,

Principally from the WEST of ENGLAND,

COLLECTED BY

THOMAS BEDDOES, M. D.

BRISTOL:
PRINTED BY BIGGS & COTTLE,
FOR T. N. LONGMAN AND O. REES, PATERNOSTER-ROW,
LONDON.
1799.

(14)

The phænomena of repulsion are not dependant on a peculiar elastic fluid for their existence, or Caloric does not exist.

Without considering the effects of the repulsive power on bodies, or endeavouring to prove from these effects that it is motion, I shall attempt to demonstrate by experiments that it is not matter; and in doing this, I shall use the method called by mathematicians, reductio ad absurdum.

Let heat be considered as matter, and let it be granted that the temperature of bodies cannot be increased, unless their capacities are diminished from some cause, or heat added to them from some bodies in contact.

Now the temperatures of bodies are uniformly raised by friction and percussion. And since an increase of temperature is consequent on friction and percussion, it must consequently be generated in one of these modes.

But soon Kelvin is getting to the main point:

6. The object of the present paper is threefold:—

(1.) To show what modifications of the conclusions arrived at by Carnot, and by others who have followed his peculiar mode of reasoning regarding the motive power of heat, must be made when the hypothesis of the dynamical theory, contrary as it is to Carnot's fundamental hypothesis, is adopted.

(2.) To point out the significance in the dynamical theory, of the numerical results deduced from Regnault's observations on steam, and communicated about two years ago to the Society, with an account of Carnot's theory, by the author of the present paper; and to show that by taking these numbers (subject to correction when accurate experimental data regarding the density of saturated steam shall have been afforded), in connexion with Joule's mechanical equivalent of a thermal unit, a complete theory of the motive power of heat, within the temperature limits of the experimental data, is obtained.

(3.) To point out some remarkable relations connecting the physical properties of all substances, established by reasoning analogous to that of Carnot, but founded in part on the contrary principle of the dynamical theory.

And then we have it: a statement of the Second Law (albeit with some hedging to which we'll come back later):

> 12. The demonstration of the second proposition is founded on the following axiom :—
>
> *It is impossible, by means of inanimate material agency, to derive mechanical effect from any portion of matter by cooling it below the temperature of the coldest of the surrounding objects*.*

And there's immediately a footnote that basically asserts the "absurdity" of a Second-Law-violating perpetual motion machine:

> * If this axiom be denied for all temperatures, it would have to be admitted that a self-acting machine might be set to work and produce mechanical effect by cooling the sea or earth, with no limit but the total loss of heat from the earth and sea, or, in reality, from the whole material world.

But by the next page we find out that Kelvin admits he's in some sense been "scooped"—by a certain Rudolf Clausius (1822–1888), who we'll be discussing soon. But what's remarkable is that Clausius's "axiom" turns out to be exactly equivalent to Kelvin's statement:

> upon an axiom (§ 12) which I think will be generally admitted. It is with no wish to claim priority that I make these statements, as the merit of first establishing the proposition upon correct principles is entirely due to Clausius, who published his demonstration of it in the month of May last year, in the second part of his paper on the motive power of heat‡. I may be allowed to add, that I have given the demonstration exactly as it occurred to me before I knew that Clausius had either enunciated or demonstrated the proposition. The following is the axiom on which Clausius' demonstration is founded :—
>
> *It is impossible for a self-acting machine, unaided by any external agency, to convey heat from one body to another at a higher temperature.*
>
> It is easily shown, that, although this and the axiom I have used are different in form, either is a consequence of the other. The reasoning in each demonstration is strictly analogous to that which Carnot originally gave.

And what this suggests is that the underlying concept—the Second Law—is something quite robust. And indeed, as Kelvin implies, it's the main thing that ultimately underlies Carnot's results. And so even though Carnot is operating on the now-outmoded idea of caloric theory, his main results are still correct, because in the end all they really depend on is a certain amount of "logical structure", together with the Second Law (and a version of the First Law, but that's a slightly trickier story).

Kelvin recognized, though, that Carnot had chosen to look at the particular ("equilibrium thermodynamics") case of processes that occur reversibly, effectively at an infinitesimal rate. And at the end of the first installment of his exposition, he explains that things will be more complicated if finite rates are considered—and that in particular the results one gets in such cases will depend on things like having a correct model for the nature of heat.

Kelvin's exposition on the "dynamical nature of heat" runs to four installments, and the next two dive into detailed derivations and attempted comparison with experiment:

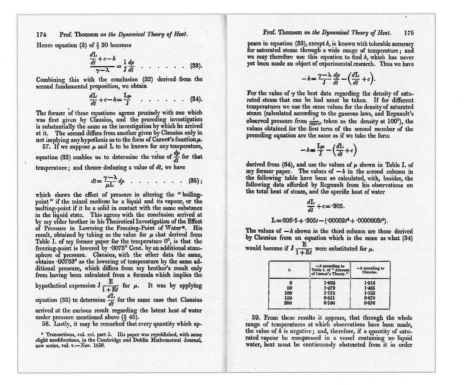

But before Kelvin gets to publish part four of his exposition he publishes two other pieces. In the first, he's talking about sources of energy for human use (now that he believes energy is conserved):

> ## 2. On the Mechanical action of Radiant Heat or Light: On the Power of Animated Creatures over Matter: On the Sources available to Man for the production of Mechanical Effect. By Professor William Thomson.
>
> *On the Mechanical Action of Radiant Heat or Light.*
>
> It is assumed in this communication that the undulatory theory

He emphasizes that the Sun is—directly or indirectly—the main source of energy on Earth (later he'll argue that coal will run out, etc.):

> the following general conclusions :——
>
> 1. *Heat radiated from the sun* (sunlight being included in this term) *is the principal source of mechanical effect available to man.** From it is derived the whole mechanical effect obtained by means of animals working, water-wheels worked by rivers, steam-engines, and galvanic engines, and part at least of the mechanical effect obtained by means of windmills and the sails of ships not driven by the trade-winds.
>
> 2. The motions of the earth, moon, and sun, and their mutual attractions, constitute an important source of available mechanical effect. From them all, but chiefly, no doubt, from the earth's motion of rotation, is derived the mechanical effect of water-wheels driven by the tides. The mechanical effect so largely used in the

But he wonders how animals actually manage to produce mechanical work, noting that "the animal body does not act as a thermo-dynamic engine; and [it is] very probable that the chemical forces produce the external mechanical effects through electrical means":

> A curious inference is pointed out, that an animal would be sensibly less warm in going up-hill than in going down-hill, were the breathing not greater in the former case than in the latter.
>
> The application of Carnot's principle, and of Joule's discoveries regarding the heat of electrolysis and the calorific effects of magneto-electricity, is pointed out; according to which it appears nearly certain that, when an animal works against resisting force, there is not a *conversion of heat into external mechanical effect*, but the full thermal equivalent of the chemical forces is *never produced ;* in other words that the animal body does not act as a *thermo-dynamic engine ;* and very probable that the chemical forces produce the external mechanical effects through electrical means.

And then, by April 1852, he's back to thinking directly about the Second Law, and he's cut through the technicalities, and is stating the Second Law in everyday (if slightly ponderous) terms:

XLVII. *On a Universal Tendency in Nature to the Dissipation of Mechanical Energy.* By Prof. W. THOMSON*.

THE object of the present communication is to call attention to the remarkable consequences which follow from Carnot's proposition, that there is an absolute waste of mechanical energy available to man when heat is allowed to pass from one body to another at a lower temperature, by any means not fulfilling his criterion of a "perfect thermo-dynamic engine," established, on a new foundation, in the dynamical theory of heat. As it is most certain that Creative Power alone can either call into existence or annihilate mechanical energy, the "waste" referred to cannot be annihilation, but must be some transformation of energy†. To explain the nature of this transformation, it is convenient, in the first place, to divide *stores* of mechanical energy into two classes—*statical* and *dynamical*. A quantity of weights at a height, ready to descend and do work when wanted, an electrified body, a quantity of fuel, contain stores of mechanical energy of the statical kind. Masses of matter in motion, a volume of space through which undulations of light or radiant heat are passing, a body having thermal motions among its particles (that is, not infinitely cold), contain stores of mechanical energy of the dynamical kind.

The following propositions are laid down regarding the *dissipation* of mechanical energy from a given store, and the *restoration* of it to its primitive condition. They are necessary consequences of the axiom, "*It is impossible, by means of inanimate material agency, to derive mechanical effect from any portion of matter by cooling it below the temperature of the coldest of the surrounding objects.*" (Dynam. Th. of Heat, § 12.)

I. When heat is created by a reversible process (so that the mechanical energy thus spent may be *restored* to its primitive condition), there is also a transference from a cold body to a hot body of a quantity of heat bearing to the quantity created a definite proportion depending on the temperatures of the two bodies.

II. When heat is created by any unreversible process (such as friction), there is a *dissipation* of mechanical energy, and a full *restoration* of it to its primitive condition is impossible.

III. When heat is diffused by *conduction*, there is a *dissipation* of mechanical energy, and perfect *restoration* is impossible.

IV. When radiant heat or light is absorbed, otherwise than in

---

It's interesting to see his apparently rather deeply held Presbyterian beliefs manifest themselves here in his mention that "Creative Power" is what must set the total energy of the universe. He ends his piece with:

> The following general conclusions are drawn from the propositions stated above, and known facts with reference to the mechanics of animal and vegetable bodies :—
>
> 1. There is at present in the material world a universal tendency to the dissipation of mechanical energy.
>
> 2. Any *restoration* of mechanical energy, without more than an equivalent of dissipation, is impossible in inanimate material processes, and is probably never effected by means of organized matter, either endowed with vegetable life or subjected to the will of an animated creature.
>
> 3. Within a finite period of time past the earth must have been, and within a finite period of time to come the earth must again be, unfit for the habitation of man as at present constituted, unless operations have been, or are to be performed, which are impossible under the laws to which the known operations going on at present in the material world are subject.

In (2) the hedging is interesting. He makes the definitive assertion that what amounts to a violation of the Second Law "is impossible in inanimate material processes". And he's pretty sure the same is true for "vegetable life" (recognizing that in his previous paper he discussed the harvesting of sunlight by plants). But what about "animal life", like us humans? Here he says that "by our will" we can't violate the Second Law—so we can't, for example, build a machine to do it. But he leaves it open whether we as humans might have some innate ("God-given"?) ability to overcome the Second Law.

And then there's his (3). It's worth realizing that his whole paper is less than 3 pages long, and right before his conclusions we're seeing triple integrals:

> If the system of thermometry adopted* be such that $\mu = \dfrac{J}{t+a}$, that is, if we agree to call $\dfrac{J}{\mu} - a$ the *temperature* of a body, for which μ is the *value of Carnot's function* (a and J being constants), the preceding expression becomes
>
> $$T = \frac{\iiint c\, dx\, dy\, dz}{\iiint \frac{c}{t+a}\, dx\, dy\, dz} - a.$$
>
> The following general conclusions are drawn from the propositions stated above, and known facts with reference to the mechanics of animal and vegetable bodies :—

So what is (3) about? It's presumably something like a Second-Law-implies-heat-death-of-the-universe statement (but what's this stuff about the past?)—but with an added twist that there's something (God?) beyond the "known operations going on at present in the material world" that might be able to swoop in to save the world for us humans.

It doesn't take people long to pick up on the "cosmic significance" of all this. But in the fall of 1852, Kelvin's colleague, the Glasgow engineering professor William Rankine (1820–1872) (who was deeply involved with the First Law of thermodynamics), is writing about a way the universe might save itself:

> LVI. *On the Reconcentration of the Mechanical Energy of the Universe.* By WILLIAM JOHN MACQUORN RANKINE, *C.E., F.R.S.E. &c.* *
>
> THE following remarks have been suggested by a paper by Professor William Thomson of Glasgow, on the tendency which exists in nature to the dissipation or indefinite diffusion of mechanical energy originally collected in stores of power.
>
> * Communicated by the Author; having been read to the British Association for the Advancement of Science, Section A, at Belfast, on the 2nd of September 1852.

After touting the increasingly solid evidence for energy conservation and the First Law

> The experimental evidence is every day accumulating, of a law which has long been conjectured to exist,—that all the different kinds of physical energy in the universe are mutually convertible, —that the total amount of physical energy, whether in the form of visible motion and mechanical power, or of heat, light, magnetism, electricity, or chemical agency, or in other forms not yet understood, is unchangeably the transformations of its different portions from one of those forms of power into another, and their transference from one portion of matter to another, constituting the phænomena which are the objects of experimental physics.

he goes on to talk about dissipation of energy and what we now call the Second Law

> the state of heat. On the other hand, all visible motion is of necessity ultimately converted entirely into heat by the agency of friction. There is thus, in the present state of the known world, a tendency towards the conversion of all physical energy into the sole form of heat.
>
> Heat, moreover, tends to diffuse itself uniformly by conduction and radiation, until all matter shall have acquired the same temperature.
>
> There is, consequently, Professor Thomson concludes, so far as we understand the present condition of the universe, a tendency towards a state in which all physical energy will be in the state of heat, and that heat so diffused that all matter will be at the same temperature; so that there will be an end of all physical phænomena.

and the fact that it implies an "end of all physical phenomena", i.e. heat death of the universe. He continues:

> Vast as this speculation may seem, it appears to be soundly based on experimental data, and to represent truly the present condition of the universe, so far as we know it.

But now he offers a "ray of hope". He believes that there must exist a "medium capable of transmitting light and heat", i.e. an aether, "[between] the heavenly bodies". And if this aether can't itself acquire heat, he concludes that all energy must be converted into a radiant form:

> My object now is to point out how it is conceivable that, at some indefinitely distant period, an opposite condition of the world may take place, in which the energy which is now being diffused may be reconcentrated into foci, and stores of chemical power again produced from the inert compounds which are now being continually formed.
>
> There must exist between the atmospheres of the heavenly bodies a material medium capable of transmitting light and heat; and it may be regarded as almost certain, that this interstellar medium is perfectly transparent and diathermanous; that is to say, that it is incapable of converting heat, or light (which is a species of heat), from the radiant into the fixed or conductible form.
>
> If this be the case, the interstellar medium must be incapable of acquiring any temperature whatsoever; and all heat which arrives in the conductible form at the limits of the atmosphere of a star or planet, will there be totally converted, partly into ordinary motion, by the expansion of the atmosphere, and partly into the radiant form. The ordinary motion will again be converted into heat, so that *radiant heat* is the ultimate form to which all physical energy tends; and in this form it is, in the present condition of the world, diffusing itself from the heavenly bodies through the interstellar medium.

Now he supposes that the universe is effectively a giant drop of aether, with nothing outside, so that all this radiant energy will get totally internally reflected from its surface, allowing the universe to "[reconcentrate] its physical energies, and [renew] its activity and life"—and save it from heat death:

> Let it now be supposed, that, in all directions round the visible world, the interstellar medium has bounds beyond which there is empty space.
>
> If this conjecture be true, then on reaching those bounds the radiant heat of the world will be totally reflected, and will ultimately be reconcentrated into foci. At each of these foci the intensity of heat may be expected to be such, that should a star (being at that period an extinct mass of inert compounds) in the course of its motions arrive at that part of space, it will be vaporized and resolved into its elements; a store of chemical power being thus reproduced at the expense of a corresponding amount of radiant heat.
>
> Thus it appears, that although, from what we can see of the known world, its condition seems to tend continually towards the equable diffusion, in the form of radiant heat, of all physical energy, the extinction of the stars, and the cessation of all phæ-nomena, yet the world, as now created, may possibly be pro-vided within itself with the means of reconcentrating its physical energies, and renewing its activity and life.
>
> For aught we know, these opposite processes may go on together; and some of the luminous objects which we see in distant regions of space may be, not stars, but foci in the inter-stellar æther.

He ends with the speculation that perhaps "some of the luminous objects which we see in distant regions of space may be, not stars, but foci in the interstellar aether".

But independent of cosmic speculations, Kelvin himself continues to study the "dynamical theory of gases". It's often a bit unclear what's being assumed. There's the First Law (energy conservation). And the Second Law. But there's also reversibility. Equilibrium. And the ideal gas law ($PV = RT$). But it soon becomes clear that that's not always correct for real gases—as the Joule–Thomson effect demonstrates:

> **LXXVI.** *On the Thermal Effects experienced by Air in rushing through small Apertures. By* J. P. JOULE *and* W. THOMSON*.
>
> THE hypothesis that the heat evolved from air compressed and kept at a constant temperature is mechanically equi-valent to the work spent in effecting the compression, assumed by Mayer as the foundation for an estimate of the numerical relation between quantities of heat and mechanical work, and adopted by Holtzmann, Clausius, and other writers, was made the subject of an experimental research by Mr. Joule†, and verified as at least approximately true for air at ordinary atmospheric temperatures. A theoretical investigation, founded on a conclu-sion of Carnot's‡, which requires no modification§ in the dyna-mical theory of heat, also leads to a verification of Mayer's hypo-

Kelvin soon returned to more cosmic speculations, suggesting that perhaps gravitation—rather than direct "Creative Power"—might "in reality [be] the ultimate created antecedent of all motion...":

> Published speculations* were referred to, by which it is shown to be possible that the motions of the earth and of the heavenly bodies, and the heat of the sun, may all be due to gravitation; or, *that the potential energy of gravitation may be in reality the ultimate created antecedent of all motion, heat, and light at present existing in the universe.*
>
> * Prof. W. Thomson, "On the Mechanical Energies of the Solar System" (*Trans. Roy. Soc. Edinburgh*, April, 1854 [Art. LXVI. above]), and "On the Mechanical Antecedents of Motion, Heat, and Light" (*British Association Report*, Liverpool, 1854 [Art. LXIX. above]).

Not long after these papers Kelvin got involved with the practical "electrical" problem of laying a transatlantic telegraph cable, and in 1858 was on the ship that first succeeded in doing this. (His commercial efforts soon allowed him to buy a 126-ton yacht.) But he continued to write physics papers, which ranged over many different areas, occasionally touching thermodynamics, though most often in the service of answering a "general science" question—like how old the Sun is (he estimated 32,000 years from thermodynamic arguments, though of course without knowledge of nuclear reactions).

Kelvin's ideas about the inevitable dissipation of "useful energy" spread quickly—by 1854, for example, finding their way into an eloquent public lecture by Hermann von Helmholtz (1821–1894). Helmholtz had trained as a doctor, becoming in 1843 a surgeon to a German military regiment. But he was also doing experiments and developing theories about "animal heat" and how muscles manage to "do mechanical work", for example publishing an 1845 paper entitled "On Metabolism during Muscular Activity". And in 1847 he was one of the inventors of the law of conservation of energy—and the First Law of thermodynamics—as well as perhaps its clearest expositor at the time (the word "force" in the title is what we now call "energy"):

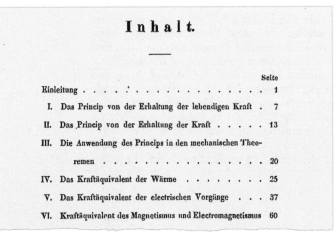

Einleitung.

Vorliegende Abhandlung musste ihrem Hauptinhalte nach hauptsächlich für Physiker bestimmt werden, ich habe es daher vorgezogen, die Grundlagen derselben unabhängig von einer philosophischen Begründung rein in der Form einer physikalischen Voraussetzung hinzustellen, deren Folgerungen zu entwickeln, und dieselben in den verschiedenen Zweigen der Physik mit den erfahrungsmässigen Gesetzen der Naturerscheinungen zu vergleichen. Die Herleitung der

By 1854 Helmholtz was a physiology professor, beginning a distinguished career in physics, psychophysics and physiology—and talking about the Second Law and its implications. He began his lecture by saying that "A new conquest of very general interest has been recently made by natural philosophy"—and what he's referring to here is the Second Law:

INTERACTION OF NATURAL FORCES.

A NEW conquest of very general interest has been recently made by natural philosophy. In the following pages, I will endeavour to give a notion of the nature of this conquest. It has reference to a new and universal natural law, which rules the action of natural forces in their mutual relations towards each other, and is as influential on our theoretic views of natural processes as it is important in their technical applications.

Among the practical arts which owe their progress to the development of the natural sciences, from the conclusion of the middle ages downwards, practical mechanics, aided by the mathematical science which bears the same name, was one of the most prominent. The character of the art was, at the time referred to, naturally very different from its present one. Surprised and stimulated by its own success, it thought no problem beyond its power, and immediately attacked some of the most difficult and complicated. Thus it was attempted to build automaton figures which should perform the functions of men and animals. The wonder of the last century was

Having discussed the inability of "automata" (he uses that word) to reproduce living systems, he starts talking about perpetual motion machines:

> From these efforts to imitate living creatures, another idea, also by a misunderstanding, seems to have developed itself, which, as it were, formed the new philosopher's stone of the seventeenth and eighteenth centuries. It was now the endeavour to construct a perpetual motion. Under this term was un-

First he disposes of the idea that perpetual motion can be achieved by generating energy from nothing (i.e. violating the First Law), charmingly including the anecdote:

> magneto-electric machine, decomposed the water, and thus continually prepared its own fuel. This would certainly have been the most splendid of all discoveries; a perpetual motion which, besides the force which kept it going, generated light like the sun, and warmed all around it. The matter was by no means badly cogitated. Each practical step in the affair was known to be possible; but those which at that time were acquainted with the physical investigations which bear upon this subject could have affirmed, on the first hearing the report, that the matter was to be numbered among the numerous stories of the fable-rich America; and indeed, a fable it remained.

And then he's on to talking about the Second Law

> capacity for heat, and the expansion by heat of all bodies. It is not yet considered as actually proved, but some remarkable deductions having been drawn from it, and afterwards proved to be facts by experiment, it has attained thereby a great degree of probability. Besides the mathematical form in which the law was first expressed by Carnot, we can give it the following more general expression :—" Only when heat passes from a warmer to a colder body, and even then only partially, can it be converted into mechanical work."

and discussing how it implies the heat death of the universe:

> But the heat of the warmer bodies strives perpetually to pass to bodies less warm by radition and conduction, and thus to establish an equilibrium of temperature. At each motion of a terrestrial body, a portion of mechanical force passes by friction or collision into heat, of which only a part can be converted back again into mechanical force. This is also generally the case in every electrical and chemical process. From this, it follows that the first portion of the store of force,

> the unchangeable heat, is augmented by every natural process, while the second portion, mechanical, electrical, and chemical force, must be diminished; so that if the universe be delivered over to the undisturbed action of its physical processes, all force will finally pass into the form of heat, and all heat come into a state of equilibrium. Then all possibility of a further change would be at an end, and the complete cessation of all natural processes must set in. The life of men, animals, and plants, could not of course continue if the sun had lost its high temperature, and with it his light,—if all the components of the earth's surface had closed those combinations which their affinities demand. In short, the universe from that time forward would be condemned to a state of eternal rest.

He notes, correctly, that the Second Law hasn't been "proved". But he's impressed at how Kelvin was able to go from a "mathematical formula" to a global fact about the fate of the universe:

> These consequences of the law of Carnot are, of course, only valid, provided that the law, when sufficiently tested, proves to be universally correct. In the mean time there is little prospect of the law being proved incorrect. At all events we must admire the sagacity of Thomson, who, in the letters of a long known little mathematical formula, which only speaks of the heat, volume, and pressure of bodies, was able to discern consequences which threatened the universe, though certainly after an infinite period of time, with eternal death.

He ends the whole lecture quite poetically:

> Thus the thread which was spun in darkness by those who sought a perpetual motion has conducted us to a universal law of nature, which radiates light into the distant nights of the beginning and of the end of the history of the universe. To our own race it permits a long but not an endless existence; it threatens it with a day of judgment, the dawn of which is still happily obscured. As each of us singly must endure the thought of his death, the race must endure the same. But above the forms of life gone by, the human race has higher moral problems before it, the bearer of which it is, and in the completion of which it fulfils its destiny.

We've talked quite a bit about Kelvin and how his ideas spread. But let's turn now to Rudolf Clausius, who in 1850 at least to some extent "scooped" Kelvin on the Second Law. At that time Clausius was a freshly minted German physics PhD. His thesis had been on an ingenious but ultimately incorrect theory of why the sky is blue. But he'd also worked on elasticity theory, and there he'd been led to start thinking about molecules and their configurations in materials. By 1850 caloric theory had become fairly elaborate, complete with concepts like "latent heat" (bound to molecules) and "free heat" (able to be transferred). Clausius's experience in elasticity theory made him skeptical, and knowing Mayer's and Joule's results he decided to break with the caloric theory—writing his career-launching paper (translated from German in 1851, with Carnot's *puissance motrice* ["motive power"] being rendered as "moving force"):

> I. *On the Moving Force of Heat, and the Laws regarding the Nature of Heat itself which are deducible therefrom.* By R. CLAUSIUS*.
>
> THE steam-engine having furnished us with a means of converting heat into a motive power, and our thoughts being thereby led to regard a certain quantity of work as an equivalent for the amount of heat expended in its production, the idea of establishing theoretically some fixed relation between a quantity of heat and the quantity of work which it can possibly produce, from which relation conclusions regarding the nature of heat itself might be deduced, naturally presents itself. Already, indeed, have many instructive experiments been made with this view; I believe, however, that they have not exhausted the subject, but that, on the contrary, it merits the continued attention of physicists; partly because weighty objections lie in the way of the conclusions already drawn, and partly because other conclusions, which might render efficient aid towards establishing and completing the theory of heat, remain either entirely unnoticed, or have not as yet found sufficiently distinct expression.
>
> The most important investigation in connexion with this subject is that of S. Carnot†. Later still, the ideas of this author

The first installment of the English version of the paper gives a clear description of the ideal gas laws and the Carnot cycle, having started from a statement of the "caloric-busting" First Law:

> ### 1. *Deductions from the principle of the equivalence of heat and work.*
>
> We shall forbear entering at present on the nature of the motion which may be supposed to exist within a body, and shall assume generally that a motion of the particles does exist, and that heat is the measure of their *vis viva*. Or yet more general, we shall merely lay down one maxim which is founded on the above assumption :—
>
> *In all cases where work is produced by heat, a quantity of heat proportional to the work done is expended; and inversely, by the expenditure of a like quantity of work, the same amount of heat may be produced.*
>
> Before passing on to the mathematical treatment of this maxim, a few of its more immediate consequences may be noticed, which

The general discussion continues in the second installment, but now there's a critical side comment that describes the "general deportment of heat, which every-where exhibits the tendency to annul differences of temperature, and therefore to pass from a warmer body to a colder one":

> thus on the whole a transmission from B to A would take place.
> Hence by repeating both these alternating processes, without expenditure of force or other alteration whatever, any quantity of heat might be transmitted from a *cold* body to a *warm* one; and this contradicts the general deportment of heat, which every-where exhibits the tendency to annul differences of temperature, and therefore to pass from a *warmer* body to a *colder* one.
>
> From this it would appear that we are *theoretically* justified in

Clausius "has" the Second Law, as Carnot basically did before him. But when Kelvin quotes Clausius he does so much more forcefully:

> which Clausius' demonstration is founded :—
> *It is impossible for a self-acting machine, unaided by any external agency, to convey heat from one body to another at a higher temperature.*

But there it is: by 1852 the Second Law is out in the open, in at least two different forms. The path to reach it has been circuitous and quite technical. But in the end, stripped of its technical origins, the law seems somehow unsurprising and even obvious. For it's a matter of common experience that heat flows from hotter bodies to colder ones, and that motion is dissipated by friction into heat. But the point is that it wasn't until basically 1850 that the overall scientific framework existed to make it useful—or even really possible—to enunciate such observations as a formal scientific law.

Of course the fact that a law "seems true" based on common experience doesn't mean it'll always be true, and that there won't be some special circumstance or elaborate construction that will evade it. But somehow the very fact that the Second Law had in a sense been "technically hard won"—yet in the end seemed so "obvious"—appears to have given it a sense of inevitability and certainty. And it didn't hurt that somehow it seemed to have emerged from Carnot's work, which had a certain air of "logical necessity". (Of course, in reality, the Second Law entered Carnot's logical structure as an "axiom".) But all this helped set the stage for some of the curious confusions about the Second Law that would develop over the century that followed.

The Concept of Entropy

In the first half of the 1850s the Second Law had in a sense been presented in two ways. First, as an almost "footnote-style" assumption needed to support the "pure thermodynamics" that had grown out of Carnot's work. And second, as an explicitly-stated-for-the-first-time—if "obvious"—"everyday" feature of nature, that was now realized as having potentially cosmic significance. But an important feature of the decade that followed was a certain progressive at-least-phenomenological "mathematicization" of the Second Law—pursued most notably by Rudolf Clausius.

In 1854 Clausius was already beginning this process. Perhaps confusingly, he refers to the Second Law as the "second fundamental theorem [*Hauptsatz*]" in the "mechanical theory of heat"—suggesting it's something that is proved, even though it's really introduced just as an empirical law of nature, or perhaps a theoretical axiom:

ON A MODIFIED FORM OF THE SECOND FUNDAMENTAL THEOREM IN THE MECHANICAL THEORY OF HEAT*.

In my memoir "On the Moving Force of Heat, &c."†, I have shown that the theorem of the equivalence of heat and work, and Carnot's theorem, are not mutually exclusive, but that, by a small modification of the latter, which does not affect its principal part, they can be brought into accordance. With the exception of this indispensable change, I allowed the theorem of Carnot to retain its original form, my chief object then being, by the application of the two theorems to special cases, to arrive at conclusions which, according as they involved known or unknown properties of bodies, might suitably serve as proofs of the truth of the theorems, or as examples of their fecundity.

This form, however, although it may suffice for the deduction of the equations which depend upon the theorem, is incomplete, because we cannot recognize therein, with sufficient clearness, the real nature of the theorem, and its connexion with the first fundamental theorem. The modified form in the following pages will, I think, better fulfil this demand, and in its applications will be found very convenient.

He starts off by discussing the "first fundamental theorem", i.e. the First Law. And he emphasizes that this implies that there's a quantity U (which we now call "internal energy") that is a pure "function of state"—so that its value depends only on the state of a system, and not the path by which that state was reached. And as an "application" of this, he then points out that the overall change in U in a cyclic process (like the one executed by Carnot's heat engine) must be zero.

And now he's ready to tackle the Second Law. He gives a statement that at first seems somewhat convoluted:

> ### *Theorem of the equivalence of transformations.*
>
> **Carnot's theorem, when brought into agreement with the first fundamental theorem, expresses a relation between two kinds of transformations, the transformation of heat into work, and the passage of heat from a warmer to a colder body, which may be regarded as the transformation of heat at a higher, into heat at a lower temperature. The theorem, as hitherto used, may be enunciated in some such manner as the following :—*In all cases where a quantity of heat is converted into work, and where the body effecting this transformation ultimately returns to its original condition, another quantity of heat must necessarily be transferred from a warmer to a colder body ; and the magnitude of the last quantity of heat, in relation to the first, depends only upon the temperatures of the bodies between which heat passes, and not upon the nature of the body effecting the transformation.***

But soon he's deriving this from a more "everyday" statement of the Second Law (which, notably, is clearly not a "theorem" in any normal sense):

> **This principle, upon which the whole of the following development rests, is as follows :—*Heat can never pass from a colder to a warmer body without some other change, connected therewith, occurring at the same time**. Everything we know concerning**

> **the interchange of heat between two bodies of different temperatures confirms this ; for heat everywhere manifests a tendency to equalize existing differences of temperature, and therefore to pass in a contrary direction, *i. e.* from warmer to colder bodies. Without further explanation, therefore, the truth of the principle will be granted.**

After giving a Carnot-style argument he's then got a new statement (that he calls "the theorem of the equivalence of transformations") of the Second Law:

> According to to this, the second fundamental theorem in the mechanical theory of heat, which in this form might appropriately be called the *theorem of the equivalence of transformations*, may be thus enunciated:
>
> If two transformations which, without necessitating any other permanent change, can mutually replace one another, be called

> equivalent, then the generation of the quantity of heat Q of the temperature t from work, has the equivalence-value
>
> $$\frac{Q}{T},$$
>
> and the passage of the quantity of heat Q from the temperature t_1 to the temperature t_2, has the equivalence-value
>
> $$Q\left(\frac{1}{T_2}-\frac{1}{T_1}\right),$$
>
> wherein T is a function of the temperature, independent of the nature of the process by which the transformation is effected.

And there it is: basically what we now call entropy (even with the same notation of Q for heat and T for temperature)—together with the statement that this quantity is a function of state, so that its differences are "independent of the nature of the process by which the transformation is effected".

Pretty soon there's a familiar expression for entropy change:

> bodies; then the foregoing equation will assume the form
>
> $$N=\int\frac{dQ}{T}, \qquad \ldots \ldots \quad (11)$$
>
> wherein the integral extends over all the quantities of heat received by the several bodies.
>
> If the process is *reversible*, then, however complicated it may be, we can prove, as in the simple process before considered, *that the transformations which occur must exactly cancel each other, so that their algebraical sum is zero.*

And by the next page he's giving what he describes as "the analytical expression" of the Second Law, for the particular case of reversible cyclic processes:

> Consequently the equation
>
> $$\int\frac{dQ}{T}=0 \qquad \ldots \ldots \quad (II)$$
>
> is the analytical expression, for all *reversible cyclical processes*, of the second fundamental theorem in the mechanical theory of heat.

A bit later he backs out of the assumption of reversibility, concluding that:

> ***The algebraical sum of all transformations occurring in a cyclical process can only be positive.***

(And, yes, with modern mathematical rigor, that should be "non-negative" rather than "positive".)

He goes on to say that if something has changed after going around a cycle, he'll call that an "uncompensated transformation"—or what we would now refer to as an irreversible change. He lists a few possible (now very familiar) examples:

> **The different kinds of operations giving rise to uncompensated transformations are, as far as external appearances are concerned, rather numerous, even though they may not differ very essentially. One of the most frequently occurring examples is that of the transmission of heat by mere conduction, when two bodies of different temperatures are brought into immediate contact; other cases are the production of heat by friction, and by an electric current when overcoming the resistance due to imperfect conductibility, together with all cases where a force, in doing mechanical work, has not to overcome an equal resistance, and**

Earlier in his paper he's careful to say that T is "a function of temperature"; he doesn't say it's actually the quantity we measure as temperature. But now he wants to determine what it is:

> **In conclusion, we must direct our attention to the function T, which hitherto has been left quite undetermined; we shall not be able to determine it entirely without hypothesis, but by means of a very probable hypothesis it will be possible so to do.**

He doesn't talk about the ultimately critical assumption (effectively the Zeroth Law of thermodynamics) that the system is "in equilibrium", with a uniform temperature. But he uses an ideal gas as a kind of "standard material", and determines that, yes, in that case T can be simply the absolute temperature.

So there it is: in 1854 Clausius has effectively defined entropy and described its relation to the Second Law, though everything is being done in a very "heat-engine" style. And pretty soon he's writing about "Theory of the Steam-Engine" and filling actual approximate steam tables into his theoretical formulas:

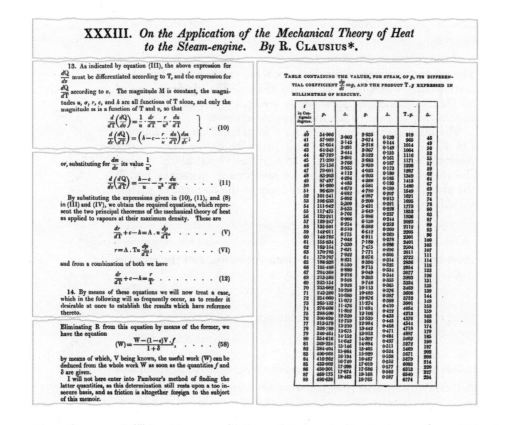

After a few years "off" (working, as we'll discuss later, on the kinetic theory of gases) Clausius is back in 1862 talking about the Second Law again, in terms of his "theorem of the equivalence of transformations":

XIII. On the Application of the Theorem of the Equivalence of Transformations to the Internal Work of a mass of Matter. By Professor R. CLAUSIUS*.

IN a memoir published in the year 1854†, wherein I sought to simplify to some extent the form of the developments I had previously published, I deduced, from my fundamental proposition *that heat cannot of itself pass from a colder into a warmer body*, a principle which is closely allied to, but does not entirely coincide with, the one first deduced by S. Carnot from considerations of a different class, based upon the older views of the nature of heat. It has reference to the circumstances under which work can be transformed into heat, and, conversely, heat converted into work; and I have called it the *Principle of the Equivalence of Transformations*. I did not, however, there communicate the entire proposition in the general form in which I had deduced it, but confined myself on that occasion to the publication of a part which can be treated separately from the rest, and is capable of more strict proof.

He's slightly tightened up his 1854 discussion, but, more importantly, he's now stating a result not just for reversible cyclic processes, but for general ones:

> The proposition respecting the equivalent values of the transformations may accordingly be stated thus:—*The algebraic sum of all the transformations occurring in a circular process can only be positive, or, as an extreme case, equal to nothing.*
>
> The mathematical expression for this proposition is as follows. Let dQ be an element of the heat given up by the body to any reservoir of heat during its modifications (heat which it may absorb from a reservoir being here reckoned as negative), and T the absolute temperature of the body at the moment of giving up this heat, then the equation
>
> $$\int \frac{d\mathrm{Q}}{\mathrm{T}} = 0 \quad \cdot \quad \cdot \quad \cdot \quad \cdot \quad \cdot \quad \cdot \quad \text{(I.)}$$
>
> must be true for every reversible circular process, and the relation
>
> $$\int \frac{d\mathrm{Q}}{\mathrm{T}} \geqq 0 \quad \cdot \quad \cdot \quad \cdot \quad \cdot \quad \cdot \quad \cdot \quad \text{(I}a\text{.)}$$
>
> must hold good for every circular process which is in any way possible.

But what does this result really mean? Clausius claims that this "theorem admits of strict mathematical proof if we start from the fundamental proposition above quoted"—though it's not particularly clear just what that proposition is. But then he says he wants to find a "physical cause":

> § 2. Although the necessity of this proposition admits of strict mathematical proof if we start from the fundamental principle above quoted, it thereby nevertheless retains an abstract form, in which it is difficultly embraced by the mind, and we feel compelled to seek for the precise physical cause, of which this proposition is a consequence. Moreover, since there is no

A little earlier in the paper he said:

> I have delayed till now the publication of the remainder of my theorem, because it leads to a consequence which is considerably at variance with the ideas hitherto generally entertained of the heat contained in bodies, and I therefore thought it desirable to make still further trial of it. But as I have become more and more convinced in the course of years that we must not attach too great weight to such ideas, which in part are founded more upon usage than upon a scientific basis, I feel that I ought to hesitate no longer, but to submit to the scientific public the theorem of the equivalence of transformations in its complete form, with the principles which attach themselves to it. I venture to

So what does he think the "physical cause" is? He says that even from his first investigations he'd assumed a general law:

> *In all cases in which the heat contained in a body does mechanical work by overcoming a resistance, the magnitude of the resistance which it is capable of overcoming is proportional to the absolute temperature.*

What are these "resistances"? He's basically saying they are the forces between molecules in a material (which from his work on the kinetic theory of gases he now imagines exist):

> In order to understand the significance of this law, we require to consider more closely the processes by which heat can perform mechanical work. These processes always admit of being reduced to the alteration in some way or another of the arrangement of the constituent molecules of a body. For instance, bodies are expanded by heat, their molecules being thus separated from each other: in this case the mutual attractions of the molecules on the one hand, and on the other external opposing forces, in so far as any such are in operation, have to be overcome. Again,

He introduces what he calls the "disgregation" to represent the microscopic effect of adding heat:

> distances from one another. In order to be able to represent this mathematically, we will express the degree in which the molecules of a body are dispersed, by introducing a new magnitude, which we will call the *disgregation* of the body, and by help

> of which we can define the effect of heat as simply *tending to increase the disgregation*. The way in which a definite measure of

For ideal gases things are straightforward, including the proportionality of "resistance" to absolute temperature. But in other cases, it's not so clear what's going on. A decade later he identifies "disgregation" with average kinetic energy per molecule—which is indeed proportional to absolute temperature. But in 1862 it's all still quite muddy, with somewhat curious statements like:

> temperature, even water. To this it might perhaps be objected that, in other cases, the effect of increased temperature is to favour the union of two substances—that, for instance, hydrogen and oxygen do not combine at low temperatures, but do so easily at higher temperatures. I believe, however, that the heat exerts here only a secondary influence, contributing to bring the atoms into such relative positions that their inherent forces, by virtue of which they strive to unite, are able to come into operation. Heat itself can never, in my opinion, tend to produce combination, but only, and in every case, decomposition.

And then the main part of the paper ends with what seems to be an anticipation of the Third Law of thermodynamics:

> If we desired to cool a body down to the absolute zero of temperature, the corresponding alteration of disgregation, as shown by the foregoing formula, in which we should then have $T=0$, would be infinitely great. Herein lies a chief argument for supposing it to be impossible to produce such a degree of cold, by any alteration of the condition of a body, as to arrive at the absolute zero.

There's an appendix entitled "On Terminology" which admits that between Clausius's own work, and other people's, it's become rather difficult to follow what's going on. He agrees that the term "energy" that Kelvin is using makes sense. He suggests "energy of the body" for what he calls U and we now call "internal energy". He suggests "heat of the body" or "thermal content of the body" for Q. But then he talks about the fact that these are measured in thermal units (say the amount of heat needed to increase the temperature of water by 1°), while mechanical work is measured in units related to kilograms and meters. He proposes therefore to introduce the concept of "ergon" for "work measured in thermal units":

> For this purpose, therefore, I will venture another proposition. Let heat and work continue to be measured each according to its most convenient unit, that is to say, heat according to the thermal unit, and work according to the mechanical one. But besides the work measured according to the mechanical unit, let another magnitude be introduced denoting *the work measured according to the thermal unit*, that is to say, *the numerical value of the work when the unit of work is that which is equivalent to the thermal unit*. For the work thus expressed a particular name is requisite. I propose to adopt for it the Greek word (ἔργον) *ergon**.

And pretty soon he's talking about the "interior ergon" and "exterior ergon", as well as concepts like "ergonized heat". (In later work he also tries to introduce the concept of "ergal" to go along with his development of what he called—in a name that did stick—the "virial theorem".)

But in 1865 he has his biggest success in introducing a term. He's writing a paper, he says, basically to clarify the Second Law, (or, as he calls it, "the second fundamental theorem"—rather confidently asserting that he will "prove this theorem"):

> ON SEVERAL CONVENIENT FORMS OF THE FUNDAMENTAL EQUATIONS
> OF THE MECHANICAL THEORY OF HEAT*.
>
> IN my former Memoirs on the Mechanical Theory of Heat,
> my chief object was to secure a firm basis for the theory,
> and I especially endeavoured to bring the second fundamental
> theorem, which is much more difficult to understand than the
> first, to its simplest and at the same time most general form,
> and to prove the necessary truth thereof. I have pursued special

Part of the issue he's trying to address is how the calculus is done:

> The more the mechanical theory of heat is acknowledged to
> be correct in its principles, the more frequently endeavours are
> made in physical and mechanical circles to apply it to different
> kinds of phenomena, and as the corresponding differential equa-
> tions must be somewhat differently treated from the ordinarily
> occurring differential equations of similar forms, difficulties of
> calculation are frequently encountered which retard progress
> and occasion errors. Under these circumstances I believe I

The partial derivative symbol ∂ had been introduced in the late 1700s. He doesn't use it, but he does introduce the now-standard-in-thermodynamics subscript notation for variables that are kept constant:

> which is supposed to be constant during differentiation. Accord-
> ingly, we will write the two differential coefficients which denote
> the specific heat at constant volume, and the specific heat at con-
> stant pressure, in the following manner :—
>
> $$\left(\frac{dQ}{dT}\right)_v \quad \text{and} \quad \left(\frac{dQ}{dT}\right)_p.$$

A little later, as part of the "notational cleanup", we see the variable S:

> present existing condition of the body, and not upon the way by
> which it reached the latter. Denoting this magnitude by S, we
> can write
>
> $$dS = \frac{dQ}{T}; \quad \cdots \cdots \cdots \quad (59)$$
>
> or, if we conceive this equation to be integrated for any re-

And then—there it is—Clausius introduces the term "entropy", "Greekifying" his concept of "transformation":

> **S is determined.**
>
> We might call S the *transformational content* of the body, just as we termed the magnitude U its *thermal and ergonal content*. But as I hold it to be better to borrow terms for important magnitudes from the ancient languages, so that they may be adopted unchanged in all modern languages, I propose to call the magnitude S the *entropy* of the body, from the Greek word τροπή, *transformation*. I have intentionally formed the word *entropy* so as to be as similar as possible to the word *energy*; for the two magnitudes to be denoted by these words are so nearly allied in their physical meanings, that a certain similarity in designation appears to be desirable.
>
> Before proceeding further, let us collect together for the sake

His paper ends with his famous crisp statements of the First and Second Laws of thermodynamics—manifesting the parallelism he's been claiming between energy and entropy:

> ception of energy, we may express in the following manner the fundamental laws of the universe which correspond to the two fundamental theorems of the mechanical theory of heat.
>
> 1. *The energy of the universe is constant.*
> 2. *The entropy of the universe tends to a maximum.*

The Kinetic Theory of Gases

We began above by discussing the history of the question of "What is heat?" Was it like a fluid—the caloric theory? Or was it something more dynamical, and in a sense more abstract? But then we saw how Carnot—followed by Kelvin and Clausius—managed in effect to sidestep the question, and come up with all sorts of "thermodynamic conclusions", by talking just about "what heat does" without ever really having to seriously address the question of "what heat is". But to be able to discuss the foundations of the Second Law—and what it says about heat—we have to know more about what heat actually is. And the crucial development that began to clarify the nature of heat was the kinetic theory of gases.

Central to the kinetic theory of gases is the idea that gases are made up of discrete molecules. And it's important to remember that it wasn't until the beginning of the 1900s that anyone knew for sure that molecules existed. Yes, something like them had been discussed ever since antiquity, and in the 1800s there was increasing "circumstantial evidence" for them. But nobody had directly "seen a molecule", or been able, for example, until about 1870, to even guess what the size of molecules might be. Still, by the mid-1800s it had become common for physicists to talk and reason in terms of ordinary matter at least effectively being made of up molecules.

But if a gas was made of molecules bouncing off each other like billiard balls according to the laws of mechanics, what would its overall properties be? Daniel Bernoulli had in 1738 already worked out the basic answer that pressure would vary inversely with volume, or in his notation, $\pi = P/s$ (and he even also gave formulas for molecules of nonzero size—in a precursor of van der Waals):

Rationi D ad *d* aliam fubftituere poffumus magis intelligibilem : nempe fi putemus operculum EF pondere infinito depreffum defcendere usque in fitum *mn*, in quo particulæ omnes fe tangunt, atque lineam *m*C vocemus *m*, erit D ad *d* ut 1 ad $\sqrt[3]{m}$, quâ ratione fubftituta, erunt tandem vires aëris naturalis E CD F & compreffi *e* CD*f* ut $s^{\frac{2}{3}} \times (\sqrt[3]{s} - \sqrt[3]{m})$ ad $1 - \sqrt[3]{m}$, feu ut $s - \sqrt[3]{m}ss$ ad $1 - \sqrt[3]{m}$. Eft igitur $\pi = \dfrac{1 - \sqrt[3]{m}}{s - \sqrt[3]{m}ss} \times P$.

§. 5. Ex omnibus phænomenis judicare poffumus aërem naturalem admodum condenfari poffe, & fere in fpatiolum infinite parvum comprimi; facta igitur *m* $= o$, fit $\pi = \dfrac{P}{s}$, ita ut pondera comprimentia fint fere in ratione inverfa fpatiorum, quæ aër diverfimode compreffus occupat;

Results like Bernouilli's would be rediscovered several times, for example in 1820 by John Herapath (1790–1868), a math teacher in England, who developed a fairly elaborate theory that purported to describe gravity as well as heat (but for example implied a $PV = aT^2$ gas law):

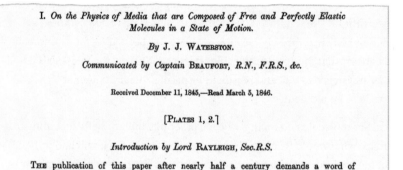

Then there was the case of John Waterston (1811–1883), a naval instructor for the East India company, who in 1843 published a book called *Thoughts on the Mental Functions*, which included results on what he called the "*vis viva* theory of heat"—that he developed in more detail in a paper he wrote in 1846. But when he submitted the paper to the Royal Society it was rejected as "nonsense", and its manuscript was "lost" until 1891 when it was finally published (with an "explanation" of the "delay"):

I. *On the Physics of Media that are Composed of Free and Perfectly Elastic Molecules in a State of Motion.*

By J. J. WATERSTON.

Communicated by Captain BEAUFORT, *R.N., F.R.S., &c.*

Received December 11, 1845,—Read March 5, 1846.

[PLATES 1, 2.]

Introduction by Lord RAYLEIGH, *Sec. R.S.*

THE publication of this paper after nearly half a century demands a word of explanation; and the opportunity may be taken to point out in what respects the

The paper had included a perfectly sensible mathematical analysis that included a derivation of the kinetic theory relation between pressure and mean-square molecular velocity:

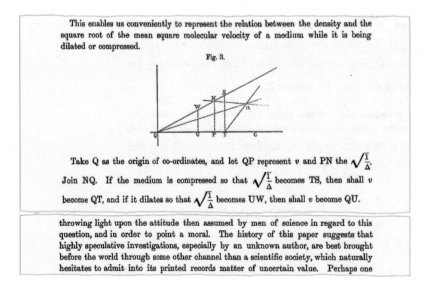

This enables us conveniently to represent the relation between the density and the square root of the mean square molecular velocity of a medium while it is being dilated or compressed.

Fig. 3.

Take Q as the origin of co-ordinates, and let QP represent v and PN the $\sqrt{\frac{I}{\Delta}}$. Join NQ. If the medium is compressed so that $\sqrt{\frac{I}{\Delta}}$ becomes TS, then shall v become QT, and if it dilates so that $\sqrt{\frac{I}{\Delta}}$ becomes UW, then shall v become QU.

throwing light upon the attitude then assumed by men of science in regard to this question, and in order to point a moral. The history of this paper suggests that highly speculative investigations, especially by an unknown author, are best brought before the world through some other channel than a scientific society, which naturally hesitates to admit into its printed records matter of uncertain value. Perhaps one

But with all these pieces of work unknown, it fell to a German high-school chemistry teacher (and sometime professor and philosophical/theological writer) named August Krönig (1822–1879) to publish in 1856 yet another "rediscovery", that he entitled "Principles of a Theory of Gases". He said it was going to analyze the "mechanical theory of heat", and once again he wanted to compute the pressure associated with colliding molecules. But to simplify the math, he assumed that molecules went only along the coordinate directions, at a fixed speed—almost anticipating a cellular automaton fluid:

VII. *Grundzüge einer Theorie der Gase;*
von Dr. A. Krönig.
(Mitgetheilt vom Hrn. Verfasser.)

Die mechanische Wärmetheorie behauptet, daß die Wärme eines Körpers in nichts anderem besteht als in einer Bewegung seiner kleinsten Theile. Es fehlt aber durchaus an einer klaren Anschauung darüber, wie diese Bewegung eigentlich beschaffen ist. Für die gasförmigen Körper, welche in Beziehung auf mechanische Wärmetheorie bis-

317

In einem rechtwinklig parallelepipedischen Gefäß mit ebenen Wänden und von den Dimensionen x, y, z seyen n Gasatome, jedes von der Masse m, enthalten. Der von dem Gefäß umschlossene Raum sey in $\frac{1}{6}n$ gleiche Würfel zerlegt. In einem bestimmten Momente seyen in jedem dieser Würfel 6 Atome befindlich, die sich bezüglich in den Richtungen $+x$, $-x$, $+y$, $-y$, $+z$, $-z$ mit der gemeinschaftlichen Geschwindigkeit c bewegen. Nehmen wir an, die Atome afficirten sich gegenseitig durchaus nicht, gingen vielmehr jedesmal bis zu einer Wand hin ungehindert fort. (Solche Atome, deren Mittelpunkte in derselben graden Linie sich bewegen, verhalten sich auch wirklich so, als ob sie einander nicht afficirten, da sie bei jedem Zusammenstoß nur ihre beiderseitigen Geschwindigkeiten vertauschen.) Es soll der Druck bestimmt werden, den z. B. eine der beiden Wände yz durch das Gas erleidet.

Dieser Druck wird hervorgebracht durch die Stöße der Gasatome gegen die Wand. Stieße gegen die Wand nur ein Atom, so würde der Druck gleich $m.c.a$ seyn, unter a die Anzahl der Stöße während der Zeiteinheit verstanden. Ein senkrecht gegen yz, also parallel x sich bewegendes Atom stößt gegen yz jedesmal, nachdem es den Raum $2x$ durchlaufen hat; folglich ist $a = \frac{c}{2x}$. Um den Gesammtdruck P gegen yz zu finden, muß $m.c.a$ noch mit der Anzahl der parallel x sich bewegenden Atome multiplicirt werden. Diese ist, da von je 6 Atomen 2 parallel x sich bewegen, gleich $\frac{n}{3}$. Folglich ist $P = m.c.\frac{c}{2x}.\frac{n}{3}$. Bezeichnen wir mit p den Druck gegen die Flächeneinheit der Wand yz, so findet sich $p = m.c.\frac{c}{2x}.\frac{n}{3}.\frac{1}{yz}$, oder wenn wir $xyz = v$ setzen und den constanten Factor fortlassen.

$$p = \frac{nmc^2}{v}.$$

Dieser Ausdruck zeigt, daß der Druck des Gases gegen die Flächeneinheit jeder der Gefäßwände gleich groß, so

What ultimately launched the subsequent development of the kinetic theory of gases, however, was the 1857 publication by Rudolf Clausius (by then an increasingly established German physics professor) of a paper entitled rather poetically "On the Nature of the Motion Which We Call Heat" ("Über die Art der Bewegung die wir Wärme nennen"):

XI. *On the Nature of the Motion which we call Heat.*
By R. CLAUSIUS*.

1. **B**EFORE writing my first memoir on heat, which was published in 1850, and in which heat is assumed to be a motion, I had already formed for myself a distinct conception of the nature of this motion, and had even employed the same in several investigations and calculations. In my former memoirs I intentionally avoided mentioning this conception, because I wished to separate the conclusions which are deducible from certain general principles from those which presuppose a particular kind of motion, and because I hoped to be able at some future time to devote a separate memoir to my notion of this motion and to the special conclusions which flow therefrom. The execution of this project, however, has been retarded longer than I at first expected, inasmuch as the difficulties of the subject, as well as other occupations, have hitherto prevented me from giving to its development that degree of completeness which I deemed necessary for publication.

A memoir has lately been published by Krönig, under the title *Grundzüge einer Theorie der Gase*†, in which I have recognized some of my own views. Seeing that Krönig has arrived at these views just as independently as I have, and has published them before me, all claim to priority on my part is of course out of the question ; nevertheless, the subject having once been mooted in this memoir, I feel myself induced to publish

---

It's a clean and clear paper, with none of the mathematical muddiness around Clausius's work on the Second Law (which, by the way, isn't even mentioned in this paper even though Clausius had recently worked on it). Clausius figures out lots of the "obvious" implications of his molecular theory, outlining for example what happens in different phases of matter:

In the *solid* state, the motion is such that the molecules move about certain positions of equilibrium without ever forsaking the same, unless acted upon by foreign forces. In solid bodies, therefore, the motion may be characterized as a vibrating one,

---

In the *liquid* state the molecules have no longer any definite position of equilibrium. They can turn completely around their centres of gravity ; and the latter, too, may be moved completely out of its place. The separating action of the motion is not, however, sufficiently strong, in comparison to the mutual attraction between the molecules, to be able to separate the latter entirely. Although a molecule no longer adheres to definite neigh-

---

Lastly, in the *gaseous* state the motion of the molecules entirely transports them beyond the spheres of their mutual attraction, causing them to recede in right lines according to the ordinary laws of motion. If two such molecules come into collision during their motion, they will in general fly asunder again with the same vehemence with which they moved towards each other ;

It takes him only a couple of pages of very light mathematics to derive the standard kinetic theory formula for the pressure of an ideal gas:

He's implicitly assuming a certain randomness to the motions of the molecules, but he barely mentions it (and this particular formula is robust enough that average values are actually all that matter):

> Let us consider one only of the two large sides; during the unit of time it is struck a certain number of times by molecules moving in all possible directions compatible with an approach towards the surface. We must first determine the number of such shocks, and how many correspond on the average to each direction.

> Lastly, there is no doubt that actually the greatest possible variety exists amongst the velocities of the several molecules. In our considerations, however, we may ascribe a certain mean velocity to all molecules. It will be evident from the following for-

But having derived the formula for pressure, he goes on to use the ideal gas law to derive the relation between average molecular kinetic energy (which he still calls "*vis viva*") and absolute temperature:

> *motion of the molecules**.   But, according to Mariotte's and Gay-Lussac's laws,
>
> $$pv = T \text{ . const.,}$$
>
> where T is the absolute temperature; hence
>
> $$\frac{nmu^2}{2} = T \text{ . const.;}$$
>
> and, as before stated, the *vis viva* of the translatory motion is proportional to the absolute temperature.

From this he can do things like work out the actual average velocities of molecules in different gases—which he does without any mention of the question of just how real or not molecules might be. By knowing experimental results about specific heats of gases he also manages to determine that not all the energy ("heat") in a gas is associated with "translatory motion": he realizes that for molecules involving several atoms there can be energy associated with other (as we would now say) internal degrees of freedom:

> Thus is corroborated what was before stated, that the *vis viva* of the translatory motion does not alone represent the whole quantity of heat in the gas, and that the difference is greater the greater the number of atoms of which the several molecules of the combination consist. We must conclude, therefore, that besides the translatory motion of the molecules as such, the constituents of these molecules perform other motions, whose *vis viva* also forms a part of the contained quantity of heat.

Clausius's paper was widely read. And it didn't take long before the Dutch meteorologist (and effectively founder of the World Meteorological Organization) Christophorus Buys Ballot (1817–1890) asked why—if molecules were moving as quickly as Clausius suggested—gases didn't mix much more quickly than they're observed to do:

> X. *On the Mean Length of the Paths described by the separate Molecules of Gaseous Bodies on the occurrence of Molecular Motion : together with some other Remarks upon the Mechanical Theory of Heat.* By R. CLAUSIUS*.
>
> (1.) THE February Number of Poggendorff's *Annalen* contains a paper by Buijs-Ballot "On the Nature of the Motion which we call Heat and Electricity." Amongst the objections which the author there makes against the views advanced by Joule, Krönig, and myself concerning the molecular motion of gaseous bodies, the following deserves especial consideration. Attention is drawn to the circumstance that, if the molecules moved in straight lines, volumes of gases in contact would necessarily speedily mix with one another,—a result which does not actually take place. To prove that such mixture does

Within a few months, Clausius published the answer: the molecules didn't just keep moving in straight lines; they were constantly being deflected, to follow what we would now call a random walk. He invented the concept of a mean free path to describe how far on average a molecule goes before it hits another molecule:

> (4.) If, now, in a given space, we imagine a great number of molecules moving irregularly about amongst one another, and if we select one of them to watch, such a one would ever and anon impinge upon one of the other molecules, and bound off from it. We have now, therefore, to solve the question as to how great is the mean length of the path between two such impacts; or more exactly expressed, *how far on an average can the molecule move, before its centre of gravity comes into the sphere of action of another molecule.*

As a capable theoretical physicist, Clausius quickly brings in the concept of probability

> If, now, a point moves through this space in a straight line, let us suppose the space to be divided into parallel layers perpendicular to the motion of the point, and let us determine *how great is the probability that the point will pass freely through a layer of the thickness* x *without encountering the sphere of action of a molecule.*

and is soon computing the average number of molecules which will survive undeflected for a certain distance:

> According to equation (5), the number of points which either reach or pass the distance x from the commencement of the motion is represented by
>
> $$Ne^{-\frac{\pi\rho^2}{\lambda^3}x};$$

Then he works out the mean free path λ (and it's often still called λ):

put:—*The mean length of path of a molecule is in the same pro*

portion to the radius of the sphere of action as the entire space occupied by the gas, to that portion of the space which is actually filled up by the spheres of action of the molecules.

And he concludes that actually there's no conflict between rapid microscopic motion and large-scale "diffusive" motion:

one is present, it is easy to convince oneself that the theory which explains the expansive force of gases does not lead to the conclusion that two quantities of gas bounding one another must mix with one another quickly and violently, but that only a comparatively small number of atoms can arrive quickly at a great distance, while the chief quantities only gradually mix at the surface of their contact.

Of course, he could have actually drawn a sample random walk, but drawing diagrams wasn't his style. And in fact it seems as if the first published drawing of a random walk was something added by John Venn (1834–1923) in the 1888 edition of his *Logic of Chance*—and, interestingly, in alignment with my computational irreducibility concept from a century later, he used the digits of π to generate his "randomness":

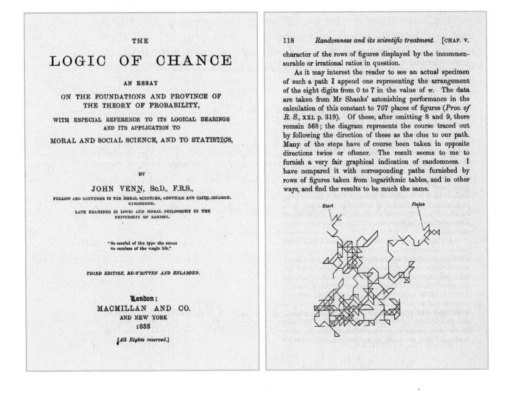

In 1859, Clausius's paper came to the attention of the then-28-year-old James Clerk Maxwell, who had grown up in Scotland, done the Mathematical Tripos in Cambridge, and was now back in Scotland as professor of "natural philosophy" at Aberdeen. Maxwell had already worked on things like elasticity theory, color vision, the mechanics of tops, the dynamics of the rings of Saturn and electromagnetism—having published his first paper (on geometry) at age 14. And, by the way, Maxwell was quite a "diagrammist"—and his early papers include all sorts of pictures that he drew:

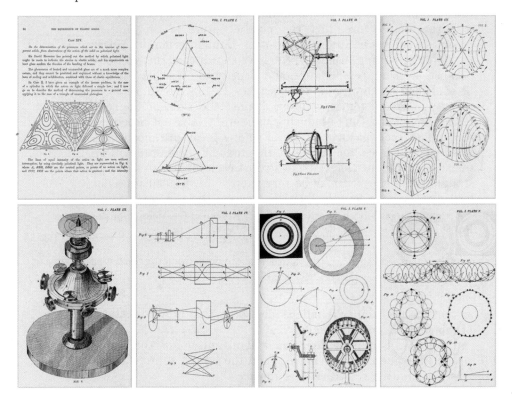

But in 1859 Maxwell applied his talents to what he called the "dynamical theory of gases":

V. *Illustrations of the Dynamical Theory of Gases.*—Part I. *On the Motions and Collisions of Perfectly Elastic Spheres. By J. C.* MAXWELL, *M.A., Professor of Natural Philosophy in Marischal College and University of Aberdeen*.*

So many of the properties of matter, especially when in the gaseous form, can be deduced from the hypothesis that their minute parts are in rapid motion, the velocity increasing with the temperature, that the precise nature of this motion becomes a subject of rational curiosity. Daniel Bernouilli, Herapath, Joule, Krönig, Clausius, &c. have shown that the relations between pressure, temperature, and density in a perfect gas can be explained by supposing the particles to move with uniform velocity in straight lines, striking against the sides of the containing vessel and thus producing pressure. It is not necessary to suppose each particle to travel to any great distance in the same straight line; for the effect in producing pressure will be the same if the particles strike against each other; so that the straight line described may be very short. M. Clausius has determined the mean length of path in terms of the average distance

* Communicated by the Author, having been read at the Meeting of the British Association at Aberdeen, September 21, 1859.

He models molecules as hard spheres, and sets about computing the "statistical" results of their collisions:

---

so that the probability is independent of ϕ, that is, all directions of rebound are equally likely.

   Prop. III. Given the direction and magnitude of the velocities of two spheres before impact, and the line of centres at impact; to find the velocities after impact.

   Let O A, O B represent the velocities before impact, so that if there had been no action between the bodies they would have been at A′ and B at the end of a second. Join A B, and let G be their

---

And pretty soon he's trying to compute distribution of their velocities:

Prop. IV. To find the average number of particles whose velocities lie between given limits, after a great number of collisions among a great number of equal particles.

Let N be the whole number of particles. Let x, y, z be the components of the velocity of each particle in three rectangular directions, and let the number of particles for which x lies between x and $x + dx$ be $Nf(x)dx$, where $f(x)$ is a function of x to be determined.

The number of particles for which y lies between y and $y + dy$ will be $Nf(y)dy$; and the number for which z lies between z and $z + dz$ will be $Nf(z)dz$, where f always stands for the same function.

Now the existence of the velocity x does not in any way affect that of the velocities y or z, since these are all at right angles to each other and independent, so that the number of particles whose velocity lies between x and $x + dx$, and also between y and $y + dy$, and also between z and $z + dz$, is

$$Nf(x)\,f(y)\,f(z)\,dx\,dy\,dz.$$

If we suppose the N particles to start from the origin at the same instant, then this will be the number in the element of volume $(dx\,dy\,dz)$ after unit of time, and the number referred to unit of volume will be

$$Nf(x)\,f(y)\,f(z).$$

But the directions of the coordinates are perfectly arbitrary, and therefore this number must depend on the distance from the origin alone, that is

$$f(x)\,f(y)\,f(z) = \phi(x^2 + y^2 + z^2).$$

Solving this functional equation, we find

$$f(x) = Ce^{Ax^2}, \quad \phi(r^2) = C^3 e^{Ar^2}.$$

If we make A positive, the number of particles will increase with the velocity, and we should find the whole number of particles infinite. We therefore make A negative and equal to $-\dfrac{1}{\alpha^2}$, so that the number between x and $x + dx$ is

$$NCe^{-\frac{x^2}{\alpha^2}}\,dx.$$

It's a somewhat unconvincing (or, as Maxwell himself later put it, "precarious") derivation (how does it work in 1D, for example?), but somehow it manages to produce what's now known as the Maxwell distribution:

> 2nd. The number whose actual velocity lies between v and $v + dv$ is
> $$N \frac{4}{\alpha^3 \sqrt{\pi}} v^2 e^{-\frac{v^2}{\alpha^2}} dv. \quad . \quad . \quad . \quad . \quad . \quad (2)$$

Maxwell observes that the distribution is the same as for "errors ... in the 'method of least squares'":

> It appears from this proposition that the velocities are distri-
> buted among the particles according to the same law as the
> errors are distributed among the observations in the theory of
> the "method of least squares." The velocities range from 0 to
> ∞, but the number of those having great velocities is compara-
> tively small. In addition to these velocities, which are in all

Maxwell didn't get back to the dynamical theory of gases until 1866, but in the meantime he was making a "dynamical theory" of something else: what he called the electromagnetic field:

> (3) The theory I propose may therefore be called a theory of the *Electromagnetic Field*, because it has to do with the space in the neighbourhood of the electric or magnetic bodies, and it may be called a *Dynamical* Theory, because it assumes that in that space there is matter in motion, by which the observed electromagnetic phenomena are produced.

Even though he'd worked extensively with the inverse square law of gravity he didn't like the idea of "action at a distance", and for example he wanted magnetic field lines to have some underlying "material" manifestation

> XXV. *On Physical Lines of Force.* By J. C. MAXWELL, Pro-
> *fessor of Natural Philosophy in King's College, London**.
> PART I.—*The Theory of Molecular Vortices applied to Magnetic*
> *Phenomena.*

imagining that they might be associated with arrays of "molecular vortices":

Fig: 2.

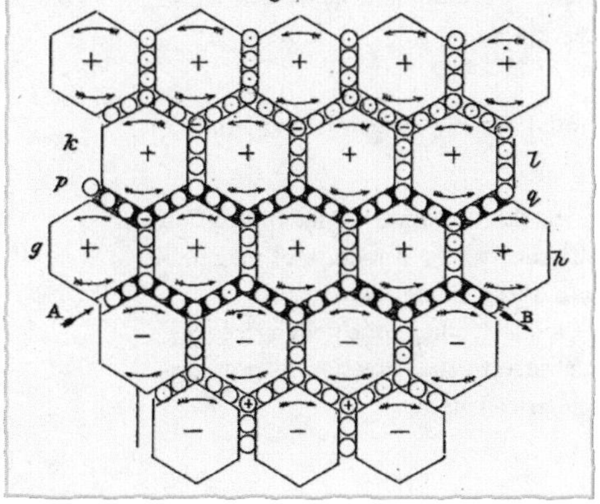

We now know, of course, that there isn't this kind of "underlying mechanics" for the electromagnetic field. But—with shades of the story of Carnot—even though the underlying framework isn't right, Maxwell successfully derives correct equations for the electromagnetic field—that are now known as Maxwell's equations:

> Between these twenty quantities we have found twenty equations, viz.
> Three equations of Magnetic Force . . . . (B)
>   ,,   Electric Currents . . . . (C)
>   ,,   Electromotive Force . . . . (D)
>   ,,   Electric Elasticity . . . . (E)
>   ,,   Electric Resistance . . . . (F)
>   ,,   Total Currents . . . . (A)
> One equation of Free Electricity . . . . (G)
>   ,,   Continuity . . . . (H)
>
> These equations are therefore sufficient to determine all the quantities which occur in them, provided we know the conditions of the problem. In many questions, however, only a few of the equations are required.

His statement of how the electromagnetic field "works" is highly reminiscent of the dynamical theory of gases:

> (74) In speaking of the Energy of the field, however, I wish to be understood literally. All energy is the same as mechanical energy, whether it exists in the form of motion or in that of elasticity, or in any other form. The energy in electromagnetic phenomena is mechanical energy. The only question is, Where does it reside? On the old theories

> it resides in the electrified bodies, conducting circuits, and magnets, in the form of an unknown quality called potential energy, or the power of producing certain effects at a distance. On our theory it resides in the electromagnetic field, in the space surrounding the electrified and magnetic bodies, as well as in those bodies themselves, and is in two different forms, which may be described without hypothesis as magnetic polarization and electric polarization, or, according to a very probable hypothesis, as the motion and the strain of one and the same medium.

But he quickly and correctly adds:

> (75) The conclusions arrived at in the present paper are independent of this hypothesis, being deduced from experimental facts of three kinds:—

And a few sections later he derives the idea of general electromagnetic waves

> This wave consists entirely of magnetic disturbances, the direction of magnetization being in the plane of the wave. No magnetic disturbance whose direction of magnetization is not in the plane of the wave can be propagated as a plane wave at all.
>
> Hence magnetic disturbances propagated through the electromagnetic field agree with light in this, that the disturbance at any point is transverse to the direction of propagation, and such waves may have all the properties of polarized light.

noting that there's no evidence that the medium through which he assumes they're propagating has elasticity:

> (100) The equations of the electromagnetic field, deduced from purely experimental evidence, show that transversal vibrations only can be propagated. If we were to go beyond our experimental knowledge and to assign a definite density to a substance which

> we should call the electric fluid, and select either vitreous or resinous electricity as the representative of that fluid, then we might have normal vibrations propagated with a velocity depending on this density. We have, however, no evidence as to the density of electricity, as we do not even know whether to consider vitreous electricity as a substance or as the absence of a substance.

By the way, when it comes to gravity he can't figure out how to make his idea of a "mechanical medium" work:

> The assumption, therefore, that gravitation arises from the action of the surrounding medium in the way pointed out, leads to the conclusion that every part of this medium possesses, when undisturbed, an enormous intrinsic energy, and that the presence of dense bodies influences the medium so as to diminish this energy wherever there is a resultant attraction.
>
> As I am unable to understand in what way a medium can possess such properties, I cannot go any further in this direction in searching for the cause of gravitation.

But in any case, after using it as an inspiration for thinking about electromagnetism, Maxwell in 1866 returns to the actual dynamical theory of gases, still feeling that he needs to justify looking at a molecular theory:

IV. *On the Dynamical Theory of Gases.* By J. CLERK MAXWELL, *F.R.S. L. & E*

Received May 16,—Read May 31, 1866.

THEORIES of the constitution of bodies suppose them either to be continuous and homogeneous, or to be composed of a finite number of distinct particles or molecules.

In certain applications of mathematics to physical questions, it is convenient to suppose bodies homogeneous in order to make the quantity of matter in each differential element a function of the coordinates, but I am not aware that any theory of this kind has been proposed to account for the different properties of bodies. Indeed the properties of a body supposed to be a uniform *plenum* may be affirmed dogmatically, but cannot be explained mathematically.

Molecular theories suppose that all bodies, even when they appear to our senses homogeneous, consist of a multitude of particles, or small parts the mechanical relations of which constitute the properties of the bodies. Those theories which suppose that the molecules are at rest relative to the body may be called statical theories, and those which suppose the molecules to be in motion, even while the body is apparently at rest, may be called dynamical theories.

And now he gives a recognizable (and correct, so far as it goes) derivation of the Maxwell distribution:

When the number of pairs of molecules which change their velocities from OA, OB to OA′ OB′ is equal to the number which change from OA′, OB′ to OA, OB, then the final distribution of velocity will be obtained, which will not be altered by subsequent exchanges. This will be the case when

$$f_1(a)f_2(b) = f_1(a')f_2(b'). \qquad \cdots \cdots \cdots (22)$$

Now the only relation between a, b and a', b' is

$$M_1 a^2 + M_2 b^2 = M_1 a'^2 + M_2 b'^2, \qquad \cdots \cdots (23)$$

whence we obtain

$$f_1(a) = C_1 e^{-\frac{a^2}{a^2}}, \quad f_2(b) = C_2 e^{-\frac{b^2}{\beta^2}}, \qquad \cdots \cdots (24)$$

where

$$M_1 \alpha^2 = M_2 \beta^2. \qquad \cdots \cdots \cdots (25)$$

By integrating $\iiint C_1 e^{-\frac{\xi^2 + \eta^2 + \zeta^2}{a^2}} d\xi\, d\eta\, d\zeta$, and equating the result to N_1, we obtain the value of C_1. If, therefore, the distribution of velocities among N_1 molecules is such that

the number of molecules whose component velocities are between ξ and $\xi + d\zeta$, η and $\eta + d\eta$, and ζ and $\zeta + d\zeta$ is

$$dN_1 = \frac{N_1}{a^3 \pi^{\frac{3}{2}}} e^{-\frac{\xi^2 + \eta^2 + \zeta^2}{a^2}} d\xi\, d\eta\, d\zeta, \qquad \cdots \cdots \cdots (26)$$

then this distribution of velocities will not be altered by the exchange of velocities among the molecules by their mutual action.

He goes on to try to understand experimental results on gases, about things like diffusion, viscosity and conductivity. For some reason, Maxwell doesn't want to think of molecules, as he did before, as hard spheres. And instead he imagines that they have "action at a distance" forces, which basically work like hard squares if it's r^{-5} force law:

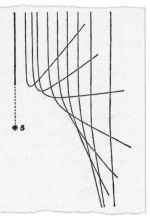

The paths described by molecules about a centre of force S, repelling inversely as the fifth power of the distance, are given in the figure.

The molecules are supposed to be originally moving with equal velocities in parallel paths, and the way in which their deflections depend on the distance of the path from S is shown by the different curves in the figure.

In the years that followed, Maxwell visited the dynamical theory of gases several more times. In 1871, a few years before he died at age 48, he wrote a textbook entitled *Theory of Heat*, which begins, in erudite fashion, discussing what "thermodynamics" should even be called:

conduction of electricity or the radiation of light. The science of heat has been called (by Dr. Whewell and others) Thermotics, and the theory of heat as a form of energy is called Thermodynamics. In the same way the theory of the equilibrium of heat might be called Thermostatics, and that of the motion of heat Thermokinematics.

Most of the book is concerned with the macroscopic "theory of heat"—though, as we'll discuss later, in the very last chapter Maxwell does talk about the "molecular theory", if in somewhat tentative terms.

"Deriving" the Second Law from Molecular Dynamics

The Second Law was in effect originally introduced as a formalization of everyday observations about heat. But the development of kinetic theory seemed to open up the possibility that the Second Law could actually be proved from the underlying mechanics of molecules. And this was something that Ludwig Boltzmann (1844–1906) embarked on towards the end of his physics PhD at the University of Vienna. In 1865 he'd published his first paper ("On the Movement of Electricity on Curved Surfaces"), and in 1866 he published his second paper, "On the Mechanical Meaning of the Second Law of Thermodynamics":

> ## Über die mechanische Bedeutung des zweiten Hauptsatzes der Wärmetheorie.
>
> ### Von Ludwig Boltzmann.
>
> #### (Vorgelegt in der Sitzung am 8. Februar 1866.)
>
> Bereits längst ist die Identität des ersten Hauptsatzes der mechanischen Wärmetheorie mit dem Princip der lebendigen Kräfte

The introduction promises "a purely analytical, perfectly general proof of the Second Law". And what he seemed to imagine was that the equations of mechanics would somehow inevitably lead to motion that would reproduce the Second Law. And in a sense what computational irreducibility, rule 30, etc. now show is that in the end that's indeed basically how things work. But the methods and conceptual framework that Boltzmann had at his disposal were very far away from being able to see that. And instead what Boltzmann did was to use standard mathematical methods from mechanics to compute average properties of cyclic mechanical motions—and then made the somewhat unconvincing claim that combinations of these averages could be related (e.g. via temperature as average kinetic energy) to "Clausius's entropy":

> wie aus dem Folgenden hervorgehen wird. Alsdann gelten ganz dieselben Schlüsse, die früher auf geschlossene Bahnen angewendet wurden, und man erhält wieder:
>
> $$\iint \frac{\partial Q}{T}\, dk = 2\Sigma \log \frac{\int_{\tau_1}^{\tau_2} \frac{mc^2}{2}\, dt}{\int_{\tau_1}^{\tau_2} \frac{mc^2}{2}\, dt},$$
>
> wenn τ_1 und τ_2 die Grenzen derjenigen Bahn sind, die dem zur untern

It's not clear how much this paper was read, but in 1871 Boltzmann (now a professor of mathematical physics in Graz) published another paper entitled simply "On the Priority of Finding the Relationship between the Second Law of Thermodynamics and the Principle of Least Action" that claimed (with some justification) that Clausius's then-newly-announced virial theorem was already contained in Boltzmann's 1866 paper.

But back in 1868—instead of trying to get all the way to Clausius's entropy—Boltzmann instead uses mechanics to get a generalization of Maxwell's law for the distribution of molecular velocities. His paper "Studies on the Equilibrium of [Kinetic Energy] between [Point Masses] in Motion" opens by saying that while analytical mechanics has in effect successfully studied the evolution of mechanical systems "from a given state to another", it's had little to say about what happens when such systems "have been left moving on their own for a long time". He intends to remedy that, and spends 47 pages—complete with elaborate diagrams and formulas about collisions between hard spheres—in deriving an exponential distribution of energies if one assumes "equilibrium" (or, more specifically, balance between forward and backward processes):

It's notable that one of the mathematical approaches Boltzmann uses is to discretize (i.e. effectively quantize) things, then look at the "combinatorial" limit. (Based on his later statements, he didn't want to trust "purely continuous" mathematics—at least in the context of discrete molecular processes—and wanted to explicitly "watch the limits happening".) But in the end it's not clear that Boltzmann's 1868 arguments do more than the few-line functional-equation approach that Maxwell had already used. (Maxwell would later complain about Boltzmann's "overly long" arguments.)

Boltzmann's 1868 paper had derived what the distribution of molecular energies should be "in equilibrium". (In 1871 he was talking about "equipartition" not just of kinetic energy, but also of energies associated with "internal motion" of polyatomic molecules.) But what about the approach to equilibrium? How would an initial distribution of molecular energies evolve over time? And would it always end up at the exponential ("Maxwell–Boltzmann") distribution? These are questions deeply related to a microscopic understanding of the Second Law.

And they're what Boltzmann addressed in 1872 in his 22nd published paper "Further Studies on the Thermal Equilibrium of Gas Molecules":

> ## Weitere Studien über das Wärmegleichgewicht unter Gasmolekülen.
>
> ### Von Ludwig Boltzmann in Graz.
>
> (Mit 1 Holzschnitten.)
>
> **(Vorgelegt in der Sitzung am 10. October 1872.)**
>
> Die mechanische Wärmetheorie setzt voraus, dass sich die Moleküle der Gase keineswegs in Ruhe, sondern in der lebhaftesten Bewegung befinden. Wenn daher auch der Körper seinen Zustand gar nicht verändert, so wird doch jedes einzelne seiner Moleküle seinen Bewegungszustand beständig verändern, und

Boltzmann explains that:

> Maxwell already found the value $Av^2 e^{-Bv^2}$ [for the distribution of velocities] ... so that the probability of different velocities is given by a formula similar to that for the probability of different errors of observation in the theory of the method of least squares. The first proof which Maxwell gave for this formula was recognized to be incorrect even by himself. He later gave a very elegant proof that, if the above distribution has once been established, it will not be changed by collisions. He also tries to prove that it is the only velocity distribution that has this property. But the latter proof appears to me to contain a false inference. It has still not yet been proved that, whatever the initial state of the gas may be, it must always approach the limit found by Maxwell. It is possible that there may be other possible limits. This proof is easily obtained, however, by the method which I am about to explain...

(He gives a long footnote explaining why Maxwell might be wrong, talking about how a sequence of collisions might lead to a "cycle of velocity states"—which Maxwell hasn't proved will be traversed with equal probability in each direction. Ironically, this is actually already an analog of where things are going to go wrong with Boltzmann's own argument.)

The main idea of Boltzmann's paper is not to assume equilibrium, but instead to write down an equation (now called the Boltzmann Transport Equation) that explicitly describes how the velocity (or energy) distribution of molecules will change as a result of collisions. He begins by defining infinitesimal changes in time:

> küle, welche zur Zeit $t+\tau$ diese lebendige Kraft haben, also $f(x, t+\tau)dx$. Wir erhalten somit:
>
> $$f(x, t+\tau)dx = f(x, t)dx - \int dn + \int d\nu. \qquad 5)$$

He then goes through a rather elaborate analysis of velocities before and after collisions, and how to integrate over them, and eventually winds up with a partial differential equation for the time variation of the energy distribution (yes, he confusingly uses x to denote energy)—and argues that Maxwell's exponential distribution is a stationary solution to this equation:

$$\frac{\partial f(x,t)}{\partial t} = \int_0^\infty \int_0^{x+x'} \left[\frac{f(\xi,t)}{\sqrt{\xi}} \frac{f(x+x'-\xi,t)}{\sqrt{x+x'-\xi}} - \frac{f(x,t)}{\sqrt{x}} \frac{f(x',t)}{\sqrt{x'}} \right] \times \quad 16)$$

$$\times \sqrt{x x'}\, \psi(x,x',\xi)\, dx'\, d\xi.$$

Dies ist die Fundamentalgleichung für die Veränderung der Function $f(x,t)$. Ich bemerke nochmal, dass die Wurzeln alle positiv zu nehmen sind, sowie auch ψ und die f wesentlich positive Grössen sind. Setzen wir für einen Augenblick

$$f(x,t) = C\sqrt{x}\, e^{-hx}, \qquad 16\,\text{a})$$

A few paragraphs further on, something important happens: Boltzmann introduces a function that here he calls E, though later he'll call it H:

$$E = \int_0^\infty f(x,t) \left\{ \log \left[\frac{f(x,t)}{\sqrt{x}} \right] - 1 \right\} dx \qquad 17)$$

Ten pages of computation follow

and finally Boltzmann gets his main result: if the velocity distribution evolves according to his equation, H can never increase with time, becoming zero for the Maxwell distribution. In other words, he is saying that he's proved that a gas will always ("monotonically") approach equilibrium—which seems awfully like some kind of microscopic proof of the Second Law.

But then Boltzmann makes a bolder claim:

> It has thus been rigorously proved that, whatever the initial distribution of kinetic energy may be, in the course of a very long time it must always necessarily approach the one found by Maxwell. The procedure used so far is of course nothing more than a mathematical artifice employed in order to give a rigorous proof of a theorem whose exact proof has not previously been found. It gains meaning by its applicability to the theory of polyatomic gas molecules. There one can again prove that a certain quantity E can only decrease as a consequence of molecular motion, or in a limiting case can remain constant. One can also prove that for the atomic motion of a system of arbitrarily many material points there always exists a certain quantity which, in consequence of any atomic motion, cannot increase, and this quantity agrees up to a constant factor with the value found for the well-known integral $\int dQ/T$ in my [1871] paper on the "Analytical proof of the 2nd law, etc.". We have therefore prepared the way for an analytical proof of the Second Law in a completely different way from those previously investigated. Up to now the object has been to show that $\int dQ/T = 0$ for reversible cyclic processes, but it has not been proved analytically that this quantity is always negative for irreversible processes, which are the only ones that occur in nature. The reversible cyclic process is only an ideal, which one can more or less closely approach but never completely attain. Here, however, we have succeeded in showing that $\int dQ/T$ is in general negative, and is equal to zero only for the limiting case, which is of course the reversible cyclic process (since if one can go through the process in either direction, $\int dQ/T$ cannot be negative).

In other words, he's saying that the quantity H that he's defined microscopically in terms of velocity distributions can be identified (up to a sign) with the entropy that Clausius defined as dQ/T. He says that he'll show this in the context of analyzing the mechanics of poly-atomic molecules.

But first he's going to take a break and show that his derivation doesn't need to assume contin-uity. In a pre-quantum-mechanics pre-cellular-automaton-fluid kind of way he replaces all the integrals by limits of sums of discrete quantities (i.e. he's quantizing kinetic energy, etc.):

$$\int_0^\infty f(x,t)\,dx = \lim \varepsilon [f(\varepsilon, t) + f(2\varepsilon, t) + f(3\varepsilon, t) + \ldots f(p\varepsilon, t)]$$
$$\text{für } \lim \varepsilon = 0, \ \lim p\varepsilon = \infty.$$

He says that this discrete approach makes everything clearer, and quotes Lagrange's derivation of vibrations of a string as an example of where this has happened before. But then he argues that everything works out fine with the discrete approach, and that H still decreases, with the Maxwell distribution as the only possible end point. As an aside, he makes a jab at Maxwell's derivation, pointing out that with Maxwell's functional equation:

> ... there are infinitely many other solutions, which are not useful however since $f(x)$ comes out negative or imaginary for some values of x. Hence, it follows very clearly that Maxwell's attempt to prove *a priori* that his solution is the only one must fail, since it is not the only one but rather it is the only one that gives purely positive probabilities, and therefore the only useful one.

But finally—after another aside about computing thermal conductivities of gases—Boltzmann digs into polyatomic molecules, and his claim about H being related to entropy. There's another 26 pages of calculations, and then we get to a section entitled "Solution of Equation (81) and Calculation of Entropy". More pages of calculation about polyatomic molecules ensue. But finally we're computing H, and, yes, it agrees with the Clausius result—but anticlimactically he's only dealing with the case of equilibrium for monatomic molecules, where we already knew we got the Maxwell distribution:

$$f^\bullet = \frac{1}{V\left(\frac{4\pi T}{3m}\right)^{\frac{3}{2}}} e^{-\frac{3m}{4T}(u^2+v^2+w^2)},$$

daher

$$E^\bullet = N \iint \cdot \cdot f^\bullet \log f^\bullet \, dx\,dy\,dz\,du\,dv\,dw =$$
$$= -N\log\left[V\left(\frac{4\pi T}{3m}\right)^{\frac{3}{2}}\right] - \frac{3}{2}N,$$

And now he decides he's not talking about polyatomic molecules anymore, and instead:

> In order to find the relation of the quantity [H] to the second law of thermodynamics in the form $\int dQ/T < 0$, we shall interpret the system of mass points not, as previously, as a gas molecule, but rather as an entire body.

But then, in the last couple of pages of his paper, Boltzmann pulls out another idea. He's discussed the concept that polyatomic molecules (or, now, whole systems) can be in many different configurations, or "phases". But now he says: "We shall replace [our] single system by a large number of equivalent systems distributed over many different phases, but which do not interact with each other". In other words, he's introducing the idea of an ensemble of states of a system. And now he says that instead of looking at the distribution just for a single velocity, we should do it for all velocities, i.e. for the whole "phase" of the system.

> [These distributions] may be discontinuous, so that they have large values when the variables are very close to certain values determined by one or more equations, and otherwise vanishingly small. We may choose these equations to be those that characterize visible external motion of the body and the kinetic energy contained in it. In this connection it should be noted that the kinetic energy of visible motion corresponds to such a large deviation from the final equilibrium distribution of kinetic energy that it leads to an infinity in H, so that from the point of view of the Second Law of thermodynamics it acts like heat supplied from an infinite temperature.

There are a bunch of ideas swirling around here. Phase-space density (*cf.* Liouville's equation). Coarse-grained variables. Microscopic representation of mechanical work. Etc. But the paper is ending. There's a discussion about H for systems that interact, and how there's an equilibrium value achieved. And finally there's a formula for entropy

$$E^\bullet = \iint f^\bullet \log f^\bullet \, ds\,d\sigma = \log A - h\frac{\int \chi e^{-h\chi}\,d\sigma}{\int e^{-h\chi}\,d\sigma} - \frac{3r}{2}$$

that Boltzmann said "agrees ... with the expression I found in my previous [1871] paper".

So what exactly did Boltzmann really do in his 1872 paper? He introduced the Boltzmann Transport Equation which allows one to compute at least certain non-equilibrium properties of gases. But is his $f \log f$ quantity really what we can call "entropy" in the sense Clausius meant? And is it true that he's proved that entropy (even in his sense) increases? A century and a half later there's still a remarkable level of confusion around both these issues.

But in any case, back in 1872 Boltzmann's "minimum theorem" (now called his "H theorem") created quite a stir. But after some time there was an objection raised, which we'll discuss below. And partly in response to this, Boltzmann (after spending time working on microscopic models of electrical properties of materials—as well as doing some actual experiments) wrote another major paper on entropy and the Second Law in 1877:

> Über die Beziehung zwischen dem zweiten Hauptsatze der mechanischen Warmetheorie und der Wahrscheinlichkeitsrechnung, respective den Sätzen über das Wärmegleichgewicht.
>
> Von dem c. M. **Ludwig Boltzmann** in Graz.
>
> Eine Beziehung des zweiten Hauptsatzes zur Wahrscheinlichkeitsrechnung zeigte sich zuerst, als ich nachwies, dass ein analytischer Beweis desselben auf keiner anderen Grundlage

The translated title of the paper is "On the Relation between the Second Law of Thermodynamics and Probability Theory with Respect to the Laws of Thermal Equilibrium". And at the very beginning of the paper Boltzmann makes a statement that was pivotal for future discussions of the Second Law: he says it's now clear to him that an "analytical proof" of the Second Law is "only possible on the basis of probability calculations". Now that we know about computational irreducibility and its implications one could say that this was the point where Boltzmann and those who followed him went off track in understanding the true foundations of the Second Law. But Boltzmann's idea of introducing probability theory was effectively what launched statistical mechanics, with all its rich and varied consequences.

Boltzmann makes his basic claim early in the paper

> mit folgenden Worten: „Es ist klar, dass jede einzelne gleichförmige Zustandsvertheilung, welche bei einem bestimmten

> Anfangszustande nach Verlauf einer bestimmten Zeit entsteht, ebenso unwahrscheinlich ist, wie eine einzelne noch so ungleichförmige Zustandsvertheilung, gerade so wie im Lottospiele jede einzelne Quinterne ebenso unwahrscheinlich ist, wie die Quinterne

with the statement (quoting from a comment in a paper he'd written earlier the same year) that "it is clear" (always a dangerous thing to say!) that in thermal equilibrium all possible states of the system—say, spatially uniform and nonuniform alike—are equally probable

> ... comparable to the situation in the game of Lotto where every single quintet is as improbable as the quintet 12345. The higher probability that the state distribution becomes uniform with time arises only because there are far more uniform than nonuniform state distributions...

He goes on:

> [Thus] it is possible to calculate the thermal equilibrium state by finding the probability of the different possible states of the system. The initial state will in most cases be highly improbable but from it the system will always rapidly approach a more probable state until it finally reaches the most probable state, i.e., that of thermal equilibrium. If we apply this to the Second Law we will be able to identify the quantity which is usually called entropy with the probability of the particular state...

He's talked about thermal equilibrium, even in the title, but now he says:

> ... our main purpose here is not to limit ourselves to thermal equilibrium, but to explore the relationship of the probabilistic formulation to the [Second Law].

He says his goal is to calculate probability distribution for different states, and he'll start with

> as simple a case as possible, namely a gas of rigid absolutely elastic spherical molecules trapped in a container with absolutely elastic walls. (Which interact with central forces only within a certain small distance, but not otherwise; the latter assumption, which includes the former as a special case, does not change the calculations in the least).

In other words, yet again he's going to look at hard-sphere gases. But, he says:

> Even in this case, the application of probability theory is not easy. The number of molecules is not infinite, in a mathematical sense, yet the number of velocities each molecule is capable of is effectively infinite. Given this last condition, the calculations are very difficult; to facilitate understanding, I will, as in earlier work, consider a limiting case.

And this is where he "goes discrete" again—allowing ("cellular-automaton-style") only discrete possible velocities for each molecule:

> **I. Die Zahl der lebendigen Kräfte ist eine discrete.**
>
> Wir wollen zunächst annehmen, jedes Molekül sei nur im Stande, eine bestimmte endliche Anzahl von Geschwindigkeiten anzunehmen, z. B. die Geschwindigkeiten
>
> $$0, \frac{1}{q}, \frac{2}{q}, \frac{3}{q}, \ldots \frac{p}{q},$$

He says that upon colliding, two molecules can exchange these discrete velocities, but nothing more. As he explains, though:

> Even if, at first sight, this seems a very abstract way of treating the problem, it rapidly leads to the desired objective, and when you consider that in nature all infinities are but limiting cases, one assumes each molecule can behave in this fashion only in the limiting case where each molecule can assume more and more values of the velocity.

But now—much like in an earlier paper—he makes things even simpler, saying he's going to ignore velocities for now, and just say that the possible energies of molecules are "in an arithmetic progression":

Molekül anzunehmen im Stande ist, eine arithmetische Progression bildet, z. B. folgende:

$$0, \varepsilon, 2\varepsilon, 3\varepsilon, \ldots p\varepsilon.$$

He plans to look at collisions, but first he just wants to consider the combinatorial problem of distributing these energies among n molecules in all possible ways, subject to the constraint of having a certain fixed total energy. He sets up a specific example, with 7 molecules, total energy 7, and maximum energy per molecule 7—then gives an explicit table of all possible states (up to, as he puts it, "immaterial permutations of molecular labels"):

aufgeführte Reihe von Zustandsvertheilungen. Die Zahlen der ersten Colonne numeriren die verschiedenen Zustandsverthei-lungen.

		$\mathfrak{P}$
1.	0000007	7
2.	0000016	42
3.	0000025	42
4.	0000034	42
5.	0000115	105
6.	0000124	210
7.	0000133	105
8.	0000223	105
9.	0001114	140
10.	0001123	420
11.	0001222	140
12.	0011113	105
13.	0011122	210
14.	0111112	42
15.	1111111	1

Tables like this had been common for nearly two centuries in combinatorial mathematics books like Jacob Bernoulli's (1655–1705) *Ars Conjectandi*

	I.	II.	III.	IV.	V.	VI.	VII.	VIII.	IX.	X.	XI.	XII.
1.	1	0	0	0	0	0	0	0	0	0	0	0
2.	1	1	0	0	0	0	0	0	0	0	0	0
3.	1	2	1	0	0	0	0	0	0	0	0	0
4.	1	3	3	1	0	0	0	0	0	0	0	0
5.	1	4	6	4	1	0	0	0	0	0	0	0
6.	1	5	10	10	5	1	0	0	0	0	0	0
7.	1	6	15	20	15	6	1	0	0	0	0	0
8.	1	7	21	35	35	21	7	1	0	0	0	0
9.	1	8	28	56	70	56	28	8	1	0	0	0
10.	1	9	36	84	126	126	84	36	9	1	0	0
11.	1	10	45	120	210	252	210	120	45	10	1	0
12.	1	11	55	165	330	462	462	330	165	55	11	1

Combinationum, seu Numerorum Figuratorum.
Exponentes Combinationum.
Numeri Rerum Combinandarum.

but this might have been the first place such a table had appeared in a paper about fundamental physics.

And now Boltzmann goes into an analysis of the distribution of states—of the kind that's now long been standard in textbooks of statistical physics, but will then have been quite unfamiliar to the pure-calculus-based physicists of the time:

Elemente unter einander gleich. Die Anzahl dieser Permutationen ist also bekanntlich

$$\mathfrak{P} = \frac{n!}{(w_0)!\,(w_1)!\ldots} \qquad 3)$$

6 b) darauf hinausläuft, dass wir an Stelle des Minimums des Ausdrucks 3) das Minimum von

$$\frac{\sqrt{2\pi}\left(\frac{n}{e}\right)^n}{\sqrt{2\pi}\left(\frac{w_0}{e}\right)^{w_0}\sqrt{2\pi}\left(\frac{w_1}{e}\right)^{w_1}\ldots}$$

He derives the average energy per molecule, as well as the fluctuations:

täls, daher $\frac{\lambda}{n} = \frac{\mu}{\varepsilon}$, also jedenfalls ausserordentlich gross, wesshalb man hat

$$(\lambda+n-1)^{\lambda+n-\frac{1}{2}} = \lambda^{\lambda+n-\frac{1}{2}}\left[\left(1+\frac{n-1}{\lambda}\right)^\lambda\right]^{1+\frac{2n-1}{2\lambda}} = \lambda^{\lambda+n-\frac{1}{2}}\cdot e^{n-1}.$$

Es ist also

$$J = \frac{1}{\sqrt{2\pi}}\frac{\lambda^{n-1}e^{n-1}}{(n-1)^{n-\frac{1}{2}}},$$

daher ist, abgesehen von verschwindenden Grössen

$$lJ = nl\frac{\lambda}{n} + n - l\lambda + \frac{1}{2}ln - 1 - \frac{1}{2}l(2\pi).$$

He says that "of course" the real interest is in the limit of an infinite number of molecules, but he still wants to show that for "moderate values" the formulas remain quite accurate. And then (even without Wolfram Language!) he's off finding (using Newton's method it seems) approximate roots of the necessary polynomials:

Dann verwandelt sich die Gleichung 12) in folgende:

$$6x^9 - 7x^5 + 2x - 1 = 0 \qquad 16)$$

woraus folgt:

$$x = \frac{1}{2} + \frac{7}{2}x^5 - 3x^9 \qquad 17$$

Da x nahe $= \frac{1}{2}$ ist, so können wir in den beiden letzten.

ohnehin sehr kleinen Gliedern der rechten Seite $x = \frac{1}{2}$ setzen.

und erhalten:

$$x = \frac{1}{2} + \frac{1}{2^9}(7-3) = \frac{1}{2} + \frac{1}{2^7} = 0{\cdot}5078125.$$

Just to show how it all works, he considers a slightly larger case as well:

$$0000011122345,$$

deren Ziffersumme in der That $= 19$ ist.

Die Anzahl der Permutationen, deren diese Complexion fähig ist, ist

$$\frac{13!}{5!\,3!\,2!} = \frac{13!}{4!\,3!\,2!} \cdot \frac{1}{5}.$$

Eine Complexion, deren Ziffernsumme ebenfalls $= 19$ ist, und von welcher man vermuthen könnte, dass sie sehr vieler Permutationen fähig sein wird, wäre folgende:

$$0000111222334.$$

Die Anzahl ihrer Permutationen ist

$$\frac{13!}{4!\,3!\,3!\,2!} = \frac{13!}{4!\,3!\,2!} \cdot \frac{1}{6},$$

also bereits kleiner als die Anzahl der Permutationen der ersten von uns aus der Annäherungsformel gefundenen Complexion. Ebenso überzeugt man sich, dass die Zahl der Permutationen der beiden Complexionen

$$0000111122335 \text{ und}$$
$$0000111122344$$

kleiner ist. Dieselbe ist nämlich für beide Complexionen

$$\frac{13!}{4!\,4!\,2!\,2!} = \frac{13!}{4!\,3!\,2!} \cdot \frac{1}{8}.$$

Now he's computing the probability that a given molecule has a particular energy

$$w_s = \frac{n^2}{n+\lambda} \cdot \left(\frac{\lambda}{n+\lambda}\right)^s.$$

and determining that in the limit it's an exponential

$$w_s = \frac{n\varepsilon}{\mu} e^{-\frac{s}{\mu}}.$$

that is, as he says, "consistent with that known from gases in thermal equilibrium".

He claims that in order to really get a "mechanical theory of heat" it's necessary to take a continuum limit. And here he concludes that thermal equilibrium is achieved by maximizing the quantity Ω (where the "l" stands for log, so this is basically $f \log f$):

welche ein Maximum werden soll, den Werth erhält:

$$\Omega = - \int_{-\infty}^{+\infty} \int_{-\infty}^{+\infty} \int_{-\infty}^{+\infty} f(u, v, w)\, lf(u, v, w)\,du\,dv\,dw \qquad 34)$$

He explains that Ω is basically the log of the number of possible permutations, and that it's "of special importance", and he'll call it the "permutability measure". He immediately notes that "the total permutability measure of two bodies is equal to the sum of the permutability measures of each body". (Note that Boltzmann's Ω isn't the modern total-number-of-states Ω; confusingly, that's essentially the exponential of Boltzmann's Ω.)

He goes through some discussion of how to handle extra degrees of freedom in polyatomic molecules, but then he's on to the main event: arguing that Ω is (essentially) the entropy. It doesn't take long:

kules, so ist für den Zustand des Warmegleichgewichtes

$$f(x, y, z, u, v, w) = \frac{N}{V\left(\frac{4\pi T}{3m}\right)^{\frac{3}{2}}} \cdot e^{-\frac{3m}{4T}(u^2 + v^2 + w^2)}.$$

Substituirt man diesen Werth in Gleichung 61), so erhält man

$$\Omega = \frac{3N}{2} + Nl\left[V\left(\frac{4\pi T}{3m}\right)^{\frac{3}{2}}\right] - NlN \qquad 62)$$

Versteht man nun unter dQ das dem Gase zugeführte Wärmedifferentiale, so ist

$$dQ = NdT + pdV \qquad 63)$$

und

$$pV = \frac{2N}{3} \cdot T \qquad 64)$$

p ist der Druck, bezogen auf die Flächeneinheit. Die Entropie des Gases ist dann:

$$\int \frac{dQ}{T} = \frac{2}{3} N . l(V . T^{\frac{3}{2}}) + C.$$

Da hier N als eine rein Constante anzusehen ist, so ist bei passender Bestimmung dieser Constante

$$\int \frac{dQ}{T} = \frac{2}{3} \Omega \qquad 65)$$

Basically he just says that in equilibrium the probability $f(...)$ for a molecule to have a particular velocity is given by the Maxwell distribution, then he substitutes this into the formula for Ω, and shows that indeed, up to a constant, Ω is exactly the "Clausius entropy" $\int dQ/T$.

So, yes, in equilibrium Ω seems to be giving the entropy. But then Boltzmann makes a bit of a jump. He says that in processes that aren't reversible both "Clausius entropy" and Ω will increase, and can still be identified—and enunciates the general principle, printed in his paper in special doubled-spaced form:

> Denken wir uns ein beliebiges System von Kör-
> pern gegeben, dasselbe mache eine beliebige Zu-
> standsveränderung durch, ohne dass nothwendig
> der Anfangs- und Endzustand Zustände des Gleich-
> gewichtes zu sein brauchen; dann wird immer das
> Permutabilitätsmass aller Körper im Verlaufe der
> Zustandsveränderungen fortwährend wachsen und
> kann höchstens constant bleiben, so lange sich
> sämmtliche Körper während der Zustandsverände-
> rung mit unendlicher Annäherung im Wärmegleich-
> gewichte befinden (umkehrbare Zustandsverände-
> rungen).
>
> Um ein Beispiel zu geben, betrachten wir ein Gefäss, wel-
> ches durch eine unendlich dünne Scheidewand in zwei Hälften

... [In] any system of bodies that undergoes state changes ... even if the initial and final states are not in thermal equilibrium ... the total permutability measure for the bodies will continually increase during the state changes, and can remain constant only so long as all the bodies during the state changes remain infinitely close to thermal equilibrium (reversible state changes).

In other words, he's asserting that Ω behaves the same way entropy is said to behave according to the Second Law. He gives various thought experiments about gases in boxes with dividers, gases under gravity, etc. And finally concludes that, yes, the relationship of entropy to Ω "applies to the general case".

There's one final paragraph in the paper, though:

> Up to this point, these propositions may be demonstrated exactly using the theory of gases. If one tries, however, to generalize to liquid drops and solid bodies, one must dispense with an exact treatment from the outset, since far too little is known about the nature of the latter states of matter, and the mathematical theory is barely developed. But I have already mentioned reasons in previous papers, in virtue of which it is likely that for these two aggregate states, the thermal equilibrium is achieved when Ω becomes a maximum, and that when thermal equilibrium exists, the entropy is given by the same expression. It can therefore be described as likely that the validity of the principle which I have developed is not just limited to gases, but that the same constitutes a general natural law applicable to solid bodies and liquid droplets, although the exact mathematical treatment of these cases still seems to encounter extraordinary difficulties.

Interestingly, Boltzmann is only saying that it's "likely" that in thermal equilibrium his permutability measure agrees with Clausius's entropy, and he's implying that actually that's really the only place where Clausius's entropy is properly defined. But certainly his definition is more general (after all, it doesn't refer to things like temperature that are only properly defined in equilibrium), and so—even though Boltzmann didn't explicitly say it— one can imagine basically just using it as the definition of entropy for arbitrary cases. Needless to say, the story is actually more complicated, as we'll see soon.

But this definition of entropy—crispened up by Max Planck (1858–1947) and with different notation—is what ended up years later "written in stone" at Boltzmann's grave:

The Concept of Ergodicity

In his 1877 paper Boltzmann had made the claim that in equilibrium all possible microscopic states of a system would be equally probable. But why should this be true? One reason could be that in its pure "mechanical evolution" the system would just successively visit all these states. And this was an idea that Boltzmann seems to have had—with increasing clarity—from the time of his very first paper in 1866 that purported to "prove the Second Law" from mechanics.

In modern times—with our understanding of discrete systems and computational rules—it's not difficult to describe the idea of "visiting all states". But in Boltzmann's time it was considerably more complicated. Did one expect to hit all the infinite possible infinitesimally separated configurations of a system? Or somehow just get close? The fact is that Boltzmann had certainly dipped his toe into thinking about things in terms of discrete quantities. But he didn't make the jump to imagining discrete rules, even though he certainly did know about discrete iterative processes, like Newton's method for finding roots.

Boltzmann knew about cases—like circular motion—where everything was purely periodic. But maybe when motion wasn't periodic, it'd inevitably "visit all states". Already in 1868 Boltzmann was writing a paper entitled "Solution to a Mechanical Problem" where he studies a single point mass moving in an $\alpha/r - \beta/r^2$ potential and bouncing elastically off a line—and manages to show that it visits every position with equal probability. In this paper he's just got traditional formulas, but by 1871, in "Some General Theorems about Thermal Equilibrium"—computing motion in the same potential as before—he's got a picture:

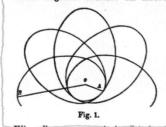

270   19. Einige allgemeine Sätze über Wärmegleichgewicht.

folgender Apsidenlinien in einem rationalen Verhältnisse zu π steht; die Gestalt der Bahn ist durch die nachstehende Fig. 1 versinnlicht. Der materielle Punkt beschreibt jetzt keine in sich zurückkehrende Linie, sondern durchwandert allmählich das ganze Stück der Ebene, welches zwischen zwei aus dem Zentrum O mit den Radien OA und OB beschriebenen Kreisen liegt, ohne je wieder exakt zu demselben Punkte der Ebene zurückzukehren. Betrachten wir irgend ein innerhalb jener Kreise liegendes Element der Ebene $dx\,dy$ und lassen den materiellen Punkt eine sehr lange Zeit T hindurch sich bewegen, so ist das Verhältnis jenes Bruchteils von T, während welches sich der Punkt innerhalb $dx \cdot dy$ befindet, zur ganzen Zeit T eine mathematisch vollständig definierte Größe.

Fig. 1.

Boltzmann probably knew about Lissajous figures—cataloged in 1857

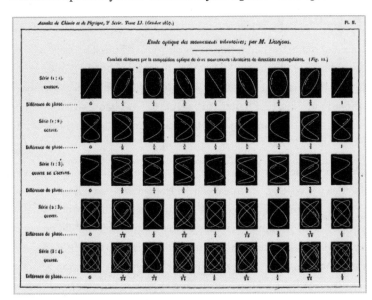

and the fact that in this case a rational ratio of x and y periods gives a periodic overall curve while an irrational one always gives a curve that visits every position might have led him to suspect that all systems would either be periodic, or would visit every possible configuration (or at least, as he identified in his paper, every configuration that had the same values of "constants of the motion", like energy).

In early 1877 Boltzmann returned to the same question, including as one section in his "Remarks on Some Problems in the Mechanical Theory of Heat" more analysis of the same potential as before, but now showing a diversity of more complicated pictures that almost seem to justify his rule-30-before-its-time idea that there could be "pure mechanics" that would lead to "Second Law" behavior:

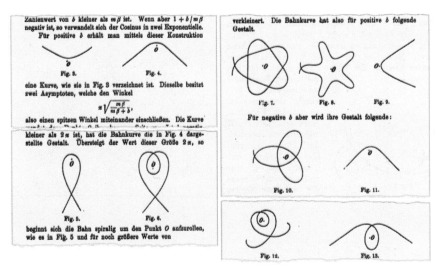

In modern times, of course, it's easy to solve those equations of motion, and typical results obtained for an array of values of parameters are:

Boltzmann returned to these questions in 1884, responding to Helmholtz's analysis of what he was calling "monocyclic systems". Boltzmann used the same potential again, but now with a name for the "visit-all-states" property: isodic. Meanwhile, Boltzmann had introduced the name "ergoden" for the collection of all possible configurations of a system with a given energy (what would now be called the microcanonical ensemble). But somehow, quite a few years later, Boltzmann's student Paul Ehrenfest (1880–1933) (along with Tatiana Ehrenfest-Afanassjewa (1876–1964)) would introduce the term "ergodic" for Boltzmann's isodic. And "ergodic" is the term that caught on. And in the twentieth century there was all sorts of development of "ergodic theory", as we'll discuss a bit later.

But back in the 1800s people continued to discuss the possibility that what would become called ergodicity was somehow generic, and would explain why all states would somehow be equally probable, why the Maxwell distribution of velocities would be obtained, and ultimately why the Second Law was true. Maxwell worked out some examples. So did Kelvin. But it remained unclear how it would all work out, as Kelvin (now with many letters after his name) discussed in a talk he gave in 1900 celebrating the new century:

> I. *Nineteenth Century Clouds over the Dynamical Theory of Heat and Light **. *By* The Right. Hon. Lord KELVIN, *G.C.V.O., D.C.L., LL.D., F.R.S., M.R.I.* †.

> § 1. THE beauty and clearness of the dynamical theory, which asserts heat and light to be modes of motion, is at present obscured by two clouds. I. The first

The dynamical theory of light didn't work out. And about the dynamical theory of heat, he quotes Maxwell (following Boltzmann) in one of his very last papers, published in 1878, as saying, in reference to what amounts to a proof of the Second Law from underlying dynamics:

> " The only assumption which is necessary for the direct
> " proof is that the system, if left to itself in its actual state of

> " motion, will, sooner or later, pass [infinitely nearly *]
> " through every phase which is consistent with the equation
> " of energy " (p. 714) and, again (p. 716).

> I have never seen validity in the demonstration ‡ on which
> Maxwell founds this statement, and it has always seemed to
> me exceedingly improbable that it can be true. If true, it
> would be very wonderful, and most interesting in pure
> mathematical dynamics. Having been published by Boltz-
> mann and Maxwell it would be worthy of most serious
> attention, even without consideration of its bearing on
> thermo-dynamics. But, when we consider its bearing

Kelvin talks about exploring test cases:

> mathematics. Ten years ago *, I suggested a number of test-
> cases, some of which have been courteously considered by
> Boltzmann ; but no demonstration either of the truth or
> untruth of the doctrine as applied to any one of them has
> hitherto been given. A year later, I suggested what seemed
> to me a decisive test-case disproving the doctrine; but my
> statement was quickly and justly criticised by Boltzmann
> and Poincaré; and more recently Lord Rayleigh† has shown
> very clearly that my simple test-case was quite indecisive.
> This last article of Rayleigh's has led me to resume the
> consideration of several classes of dynamical problems, which
> had occupied me more or less at various times during the last
> twenty years, each presenting exceedingly interesting features
> in connection with the double question: Is this a case which
> admits of the application of the Boltzmann-Maxwell doctrine;
> and if so, is the doctrine true for it?

When, for example, is the motion of a single particle bouncing around in a fixed region ergodic? He considers first an ellipse, and proves that, no, there isn't in general ergodicity there:

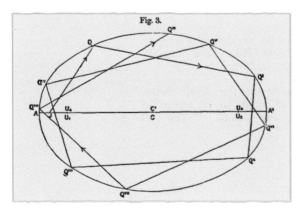

Fig. 3.

Then he goes on to the much more complicated case

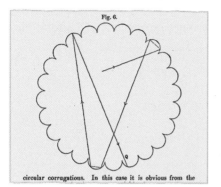

Fig. 6.

circular corrugations. In this case it is obvious from the

and now he does an "experiment" (with a rather Monte Carlo flavor):

circular corrugations. In this case it is obvious from the symmetry that the time-integral of kinetic energy of component motion parallel to any straight line must, in the long run, be equal to that parallel to any other. But the Boltzmann-Maxwell doctrine asserts, that the time-integrals of the kinetic energies of the two components, radial and transversal, according to polar coordinates, would be equal. To test this, I have taken the case of an infinite number of the semicircular corrugations, so that in the time-integral it is not necessary to include the times between successive impacts of the particle on any one of the semicircles. In this case the geometrical construction would, of course, fail to show the precise point Q at which the free path would cut the diameter AB of the semicircular hollow to which it is approaching; and I have evaded the difficulty in a manner thoroughly suitable for thermodynamic application such as the kinetic theory of gases. I arranged to draw lots for 1

out of the 199 points dividing AB into 200 equal parts. This was done by taking 100 cards*, 0, 1 . . . . . 98, 99, to represent distances from the middle point, and, by the toss of a coin, determining on which side of the middle point it was to be (plus or minus for head or tail, frequently changed to avoid possibility of error by bias). The draw for one of the hundred numbers (0 . . . . 99) was taken after very thorough shuffling of the cards in each case. The point of entry having been found, a large-scale geometrical construction was used to determine the successive points of impact and the inclination θ of the emergent path to the diameter AB. The inclination of the entering path to the diameter of the semicircular hollow struck at the end of the flight, has the same value θ. If we call the diameter of the large circle unity, the length of each flight is $\sin\theta$. Hence, if the velocity is unity and the mass of the particle 2, the time-integral of the whole kinetic energy is $\sin\theta$; and it is easy to prove that the time-integrals of the components of the velocity, along and perpendicular to the line from each point of the path to the centre of the large circle, are respectively $\theta\cos\theta$, and $\sin\theta - \theta\cos\theta$. The excess of the latter above the former is $\sin\theta - 2\theta\cos\theta$. By summation for 143 flights we have found,

$$\Sigma \sin\theta = 121\cdot3 \; ; \; 2\Sigma\theta\cos\theta = 108\cdot3 ;$$

whence,

$$\Sigma \sin\theta - 2\Sigma\theta\cos\theta = 13\cdot0.$$

This is a notable deviation from the Boltzmann-Maxwell doctrine, which makes $\Sigma(\sin\theta - \theta\cos\theta)$ equal to $\Sigma\theta\cos\theta$. We have found the former to exceed the latter by a difference which amounts to $10\cdot7$ of the whole $\Sigma\sin\theta$.

Out of fourteen sets of ten flights, I find that the time-integral of the transverse component is less than half the whole in twelve sets, and greater in only two. This seems to prove beyond doubt that the deviation from the Boltzmann-Maxwell doctrine is genuine; and that the time-integral of the transverse component is certainly smaller than the time-integral of the radial component.

Kelvin considers a few other examples

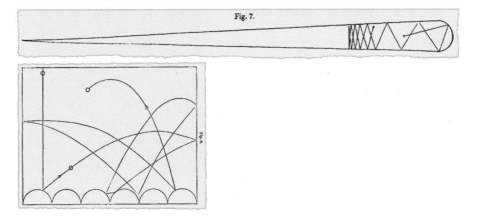

but mostly concludes that he can't tell in general about ergodicity—and that probably something else is needed, or as he puts it (somehow wrapping the theory of light into the story as well):

> The simplest way of arriving at this desired result is to deny the conclusion; and so, in the beginning of the twentieth century, to lose sight of a cloud which has obscured the brilliance of the molecular theory of heat and light during the last quarter of the nineteenth century.

But What about Reversibility?

Had Boltzmann's 1872 *H* theorem proved the Second Law? Was the Second Law—with its rather downbeat implication about the heat death of the universe—even true? One skeptic was Boltzmann's friend and former teacher, the chemist Josef Loschmidt (1821–1895), who in 1866 had used kinetic theory to (rather accurately) estimate the size of air molecules. And in 1876 Loschmidt wrote a paper entitled "On the State of Thermal Equilibrium in a System of Bodies with Consideration of Gravity" in which he claimed to show that when gravity was taken into account, there wouldn't be uniform thermal equilibrium, the Maxwell distribution, or the Second Law—and thus, as he poetically explained:

> The terroristic nimbus of the Second Law is destroyed, a nimbus which makes that Second Law appear as the annihilating principle of all life in the universe—and at the same time we are confronted with the comforting perspective that, as far as the conversion of heat into work is concerned, mankind will not solely be dependent on the intervention of coal or of the Sun, but will have available an inexhaustible resource of convertible heat at all times.

His main argument revolves around a thought experiment involving molecules in a gravitational field:

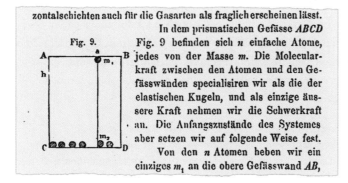

Over the next couple of years, despite Loschmidt's progressively more elaborate constructions

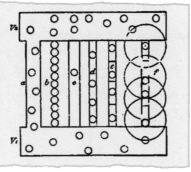

Boltzmann and Maxwell will debunk this particular argument—even though to this day the role of gravity in relation to the Second Law remains incompletely resolved.

But what's more important for our narrative about Loschmidt's original paper are a couple of paragraphs tucked away at the end of one section (that in fact Kelvin had basically anticipated in 1874):

> [Consider what would happen if] after a time *t* sufficiently long for the stationary state to obtain, we suddenly reversed the velocities of all atoms. Initially we would be in a state that would look like the stationary state. This would be true for some time, but in the long run the stationary state would deteriorate and after the time *t* we would inevitably return to the initial state...

> It is clear that in general in any system one can revert the entire course of events by suddenly inverting the velocities of all the elements of the system. This doesn't give a solution to the problem of undoing everything that happens [in the universe] but it does give a simple prescription: just suddenly revert the instantaneous velocities of all atoms of the universe.

How did this relate to the *H* theorem? The underlying molecular equations of motion that Boltzmann had assumed in his proof were reversible in time. Yet Boltzmann claimed that *H* was always going to a minimum. But why couldn't one use Loschmidt's argument to construct an equally possible "reverse evolution" in which *H* was instead going to a maximum?

It didn't take Boltzmann long to answer, in print, tucked away in a section of his paper "Remarks on Some Problems in the Mechanical Theory of Heat". He admits that Loschmidt's argument "has great seductiveness". But he claims it is merely "an interesting sophism"—and then says he will "locate the source of the fallacy". He begins with a classic setup: a collection of hard spheres in a box.

> Suppose that at time zero the distribution of spheres in the box is not uniform; for example, suppose that the density of spheres is greater on the right than on the left ... The sophism now consists in saying that, without reference to the initial conditions, it cannot be proved that the spheres will become uniformly mixed in the course of time.

But then he rather boldly claims that with the actual initial conditions described, the spheres will "almost always [become] uniform" at a future time *t*. Now he imagines (following Loschmidt) reversing all the velocities in this state at time *t*. Then, he says:

> ... the spheres would sort themselves out as time progresses, and at [the analog of] time 0, they would have a completely nonuniform distribution, even though the [new] initial distribution [one had used] was almost uniform.

But now he says that, yes—given this counterexample—it won't be possible to prove that the final distribution of spheres will always be uniform.

This is in fact a consequence of probability theory, for any nonuniform distribution, no matter how improbable it may be, is still not absolutely impossible. Indeed it is clear that any individual uniform distribution, which might arise after a certain time from some particular initial state, is just as improbable as an individual nonuniform distribution; just as in the game of Lotto, any individual set of five numbers is as improbable as the set 1, 2, 3, 4, 5. It is only because there are many more uniform distributions than nonuniform ones that the distribution of states will become uniform in the course of time. One therefore cannot prove that, whatever may be the positions and velocities of the spheres at the beginning, the distribution must become uniform after a long time; rather one can only prove that infinitely many more initial states will lead to a uniform one after a definite length of time than to a nonuniform one.

He adds:

One could even calculate, from the relative numbers of the different state distributions, their probabilities, which might lead to an interesting method for the calculation of thermal equilibrium.

And indeed within a few months Boltzmann has followed up on that "interesting method" to produce his classic paper on the probabilistic interpretation of entropy.

But in his earlier paper he goes on to argue:

Since there are infinitely many more uniform than nonuniform distributions of states, the latter case is extraordinarily improbable [to arise] and can be considered impossible for practical purposes; just as it may be considered impossible that if one starts with oxygen and nitrogen mixed in a container, after a month one will find chemically pure oxygen in the lower half and nitrogen in the upper half, although according to probability theory this is merely very improbable but not impossible.

He talks about how interesting it is that the Second Law is intimately connected with probability while the First Law is not. But at the end he does admit:

Perhaps this reduction of the Second Law to the realm of probability makes its application to the entire universe appear dubious, but the laws of probability theory are confirmed by all experiments carried out in the laboratory.

At this point it's all rather unconvincing. The H theorem had purported to prove the Second Law. But now he's just talking about probability theory. He seems to have given up on proving the Second Law. And he's basically just saying that the Second Law is true because it's observed to be true—like other laws of nature, but not like something that can be "proved", say from underlying molecular dynamics.

For many years not much attention was paid to these issues, but by the late 1880s there were attempts to clarify things, particularly among the rather active British circle of kinetic theorists. A published 1894 letter from the Irish mathematician Edward Culverwell (1855–1931) (who also wrote about ice ages and Montessori education) summed up some of the confusions that were circulating:

Boltzmann's Minimum Theorem.

THE remarkable differences of opinion as to what the H-theorem *is*, and how it can be proved, show how necessary is the discussion elicited by my letter on the oversight in Dr. Wa'son's proof. Each of the four authorities who have replied takes a different view.

Dr. Larmor enforces the view I put forward at the close of my letter, and says that the theorem *is* what I said appeared an *à priori* possibility ; and I may here point out that his letter is a complete answer to the argument I used in the *Phil. Mag.* 1890, p. 95, urging that, as there were as many configurations which receded from the permanent state as approached it, there was an *à priori* improbability that a permanent state would ever be reached. This argument was criticised at some length, not really answered, in Messrs. Larmor and Bryan's Report on Thermodynamics (British Association Report, 1891), but the suggestive remarks there given helped me, I think, to arrive (independently) at the complete answer given in Dr. Larmor's recent letter. But my present use of the argument is not that which Dr. Larmor criticises ; I now use it as a test of a particular proof of the H-theorem. I say that if that proof does not somewhere or other introduce some assumption about averages, probability, or irreversibility, it cannot be valid.

Mr. Burbury appears to consider that the theorem can only be proved if we assume that some element of the distribution does tend to an average (quite a different position from Dr. Larmor's), and he is as yet unable to state the appropriate assumption except for the case of hard elastic spherical particles colliding or "encountering" (for since *a* is constant in his last letter, it seems as if the $q_1 \ldots q_{n-3}$ coordinates are really dummies). Yet Mr. Burbury has already given what purports to be a *general* proof of the theorem for any number of degrees of freedom.

Mr. Bryan thinks that a condition which excludes the reversed motion is implied in Dr. Watson's proof, for he says that in taking unaccented letters F*f* as proportional to the number of molecules passing from one configuration to another in the reversed motion, I make a less "natural" supposition than Dr. Watson, who takes accented letters F'*f*'. I cannot see what virtue there is in putting accents on or leaving them off, and after a very careful study of Mr. Bryan's letter, I can only think that he has fallen into some confusion owing to the way in which he uses at one time *accented* and at another time *unaccented differentials*, although (as he himself remarks) there is no difference whatever between their accented and unaccented *products*. But even if Mr. Bryan be right, would he put us any "forrarder"? What we want is a *proof* that the collisions will make H decrease, and we can hardly be satisfied with a proof

which depends on the previous assumption that the particles do "naturally" tend to move in the desired way.

Dr. Watson meets my reversibility argument by saying that H decreases even in the reversed motion, when the system is confessedly *receding* from its permanent state. No other correspondent agrees with him in this view, which would indeed *take away all physical meaning from the H theorem*, for the decrease of H would then be quite unconnected with the approach to a permanent state. As to the other point, Dr. Watson does not amend his proof himself, but says it is "easy" to do, and so does Mr. Bryan. Yet one has an instinctive distrust of things which are said to be "easily seen," and at all events Dr. Watson's reference to the case in which the theorem is *applied* does not help one in the *proof*, where it is necessary to express *separately* the products of the differentials expressed by the small and capital letters respectively in his "Kinetic Theory."

Mr. Burbury asks why I say the error law has been proved for the case of hard spheres without the use of Boltzmann's Minimum Theorem. I thought that Tait had done so (*Trans.* R.S.E. 1886), and at all events I thought the ordinary investigation showed that there was but *one* solution, that of the error law, in that case ; but perhaps I am mistaken.

Mr. Bryan says Lorenz gives the clearest account of the assumptions in Boltzmann's theorem. He would earn our gratitude if he would state them in his next letter.

EDW. P. CULVERWELL.

Trinity College, Dublin, December 29, 1894.

At a lecture in England the next year, Boltzmann countered (conveniently, in English):

§ 2. Mr. Culverwell's objections against my Minimum Theorem bear the closest connections to what I pointed out in the second part of my paper „Bemerkungen über einige Probleme der physikalischen Wärmetheorie", Wien. Ber. 75. 1877.¹) There I pointed out that my Minimum Theorem, as well as the so-called Second Law of Thermodynamics, are only theorems of probability. The Second Law can never be proved mathematically by means of the equations of dynamics alone.

He goes on, but doesn't get much more specific:

Though interesting and striking at the first moment, Mr. Culverwell's arguments rest, as I think, only upon a mistake of my assumptions. It can never be proved from the equations of motion alone, that the minimum function *H* must always decrease. It can only be deduced from the laws of probability, that if the initial state is not specially arranged for a certain purpose, but haphazard governs freely, the probability that *H* decreases is always greater than that it increases. It is well known that the theory of probability is

He then makes an argument that will be repeated many times in different forms, saying that there will be fluctuations, where H deviates temporarily from its minimum value, but these will be rare:

> shall beexcluded. During the greater part of this time $\dot{H}$ will be very nearly equal to its minimum value H (min.). Let us construct the H-curve, *i. e.* let us take the time as axis of abscissæ and draw the curve, whose ordinates are the corresponding values of H. The greater majority of the ordinates of this curve are very nearly equal to H (min.). But because greater values of H are not mathematically impossible, but only very improbable, the curve has certain, though very few, summits or maximum ordinates which rise to a greater height than H (min.).

Later he's talking about what he calls the "H curve" (a plot of H as a function of time), and he's trying to describe its limiting form:

> remark. Not for every curve, but only for the particular form of the H-curve, disymmetrical in the upward and downward direction, can it be proved that H has a tendency to decrease. This particular form is very well illustrated by Mr. Culverwell's suggestion of an inverted tree. The H-curve is composed of a succession of such trees. Almost all these trees are extremely low, and have branches very nearly horizontal. Here H has nearly the minimum value. Only very few trees are higher, and have branches inclined to the axis of abscissæ, and the improbability of such a tree increases enormously with its height. The difficulty consists only in imagining all these branches infinitely short.
>
> Finally there is the difference between the ordinary cases, where H decreases or is near to its minimum value, and the very rare cases, where H is far from the minimum value and still increasing. In the last cases, H will reach, probably in a very short time, a maximum value. Then it will decrease from that value to the well-known minimum value.

And he even refers to Weierstrass's recent work on nondifferentiable functions:

> increase of x for all points, whose ordinates are $= 1$. The P-curve belongs to the large class of curves which have nowhere a uniquely defind tangent. Even at the top of each summit the tangent is not parallel to the x-axis, but is undefined. In other words, the chord joining two points of the curve does not tend toward a difinite limiting position when one of the two points approaches and ultimately coincides with the other.[1]) The same applies to the H-curve in the Theory of Gases. If I find a certain negative value for dH/dt, that does not defiue the tangent of the curve in the ordinary sense, but it is only an average value.

But he doesn't pursue this, and instead ends his "rebuttal" with a more philosophical—and in some sense anthropic—argument that he attributes to his former assistant Ignaz Schütz (1867–1927):

> § 3. Mr. Culverwell says that my theorem cannot be true because if it were true every atom of the universe would have the same average *vis viva*, and all energy would be dissipated. I find, on the contrary, that this argument only tends to confirm my theorem, which requires only that in the course of time the universe must tend to a state where the average *vis viva* of every atom is the same and all energy is dissipated, and that is indeed the case. But if we ask why this state is not yet reached, we again come to a „Salisburian mystery".
>
> I will conclude this paper with an idea of my old assistant, Dr. Schuetz.
>
> We assume that the whole universe is, and rests for ever, in thermal equilibrium. The probability that one (only one) part of the universe is in a certain state, is the smaller the further this state is from thermal equilibrium; but this probability is greater, the greater the universe itself is. If we assume the universe great enough we can make the probability of one relatively small part being in any given state (however far from the state of thermal equilibrium), as great as we please. We can also make the probability great that, though the whole universe is in thermal equilibrium, our world is in its present state. It may be sayd that the world is so far from thermal equilibrium that we cannot imagine the improbability of such a state. But can we imagine, on the other side, how small a part of the whole universe this world is? Assuming the universe great enough, the probability that such a small part of it as our world should be in its present state, is no longer small.
>
> If this assumption were correct, our world would return more and more to thermal equilibrium; but because the whole universe is so great, it might be probable that at some future time some other world might deviate as far from thermal equilibrium as our world does at present. Then the afore-mentioned *H*-curve would form a representation of what takes place in the universe. The summits of the curve would represent the worlds where visible motion and life exist.

It's an argument that we'll see in various forms repeated over the century and a half that follows. In essence what it's saying is that, yes, the Second Law implies that the universe will end up in thermal equilibrium. But there'll always be fluctuations. And in a big enough universe there'll be fluctuations somewhere that are large enough to correspond to the world as we experience it, where "visible motion and life exist".

But regardless of such claims, there's a purely formal question about the H theorem. How exactly is it that from the Boltzmann transport equation—which is supposed to describe reversible mechanical processes—the H theorem manages to prove that the H function irreversibly decreases? It wasn't until 1895—fully 25 years after Boltzmann first claimed to prove the H theorem—that this issue was even addressed. And it first came up rather circuitously through Boltzmann's response to comments in a textbook by Gustav Kirchhoff (1824–1887) that had been completed by Max Planck.

The key point is that Boltzmann's equation makes an implicit assumption, that's essentially the same as Maxwell made back in 1860: that before each collision between molecules, the molecules are statistically uncorrelated, so that the probability for the collision has the factored form $f(v_1) f(v_2)$. But what about after the collision? Inevitably the collision itself will lead to correlations. So now there's an asymmetry: there are no correlations before each collision, but there are correlations after. And that's why the behavior of the system doesn't have to be symmetrical—and the H theorem can prove that H irreversibly decreases.

In 1895 Boltzmann wrote a 3-page paper (after half in footnotes) entitled "More about Maxwell's Distribution Law for Speeds" where he explained what he thought was going on:

> [The reversibility of the laws of mechanics] has been recently applied in judging the assumptions necessary for a proof of [the H theorem]. This proof requires the hypothesis that the state of the gas is and remains molecularly disordered, namely, that the molecules of a given class do not always or predominantly collide in a specific manner and that, on the contrary, the number of collisions of a given kind can be found by the laws of probability.
>
> Now, if we assume that in general a state distribution never remains molecularly ordered for an unlimited time and also that for a stationary state-distribution every velocity is as probable as the reversed velocity, then it follows that by inversion of all the velocities after an infinitely long time every stationary state-distribution remains unchanged. After the reversal, however, there are exactly as many collisions occurring in the reversed way as there were collisions occurring in the direct way. Since the two state distributions are identical, the probability of direct and indirect collisions must be equal for each of them, whence follows Maxwell's distribution of velocities.

Boltzmann is introducing what we'd now call the "molecular chaos" assumption (and what Ehrenfest would call the *Stosszahlansatz*)—giving a rather self-fulfilling argument for why the assumption should be true. In Boltzmann's time there wasn't really anything better to do. By the 1940s the BBGKY hierarchy at least let one organize the hierarchy of correlations between molecules—though it still didn't give one a tractable way to assess what correlations should exist in practice, and what not.

Boltzmann knew these were all complicated issues. But he wrote about them at a technical level only a few more times in his life. The last time was in 1898 when, responding to a request from the mathematician Felix Klein (1849–1925), he wrote a paper about the H curve for mathematicians. He begins by saying that although this curve comes from the theory of gases, the essence of it can be reproduced by a process based on accumulating balls randomly picked from an urn. He then goes on to outline what amounts to a story of random walks and fractals. In another paper, he actually sketches the curve

große, endliche Zahl von unendlich wenig deformierbaren Gasmolekülen handelt. Verschwindend wenige spezielle Anfangs-
zustände ausgenommen wird dann allerdings auch der wahr-

scheinlichste Zustand am häufigsten vorkommen (wenigstens bei
einer sehr großen Zahl von Molekülen). Die Ordinaten dieser

saying that his drawing "should be taken with a large grain of salt", noting—in a remarkably fractal-reminiscent way—that "a zincographer [i.e. an engraver of printing plates] would not have been able to produce a real figure since the H-curve has a very large number of maxima and minima on each finite segment, and hence defies representation as a line of continuously changing direction."

Of course, in modern times it's easy to produce an approximation to the H curve according to his prescription:

But at the end of his "mathematical" paper he comes back to talking about gases. And first he makes the claim that the effective reversibility seen in the H curve will never be seen in actual physical systems because, in essence, there are always perturbations from outside. But then he ends, in a statement of ultimate reversibility that casts our everyday observation of irreversibility as tautological:

> There is no doubt that it is just as conceivable to have a world in which all natural processes take place in the wrong chronological order. But a person living in this upside-down world would have feelings no different than we do: they would just describe what we call the future as the past and vice versa.

The Recurrence Objection

Probably the single most prominent research topic in mathematical physics in the 1800s was the three-body problem—of solving for the motion under gravity of three bodies, such as the Earth, Moon and Sun. And in 1890 the French mathematician Henri Poincaré (1854–1912) (whose breakout work had been on the three-body problem) wrote a paper entitled "On the Three-Body Problem and the Equations of Dynamics" in which, as he said:

> It is proved that there are infinitely many ways of choosing the initial conditions such that the system will return infinitely many times as close as one wishes to its initial position. There are also an infinite number of solutions that do not have this property, but it is shown that these unstable solutions can be regarded as "exceptional" and may be said to have zero probability.

This was a mathematical result. But three years later Poincaré wrote what amounted to a philosophy paper entitled "Mechanism and Experience" which expounded on its significance for the Second Law:

> In the mechanistic hypothesis, all phenomena must be reversible; for example, the stars might traverse their orbits in the retrograde sense without violating Newton's law; this would be true for any law of attraction whatever. This is therefore not a fact peculiar to astronomy; reversibility is a necessary consequence of all mechanistic hypotheses.
>
> Experience provides on the contrary a number of irreversible phenomena. For example, if one puts together a warm and a cold body, the former will give up its heat to the latter; the opposite phenomenon never occurs. Not only will the cold body not return to the warm one the heat which it has taken away when it is in direct contact with it; no matter what artifice one may employ, using other intervening bodies, this restitution will be impossible, at least unless the gain thereby realized is compensated by an equivalent or large loss. In other words, if a system of bodies can pass from state A to state B by a certain path, it cannot return from B to A, either by the same path or by a different one. It is this circumstance that one describes by saying that not only is there not direct reversibility, but also there is not even indirect reversibility.

But then he continues:

> A theorem, easy to prove, tells us that a bounded world, governed only by the laws of mechanics, will always pass through a state very close to its initial state. On the other hand, according to accepted experimental laws (if one attributes absolute validity to them, and if one is willing to press their consequences to the extreme), the universe tends toward a certain final state, from which it will never depart. In this final state, which will be a kind of death, all bodies will be at rest at the same temperature.

But in fact, he says, the recurrence theorem shows that:

> This state will not be the final death of the universe, but a sort of slumber, from which it will awake after millions of millions of centuries. According to this theory, to see heat pass from a cold body to a warm one … it will suffice to have a little patience. [And we may] hope that some day the telescope will show us a world in the process of waking up, where the laws of thermodynamics are reversed.

By 1903, Poincaré was more strident in his critique of the formalism around the Second Law, writing (in English) in a paper entitled "On Entropy":

8. *Conclusions.*
A. The entropy is a function of the co-ordinates.
B. Is not defined by the equation

$$d\gamma = \int^{\cdot} \frac{dH}{\theta}.$$

This equation, arising from *another* definition of entropy, can be demonstrated for reversible changes.

C. It is wrong for *all* irreversible changes, and not only for those where there is exchange of heat in the narrow sense of the word.

D. In an irreversible change in which there is no exchange of heat the entropy increases.

E. It increases, for instance, in the case of a mixture of gases, and the increase can be calculated by the artifice I have discussed at the end of section 3.

F. If the universe is regarded as an isolated system, it can never come back to its original state; for its entropy is always growing, and this entropy being a function of the co-ordinates, would come back to its original value if the universe came back to its original state.

But back in 1896, Boltzmann and the *H* theorem had another critic: Ernst Zermelo (1871–1953), a recent German math PhD who was then working with Max Planck on applied mathematics—though would soon turn to foundations of mathematics and become the "Z" in ZFC set theory. Zermelo's attack on the *H* theorem began with a paper entitled "On a Theorem of Dynamics and the Mechanical Theory of Heat". After explaining Poincaré's recurrence theorem, Zermelo gives some "mathematician-style" conditions (the gas must be in a finite region, must have no infinite energies, etc.), then says that even though there must exist states that would be non-recurrent and could show irreversible behavior, there would necessarily be infinitely more states that "would periodically repeat themselves … with arbitrarily small variations". And, he argues, such repetition would affect macroscopic quantities discernable by our senses. He continues:

> In order to retain the general validity of the Second Law, we therefore would have to assume that just those initial states leading to irreversible processes are realized in nature, their small number notwithstanding, while the other ones, whose probability of existence is higher, mathematically speaking, do not actually occur.

And he concludes that the Poincaré recurrence phenomenon means that:

> ... it is certainly impossible to carry out a mechanical derivation of the Second Law on the basis of the existing theory without specializing the initial states.

Boltzmann responded promptly but quite impatiently:

> I have pointed out particularly often, and as clearly as I possibly could ... that the Second Law is but a principle of probability theory as far as the molecular-theoretic point of view is concerned. ... While the theorem by Poincaré that Zermelo discusses in the beginning of his paper is of course correct, its application to heat theory is not.

Boltzmann talks about the H curve, and first makes rather a mathematician-style point about the order of limits:

> If we first take the number of gas molecules to be infinite, as was clearly done in [my 1896 proof], and only then let the time grow very large, then, in the vast majority of cases, we obtain a curve asymptotically [always close to zero]. Moreover, as can easily be seen, Poincaré's theorem is not applicable in this case. If, however, we take the time [span] to be infinitely great and, in contrast, the number of molecules to be very great but not absolutely infinite, then the H-curve has a different character. It almost always runs very close to [zero], but in rare cases it rises above that, in what we shall call a "hump" ... at which significant deviations from the Maxwell velocity distribution can occur ...

Boltzmann then argues that even if you start "at a hump", you won't stay there, and "over an enormously long period of time" you'll see something infinitely close to "equilibrium behavior". But, he says:

> ... it is [always] possible to reach again a greater hump of the H-curve by further extending the time ... In fact, it is even the case that the original state must return, provided only that we continue to sufficiently extend the time...

He continues:

> Mr. Zermelo is therefore right in claiming that, mathematically speaking, the motion is periodic. He has by no means succeeded, however, in refuting my theorems, which, in fact, are entirely consistent with this periodicity.

After giving arguments about the probabilistic character of his results, and (as we would now say it) the fact that a 1D random walk is certain to repeatedly return to the origin, Boltzmann says that:

> ... we must not conclude that the mechanical approach has to be modified in any way. This conclusion would be justified only if the approach had a consequence that runs contrary to experience. But this would be the case only if Mr. Zermelo were able to prove that the duration of the period within which the old state of the gas must recur in accordance with Poincaré's theorem has an observable length...

He goes on to imagine "a trillion tiny spheres, each with a [certain initial velocity] … in the one corner of a box" (and by "trillion" he means million million million, or today's quintillion) and then says that "after a short time the spheres will be distributed fairly evenly in the box", but the period for a "Poincaré recurrence" in which they all will return to their original corner is "so great that nobody can live to see it happen". And to make this point more forcefully, Boltzmann has an appendix in which he tries to get an actual approximation to the recurrence time, concluding that its numerical value "has many trillions of digits".

He concludes:

> If we consider heat as a motion of molecules that occurs in accordance with the general equations of mechanics and assume that the arrangement of bodies that we perceive is currently in a highly improbable state, then a theorem follows that is in agreement with the Second Law for all phenomena so far observed.

> Of course, this theorem can no longer hold once we observe bodies of so small a scale that they only contain a few molecules. Since, however, we do not have at hand any experimental results on the behavior of bodies so small, this assumption does not run counter to previous experience. In fact, certain experiments conducted on very small bodies in gases seem rather to support the assumption, although we are still far from being able to assert its correctness on the basis of experimental proof.

But then he gives an important caveat—with a small philosophical flourish:

> Of course, we cannot expect natural science to answer the question as to why the bodies surrounding us currently exist in a highly improbable state, just as we cannot expect it to answer the question as to why there are any phenomena at all and why they adhere to certain given principles.

Unsurprisingly—particularly in view of his future efforts in the foundations of mathematics— Zermelo is unconvinced by all of this. And six months later he replies again in print. He admits that a full Poincaré recurrence might take astronomically long, but notes that (where, by "physical state", he means one that we perceive):

> … we are after all always concerned only with the "physical state", which can be realized by many different combinations, and hence can recur much sooner.

Zermelo zeroes in on many of the weaknesses in Boltzmann's arguments, saying that the thing he particularly "contests … is the analogy that is supposed to exist between the properties of the H curve and the Second Law". He claims that irreversibility cannot be explained from "mechanical suppositions" without "new physical assumptions"—and in particular criteria for choosing appropriate initial states. He ends by saying that:

> From the great successes of the kinetic theory of gases in explaining the relationships among states we must not deduce its … applicability also to temporal processes. … [For in this case I am] convinced that it necessarily fails in the absence of entirely new assumptions.

Boltzmann replies again—starting off with the strangely weak argument:

> The Second Law receives a mechanical explanation by virtue of the assumption, which is of course unprovable, that the universe, when considered as a mechanical system, or at least a very extensive part thereof surrounding us, started out in a highly improbable state and still is in such a state.

And, yes, there's clearly something missing in the understanding of the Second Law. And even as Zermelo pushes for formal mathematician-style clarity, Boltzmann responds with physicist-style "reasonable arguments". There's lots of rhetoric:

> The applicability of the calculus of probabilities to a particular case can of course never be proved with precision. If 100 out of 100,000 objects of a particular sort are consumed by fire per year, then we cannot infer with certainty that this will also be the case next year. On the contrary! If the same conditions continue to obtain for 10^{10} years, then it will often be the case during this period that the 100,000 objects are all consumed by fire at once on a single day, and even that not a single object suffers damage over the course of an entire year. Nevertheless, every insurance company places its faith in the calculus of probabilities.

Or, in justification of the idea that we live in a highly improbable "low-entropy" part of the universe:

> I refuse to grant the objection that a mental picture requiring so great a number of dead parts of the universe for the explanation of so small a number of animated parts is wasteful, and hence inexpedient. I still vividly remember someone who adamantly refused to believe that the Sun's distance from the Earth is 20 million miles on the ground that it would simply be foolish to assume so vast a space only containing luminiferous aether alongside so small a space filled with life.

Curiously—given his apparent reliance on "commonsense" arguments—Boltzmann also says:

> I myself have repeatedly cautioned against placing excessive trust in the extension of our mental pictures beyond experience and issued reminders that the pictures of contemporary mechanics, and in particular the conception of the smallest particles of bodies as material points, will turn out to be provisional.

In other words, we don't know that we can think of atoms (even if they exist at all) as points, and we can't really expect our everyday intuition to tell us about how they work. Which presumably means that we need some kind of solid, "formal" argument if we're going to explain the Second Law.

Zermelo didn't respond again, and moved on to other topics. But Boltzmann wrote one more paper in 1897 about "A Mechanical Theorem of Poincaré" ending with two more why-it-doesn't-apply-in-practice arguments:

> Poincaré's theorem is of course never applicable to terrestrial bodies which we can hold in our hands as none of them is entirely closed. Nor it is applicable to an entirely closed gas of the sort considered by the kinetic theory if first the number of molecules and only then the quotients of the intervals between two neighboring collisions in the observation time is allowed to become infinite.

Ensembles, and an Effort to Make Things Rigorous

Boltzmann—and Maxwell before him—had introduced the idea of using probability theory to discuss the emergence of thermodynamics and potentially the Second Law. But it wasn't until around 1900—with the work of J. Willard Gibbs (1839–1903)—that a principled mathematical framework for thinking about this developed. And while we can now see that this framework distracts in some ways from several of the key issues in understanding the foundations of the Second Law, it's been important in framing the discussion of what the Second Law really says—as well as being central in defining the foundations for much of what's been done over the past century or so under the banner of "statistical mechanics".

Gibbs seems to have first gotten involved with thermodynamics around 1870. He'd finished his PhD at Yale on the geometry of gears in 1863—getting the first engineering PhD awarded in the US. After traveling in Europe and interacting with various leading mathematicians and physicists, he came back to Yale (where he stayed for the remaining 34 years of his life) and in 1871 became professor of mathematical physics there.

His first papers (published in 1873 when he was already 34 years old) were in a sense based on taking seriously the formalism of equilibrium thermodynamics defined by Clausius and Maxwell—treating entropy and internal energy, just like pressure, volume and temperature, as variables that defined properties of materials (and notably whether they were solids, liquids or gases). Gibbs's main idea was to "geometrize" this setup, and make it essentially a story of multivariate calculus:

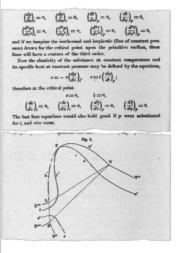

Unlike the European developers of thermodynamics, Gibbs didn't interact deeply with other scientists—with the possible exception of Maxwell, who (a few years before his death in 1879) made a 3D version of Gibbs's thermodynamic surface out of clay—and supplemented his 2D thermodynamic diagrams after the first edition of his textbook *Theory of Heat* with renderings of 3D versions:

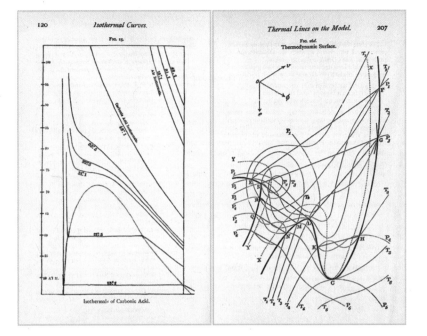

Three years later, Gibbs began publishing what would be a 300-page work defining what has become the standard formalism for equilibrium chemical thermodynamics. He began with a quote from Clausius:

V. On the Equilibrium of Heterogeneous Substances.
By J. Willard Gibbs.

"Die Energie der Welt ist constant.
Die Entropie der Welt strebt einem Maximum zu."
Clausius.*

THE comprehension of the laws which govern any material system is greatly facilitated by considering the energy and entropy of the system in the various states of which it is capable. As the difference of the values of the energy for any two states represents the combined amount of work and heat received or yielded by the system when it is brought from one state to the other, and the difference of entropy is the limit of all the possible values of the integral $\int \frac{dQ}{t}$, (dQ denoting the element of the heat received from external sources, and t the temperature of the part of the system receiving it,) the

In the years that followed, Gibbs's work—stimulated by Maxwell—mostly concentrated on electrodynamics, and later quaternions and vector analysis. But Gibbs published a few more small papers on thermodynamics—always in effect taking equilibrium (and the Second Law) for granted.

In 1882—a certain Henry Eddy (1844–1921) (who in 1879 had written a book on thermodynamics, and in 1890 would become president of the University of Cincinnati), claimed that "radiant heat" could be used to violate the Second Law:

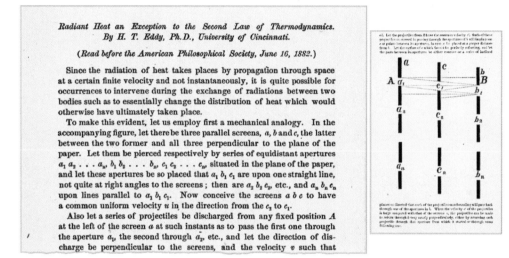

Gibbs soon published a 2-page rebuttal (in the 6th-ever issue of *Science* magazine):

ON AN ALLEGED EXCEPTION TO THE SECOND LAW OF THERMODYNAMICS.

According to the received doctrine of radiation, heat is transmitted with the same intensity in all directions and at all points within any space which is void of ponderable matter and entirely surrounded by stationary bodies of the same temperature. We may apply this principle to the arrangement recently proposed by Prof. H. T. Eddy[1] for transferring heat from a colder body A to a warmer B without expenditure of work.

Then in 1889 Clausius died, and Gibbs wrote an obituary—praising Clausius but making it clear he didn't think the kinetic theory of gases was a solved problem:

The origin of the kinetic theory of gases is lost in remote antiquity, and its completion the most sanguine cannot hope to see. But a single generation has seen it advance from the stage of vague surmises to an extensive and well established body of doctrine. This is mainly the work of three men, Clausius, Maxwell, and Boltzmann, of whom Clausius was the earliest in the field, and has been called by Maxwell the principal founder of the science.[*] We may regard his

That same year Gibbs announced a short course that he would teach at Yale on "The *a priori* Deduction of Thermodynamic Principles from the Theory of Probabilities". After a decade of work, this evolved into Gibbs's last publication—an original and elegant book that's largely defined how the Second Law has been thought about ever since:

ELEMENTARY PRINCIPLES

IN

STATISTICAL MECHANICS

DEVELOPED WITH ESPECIAL REFERENCE TO

THE RATIONAL FOUNDATION OF
THERMODYNAMICS

BY

J. WILLARD GIBBS
Professor of Mathematical Physics in Yale University

NEW YORK: CHARLES SCRIBNER'S SONS
LONDON: EDWARD ARNOLD
1902

The book begins by explaining that mechanics is about studying the time evolution of single systems:

THE usual point of view in the study of mechanics is that where the attention is mainly directed to the changes which take place in the course of time in a given system. The principal problem is the determination of the condition of the system with respect to configuration and velocities at any required time, when its condition in these respects has been given for some one time, and the fundamental equations are those which express the changes continually taking place in the system. Inquiries of this kind are often simplified by

But Gibbs says he is going to do something different: he is going to look at what he'll call an ensemble of systems, and see how the distribution of their characteristics changes over time:

> For some purposes, however, it is desirable to take a broader view of the subject. We may imagine a great number of systems of the same nature, but differing in the configurations and velocities which they have at a given instant, and differing not merely infinitesimally, but it may be so as to embrace every conceivable combination of configuration and velocities. And here we may set the problem, not to follow a particular system through its succession of configurations, but to determine how the whole number of systems will be distributed among the various conceivable configurations and velocities at any required time, when the distribution has been given for some one time. The fundamental equation

He explains that these "inquiries" originally arose in connection with deriving the laws of thermodynamics:

> Such inquiries have been called by Maxwell *statistical*. They belong to a branch of mechanics which owes its origin to the desire to explain the laws of thermodynamics on mechanical principles, and of which Clausius, Maxwell, and Boltzmann are to be regarded as the principal founders. The first inquiries in this field were indeed somewhat narrower in their scope than that which has been mentioned, being applied to the particles of a system, rather than to independent systems.

But he argues that this area—which he's calling statistical mechanics—is worth investigating even independent of its connection to thermodynamics:

> But although, as a matter of history, statistical mechanics owes its origin to investigations in thermodynamics, it seems eminently worthy of an independent development, both on account of the elegance and simplicity of its principles, and because it yields new results and places old truths in a new light in departments quite outside of thermodynamics. More-

Still, he expects this effort will be relevant to the foundations of thermodynamics:

> light in departments quite outside of thermodynamics. Moreover, the separate study of this branch of mechanics seems to afford the best foundation for the study of rational thermodynamics and molecular mechanics.

He immediately then goes on to what he'll claim is the way to think about the relation of "observed thermodynamics" to his exact statistical mechanics:

> The laws of thermodynamics, as empirically determined, express the approximate and probable behavior of systems of a great number of particles, or, more precisely, they express the laws of mechanics for such systems as they appear to beings who have not the fineness of perception to enable them to appreciate quantities of the order of magnitude of those which relate to single particles, and who cannot repeat their experiments often enough to obtain any but the most probable results. The laws of statistical mechanics apply to

Soon he makes the interesting—if, in the light of history, very overly optimistic—claim that "the laws of thermodynamics may be easily obtained from the principles of statistical mechanics":

> that the effect of the quantities and circumstances neglected will be negligible in the result. The laws of thermodynamics may be easily obtained from the principles of statistical mechanics, of which they are the incomplete expression, but they make a somewhat blind guide in our search for those laws. This is perhaps the principal cause of the slow progress of rational thermodynamics, as contrasted with the rapid deduction of the consequences of its laws as empirically established. To this must be added that the rational foundation

At first the text of the book reads very much like a typical mathematical work on mechanics:

> ## CHAPTER I.
>
> ### GENERAL NOTIONS. THE PRINCIPLE OF CONSERVATION OF EXTENSION-IN-PHASE.
>
> WE shall use Hamilton's form of the equations of motion for a system of n degrees of freedom, writing $q_1, \ldots q_n$ for the (generalized) coördinates, $\dot{q}_1, \ldots \dot{q}_n$ for the (generalized) velocities, and
>
> $$F_1 \, dq_1 + F_2 \, dq_2 \ldots + F_n \, dq_n \qquad (1)$$
>
> for the moment of the forces. We shall call the quantities $F_1, \ldots F_n$ the (generalized) forces, and the quantities $p_1 \ldots p_n$, defined by the equations
>
> $$p_1 = \frac{d\epsilon_p}{d\dot{q}_1}, \quad p_2 = \frac{d\epsilon_p}{d\dot{q}_2}, \quad \text{etc.,} \qquad (2)$$
>
> where ϵ_p denotes the kinetic energy of the system, the (generalized) momenta. The kinetic energy is here regarded as a function of the velocities and coördinates. We shall usually

But soon it's "going statistical", talking about the "density" of systems in "phase" (i.e. with respect to the variables defining the configuration of the system). And a few pages in, he's proving the fundamental result that the density of "phase fluid" satisfies a continuity equation (which we'd now call the Liouville equation):

> *In an ensemble of mechanical systems identical in nature and subject to forces determined by identical laws, but distributed in phase in any continuous manner, the density-in-phase is constant in time for the varying phases of a moving system; provided, that the forces of a system are functions of its co-ordinates, either alone or with the time.**
>
> This may be called the principle of *conservation of density-in-phase*. It may also be written
>
> $$\left(\frac{dD}{dt}\right)_{a,\ldots\mathbf{h}} = 0, \tag{22}$$

It's all quite elegant, and all very rooted in the calculus-based mathematics of its time. He's thinking about a collection of instances of a system. But while with our modern computational paradigm we'd readily be able to talk about a discrete list of instances, with his calculus-based approach he has to consider a continuous collection of instances—whose treatment inevitably seems more abstract and less explicit.

He soon makes contact with the "theory of errors", discussing in effect how probability distributions over the space of possible states evolve. But what probability distributions should one consider? By chapter 4, he's looking at what he calls (and is still called) the "canonical distribution":

> ## CHAPTER IV.
>
> ### ON THE DISTRIBUTION IN PHASE CALLED CANONICAL, IN WHICH THE INDEX OF PROBABILITY IS A LINEAR FUNCTION OF THE ENERGY.
>
> LET us now give our attention to the statistical equilibrium of ensembles of conservation systems, especially to those cases and properties which promise to throw light on the phenomena of thermodynamics.

He gives a now-classic definition for the probability as a function of energy ϵ:

or

$$P = e^{\frac{\psi - \epsilon}{\Theta}}, \qquad (91)$$

where Θ and ψ are constants, and Θ positive, seems to represent the most simple case conceivable, since it has the property that when the system consists of parts with separate energies, the laws of the distribution in phase of the separate parts are of the same nature,— a property which enormously simplifies the discussion, and is the foundation of extremely important relations to thermodynamics. The case is not rendered less

He observes that this distribution combines nicely when independent parts of a system are brought together, and soon he's noting that:

The modulus Θ has properties analogous to those of temperature in thermodynamics. Let the system A be defined as

But so far he's careful to just talk about how things are "analogous", without committing to a true connection:

How far, or in what sense, the similarity of these equations constitutes any demonstration of the thermodynamic equations, or accounts for the behavior of material systems, as described in the theorems of thermodynamics, is a question of which we shall postpone the consideration until we have further investigated the properties of an ensemble of systems distributed in phase according to the law which we are considering. The analogies which have been pointed out will at least supply the motive for this investigation, which will naturally commence with the determination of the average values in the ensemble of the most important quantities relating to the systems, and to the distribution of the ensemble with respect to the different values of these quantities.

More than halfway through the book he's defined certain properties of his probability distributions that "may ... correspond to the thermodynamic notions of entropy and temperature":

Now we have already noticed a certain correspondence between the quantities Θ and $\bar{\eta}$ and those which in thermodynamics are called temperature and entropy. The property just demonstrated, with those expressed by equation (336), therefore suggests that the quantities ϕ and $d\epsilon/d\phi$ may also correspond to the thermodynamic notions of entropy and temperature. We leave the discussion of this point to a subsequent chapter, and only mention it here to justify the somewhat detailed investigation of the relations of these quantities.

Next he's on to the concept of a "microcanonical ensemble" that includes only states of a given energy. For him—with his continuum-based setup—this is a slightly elaborate thing to define; in our modern computational framework it actually becomes more straightforward than his "canonical ensemble". Or, as he already says:

> From a certain point of view, the microcanonical distribution may seem more simple than the canonical, and it has perhaps been more studied, and been regarded as more closely related to the fundamental notions of thermodynamics. To this last point we shall return in a subsequent chapter. It is sufficient here to remark that analytically the canonical distribution is much more manageable than the microcanonical.

But what about the Second Law? Now he's getting a little closer:

> ## CHAPTER XI.
>
> ### MAXIMUM AND MINIMUM PROPERTIES OF VARIOUS DISTRIBUTIONS IN PHASE.

When he says "index of probability" he's talking about the log of a probability in his ensemble, so this result is about the fact that this quantity is extremized when all the elements of the ensemble have equal probability:

> original ensembles are all identical.
>   *Theorem IX.* A uniform distribution of a given number of systems within given limits of phase gives a less average index of probability of phase than any other distribution.

Soon he's discussing whether he can use his index as a way—like Boltzmann tried to do with his version of entropy—to measure deviations from "statistical equilibrium":

> It would seem, therefore, that we might find a sort of measure of the deviation of an ensemble from statistical equilibrium in the excess of the average index above the minimum which is consistent with the condition of the invariability of the distribution with respect to the constant functions of phase. But we have seen that the index of probability is constant in time for each system of the ensemble. The average index is therefore constant, and we find by this method no approach toward statistical equilibrium in the course of time.

But now Gibbs has hit one of the classic gotchas of his approach: if you look in perfect detail at the evolution of an ensemble of systems, there'll never be a change in the value of his index—essentially because of the overall conservation of probability. Gibbs brings in what amounts to a commonsense physics argument to handle this. He says to consider putting "coloring matter" in a liquid that one stirs. And then he says that even though the liquid (like his phase fluid) is microscopically conserved, the coloring matter will still end up being "uniformly mixed" in the liquid:

distributed with a variable density. If we give the liquid any motion whatever, subject only to the hydrodynamic law of incompressibility, — it may be a steady flux, or it may vary with the time, — the density of the coloring matter at any same point of the liquid will be unchanged, and the average square of this density will therefore be unchanged. Yet no fact is more familiar to us than that stirring tends to bring a liquid to a state of uniform mixture, or uniform densities of its components, which is characterized by minimum values of the average squares of these densities. It is quite true that

He talks about how the conclusion about whether mixing happens in effect depends on what order one takes limits in. And while he doesn't put it quite this way, he's essentially realized that there's a competition between the system "mixing things up more and more finely" and the observer being able to track finer and finer details. He realizes, though, that not all systems will show this kind of mixing behavior, noting for example that there are mechanical systems that'll just keep going in simple cycles forever.

He doesn't really resolve the question of why "practical systems" should show mixing, more or less ending with a statement that even though his underlying mechanical systems are reversible, it's somehow "in practice" difficult to go back:

But while the distinction of prior and subsequent events may be immaterial with respect to mathematical fictions, it is quite otherwise with respect to the events of the real world. It should not be forgotten, when our ensembles are chosen to illustrate the probabilities of events in the real world, that

while the probabilities of subsequent events may often be determined from the probabilities of prior events, it is rarely the case that probabilities of prior events can be determined from those of subsequent events, for we are rarely justified in excluding the consideration of the antecedent probability of the prior events.

Despite things like this, Gibbs appears to have been keen to keep the majority of his book "purely mathematical", in effect proving theorems that necessarily followed from the setup he had given. But in the penultimate chapter of the book he makes what he seems to have viewed as a less-than-satisfactory attempt to connect what he's done with "real thermodynamics". He doesn't really commit to the connection, though, characterizing it more as an "analogy":

CHAPTER XIV.

DISCUSSION OF THERMODYNAMIC ANALOGIES.

If we wish to find in rational mechanics an *a priori* foundation for the principles of thermodynamics, we must seek mechanical definitions of temperature and entropy. The

But he soon starts to be pretty clear that he actually wants to prove the Second Law:

> We have also to enunciate in mechanical terms, and to prove, what we call the tendency of heat to pass from a system of higher temperature to one of lower, and to show that this tendency vanishes with respect to systems of the same temperature.

He quickly backs off a little, in effect bringing in the observer to soften the requirements:

> At least, we have to show by *a priori* reasoning that for such systems as the material bodies which nature presents to us, these relations hold with such approximation that they are sensibly true for human faculties of observation. This indeed is all that is really necessary to establish the science of thermodynamics on an *a priori* basis. Yet we will naturally desire to find the exact expression of those principles of which the laws of thermodynamics are the approximate expression.

But then he fires his best shot. He says that the quantities he's defined in connection with his canonical ensemble satisfy the same equations as Clausius originally set up for temperature and entropy:

> Now we have defined what we have called the *modulus* (Θ) of an ensemble of systems canonically distributed in phase, and what we have called the index of probability (η) of any phase in such an ensemble. It has been shown that between

> the modulus (Θ), the external coördinates (a_1, etc.), and the average values in the ensemble of the energy (ϵ), the index of probability (η), and the external forces (A_1, etc.) exerted by the systems, the following differential equation will hold:
> $$d\bar{\epsilon} = -\Theta\, d\bar{\eta} - \bar{A_1}\, da_1 - \bar{A_2}\, da_2 - \text{etc.} \qquad (483)$$
> This equation, if we neglect the sign of averages, is identical in form with the thermodynamic equation (482), the modulus (Θ) corresponding to temperature, and the index of probability of phase with its sign reversed corresponding to entropy.*

He adds that fluctuations (or "anomalies", as he calls them) become imperceptible in the limit of a large system:

> We have also shown that the average square of the anomalies of ϵ, that is, of the deviations of the individual values from the average, is in general of the same order of magnitude as the reciprocal of the number of degrees of freedom, and therefore to human observation the individual values are indistinguishable from the average values when the number of degrees of freedom is very great.† In this case also the anomalies of η

But in physical reality, why should one have a whole collection of systems as in the canonical ensemble? Gibbs suggests it would be more natural to look at the microcanonical ensemble—and in fact to look at a "time ensemble", i.e. an averaging over time rather than an averaging over different possible states of the system:

> The definitions and propositions which we have been considering relate essentially to what we have called a canonical ensemble of systems. This may appear a less natural and simple conception than what we have called a microcanonical ensemble of systems, in which all have the same energy, and which in many cases represents simply the *time-ensemble*, or ensemble of phases through which a single system passes in the course of time.

Gibbs has proved some results (e.g. related to the virial theorem) about the relation between time and ensemble averages. But as the future of the subject amply demonstrates, they're not nearly strong enough to establish any general equivalence. Still, Gibbs presses on.

In the end, though, as he himself recognized, things weren't solved—and certainly the canonical ensemble wasn't the whole story:

> It is certainly in the quantities relating to a canonical ensemble, $\bar{\epsilon}$, Θ, $\bar{\eta}$, $\bar{A}_1$, etc., a_1, etc., that we find the most complete correspondence with the quantities of the thermodynamic equation (482). Yet the conception itself of the canonical ensemble may seem to some artificial, and hardly germane to a natural exposition of the subject; and the quantities ϵ, $\frac{d\epsilon}{d \log V}$, $\log V$, $\overline{A_1}_\epsilon$, etc., a_1, etc., or ϵ, $\frac{d\epsilon}{d\phi}$, ϕ, $\left(\frac{d\epsilon}{da_1}\right)_{\phi, a}$, etc., a_1, etc., which are closely related to ensembles of constant energy, and to average and most probable values in such ensembles, and most of which are defined without reference to any ensemble, may appear the most natural analogues of the thermodynamic quantities.

He discusses the tradeoff between having a canonical ensemble "heat bath" of a known temperature, and having a microcanonical ensemble with known energy. At one point he admits that it might be better to consider the time evolution of a single state, but basically decides that—at least in his continuous-probability-distribution-based formalism—he can't really set this up:

> arbitrary, will represent better than any one time-ensemble the effect of the bath. Indeed a single time-ensemble, when it is not also a microcanonical ensemble, is too ill-defined a notion to serve the purposes of a general discussion. We will therefore direct our attention, when we suppose the body placed in a bath, to the microcanonical ensemble of phases thus obtained.

Gibbs definitely encourages the idea that his "statistical mechanics" has successfully "derived" thermodynamics. But he's ultimately quite careful and circumspect in what he actually says. He mentions the Second Law only once in his whole book—and then only to note that he can get the same "mathematical expression" from his canonical ensemble as Clausius's form of the Second Law. He doesn't mention Boltzmann's H theorem anywhere in the book, and—apart from one footnote concerning "difficulties long recognized by physicists"—he mentions only Boltzmann's work on theoretical mechanics.

One can view the main achievement of Gibbs's book as having been to define a framework in which precise results about the statistical properties of collections of systems could be defined and in some cases derived. Within the mathematics and other formalism of the time, such ensemble results represented in a sense a distinctly "higher-order" description of things. Within our current computational paradigm, though, there's much less of a distinction to be made: whether one's looking at a single path of evolution, or a whole collection, one's ultimately still just dealing with a computation. And that makes it clearer that— ensembles or not—one's thrown back into the same kinds of issues about the origin of the Second Law. But even so, Gibbs provided a language in which to talk with some clarity about many of the things that come up.

Maxwell's Demon

In late 1867 Peter Tait (1831–1901)—a childhood friend of Maxwell's who was by then a professor of "natural philosophy" in Edinburgh—was finishing his sixth book. It was entitled *Sketch of Thermodynamics* and gave a brief, historically oriented and not particularly conceptual outline of what was then known about thermodynamics. He sent a draft to Maxwell, who responded with a fairly long letter:

The letter begins:

> I do not know in a controversial manner the history of thermodynamics ... [and] I could make no assertions about the priority of authors ...

> Any contributions I could make ... [involve] picking holes here and there to ensure strength and stability.

Then he continues (with "ΘΔcs" being his whimsical Greekified rendering of the word "thermodynamics"):

> To pick a hole—say in the 2nd law of ΘΔcs, that if two things are in contact the hotter cannot take heat from the colder without external agency.

Now let A and B be two vessels divided by a diaphragm ... Now conceive a finite being who knows the paths and velocities of all the molecules by simple inspection but who can do no work except open and close a hole in the diaphragm by means of a slide without mass. Let him ... observe the molecules in A and when he sees one coming ... whose velocity is less than the mean [velocity] of the molecules in B let him open the hole and let it go into B [and vice versa].

Then the number of molecules in A and B are the same as at first, but the energy in A is increased and that in B diminished, that is, the hot system has got hotter and the cold colder and yet no work has been done, only the intelligence of a very observant and neat-fingered being has been employed.

Or in short [we can] ... restore a uniformly hot system to unequal temperatures.... Only we can't, not being clever enough.

And so it was that the idea of "Maxwell's demon" was launched. Tait must at some point have shown Maxwell's letter to Kelvin, who wrote on it:

Very good. Another way is to reverse the motion of every particle of the Universe and to preside over the unstable motion thus produced.

But the first place Maxwell's demon idea appeared in print was in Maxwell's 1871 textbook *Theory of Heat*:

THEORY OF HEAT.

BY

J. CLERK MAXWELL, M.A.

LL.D. Edin., F.R.SS. L. & E.

Professor of Experimental Physics in the University of Cambridge.

LONDON:

LONGMANS, GREEN, AND CO.

1871.

CONTENTS.

Much of the book is devoted to what was by then quite traditional, experimentally oriented thermodynamics. But Maxwell included one final chapter:

CHAPTER XXII.

MOLECULAR THEORY OF THE CONSTITUTION OF BODIES.

Even in 1871, after all his work on kinetic theory, Maxwell is quite circumspect in his discussion of molecules:

> We have now arrived at the conception of a body as consisting of a great many small parts, each of which is in motion. We shall call any one of these parts a molecule of the substance. A molecule may therefore be defined as a small mass of matter the parts of which do not part company during the excursions which the molecule makes when the body to which it belongs is hot.
>
> The doctrine that visible bodies consist of a determinate number of molecules is called the molecular theory of matter. The opposite doctrine is that, however small the parts may be into which we divide a body, each part retains all the properties of the substance. This is the theory of the infinite divisibility of bodies. We do not assert that there is an absolute limit to the divisibility of matter : what we assert is, that after we have divided a body into a certain finite number of constituent parts called molecules, then any further division of these molecules will deprive them of the properties which give rise to the phenomena observed in the substance.
>
> The opinion that the observed properties of visible bodies apparently at rest are due to the action of invisible molecules in rapid motion is to be found in Lucretius.

But Maxwell's textbook goes through a series of standard kinetic theory results, much as a modern textbook would. The second-to-last section in the whole book sounds a warning, however:

LIMITATION OF THE SECOND LAW OF THERMODYNAMICS.

Before I conclude, I wish to direct attention to an aspect of the molecular theory which deserves consideration.

One of the best established facts in thermodynamics is that it is impossible in a system enclosed in an envelope which permits neither change of volume nor passage of heat, and in which both the temperature and the pressure are everywhere the same, to produce any inequality of temperature or of pressure without the expenditure of work. This is the second law of thermodynamics, and it is undoubtedly true as long as we can deal with bodies only in mass, and have no power of perceiving or handling the separate molecules of which they are made up. But if we conceive a being whose faculties are so sharpened that he can follow every molecule in its course, such a being, whose attributes are still as essentially finite as our own, would be able to do what is at present impossible to us. For we have seen that the molecules in a vessel full of air at uniform temperature are moving with velocities by no means uniform, though the mean velocity of any great number of them, arbitrarily selected, is almost exactly uniform. Now let us suppose that such a vessel is divided into two portions, A and B, by a division in which there is a small hole, and that a being, who can see the individual molecules, opens and closes this hole, so as to allow only the swifter molecules to pass from A to B, and only the slower ones to pass from B to A. He will thus, without expenditure of work, raise the temperature of B and lower that of A, in contradiction to the second law of thermodynamics.

Interestingly, Maxwell continues, somewhat in anticipation of what Gibbs will say 30 years later:

This is only one of the instances in which conclusions which we have drawn from our experience of bodies consisting of an immense number of molecules may be found not to be applicable to the more delicate observations and experiments which we may suppose made by one who can perceive and handle the individual molecules which we deal with only in large masses.

In dealing with masses of matter, while we do not perceive the individual molecules, we are compelled to adopt what I have described as the statistical method of calculation, and to abandon the strict dynamical method, in which we follow every motion by the calculus.

But then there's a reminder that this is being written in 1871, several decades before any clear observation of molecules was made. Maxwell says:

> I do not think, however, that the perfect identity which we observe between different portions of the same kind of matter can be explained on the statistical principle of the stability of the averages of large numbers of quantities each of which may differ from the mean. For if of the

In other words, if there are water molecules, there must be something other than a law of averages that makes them all appear the same. And, yes, it's now treated as a fundamental fact of physics that, for example, all electrons have exactly—not just statistically—the same properties such as mass and charge. But back in 1871 it was much less clear what characteristics molecules—if they existed as real entities at all—might have.

Maxwell included one last section in his book that to us today might seem quite wild:

> ### NATURE AND ORIGIN OF MOLECULES.
>
> We have thus been led by our study of visible things to a theory that they are made up of a finite number of parts or molecules, each of which has a definite mass, and possesses other properties. The molecules of the same substance are all exactly alike, but different from those of other substances. There is not a regular gradation in the mass of molecules from that of hydrogen, which is the least of those known to us, to that of bismuth ; but they all fall into a limited number of classes or species, the individuals of each species being exactly similar to each other, and no inter-mediate links are found to connect one species with another by a uniform gradation.
>
> We are here reminded of certain speculations concerning the relations between the species of living things. We find that in these also the individuals are naturally grouped into species, and that intermediate links between the species are wanting. But in each species variations occur, and there is a perpetual generation and destruction of the individuals of which the species consist.
>
> Hence it is possible to frame a theory to account for the present state of things by means of generation, variation, and discriminative destruction.
>
> In the case of the molecules, however, each individual is permanent ; there is no generation or destruction, and no variation, or rather no difference, between the individuals of each species.
>
> Hence the kind of speculation with which we have become so familiar under the name of theories of evolution is quite inapplicable to the case of molecules.

In other words, aware of Darwin's (1809–1882) 1859 *Origin of Species*, he's considering a kind of "speciation" of molecules, along the lines of the discrete species observed in biology. But then he notes that unlike biological organisms, molecules are "permanent", so their "selection" must come from some kind of pure separation process:

> In speculating on the cause of this equality we are debarred from imagining any cause of equalization, on account of the immutability of each individual molecule. It is difficult, on the other hand, to conceive of selection and elimination of intermediate varieties, for where can these eliminated molecules have gone to if, as we have reason to believe, the hydrogen, &c., of the fixed stars is composed of molecules identical in

> all respects with our own? The time required to eliminate from the whole of the visible universe every molecule whose mass differs from that of some one of our so-called elements, by processes similar to Graham's method of dialysis, which is the only method we can conceive of at present, would exceed the utmost limits ever demanded by evolutionists as many times as these exceed the period of vibration of a molecule.

And at the very end he suggests that if molecules really are all identical, that suggests a level of fundamental order in the world that we might even be able to flow through to "exact principles of distributive justice" (presumably for people rather than molecules):

> But if we suppose the molecules to be made at all, or if we suppose them to consist of something previously made, why should we expect any irregularity to exist among them? If they are, as we believe, the only material things which still remain in the precise condition in which they first began to exist, why should we not rather look for some indication of that spirit of order, our scientific confidence in which is never shaken by the difficulty which we experience in tracing it in the complex arrangements of visible things, and of which our moral estimation is shown in all our attempts to think and speak the truth, and to ascertain the exact principles of distributive justice?

Maxwell has described rather clearly his idea of demons. But the actual name "demon" first appears in print in a paper by Kelvin in 1874:

KINETIC THEORY OF THE DISSIPATION OF ENERGY

IN abstract dynamics an instantaneous reversal of the motion of every moving particle of a system causes the system to move backwards, each particle of it along its old path, and at the same speed as before when again in the same position—that is to say, in mathematical language, any solution remains a solution when t is changed into $-t$. In physical dynamics, this simple and

This process of diffusion could be perfectly prevented by an army of Maxwell's "intelligent demons"* stationed at the surface, or interface as we may call it with Prof. James Thomson, separating the hot from the cold part of the bar.

* The definition of a "demon." according to the use of this word by Maxwell, is an intelligent being endowed with free will, and fine enough tactile and perceptive organisation to give him the faculty of observing and influencing individual molecules of matter

It's a British paper, so—in a nod to future nanomachinery—it's talking about (molecular) cricket bats:

Now, suppose the weapon of the ideal army to be a club,

or, as it were, a molecular cricket-bat ; and suppose for convenience the mass of each demon with his weapon to be several times greater than that of a molecule. Every time he strikes a molecule he is to send it away with the same energy as it had immediately before. Each demon is to keep as nearly as possible to a certain station, making only such excursions from it as the execution of his orders requires. He is to experience no forces except such as result from collisions with molecules, and mutual forces between parts of his own mass, including his weapon : thus his voluntary movements cannot influence the position of his centre of gravity, otherwise than by producing collision with molecules.

The whole interface between hot and cold is to be divided into small areas, each allotted to a single demon. The duty of each demon is to guard his allotment, turning molecules back or allowing them to pass through from either side, according to certain definite orders. First,

Kelvin's paper—like his note written on Maxwell's letter—imagines that the demons don't just "sort" molecules; they actually reverse their velocities, thus in effect anticipating Loschmidt's 1876 "reversibility objection" to Boltzmann's H theorem.

In an undated note, Maxwell discusses demons, attributing the name to Kelvin—and then starts considering the "physicalization" of demons, simplifying what they need to do:

> Concerning Demons.
>
> I. Who gave them this name? Thomson.
>
> 2. What were they by nature? Very small BUT lively beings incapable of doing work but able to open and shut valves which move without friction or inertia.
>
> 3. What was their chief end? To show that the 2nd Law of Thermodynamics has only a statistical certainty.
>
> 4. Is the production of an inequality of temperature their only occupation? No, for less intelligent demons can produce a difference in pressure as well as temperature by merely allowing all particles going in one direction while stopping all those going the other way. This reduces the demon to a valve. As such value him. Call him no more a demon but a valve like that of the hydraulic ram, suppose.

It didn't take long for Maxwell's demon to become something of a fixture in expositions of thermodynamics, even if it wasn't clear how it connected to other things people were saying about thermodynamics. And in 1879, for example, Kelvin gave a talk all about Maxwell's "sorting demon" (like other British people of the time he referred to Maxwell as "Clerk Maxwell"):

87. THE SORTING DEMON OF MAXWELL.

[Abstract of Royal Institution Lecture, Feb. 28, 1879, *Roy. Institution Proc.* Vol. IX. p. 113. Reprinted in *Popular Lectures and Addresses*, Vol. I. pp. 137—141.]

THE word "demon," which originally in Greek meant a supernatural being, has never been properly used to signify a real or ideal personification of malignity.

Clerk Maxwell's "demon" is a creature of imagination having certain perfectly well defined powers of action, purely mechanical in their character, invented to help us to understand the "Dissipation of Energy" in nature.

He is a being with no preternatural qualities, and differs from real living animals only in extreme smallness and agility. He can at pleasure stop, or strike, or push, or pull any single atom of matter, and so moderate its natural course of motion. Endowed ideally with arms and hands and fingers—two hands and ten fingers suffice—he can do as much for atoms as a pianoforte player can do for the keys of the piano—just a little more, he can push or pull each atom *in any direction*.

Kelvin describes—without much commentary, and without mentioning the Second Law—some of the feats of which the demon would be capable. But he adds:

> The conception of the "sorting demon" is purely mechanical, and is of great value in purely physical science. It was not invented to help us to deal with questions regarding the influence of life and of mind on the motions of matter, questions essentially beyond the range of mere dynamics.

The description of the lecture ends:

> The discourse was illustrated by a series of experiments.

Presumably no actual Maxwell's demon was shown—or Kelvin wouldn't have continued for the rest of his life to treat the Second Law as an established principle.

But in any case, Maxwell's demon has always remained something of a fixture in discussions of the foundations of the Second Law. One might think that the observability of Brownian motion would make something like a Maxwell's demon possible. And indeed in 1912 Marian Smoluchowski (1872–1917) suggested experiments that one could imagine would "systematically harvest" Brownian motion—but showed that in fact they couldn't. In later years, a sequence of arguments were advanced that the mechanism of a Maxwell's demon just couldn't work in practice—though even today microscopic versions of what amount to Maxwell's demons are routinely being investigated.

What Happened to Those People?

We've finally now come to the end of the story of how the original framework for the Second Law came to be set up. And, as we've seen, only a fairly small number of key players were involved:

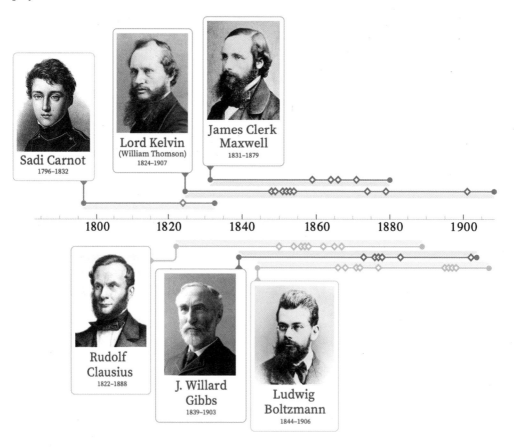

So what became of these people? Carnot lived a generation earlier than the others, never made a living as a scientist, and was all but unknown in his time. But all the others had distinguished careers as academic scientists, and were widely known in their time. Clausius, Boltzmann and Gibbs are today celebrated mainly for their contributions to thermodynamics; Kelvin and Maxwell also for other things. Clausius and Gibbs were in a sense "pure professors"; Boltzmann, Maxwell and especially Kelvin also had engagement with the more general public.

All of them spent the majority of their lives in the countries of their birth—and all (with the exception of Carnot) were able to live out the entirety of their lives without time-consuming disruptions from war or other upheavals:

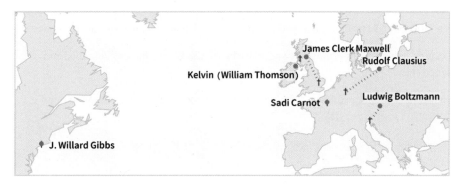

Sadi Carnot (1796–1832)

Almost all of what is known about Sadi Carnot as a person comes from a single biographical note written nearly half a century after his death by his younger brother Hippolyte Carnot (who was a distinguished French politician—and sometime education minister—and father of the Sadi Carnot who would become president of France). Hippolyte Carnot began by saying that:

> As the life of Sadi Carnot was not marked by any notable event, his biography would have occupied only a few lines; but a scientific work by him, after remaining long in obscurity, brought again to light many years after his death, has caused his name to be placed among those of great inventors.

The Carnots' father was close to Napoleon, and Hippolyte explains that when Sadi was a young child he ended up being babysat by "Madame Bonaparte"—but one day wandered off, and was found inspecting the operation of a nearby mill, and quizzing the miller about it. For the most part, however, throughout his life, Sadi Carnot apparently kept very much to himself—while with quiet intensity showing a great appetite for intellectual pursuits from mathematics and science to art, music and literature, as well as practical engineering and the science of various sports.

Even his brother Hippolyte can't explain quite how Sadi Carnot—at the age of 28—suddenly "came out" and in 1824 published his book on thermodynamics. (As we discussed above, it no doubt had something to do with the work of his father, who died two years earlier.) Sadi Carnot funded the publication of the book himself—having 600 copies printed (at least some of which remained unsold a decade later). But after the book was published, Carnot appears to have returned to just privately doing research, living alone, and never publishing again in his lifetime. And indeed he lived only another eight years, dying (apparently after some months of ill health) in the same Paris cholera outbreak that claimed General Lamarque of *Les Misérables* fame.

Twenty-three pages of unpublished personal notes survive from the period after the publication of Carnot's book. Some are general aphorisms and life principles:

> Speak little of what you know, and not at all of what you do not know.

> Why try to be witty? I would rather be thought stupid and modest than witty and pretentious.

> God cannot punish man for not believing when he could so easily have enlightened and convinced him.

> The belief in an all-powerful Being, who loves us and watches over us, gives to the mind great strength to endure misfortune.

> When walking, carry a book, a notebook to preserve ideas, and a piece of bread in order to prolong the walk if need be.

But others are more technical—and in fact reveal that Carnot, despite having based his book on caloric theory, had realized that it probably wasn't correct:

> When a hypothesis no longer suffices to explain phenomena, it should be abandoned. This is the case with the hypothesis which regards caloric as matter, as a subtile fluid.

> The experimental facts tending to destroy this theory are as follows: The development of heat by percussion or friction of bodies ... The elevation of temperature which takes place [when] air [expands into a] vacuum ...

He continues:

> At present, light is generally regarded as the result of a vibratory movement of the ethereal fluid. Light produces heat, or at least accompanies radiating heat, and moves with the same velocity as heat. Radiating heat is then a vibratory movement. It would be ridiculous to suppose that it is an emission of matter while the light which accompanies it could be only a movement.

> Could a motion (that of radiating heat) produce matter (caloric)? No, undoubtedly; it can only produce a motion. Heat is then the result of a motion.

And then—in a rather clear enunciation of what would become the First Law of thermodynamics:

> Heat is simply motive power, or rather motion which has changed form. It is a movement among the particles of bodies. Wherever there is destruction of motive power there is, at the same time, production of heat in quantity exactly proportional to the quantity of motive power destroyed. Reciprocally, wherever there is destruction of heat, there is production of motive power.

Carnot also wonders:

> Liquefaction of bodies, solidification of liquids, crystallization—are they not forms of combinations of integrant molecules? Supposing heat due to a vibratory movement, how can the passage from the solid or the liquid to the gaseous state be explained?

There is no indication of how Carnot felt about this emerging rethinking of thermodynamics, or of how it might affect the results in his book. But Carnot clearly hoped to do experiments (as outlined in his notes) to test what was really going on. But as it was, he presumably didn't get around to any of them—and his notes, ahead of their time as they were, did not resurface for many decades, by which time the ideas they contained had already been discovered by others.

Rudolf Clausius (1822–1888)

Rudolf Clausius was born in what's now Poland (and was then Prussia), one of more than 14 children of an education administrator and pastor. He went to university in Berlin, and, after considering doing history, eventually specialized in math and physics. After graduating in 1844 he started teaching at a top high school in Berlin (which he did for 6 years), and meanwhile earned his PhD in physics. His career took off after his breakout paper on thermodynamics appeared in 1850. For a while he was a professor in Berlin, then for 12 years in Zürich, then briefly in Würzburg, then—for the remaining 19 years of his life—in Bonn.

He was a diligent—if, one suspects, somewhat stiff—professor, notable for the clarity of his lectures, and his organizational care with students. He seems to have been a competent administrator, and late in his career he spent a couple of years as the president ("rector") of his university. But first and foremost, he was a researcher, writing about a hundred papers over the course of his career. Most physicists of the time devoted at least some of their efforts to doing actual physics experiments. But Clausius was a pioneer in the idea of being a "pure theoretical physicist", inspired by experiments and quoting their results, but not doing them himself.

The majority of Clausius's papers were about thermodynamics, though late in his career his emphasis shifted more to electrodynamics. Clausius's papers were original, clear, incisive and often fairly mathematically sophisticated. But from his very first paper on thermodynamics in 1850, he very much adopted a macroscopic approach, talking about what he considered to be "bulk" quantities like energy, and later entropy. He did explore some of the potential mechanics of molecules, but he never really made the connection between molecular phenomena and entropy—or the Second Law. He had a number of run-ins about academic credit with Kelvin, Tait, Maxwell and Boltzmann, but he didn't seem to ever pay much attention to, for example, Boltzmann's efforts to find molecular-based probabilistic derivations of Clausius's results.

It probably didn't help that after two decades of highly productive work, two misfortunes befell Clausius. First, in 1870, he had volunteered to lead an ambulance corps in the Franco-Prussian war, and was wounded in the knee, leading to chronic pain (as well as to his habit of riding to class on horseback). And then, in 1875, Clausius's wife died in the birth of their sixth child—leaving him to care for six young children (which apparently he did with great conscientiousness). Clausius nevertheless continued to pursue his research—even to the end of his life—receiving many honors along the way (like election to no less than 40 professional societies), but it never again rose to the level of significance of his early work on thermodynamics and the Second Law.

Kelvin (William Thomson) (1824–1907)

Of the people we're discussing here, by far the most famous during their lifetime was Kelvin. In his long career he wrote more than 600 scientific papers, received dozens of patents, started several companies and served in many administrative and governmental roles. His father was a math professor, ultimately in Glasgow, who took a great interest in the education of his children. Kelvin himself got an early start, effectively going to college at the age of 10, and becoming a professor in Glasgow at the age of 22—a position in which he continued for 53 years.

Kelvin's breakout work, done in his twenties, was on thermodynamics. But over the years he also worked on many other areas of physics, and beyond, mixing theory, experiment and engineering. Beginning in 1854 he became involved in a technical megaproject of the time: the attempt to lay a transatlantic telegraph cable. He wound up very much on the front lines, helping out as a just-in-time physicist + engineer on the cable-laying ship. The first few attempts didn't work out, but finally in 1866—in no small part through Kelvin's contributions—a cable was successfully laid, and Kelvin (or William Thomson, as he then was) became something of a celebrity. He was made "Sir William Thomson" and—along with two other techies—formed his first company, which had considerable success in exploiting telegraph-cable-related engineering innovations.

Kelvin's first wife died after a long illness in 1870, and Kelvin, with no children and already enthusiastic about the sea, bought a fairly large yacht, and pursued a number of nautical-related projects. One of these—begun in 1872—was the construction of an analog computer for calculating tides (basically with 10 gears for adding up 10 harmonic tide components), a device that, with progressive refinements, continued to be used for close to a century.

Being rather charmed by Kelvin's physicist-with-a-big-yacht persona, I once purchased a letter that Kelvin wrote in 1877 on the letterhead of "Yacht Lalla Rookh":

The letter—in true academic style—promises that Kelvin will soon send an article he's been asked to write on elasticity theory. And in fact he did write the article, and it was an expository one that appeared in the 9th edition of the *Encyclopedia Britannica*.

Kelvin was a prolific (if, to modern ears, sometimes rather pompous) writer, who took exposition seriously. And indeed—finding the textbooks available to him as a professor inadequate—he worked over the course of a dozen years (1855–1867) with his (and Maxwell's) friend Peter Guthrie Tait to produce the influential *Treatise on Natural Philosophy*.

Kelvin explored many topics and theories, some more immediately successful than others. In the 1870s he suggested that perhaps atoms might be knotted vortices in the (luminiferous) aether (causing Tait to begin developing knot theory)—a hypothesis that's in some sense a Victorian prelude to modern ideas about particles in our Physics Project.

Throughout his life, Kelvin was a devout Christian, writing that "The more thoroughly I conduct scientific research, the more I believe science excludes atheism." And indeed this belief seems to make an appearance in his implication that humans—presumably as a result of their special relationship with God—might avoid the Second Law. But more significant at the time was Kelvin's skepticism about Charles Darwin's 1859 theory of natural selection, believing that there must in the end be a "continually guiding and controlling intelligence". Despite being somewhat ridiculed for it, Kelvin talked about the possibility that life might have come to Earth from elsewhere via meteorites, believing that his estimates of the age of the Earth (which didn't take into account radioactivity) made it too young for the things Darwin described to have occurred.

By the 1870s, Kelvin had become a distinguished man of science, receiving all sorts of honors, assignments and invitations. And in 1876, for example, he was invited to Philadelphia to chair the committee judging electrical inventions at the US Centennial International Exhibition, notably reporting, in the terms of the time:

> In addition to his electro-phonetic multiple telegraph, Mr. Graham Bell exhibits apparatus by which he has achieved a result of transcendent scientific interest—the transmission of spoken words by electric currents through a telegraph wire. To obtain this result, or

Then in 1892 a "peerage of the realm" was conferred on him by Queen Victoria. His wife (he had remarried) and various friends (including Charles Darwin's son George) suggested he pick the title "Kelvin", after the River Kelvin that flowed by the university in Glasgow. And by the end of his life "Lord Kelvin" had accumulated enough honorifics that they were just summarized with "..." (the MD was an honorary degree conferred by the University of Heidelberg because "it was the only one at their disposal which he did not already possess"):

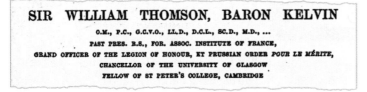

> ### SIR WILLIAM THOMSON, BARON KELVIN
> O.M., P.C., G.C.V.O., LL.D., D.C.L., SC.D., M.D., ...
> PAST PRES. R.S., FOR. ASSOC. INSTITUTE OF FRANCE,
> GRAND OFFICER OF THE LEGION OF HONOUR, ET PRUSSIAN ORDER *POUR LE MÉRITE*,
> CHANCELLOR OF THE UNIVERSITY OF GLASGOW
> FELLOW OF ST PETER'S COLLEGE, CAMBRIDGE

And when Kelvin died in 1907 he was given a state funeral and buried in Westminster Abbey near Newton and Darwin.

James Clerk Maxwell (1831–1879)

James Clerk Maxwell lived only 48 years but in that time managed to do a remarkable amount of important science. His early years were spent on a 1500-acre family estate (inherited by his father) in a fairly remote part of Scotland—to which he would return later. He was an only child and was homeschooled—initially by his mother, until she died, when he was 8. At 10 he went to an upscale school in Edinburgh, and by the age of 14 had written his first scientific paper. At 16 he went as an undergraduate to the University of Edinburgh, then, effectively as a graduate student, to Cambridge—coming second in the final exams ("Second Wrangler") to a certain Edward Routh, who would spend most of his life coaching other students on those very same exams.

Within a couple of years, Maxwell was a professor, first in Aberdeen, then in London. In Aberdeen he married the daughter of the university president, who would soon be his "Observer K" (for "Katherine") in his classic work on color vision. But after nine fairly strenuous years as a professor, Maxwell in 1865 "retired" to his family estate, supervising a house renovation, and in "rural solitude" (recreationally riding around his estate on horseback with his wife) having the most scientifically productive time of his life. In addition to his work on things like the kinetic theory of gases, he also wrote his 2-volume *Treatise on Electricity and Magnetism,* which ultimately took 7 years to finish, and which, with considerable clarity, described his approach to electromagnetism and what are now called "Maxwell's Equations". Occasionally, there were hints of his "country life"—like his 1870 "On Hills and Dales" that in his characteristic mathematicize-everything way gave a kind of "pre-topological" analysis of contour maps (perhaps conceived as he walked half a mile every day down to the mailbox at which journals and correspondence would arrive):

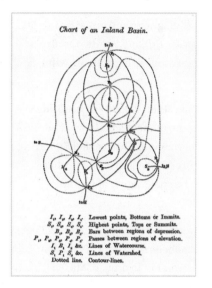

As a person, Maxwell was calm, reserved and unassuming, yet cheerful and charming—and given to writing (arguably sometimes sophomoric) poetry:

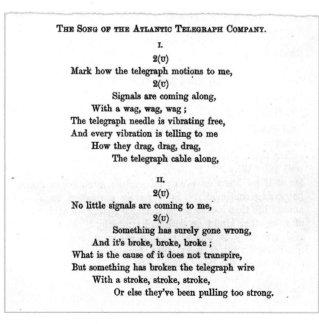

With a certain sense of the absurd, he would occasionally publish satirical pieces in *Nature*, signing them *dp/dt*, which in the thermodynamic notation created by his friend Tait was equal to JCM, which were his initials. Maxwell liked games and tricks, and spinning tops featured prominently in some of his work. He enjoyed children, though never had any of his own. As a lecturer, he prepared diligently, but often got too sophisticated for his audience. In writing, though, he showed both great clarity and great erudition, for example freely quoting Latin and Greek in articles he wrote for the 9th edition of the *Encyclopedia Britannica* (of which he was scientific co-editor) on topics such as "Atom" and "Ether".

As we mentioned above, Maxwell was quite an enthusiast of diagrams and visual presentation (even writing an article on "Diagrams" for the *Encyclopedia Britannica*). He was also a capable experimentalist, making many measurements (sometimes along with his wife), and in 1861 creating the first color photograph.

In 1871 William Cavendish, 7th Duke of Devonshire, who had studied math in Cambridge, and was now chancellor of the university, agreed to put up the money to build what became the Cavendish Laboratory and to endow a new chair of experimental physics. Kelvin having turned down the job, it was offered to the still-rather-obscure Maxwell, who somewhat reluctantly accepted—with the result that for several years he spent much of his time supervising the design and building of the lab.

The lab was finished in 1874, but then William Cavendish dropped on Maxwell a large collection of papers from his great uncle Henry Cavendish, who had been a wealthy "gentleman scientist" of the late 1700s and (among other things) the discoverer of hydrogen. Maxwell liked history (as some of us do!), noticed that Cavendish had discovered Ohm's law 50 years before Ohm, and in the end spent several years painstakingly editing and annotating the papers into a 500-page book. By 1879 Maxwell was finally ready to energetically concentrate on physics research again, but, sadly, in the fall of that year his health failed, and he died at the age of 48—having succumbed to stomach cancer, as his mother also had at almost the same age.

J. Willard Gibbs (1839–1903)

Gibbs was born near the Yale campus, and died there 64 years later, in the same house where he had lived since he was 7 years old (save for three years spent visiting European universities as a young man, and regular summer "out-in-nature" vacations). His father (who, like, "our Gibbs" was named "Josiah Willard"—making "our Gibbs" be called "Willard") came from an old and distinguished intellectual and religious New England family, and was a professor of sacred languages at Yale. Willard Gibbs went to college and graduate school at Yale, and then spent his whole career as a professor at Yale.

He was, it seems, a quiet, modest and rather distant person, who radiated a certain serenity, regularly attended church, had a small circle of friends and lived with his two sisters (and the husband and children of one of them). He diligently discharged his teaching responsibilities, though his lectures were very sparsely attended, and he seems not to have been thought forceful enough in dealing with people to have been called on for many administrative tasks—though he became the treasurer of his former high school, and himself was careful enough with money that by the end of his life he had accumulated what would now be several million dollars.

He had begun his academic career in practical engineering, for example patenting an "improved [railway] car-brake", but was soon drawn in more mathematical directions, favoring a certain clarity and minimalism of formulation, and a cleanliness, if not brevity, of exposition. His work on thermodynamics (initially published in the rather obscure *Transactions of the Connecticut Academy*) was divided into two parts: the first, in the 1870s, concentrating on macroscopic equilibrium properties, and the second, in the 1890s, concentrating on microscopic "statistical mechanics" (as Gibbs called it). Even before he started on thermodynamics, he'd been interested in electromagnetism, and between his two "thermodynamic periods", he again worked on electromagnetism. He studied Maxwell's work, and was at first drawn to the then-popular formalism of quaternions—but soon decided to invent his own approach and notation for vector analysis, which at first he presented only in notes for his students, though it later became widely adopted.

And while Gibbs did increasingly mathematical work, he never seems to have identified as a mathematician, modestly stating that "If I have had any success in mathematical physics, it is, I think, because I have been able to dodge mathematical difficulties." His last work was his book on statistical mechanics, which—with considerable effort and perhaps damage to his health—he finished in time for publication in connection with the Yale bicentennial in 1901 (an event which notably also brought a visit from Kelvin), only to die soon thereafter.

Gibbs had a few graduate students at Yale, a notable one being Lee de Forest, inventor of the vacuum tube (triode) electronic amplifier, and radio entrepreneur. (de Forest's 1899 PhD thesis was entitled "Reflection of Hertzian Waves from the Ends of Parallel Wires".) Another student of Gibbs's was Lynde Wheeler, who became a government radio scientist, and who wrote a biography of Gibbs, of which I have a copy bought years ago at a used bookstore—that I was now just about to put back on a shelf when I opened its front cover and found an inscription:

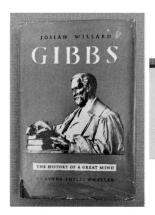

And, yes, it's a small world, and "To Willard" refers to Gibbs's sister's son (Willard Gibbs Van Name, who became a naturalist and wrote a 1929 book about national park deforestation).

Ludwig Boltzmann (1844–1906)

Of the people we're discussing, Boltzmann is the one whose career was most focused on the Second Law. Boltzmann grew up in Austria, where his father was a civil servant (who died when Boltzmann was 15) and his mother was something of an heiress. Boltzmann did his PhD at the University of Vienna, where his professor notably gave him a copy of some of Maxwell's papers, together with an English grammar book. Boltzmann started publishing his own papers near the end of his PhD, and soon landed a position as a professor of mathematical physics in Graz. Four years later he moved to Vienna as a professor of mathematics, soon moving back to Graz as a professor of "general and experimental physics"—a position he would keep for 14 years.

He'd married in 1876, and had 5 children, though a son died in 1889, leaving 3 daughters and another son. Boltzmann was apparently a clear and lively lecturer, as well as a spirited and eager debater. He seems, at least in his younger years, to have been a happy and gregarious person, with a strong taste for music—and some charming do-it-your-own-way tendencies. For example, wanting to provide fresh milk for his children, he decided to just buy a cow, which he then led from the market through the streets—though had to consult his colleague, the professor of zoology, to find out how to milk it. Boltzmann was a capable experimental physicist, as well as a creator of gadgets, and a technology enthusiast—promoting the idea of airplanes (an application for gas theory!) and noting their potential power as a means of transportation.

Boltzmann had always had mood swings, but by the early 1890s he claimed they were getting worse. It didn't help that he was worn down by administrative work, and had worsening asthma and increasing nearsightedness (that he'd thought might be a sign of going blind). He moved positions, but then came back to Vienna, where he embarked on writing what would become a 2-volume book on *Gas Theory*—in effect contextualizing his life's work. The introduction to the first volume laments that "gas theory has gone out of fashion in Germany". The introduction to the second volume, written in 1898 when Boltzmann was 54, then says that "attacks on the theory of gases have begun to increase", and continues:

> ... it would be a great tragedy for science if the theory of gases were temporarily thrown into oblivion because of a momentary hostile attitude toward it, as, for example, was the wave theory [of light] because of Newton's authority.

> I am conscious of being only an individual struggling weakly against the stream of time. But it still remains in my power to contribute in such a way that, when the theory of gases is again revived, not too much will have to be rediscovered.

But even as he was writing this, Boltzmann had pretty much already wound down his physics research, and had basically switched to exposition, and to philosophy. He moved jobs again, but in 1902 again came back to Vienna, but now also as a professor of philosophy. He gave an inaugural lecture, first quoting his predecessor Ernst Mach (1838–1916) as saying "I do not believe that atoms exist", then discussing the philosophical relations between reality, perception and models. Elsewhere he discussed things like his view of the different philosophical character of models associated with differential equations and with atomism—and he even wrote an article on the general topic of "Models" for *Encyclopedia Britannica* (which curiously also talks about "in pure mathematics, especially geometry, models constructed of papier-mâché and plaster"). Sometimes Boltzmann's philosophy could be quite polemical, like his attack on Schopenhauer, that ends by saying that "men [should] be freed from the spiritual migraine that is called metaphysics".

Then, in 1904, Boltzmann addressed the Vienna Philosophical Society (a kind of predecessor of the Vienna Circle) on the subject of a "Reply to a Lecture on Happiness by Professor Ostwald". Wilhelm Ostwald (1853–1932) (a chemist and social reformer, who was a personal friend of Boltzmann's, but intellectual adversary) had proposed the concept of "energy of will" to apply mathematical physics ideas to psychology. Boltzmann mocked this, describing its faux formalism as "dangerous for science". Meanwhile, Boltzmann gives his own Darwinian theory for the origin of happiness, based essentially on the idea that unhappiness is needed as a way to make organisms improve their circumstances in the struggle for survival.

Boltzmann himself was continuing to have problems that he attributed to the then-popular but very vague "diagnosis" of "neurasthenia", and had even briefly been in a psychiatric hospital. But he continued to do things like travel. He visited the US three times, in 1905 going to California (mainly Berkeley)—which led him to write a witty piece entitled "A German Professor's Trip to El Dorado" that concluded:

> Yes, America will achieve great things. I believe in these people, even after seeing them at work in a setting where they're not at their best: integrating and differentiating at a theoretical physics seminar...

In 1905 Einstein published his Boltzmann-and-atomism-based results on Brownian motion and on photons. But it's not clear Boltzmann ever knew about them. For Boltzmann was sinking further. Perhaps he'd overexerted himself in California, but by the spring of 1906 he said he was no longer able to teach. In the summer he went with his family to an Italian seaside resort in an attempt to rejuvenate. But a day before they were to return to Vienna he failed to join his family for a swim, and his youngest daughter found him hanged in his hotel room, dead at the age of 62.

Coarse Graining and the "Modern Formulation"

After Gibbs's 1902 book introducing the idea of ensembles, most of the language used (at least until now!) to discuss the Second Law was basically in place. But in 1912 one additional term—representing a concept already implicit in Gibbs's work—was added: coarse graining. Gibbs had discussed how the phase fluid representing possible states of a system could be elaborately mixed by the mechanical time evolution of the system. But realistic practical measurements could not be expected to probe all the details of the distribution of phase fluid; instead one could say that they would only sample "coarse-grained" aspects of it.

The term "coarse graining" first appeared in a survey article entitled "The Conceptual Foundations of the Statistical Approach in Mechanics", written for the German-language *Encyclopaedia of the Mathematical Sciences* by Boltzmann's former student Paul Ehrenfest, and his wife Tatiana Ehrenfest-Afanassjewa:

> Let us divide the Γ-space somehow into very small, but finite cells Ω: Ω_1, Ω_2, $\cdots$, Ω_λ, $\cdots$, which might be, for instance, cubes of equal size. The average value which the "fine-grained" density $\rho(q, p, t)$ has at time t over the cell Ω_λ we will call "coarse-grained" density $P_\lambda(t)$ (read: capital ρ) of this cell. Because of Eq. (54) we have

The article also introduced all sorts of now-standard notation, and in many ways can be read as a final summary of what was achieved in the original development around the foundations of thermodynamics and the Second Law. (And indeed the article was sufficiently "final" that when it was republished as a book in 1959 it could still be presented as usefully summarizing the state of things.)

Looking at the article now, though, it's notable how much it recognized was not at all settled about the Second Law and its foundations. It places Boltzmann squarely at the center, stating in its preface:

> Since 1876 numerous papers have called attention to these foundations. In these papers the Boltzmann H-theorem, a central theorem of the kinetic theory of gases, was attacked. Without exception all studies so far published dealing with the connection of mechanics with probability theory grew out of the synthesis of these polemics and of Boltzmann's replies. These discussions will therefore be referred to frequently in our report.

The section titles are already revealing:

And soon they're starting to talk about "loose ends", and lots of them. Ergodicity is something one can talk about, but there's no known example (and with this definition it was later proved that there couldn't be):

> However, the existence of ergodic systems (i.e., the consistency of their definition) is doubtful. So far, not even one example is known of a mechanical system for which the single G-path approaches arbitrarily closely each point of the corresponding energy surface.[97] More-

But, they point out, it's something Boltzmann needed in order to justify his results:

> In order to get from this ensemble average to the time average, which is what Boltzmann wants, one needs the following chain of equalities:
>
> (33) Ensemble average = the time average of the ensemble average
>   = the ensemble average of the time average
>   = time average.
>
> The first equality follows from the stationary character of the phase distribution,[110] the second because the forming of averages is commutative. The third equality, however, is based on the statement that all motions of the set in question give the same time average for $\psi(q, p)$.
> It is at this point that the hypothesis about the gas model being ergodic enters. Because of the doubts about the internal consistency of the ergodic hypothesis, this investigation cannot be considered free of objection

Soon they're talking about Boltzmann's sloppiness in his discussion of the H curve:

> 3. In particular, he promoted this misunderstanding by always calling these maxima of the H-curves "humps," which makes one think almost necessarily of a maximum with a horizontal tangent.

And then they're on to talking about Gibbs, and the gaps in his reasoning:

> realize the following: While in all these cases Chapters XI and XIII prove at most a certain change in the direction of the canonical distribution, in Chapter XIV the analogies with thermodynamics are discussed as if it had been proved that the canonical distribution will be reached, at least approximately, in time. Such a jump

In the end they conclude:

> The foregoing account dealt chiefly with the conceptual foundations of statistico-mechanical investigations. Accordingly we had to emphasize that in these investigations a large number of loosely formulated and perhaps even inconsistent statements occupy a central position. In fact, we encounter here an incompleteness which from the logical point of view is serious and which appears in other branches of mechanics to a much smaller extent. This incompleteness, however, does not seem to have influenced the physicists in their evaluation of the

> statistico-mechanical investigations. In particular, the last few years have seen a sudden and wide dissemination of Boltzmann's ideas (the H-theorem, the Maxwell-Boltzmann distribution, the equipartition of energy, the relationship between entropy and probability, etc.). However, one cannot point at a corresponding progress in the conceptual clarification of Boltzmann's system to which one can ascribe this turn of affairs.

In other words, even though people now seem to be buying all these results, there are still plenty of issues with their foundations. And despite people's implicit assumptions, we can in no way say that the Second Law has been "proved".

Radiant Heat, the Second Law and Quantum Mechanics

It was already realized in the 1600s that when objects get hot they emit "heat radiation"—which can be transferred to other bodies as "radiant heat". And particularly following Maxwell's work in the 1860s on electrodynamics it came to be accepted that radiant heat was associated with electromagnetic waves propagating in the "luminiferous aether". But unlike the molecules from which it was increasingly assumed that one could think of matter as being made, these electromagnetic waves were always treated—particularly on the basis of their mathematical foundations in calculus—as fundamentally continuous.

But how might this relate to the Second Law? Could it be, perhaps, that the Second Law should ultimately be attributed not to some property of the large-scale mechanics of discrete molecules, but rather to a feature of continuous radiant heat?

The basic equations assumed for mechanics—originally due to Newton—are reversible. But what about the equations for electrodynamics? Maxwell's equations are in and of themselves also reversible. But when one thinks about their solutions for actual electromagnetic radiation, there can be fundamental irreversibility. And the reason is that it's natural to describe the emission of radiation (say from a hot body), but then to assume that, once emitted, the radiation just "escapes to infinity"—rather than ever reversing the process of emission by being absorbed by some other body.

All the various people we've discussed above, from Clausius to Gibbs, made occasional remarks about the possibility that the Second Law—whether or not it could be "derived mechanically"—would still ultimately work, if nothing else, because of the irreversible emission of radiant heat.

But the person who would ultimately be most intimately connected to these issues was Max Planck—though in the end the somewhat-confused connection to the Second Law would recede in importance relative to what emerged from it, which was basically the raw material that led to quantum theory.

As a student of Helmholtz's in Berlin, Max Planck got interested in thermodynamics, and in 1879 wrote a 61-page PhD thesis entitled "On the Second Law of Mechanical Heat Theory". It was a traditional (if slightly streamlined) discussion of the Second Law, very much based on Clausius's approach (and even with the same title as Clausius's 1867 paper)—and without any mention whatsoever of Boltzmann:

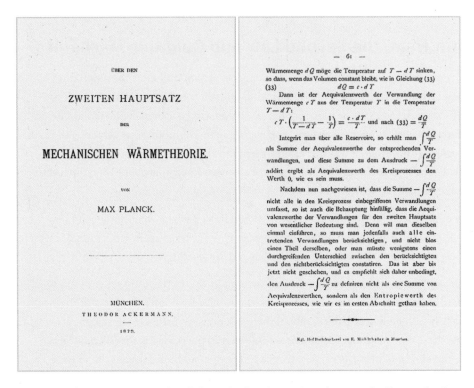

For most of the two decades that followed, Planck continued to use similar methods to study the Second Law in various settings (e.g. elastic materials, chemical mixtures, etc.)—and meanwhile ascended the German academic physics hierarchy, ending up as a professor of theoretical physics in Berlin. Planck was in many ways a physics traditionalist, not wanting to commit to things like "newfangled" molecular ideas—and as late as 1897 (with his assistant Zermelo having made his "recurrence objection" to Boltzmann's work) still saying that he would "abstain completely from any definite assumption about the nature of heat". But regardless of its foundations, Planck was a true believer in the Second Law, for example in 1891 asserting that it "must extend to all forces of nature ... not only thermal and chemical, but also electrical and other".

And in 1895 he began to investigate how the Second Law applied to electrodynamics—and in particular to the "heat radiation" that it had become clear (particularly through Heinrich Hertz's (1857–1894) experiments) was of electromagnetic origin. In 1896 Wilhelm Wien (1864–1928) suggested that the heat radiation (or what we now call blackbody radiation) was in effect produced by tiny Hertzian oscillators with velocities following a Maxwell distribution.

Planck, however, had a different viewpoint, instead introducing the concept of "natural radiation"—a kind of intrinsic thermal equilibrium state for radiation, with an associated intrinsic entropy. He imagined "resonators" interacting through Maxwell's equations with this radiation, and in 1899 invented a (rather arbitrary) formula for the entropy of these resonators, that implied (through the laws of electrodynamics) that overall entropy would increase—just like the Second Law said—and when the entropy was maximized it gave the same result as Wien for the spectrum of blackbody radiation. In early 1900 he sharpened his treatment and began to suggest that with his approach Wien's form of the blackbody spectrum would emerge as a provable consequence of the universal validity of the Second Law.

But right around that time experimental results arrived that disagreed with Wien's law. And by the end of 1900 Planck had a new hypothesis, for which he finally began to rely on ideas from Boltzmann. Planck started from the idea that he should treat the behavior of his resonators statistically. But how then could he compute their entropy? He quotes (for the first time ever) his simplification of Boltzmann's formula for entropy:

auf eine willkürlich bleibende additive Constante, proportional
dem Logarithmus der Wahrscheinlichkeit W dafür, dass die
N Resonatoren insgesamt die Energie U_N besitzen, also:

(3) $$S_N = k \log W + \text{const.}$$

Diese Festsetzung kommt nach meiner Meinung im Grunde
auf eine Definition der genannten Wahrscheinlichkeit W hinaus;

As he explains it—claiming now, after years of criticizing Boltzmann, that this is a "theorem":

> We now set the entropy S of the system proportional to the logarithm of its probability W ... In my opinion this actually serves as a definition of the probability W, since in the basic assumptions of electromagnetic theory there is no definite evidence for such a probability. The suitability of this expression is evident from the outset, in view of its simplicity and close connection with a theorem from kinetic gas theory.

But how could he figure out the probability for a resonator to have a certain energy, and thus a certain entropy? For this he turns directly to Boltzmann—who, as a matter of convenience in his 1877 paper had introduced discrete values of energy for molecules. Planck simply states that it's "necessary" (i.e. to get the experimentally right answer) to treat the resonator energy "not as a continuous, infinitely divisible quantity, but as a discrete quantity composed of an integral number of finite equal parts". As an example of how this works he gives a table just like the one in Boltzmann's paper from nearly a quarter of a century earlier:

elemente, so erhält man für jede Complexion ein Symbol von
folgender Form:

1	2	3	4	5	6	7	8	9	10
7	38	11	0	9	2	20	4	4	5

Hier ist $N = 10$, $P = 100$ angenommen. Die Anzahl $\Re$ aller

Pretty soon he's deriving the entropy of a resonator as a function of its energy, and its discrete energy unit ϵ:

Also nach (2) die Entropie S eines Resonators als Function
seiner Energie U:

(6) $$S = k \left\{ \left(1 + \frac{U}{\varepsilon} \right) \log \left(1 + \frac{U}{\varepsilon} \right) - \frac{U}{\varepsilon} \log \frac{U}{\varepsilon} \right\}.$$

Connecting this to blackbody radiation he claims that each resonator's energy unit is connected to its frequency according to

der Schwingungszahl ν sein muss, also:
$$\varepsilon = h \cdot \nu$$

so that its entropy is

und somit:
$$S = k \left\{ \left(1 + \frac{U}{h\nu} \right) \log \left(1 + \frac{U}{h\nu} \right) - \frac{U}{h\nu} \log \frac{U}{h\nu} \right\}.$$
Hierbei sind h und k universelle Constante.

"[where] h and k are universal constants".

In a similar situation Boltzmann had effectively taken the limit $\epsilon \to 0$, because that's what he believed corresponded to ("calculus-based") physical reality. But Planck—in what he later described as an "act of desperation" to fit the experimental data—didn't do that. So in computing things like average energies he's evaluating Sum[x Exp[-a x], {x, 0, ∞}] rather than Integrate[x Exp[-a x], {x, 0, Infinity}]. And in doing this it takes him only a few lines to derive what's now called the Planck spectrum for blackbody radiation (i.e. for "radiation in equilibrium"):

und aus (8) folgt dann das gesuchte Energieverteilungsgesetz:

$$(12) \qquad \mathfrak{u} = \frac{8 \pi h \nu^3}{c^3} \cdot \frac{1}{e^{\frac{h\nu}{k\vartheta}} - 1}$$

And then by fitting this result to the data of the time he gets "Planck's constant" (the correct result is 6.62):

$$(15) \qquad h = 6{,}55 \cdot 10^{-27} \text{ erg} \cdot \text{sec},$$

And, yes, this was essentially the birth of quantum mechanics—essentially as a spinoff from an attempt to extend the domain of the Second Law. Planck himself didn't seem to internalize what he'd done for at least another decade. And it was really Albert Einstein's 1905 analysis of the photoelectric effect that made the concept of the quantization of energy that Planck had assumed (more as a calculational hypothesis than anything else) seem to be something of real physical significance—that would lead to the whole development of quantum mechanics, notably in the 1920s.

Are Molecules Real? Continuous Versus Discrete

As we discussed previously in "The Basic Arc of the Story" (page 237), already in antiquity there was a notion that at least things like solids and liquids might not ultimately be continuous (as they seemed), but might instead be made of large numbers of discrete "atomic" elements. By the 1600s there was also the idea that light might be "corpuscular"—and, as we discussed above, gases too. But meanwhile, there were opposing theories that espoused continuity—like the caloric theory of heat. And particularly with the success of calculus, there was a strong tendency to develop theories that showed continuity—and to which calculus could be applied.

But in the early 1800s—notably with the work of John Dalton (1766–1844)—there began to be evidence that there were discrete entities participating in chemical reactions. Meanwhile, as we discussed above, the success of the kinetic theory of gases gave increasing evidence for some kind of—at least effectively—discrete elements in gases. But even people like Boltzmann and Maxwell were reluctant to assert that gases really were made of molecules. And there were plenty of well-known scientists (like Ernst Mach) who "opposed atomism", often effectively on the grounds that in science one should only talk about things one can actually see or experience—not things like atoms that were too small for that.

But there was something else too: with Newton's theory of gravitation as a precursor, and then with the investigation of electromagnetic phenomena, there emerged in the 1800s the idea of a "continuous field". The interpretation of this was fairly clear for something like an elastic solid or a fluid that exhibited continuous deformations.

Mathematically, things like gravity, magnetism—and heat—seemed to work in similar ways. And it was assumed that this meant that in all cases there had to be some fluid-like "carrier" for the field. And this is what led to ideas like the luminiferous aether as the "carrier" of electromagnetic waves. And, by the way, the idea of an aether wasn't even obviously incompatible with the idea of atoms; Kelvin, for example, had a theory that atoms were vortices (perhaps knotted) in the aether.

But how does this all relate to the Second Law? Well, particularly through the work of Boltzmann there came to be the impression that given atomism, probability theory could essentially "prove" the Second Law. A few people tried to clarify the formal details (as we discussed above), but it seemed like any final conclusion would have to await the validation (or not) of atomism, which in the late 1800s was still a thoroughly controversial theory.

By the first decade of the 1900s, however, the fortunes of atomism began to change. In 1897 J. J. Thomson (1856–1940) discovered the electron, showing that electricity was fundamentally "corpuscular". And in 1900 Planck had (at least calculationally) introduced discrete quanta of energy. But it was the three classic papers of Albert Einstein in 1905 that—in their different ways—began to secure the ultimate success of atomism.

First there was his paper "On a Heuristic View about the Production and Transformation of Light", which began:

> Maxwell's theory of electromagnetic [radiation] differs in a profound, essential way from the current theoretical models of gases and other matter. We consider the state of a material body to be completely determined by the positions and velocities of a finite number of atoms and electrons, albeit a very large number. But the electromagnetic state of a region of space is described by continuous functions ...

He then points out that optical experiments look only at time-averaged electromagnetic fields, and continues:

> In particular, blackbody radiation, photoluminescence, [the photoelectric effect] and other phenomena associated with the generation and transformation of light seem better modeled by assuming that the energy of light is distributed discontinuously in space. According to this picture, the energy of a light wave emitted from a point source is not spread continuously over ever larger volumes, but consists of a finite number of energy quanta that are spatially localized at points of space, move without dividing and are absorbed or generated only as a whole.

In other words, he's suggesting that light is "corpuscular", and that energy is quantized. When he begins to get into details, he's soon talking about the "entropy of radiation"—and, then, in three core sections of his paper, he's basing what he's doing on "Boltzmann's principle":

Two months later, Einstein produced another paper: "Investigations on the Theory of Brownian Motion". Back in 1827 the British botanist Robert Brown (1773–1858) had seen under a microscope tiny grains (ejected by pollen) randomly jiggling around in water. Einstein began his paper:

> In this paper it will be shown that according to the molecular-kinetic theory of heat, bodies of microscopically visible size suspended in a liquid will perform movements of such magnitude that they can be easily observed in a microscope, on account of the molecular motions of heat.

He doesn't explicitly mention Boltzmann in this paper, but there's Boltzmann's formula again:

Sind $p_1 \, p_2 \ldots p_l$ Zustandsvariable, eines physikalischen Systems, welche den momentanen Zustand desselben vollkommen bestimmen (z. B. die Koordinaten und Geschwindigkeitskomponenten aller Atome des Systems) und ist das vollständige System der Veränderungsgleichungen dieser Zustandsvariabeln von der Form

$$\frac{\partial p_\nu}{\partial t} = \varphi_\nu(p_1 \ldots p_l)(\nu = 1, 2 \ldots l) \ .$$

gegeben, wobei $\Sigma \frac{\partial \varphi_\nu}{\partial p_\nu} = 0$, so ist die Entropie des Systems durch den Ausdruck gegeben:

$$S = \frac{E}{T} + 2\varkappa \lg \int e^{-\frac{E}{2\varkappa T}} dp_1 \ldots dp_l.$$

Hierbei bedeutet T die absolute Temperatur, E die Energie des Systems, E die Energie als Funktion der p_ν. Das Integral ist über alle mit den Bedingungen des Problems vereinbaren Wertekombinationen der p_ν zu erstrecken. $\varkappa$ ist mit

And by the next year it's become clear experimentally that, yes, the jiggling Robert Brown had seen was in fact the result of impacts from discrete, real water molecules.

Einstein's third 1905 paper, "On the Electrodynamics of Moving Bodies"—in which he introduced relativity theory—wasn't so obviously related to atomism. But in showing that the luminiferous aether will (as Einstein put it) "prove superfluous" he was removing what was (almost!) the last remaining example of something continuous in physics.

In the years after 1905, the evidence for atomism mounted rapidly, segueing in the 1920s into the development of quantum mechanics. But what happened with the Second Law? By the time atomism was generally accepted, the generation of physicists that had included Boltzmann and Gibbs was gone. And while the Second Law was routinely invoked in expositions of thermodynamics, questions about its foundations were largely forgotten. Except perhaps for one thing: people remembered that "proofs" of the Second Law had been controversial, and had depended on the controversial hypothesis of atomism. But—they appear to have reasoned—now that atomism isn't controversial anymore, it follows that the Second Law is indeed "satisfactorily proved". And, after all, there were all sorts of other things to investigate in physics.

There are a couple of "footnotes" to this story. The first has to do with Einstein. Right before Einstein's remarkable series of papers in 1905, what was he working on? The answer is: the Second Law! In 1902 he wrote a paper entitled "Kinetic Theory of Thermal Equilibrium and of the Second Law of Thermodynamics". Then in 1903: "A Theory of the Foundations of Thermodynamics". And in 1904: "On the General Molecular Theory of Heat". The latter paper claims:

> I derive an expression for the entropy of a system, which is completely analogous to the one found by Boltzmann for ideal gases and assumed by Planck in his theory of radiation. Then I give a simple derivation of the Second Law.

But what's actually there is not quite what's advertised:

§ 2. Herleitung des zweiten Hauptsatzes.

Befindet sich ein System in einer Umgebung von bestimmter konstanter Temperatur T_0 und steht es mit dieser Umgebung in thermischer Wechselwirkung („Berührung"), so nimmt es ebenfalls erfahrungsgemäß die Temperatur T_0 an

(b)
$$\mathfrak{W}' = C_1 . C_2 \ldots C_i\, e^{\frac{1}{2\varkappa}\left(\sum\limits_{1}^{i} s' - \frac{\sum\limits_{1}^{i} E'}{T_0}\right)}.$$

Bei dem Vorgange hat sich weder der Zustand der Umgebung noch der Zustand der Maschine geändert, da letztere einen Kreisprozeß durchlief.

Nehmen wir nun an, daß nie unwahrscheinlichere Zustände auf wahrscheinlichere folgen, so ist:

[13]
$$\mathfrak{W}' \geqq \mathfrak{W}.$$

Es ist aber auch nach dem Energieprinzip:

$$\sum\limits_{1}^{i} E = \sum\limits_{1}^{i} E'.$$

Berücksichtigt man dies, so folgt aus Gleichungen (a) und (b):

$$\sum s' \geqq \sum s.$$

It's a short argument—about interactions between a collection of heat reservoirs. But in a sense it already assumes its answer, and certainly doesn't provide any kind of fundamental "derivation of the Second Law". And this was the last time Einstein ever explicitly wrote about deriving the Second Law. Yes, in those days it was just too hard, even for Einstein.

There's another footnote to this story too. As we said, at the beginning of the twentieth century it had become clear that lots of things that had been thought to be continuous were in fact discrete. But there was an important exception: space. Ever since Euclid (~300 BC), space had almost universally been implicitly assumed to be continuous. And, yes, when quantum mechanics was being built, people did wonder about whether space might be discrete too (and even in 1917 Einstein expressed the opinion that eventually it would turn out to be). But over time the idea of continuous space (and time) got so entrenched in the fabric of physics that when I started seriously developing the ideas that became our Physics Project based on space as a discrete network (or what—in homage to the dynamical theory of heat one might call the "dynamical theory of space") it seemed to many people quite shocking. And looking back at the controversies of the late 1800s around atomism and its application to the Second Law it's charming how familiar many of the arguments against atomism seem. Of course it turns out they were wrong—as they seem again to be in the case of space.

The Twentieth Century

The foundations of thermodynamics were a hot topic in physics in the latter half of the nineteenth century—worked on by many of the most prominent physicists of the time. But by the early twentieth century it'd been firmly eclipsed by other areas of physics. And going forward it'd receive precious little attention—with most physicists just assuming it'd "somehow been solved", or at least "didn't need to be worried about".

As a practical matter, thermodynamics in its basic equilibrium form nevertheless became very widely used in engineering and in chemistry. And in physics, there was steadily increasing interest in doing statistical mechanics—typically enumerating states of systems (quantum or otherwise), weighted as they would be in idealized thermal equilibrium. In mathematics, the field of ergodic theory developed, though for the most part it concerned itself with systems (such as ordinary differential equations) involving few variables—making it relevant to the Second Law essentially only by analogy.

There were a few attempts to "axiomatize" the Second Law, but mostly only at a macroscopic level, not asking about its microscopic origins. And there were also attempts to generalize the Second Law to make robust statements not just about equilibrium and the fact that it would be reached, but also about what would happen in systems driven to be in some manner away from equilibrium. The fluctuation-dissipation theorem about small perturbations from equilibrium—established in the mid-1900s, though anticipated in Einstein's work on Brownian motion—was one example of a widely applicable result. And there were also related ideas of "minimum entropy production"—as well as "maximum entropy production". But for large deviations from equilibrium there really weren't convincing general results, and in practice most investigations basically used phenomenological models that didn't have obvious connections to the foundations of thermodynamics, or derivations of the Second Law.

Meanwhile, through most of the twentieth century there were progressively more elaborate mathematical analyses of Boltzmann's equation (and the *H* theorem) and their relation to rigorously derivable but hard-to-manage concepts like the BBGKY hierarchy. But despite occasional claims to the contrary, such approaches ultimately never seem to have been able to make much progress on the core problem of deriving the Second Law.

And then there's the story of entropy. And in a sense this had three separate threads. The first was the notion of entropy—essentially in the original form defined by Clausius—being used to talk quantitatively about heat in equilibrium situations, usually for either engineering or chemistry. The second—that we'll discuss a little more below—was entropy as a qualitative characterization of randomness and degradation. And the third was entropy as a general and formal way to measure the "effective number of degrees of freedom" in a system, computed from the log of the number of its achievable states.

There are definitely correspondences between these different threads. But they're in no sense "obviously equivalent". And much of the mystery—and confusion—that developed around entropy in the twentieth century came from conflating them.

Another piece of the story was information theory, which arose in the 1940s. And a core question in information theory is how long an "optimally compressed" message will be. And (with various assumptions) the average such length is given by a $\sum p \log p$ form that has essentially the same structure as Boltzmann's expression for entropy. But even though it's "mathematically like entropy" this has nothing immediately to do with heat—or even physics; it's just an abstract consequence of needing $\log \Omega$ bits (i.e. $\log \Omega$ degrees of freedom) to specify one of Ω possibilities. (Still, the coincidence of definitions led to an "entropy branding" for various essentially information-theoretic methods, with claims sometimes being made that, for example, the thing called entropy must always be maximized "because we know that from physics".)

There'd been an initial thought in the 1940s that there'd be an "inevitable Second Law" for systems that "did computation". The argument was that logical gates (like And and Or) take 2 bits of input (with 4 overall states 11, 10, 01, 00) but give only 1 bit of output (1 or 0), and are therefore fundamentally irreversible. But in the 1970s it became clear that it's perfectly possible to do computation reversibly (say with 2-input, 2-output gates)—and indeed this is what's used in the typical formalism for quantum circuits.

As I've mentioned elsewhere, there were some computer experiments in the 1950s and beyond on model systems—like hard-sphere gases and nonlinear springs—that showed some sign of Second Law behavior (though less than might have been expected). But the analysis of these systems very much concentrated on various regularities, and not on the effective randomness associated with Second Law behavior.

In another direction, the 1970s saw the application of thermodynamic ideas to black holes. At first, it was basically a pure analogy. But then quantum field theory calculations suggested that black holes should produce thermal radiation as if they had a certain effective temperature. By the late 1990s there were more direct ways to "compute entropy" for black holes, by enumerating possible (quantum) configurations consistent with the overall characteristics of the black hole. But such computations in effect assume (time-invariant) equilibrium, and so can't be expected to shed light directly on the Second Law.

Talking about black holes brings up gravity. And in the course of the twentieth century there were scattered efforts to understand the effect of gravity on the Second Law. Would a self-gravitating gas achieve "equilibrium" in the usual sense? Does gravity violate the Second Law? It's been difficult to get definitive answers. Many specific simulations of n-body gravitational systems were done, but without global conclusions for the Second Law. And there were cosmological arguments, particularly about the role of gravity in accounting for entropy in the early universe—but not so much about the actual evolution of the universe and the effect of the Second Law on it.

Yet another direction has involved quantum mechanics. The standard formalism of quantum mechanics—like classical mechanics—is fundamentally reversible. But the formalism for measurement introduced in the 1930s—arguably as something of a hack—is fundamentally irreversible, and there've been continuing arguments about whether this could perhaps "explain the Second Law". (I think our Physics Project finally provides more clarity about what's going on here—but also tells us this isn't what's "needed" for the Second Law.)

From the earliest days of the Second Law, there had always been scattered but ultimately unconvincing assertions of exceptions to the Second Law—usually based on elaborately constructed machines that were claimed to be able to achieve perpetual motion "just powered by heat". Of course, the Second Law is a claim about large numbers of molecules, etc.—and shouldn't be expected to apply to very small systems. But by the end of the twentieth century it was starting to be possible to make micromachines that could operate on small numbers of molecules (or electrons). And with the right control systems in place, it was argued that such machines could—at least in principle—effectively be used to set up Maxwell's demons that would systematically violate the Second Law, albeit on a very small scale.

And then there was the question of life. Early formulations of the Second Law had tended to talk about applying only to "inanimate matter"—because somehow living systems didn't seem to follow the same process of inexorable "dissipation to heat" as inanimate, mechanical systems. And indeed, quite to the contrary, they seemed able to take disordered input (like food) and generate ordered biological structures from it. And indeed, Erwin Schrödinger (1887–1961), in his 1944 book *What Is Life?* talked about "negative entropy" associated with life. But he—and many others since—argue that life doesn't really violate the Second Law because it's not operating in a closed environment where one should expect evolution to equilibrium. Instead, it's constantly being driven away from equilibrium, for example by "organized energy" ultimately coming from the Sun.

Still, the concept of at least locally "antithermodynamic" behavior is often considered to be a potential general signature of life. But already by the early part of the 1900s, with the rise of things like biochemistry, and the decline of concepts like "life force" (which seemed a little like "caloric"), there developed a strong belief that the Second Law must at some level always apply, even to living systems. But, yes, even though the Second Law seemed to say that one can't "unscramble an egg", there was still the witty rejoinder: "unless you feed it to a chicken".

What about biological evolution? Well, Boltzmann had been an enthusiast of Darwin's idea of natural selection. And—although it's not clear he made this connection—it was pointed out many times in the twentieth century that just as in the Second Law reversible underlying dynamics generate an irreversible overall effect, so also in Darwinian evolution effectively reversible individual changes aggregate to what at least Darwin thought was an "irreversible" progression to things like the formation of higher organisms.

The Second Law also found its way into the social sciences—sometimes under names like "entropy pessimism"—most often being used to justify the necessity of "Maxwell's-demon-like" active intervention or control to prevent the collapse of economic or social systems into random or incoherent states.

But despite all these applications of the Second Law, the twentieth century largely passed without significant advances in understanding the origin and foundations of the Second Law. Though even by the early 1980s I was beginning to find results—based on computational ideas—that seemed as if they might finally give a foundational understanding of what's really happening in the Second Law, and the extent to which the Second Law can in the end be "derived" from underlying "mechanical" rules.

What the Textbooks Said: The Evolution of Certainty

Ask a typical physicist today about the Second Law and they're likely to be very sure that it's "just true". Maybe they'll consider it "another law of nature" like the conservation of energy, or maybe they'll think it is something that was "proved long ago" from basic principles of mathematics and mechanics. But as we've discussed here, there's really nowhere in the history of the Second Law that should give us this degree of certainty. So where did all the certainty come from? I think in the end it's a mixture of a kind of don't-question-this-it-comes-from-sophisticated-science mystique about the Second Law, together with a century and a half of "increasingly certain" textbooks. So let's talk about the textbooks.

While early contributions to what we now call thermodynamics (and particularly those from continental Europe) often got published as monographs, the first "actual textbooks" of thermodynamics already started to appear in the 1860s, with three examples (curiously, all in French) being:

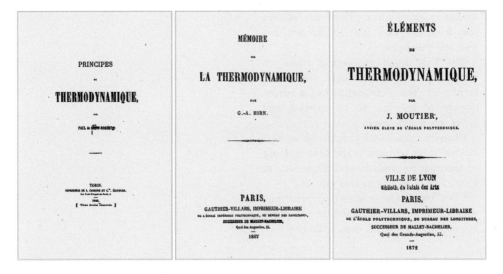

And in these early textbooks what one repeatedly sees is that the Second Law is simply cited—without much comment—as a "principle" or "axiom" (variously attributed to Carnot, Kelvin or Clausius, and sometimes called "the Principle of Carnot"), from which theory will be developed. By the 1870s there's a bit of confusion starting to creep in, because people are talking about the "Theorem of Carnot". But, at least at first, by this they mean not the Second Law, but the result on the efficiency of heat engines that Carnot derived from this.

Occasionally, there are questions in textbooks about the validity of the Second Law. A notable one, that we discussed above when we talked about Maxwell's demon, shows up under the title "Limitation of the Second Law of Thermodynamics" at the end of Maxwell's 1871 *Theory of Heat*.

Tait's largely historical 1877 *Sketch of Thermodynamics* notes that, yes, the Second Law hasn't successfully been proved from the laws of mechanics:

> modynamics. A great many attempts have of late been made to show that this second law is merely a form of Hamilton's Principle of *Varying Action*—a pure principle of dynamics. It is obvious, from the fact that if we could at any moment exactly reverse the motion of every particle, we should make a dynamical system (however complex) go back through all its previous states of motion, that such a deduction from Hamilton's principle can only be made by a method of averages which virtually *assumes* the degradation of energy, a consequence of the law to be proved.

In 1879, Eddy's *Thermodynamics* at first shows even more skepticism

> Various attempts* have been made to derive the fundamental equation of thermodynamics resulting from the second law from the first law, by known mechanical principles.
>
> But Tait says,† respecting these attempts, that they virtually assume, in course of the demonstration, a consequence of the law to be proved, and hence are inconclusive.
>
> Clausius incidentally referred to this matter in his address at the 41st meeting of German Naturalists and Physicists at Frankfort in words to this effect:
>
> "Besides, there is a second principle, which is not yet contained in the first, but requires a special demonstration."

but soon he's talking about how "Rankine's theory of molecular vortices" has actually "proved the Second Law":

> Rankine however states‡ that "Carnot's principle (*i.e.*, the second principle) is not an independent principle in the theory of heat, but is deducible as a consequence from the equations of the mutual conversion of heat and expansive power" (*i.e.*, from the first principle). The demonstration of this which he has given rests upon his hypothesis of molecular vortices.

He goes on to give some standard "phenomenological" statements of the Second Law, but then talks about "molecular hypotheses from which Carnot's principle has been derived":

> Besides these axioms, various molecular hypotheses have been advanced, from which Carnot's principle has been derived.
>
> 1°. Hypothesis of Molecular Vortices. (Rankine, 1849).
>
> 2°. Hypothesis of Circulating Streams of any figure whatever. (Rankine, 1851 and 1855) †.
>
> 3°. Hypothesis of Periodic Motions. (Boltzmann, 1866.‡)
>
> 4°. Hypothesis of Quasi-Periodic Mo-

Pretty soon there's confusion like the section in Alexandre Gouilly's (1842–1906) 1877 *Mechanical Theory of Heat* that's entitled "Second Fundamental Theorem of Thermodynamics or the Theorem of Carnot":

> **Deuxième théorème fondamental de la thermodynamique**
> **ou théorème de Carnot.**

More textbooks on thermodynamics follow, but the majority tend to be practical expositions (that are often incredibly similar to each other) with no particular theoretical discussion of the Second Law, its origins or validity.

In 1891 there's an "official report about the Second Law" commissioned by the British Association for the Advancement of Science (and written by a certain George Bryan (1864–1928) who would later produce a thermodynamics textbook):

> *Report of a Committee, consisting of* Messrs. J. LARMOR *and* G. H. BRYAN, *on the present state of our knowledge of Thermodynamics, specially with regard to the Second Law.*
>
> [Ordered by the General Committee to be printed among the Reports.]
>
> PART I.—RESEARCHES RELATING TO THE CONNECTION OF THE SECOND LAW WITH DYNAMICAL PRINCIPLES. DRAWN UP BY G. H. BRYAN.
>
> *Introduction.*
>
> 1. The present report treats exclusively of the attempts that have been made to deduce the Second Law of Thermodynamics from purely mechanical principles.

There's an enumeration of approaches so far:

> I. The Hypothesis of 'Stationary' or 'Quasi-Periodic' Motions as adopted by Clausius and Szily.
> II. The Hypothesis of 'Monocyclic Systems' of von Helmholtz, and similar hypotheses.
> III. The Statistical Hypothesis of Boltzmann, Clerk Maxwell, and other writers on the Kinetic Theory of Gases.

Somewhat confusingly it talks about a "proof of the Second Law"—actually referring to an already-in-equilibrium result:

> 46. *Applications to the Second Law.*—The simplest proof of the Second Law of Thermodynamics based on the hypothesis of the Boltzmann-Maxwell law of distribution of speed is that due to Mr. S. H. Burbury.[3] The proof is too well known to need description here. It leads to the same form for the entropy as Boltzmann's original investigation for the case of a system of point-atoms.[4] Although Watson and Burbury

There's talk of mechanical instability leading to irreversibility:

> 50. If we regard the whole matter as one of probabilities, the argument derived from reversing the system may be met without an appeal to the luminiferous æther. Although a conservative dynamical system is always reversible, the reversed motion may not unfrequently be dynamically unstable in the highest degree. One of the best illustrations in point is afforded by the impossibility of riding a bicycle backwards (*i.e.* with the steering wheel behind); here the forward motion is stable, but the reversed motion is highly unstable.

The conclusions say that, yes, the Second Law isn't proved "yet"

> 51. Although many of the researches mentioned in this report are not unfrequently called dynamical proofs of the Second Law, yet to prove the Second Law, about which we know something, by means of molecules, about which we know much less, would not be in consonance with the sentiments expressed at the end of the last paragraph. The most conclusive evidence for regarding Carnot's principle as a theorem in molecular dynamics lies in the remarkable agreement between the results obtained by the methods described in the three different sections of this report, all of which are based on different fundamental hypotheses. It

but imply that if only we knew more about molecules that might be enough to nail it:

> In conclusion we may reasonably hope that future researches in the domain of molecular science will still further strengthen the bond of

> connection which we suppose to exist between the Second Law of Thermodynamics and Newton's Laws of Motion.

But back to textbooks. In 1895 Boltzmann published his *Lectures on Gas Theory*, which includes a final chapter about the *H* theorem and its relation to the Second Law. Boltzmann goes through his mathematical derivations for gases, then (rather over-optimistically) asserts that they'll also work for solids and liquids:

We have looked mainly at processes in gases and have calculated the function H for this case. Yet the laws of probability that govern atomic motion in the solid and liquid states are clearly not qualitatively different ... from those for gases, so that the calculation of the function H corresponding to the entropy would not be more difficult in principle, although to be sure it would involve greater mathematical difficulties.

But soon he's discussing the more philosophical aspects of things (and by the time Boltzmann wrote this book, he was a professor of philosophy as well as physics). He says that the usual statement of the Second Law is "asserted phenomenologically as an axiom" (just as he says the infinite divisibility of matter also is at that time):

> ... the Second Law is formulated in such a way that the unconditional irreversibility of all natural processes is asserted as an axiom, just as general physics based on a purely phenomenological standpoint asserts the unconditional divisibility of matter without limit as an axiom.

One might then expect him to say that actually the Second Law is somehow provable from basic physical facts, such as the First Law. But actually his claims about any kind of "general derivation" of the Second Law are rather subdued:

> Since however the probability calculus has been verified in so many special cases, I see no reason why it should not also be applied to natural processes of a more general kind. The applicability of the probability calculus to the molecular motion in gases cannot of course be rigorously deduced from the differential equations for the motion of the molecules. It follows rather from the great number of the gas molecules and the length of their paths, by virtue of which the properties of the position in the gas where a molecule undergoes a collision are completely independent of the place where it collided the previous time.

But he still believes in the ultimate applicability of the Second Law, and feels he needs to explain why—in the face of the Second Law—the universe as we perceive it "still has interesting things going on":

> ... small isolated regions of the universe will always find themselves "initially" in an improbable state. This method seems to me to be the only way in which one can understand the Second Law—the heat death of each single world—without a unidirectional change of the entire universe from a definite initial state to a final state.

Meanwhile, he talks about the idea that elsewhere in the universe things might be different—and that, for example, entropy might be systematically decreasing, making (he suggests) perceived time run backwards:

> In the entire universe, the aggregate of all individual worlds, there will however in fact occur processes going in the opposite direction. But the beings who observe such processes will simply reckon time as proceeding from the less probable to the more probable states, and it will never be discovered whether they reckon time differently from us, since they are separated from us by eons of time and spatial distances $10^{10^{10}}$ times the distance of Sirius—and moreover their language has no relation to ours.

Most other textbook discussions of thermodynamics are tamer than this, but the rather anthropic-style argument that "we live in a fluctuation" comes up over and over again as an ultimate way to explain the fact that the universe as we perceive it isn't just a featureless maximum-entropy place.

It's worth noting that there are roughly three general streams of textbooks that end up discussing the Second Law. There are books about rather practical thermodynamics (of the type pioneered by Clausius), that typically spend most of their time on the equilibrium case. There are books about kinetic theory (effectively pioneered by Maxwell), that typically spend most of their time talking about the dynamics of gas molecules. And then there are books about statistical mechanics (as pioneered by Gibbs) that discuss with various degrees of mathematical sophistication the statistical characteristics of ensembles.

In each of these streams, many textbooks just treat the Second Law as a starting point that can be taken for granted, then go from there. But particularly when they are written by physicists with broader experience, or when they are intended for a not-totally-specialized audience, textbooks will quite often attempt at least a little justification or explanation for the Second Law—though rather often with a distinct sleight of hand involved.

For example, when Planck in 1903 wrote his *Treatise on Thermodynamics* he had a chapter in his discussion of the Second Law, misleadingly entitled "Proof". Still, he explains that:

> The second fundamental principle of thermodynamics [Second Law] being, like the first, an empirical law, we can speak of its proof only in so far as its total purport may be deduced from a single self-evident proposition. We, therefore, put forward the following proposition as being given directly by experience. It is impossible to construct an engine which will work in a complete cycle, and produce no effect except the raising of a weight and the cooling of a heat-reservoir.

In other words, his "proof" of the Second Law is that nobody has ever managed to build a perpetual motion machine that violates it. (And, yes, this is more than a little reminiscent of $P \neq NP$, which, through computational irreducibility, is related to the Second Law.) But after many pages, he says:

> In conclusion, we shall briefly discuss the question of the possible limitations to the Second Law. If there exist any such limitations—a view still held by many scientists and philosophers—then this [implies an error] in our starting point: the impossibility of perpetual motion ...

(In the 1905 edition of the book he adds a footnote that frankly seems bizarre in view of his—albeit perhaps initially unwilling—role in the initiation of quantum theory five years earlier: "The following discussion, of course, deals with the meaning of the Second Law only insofar as it can be surveyed from the points of view contained in this work avoiding all atomic hypotheses.")

He ends by basically saying "maybe one day the Second Law will be considered necessarily true; in the meantime let's assume it and see if anything goes wrong":

> Presumably the time will come when the principle of the increase of the entropy will be presented without any connection with experiment. Some metaphysicians may even put it forward as being *a priori* valid. In the meantime, no more effective weapon can be used by both champions and opponents of the Second Law than the indefatigable endeavour to follow the real purport of this law to the utmost consequences, taking the latter one by one to the highest court of appeal experience. Whatever the decision may be, lasting gain will accrue to us from such a proceeding, since thereby we serve the chief end of natural science—the enlargement of our stock of knowledge.

Planck's book came in a sense from the Clausius tradition. James Jeans's (1877–1946) 1904 book *The Dynamical Theory of Gases* came instead from the Maxwell + Boltzmann tradition. He says at the beginning—reflecting the fact the existence of molecules had not yet been firmly established in 1904—that the whole notion of the molecular basis of heat "is only a hypothesis":

> The essential feature of the Kinetic Theory is that it interprets heat in matter as a manifestation of a motion of the molecules which compose the matter. It need hardly be said that this identification of heat and motion is only a hypothesis: it never has been, and from the nature of things never can be, proved. At the same time this hypothesis shews an ability to explain and even to predict natural phenomena, such that there can be little doubt that it rests upon a foundation of truth.

Later he argues that molecular-scale processes are just too "fine-grained" to ever be directly detected:

> The principal lesson to be learned from the foregoing figures is that the mechanism of the Kinetic Theory is extremely "fine-grained" when measured by ordinary standards. Molecules are, in fact, not infinitely small,

> human observation go, from those of a continuous medium. It is for this reason that the hypothesis upon which the Kinetic Theory rests is, and probably will always remain, an unproved hypothesis.

But soon Jeans is giving a derivation of Boltzmann's H theorem, though noting some subtleties:

> 67. The assumption of molecular chaos (corrected, if necessary, in accordance with § 66) will therefore give correct results, provided it is interpreted with reference to the basis of probability supplied by our generalised space, and provided it is understood that it gives probable, and not certain, results. If we wish to obtain strictly accurate results the

His take on the "reversibility objection" is that, yes, the H function will be symmetric at every maximum, but, he argues, it'll also be discontinuous there:

> **70.** It may, perhaps, still be thought paradoxical that dH/dt is not zero at each of these maxima. The explanation is that the variation of H is not governed by the laws of the differential calculus, since this variation is not, strictly speaking, continuous. The value of H is constant between collisions of the molecules, and changes abruptly at every collision. When the number

And in the time-honored tradition of saying "it is clear" right when an argument is questionable, he then claims that an "obvious averaging" will give irreversibility and the Second Law:

> It is therefore clear that, averaged over all systems which have a given f, dH/dt will be negative except when f is the law of distribution for the normal state, in which case it is zero. This result is now in agreement with that of Chapter II.

Later in his book Jeans simply quotes Maxwell and mentions his demon:

> of work. This is the second law of thermodynamics, and it is undoubtedly true so long as we can deal with bodies only in mass and have no power of perceiving or handling the separate molecules of which they are made up. But if we conceive a being whose faculties are so sharpened that he can

Then effectively just tells readers to go elsewhere:

> The reader who wishes to study the question of irreversibility further is referred to the following works:
>
> (i) "Report on the Present State of our knowledge of Thermodynamics, specially with regard to the Second Law," by J. Larmor and G. H. Bryan, *British Association Report*, 1891 (Cardiff), p. 85.
>
> (ii) *Elementary Principles of Statistical Mechanics*, J. Willard Gibbs (Scribners, New York), 1902.
>
> (iii) *Vorlesungen über Gastheorie*, Boltzmann.

In 1907 George Bryan (whose 1891 report we mentioned earlier) published *Thermodynamics, an Introductory Treatise Dealing Mainly with First Principles and Their Direct Applications*. But despite its title, Bryan has now "walked back" the hopes of his earlier report and is just treating the Second Law as an "axiom":

> We are thus led to assume the following axiom which may be regarded as the simplest form of the Second Law of Thermodynamics:
> *Energy in the form of mechanical work is always wholly convertible into any other forms of energy to which the present theory is applicable, but the converse processes are not in general possible.*

And—presumably from his interactions with Boltzmann—is saying that the Second Law is basically an empirical fact of our particular experience of the universe, and thus not something fundamentally derivable:

> The necessity of the appeal to experience is manifest from the following considerations: If in our Universe events occur in a certain definite sequence, it is possible to conceive a universe in which events occur in the opposite sequence, by merely reversing the scale of *time*. In such a universe the transformations of energy would be exactly the opposite to those of which we have experience, and the forms of energy which are least capable of being converted into other forms in our Universe would become the most convertible. In stating this it is assumed that the individuals living in either universe possess the power of influencing the progress only of future events and possess a knowledge only of past events. This assumption is implicitly involved in all our ideas relating to irreversibility.

As the years went by, many thermodynamics textbooks appeared, increasingly with an emphasis on applications, and decreasingly with a mention of foundational issues—typically treating the Second Law essentially just as an absolute empirical "law of nature" analogous to the First Law.

But in other books—including some that were widely read—there were occasional mentions of the foundations of the Second Law. A notable example was in Arthur Eddington's (1882–1944) 1929 *The Nature of the Physical World*—where now the Second Law is exalted as having the "supreme position among the laws of Nature":

> dence. The law that entropy always increases—the second law of thermodynamics—holds, I think, the supreme position among the laws of Nature. If someone points out to you that your pet theory of the universe is in disagreement with Maxwell's equations—then so much the worse for Maxwell's equations. If it is found to be contradicted by observation—well, these experimentalists do bungle things sometimes. But if your theory is found to be against the second law of thermodynamics I can give you no hope; there is nothing for it but to collapse in deepest humiliation. This exaltation

Although Eddington does admit that the Second Law is probably not "mathematically derivable":

> The question whether the second law of thermodynamics and other statistical laws are mathematical deductions from the primary laws, presenting their results in a conveniently usable form, is difficult to answer; but I think it is generally considered that there is an unbridgeable hiatus. At the bottom of all the questions settled by secondary law there is an elusive conception of "*a priori* probability of states of the world" which involves an essentially different attitude to knowledge from that presupposed in the construction of the scheme of primary law.

And even though in the twentieth century questions about thermodynamics and the Second Law weren't considered "top physics topics", some top physicists did end up talking about them, if nothing else in general textbooks they wrote. Thus, for example, in the 1930s and 1940s people like Enrico Fermi (1901–1954) and Wolfgang Pauli (1900–1958) wrote in some detail about the Second Law—though rather strenuously avoided discussing foundational issues about it.

Lev Landau (1908–1968), however, was a different story. In 1933 he wrote a paper "On the Second Law of Thermodynamics and the Universe" which basically argues that our everyday experience is only possible because "the world as a whole does not obey the laws of thermodynamics"—and suggests that perhaps relativistic quantum mechanics (which he says, quoting Niels Bohr (1885–1962), could be crucial in the center of stars) might fundamentally violate the Second Law. (And yes, even today it's not clear how "relativistic temperature" works.)

But this kind of outright denial of the Second Law had disappeared by the time Lev Landau and Evgeny Lifshitz (1915–1985) wrote the 1951 version of their book *Statistical Mechanics*—though they still showed skepticism about its origins:

> There is no doubt that the foregoing simple formulations [of the Second Law] accord with reality; they are confirmed by all our everyday observations. But when we consider more closely the problem of the physical nature and origin of these laws of behaviour, substantial difficulties arise, which to some extent have not yet been overcome.

Their book continues, discussing Boltzmann's fluctuation argument:

> Firstly, if we attempt to apply statistical physics to the entire universe ... we immediately encounter a glaring contradiction between theory and experiment. According to the results of statistics, the universe ought to be in a state of complete statistical equilibrium. ... Everyday experience shows us, however, that the properties of Nature bear no resemblance to those of an equilibrium system; and astronomical results show that the same is true throughout the vast region of the Universe accessible to our observation.
>
> We might try to overcome this contradiction by supposing that the part of the Universe which we observe is just some huge fluctuation in a system which is in equilibrium as a whole. The fact that we have been able to observe this huge fluctuation might be explained by supposing that the existence of such a fluctuation is a necessary condition for the existence of an observer (a condition for the occurrence of biological evolution). This argument, however, is easily disproved, since a fluctuation within, say, the volume of the solar system only would be very much more probable, and would be sufficient to allow the existence of an observer.

What do they think is the way out? The effect of gravity:

> ... in the general theory of relativity, the Universe as a whole must be regarded not as a closed system but as a system in a variable gravitational field. Consequently the application of the law of increase of entropy does not prove that statistical equilibrium must necessarily exist.

But they say this isn't the end of the problem, essentially noting the reversibility objection. How should this be overcome? First, they suggest the solution might be that the observer somehow "artificially closes off the history of a system", but then they add:

> Such a dependence of the laws of physics on the nature of an observer is quite inadmissible, of course.

They continue:

> At the present time it is not certain whether the law of increase of entropy thus formulated can be derived on the basis of classical mechanics. ... It is more reasonable to suppose that the law of increase of entropy in the above general formulation arises from quantum effects.

They talk about the interaction of classical and quantum systems, and what amounts to the explicit irreversibility of the traditional formalism of quantum measurement, then say that if quantum mechanics is in fact the ultimate source of irreversibility:

> ... there must exist an inequality involving the quantum constant $\hbar$ which ensures the validity of the law and is satisfied in the real world...

What about other textbooks? Joseph Mayer (1904–1983) and Maria Goeppert Mayer's (1906–1972) 1940 *Statistical Mechanics* has the rather charming

4e. The Limits of Validity of the Second Law of Thermodynamics

> " What, never?" " No, never!"
> " What, never?" " Well, hardly ever!"
> *H. M. S. Pinafore.*

The second law of thermodynamics can be stated in the form: the entropy of an isolated system never decreases. We have now claimed that this fundamental law is a consequence of the theorems of mechanics. It is appropriate at this time to investigate the extent to which exceptions to this law might conceivably be observed.

though in the end they sidestep difficult questions about the Second Law by basically making convenient definitions of what S and Ω mean in $S = k \log \Omega$.

For a long time one of the most cited textbooks in the area was Richard Tolman's (1881–1948) 1938 *Principles of Statistical Mechanics*. Tolman (basically following Gibbs) begins by explaining that statistical mechanics is about making predictions when all you know are probabilistic statements about initial conditions:

theoretically possible. The principles of ordinary mechanics may be regarded as allowing us to make precise predictions as to the future state of a mechanical system from a precise knowledge of its initial state.†
On the other hand, the principles of statistical mechanics are to be regarded as permitting us to make reasonable predictions as to the future condition of a system, which may be expected to hold on the average, starting from an incomplete knowledge of its initial state.

Tolman continues:

> Since our actual contacts with the physical world are such that we never do have the maximal knowledge of systems regarded as theoretically allowable, the idea of the precise state of a system is in any case an abstract limiting concept. Hence the methods of ordinary mechanics really apply to somewhat highly idealized situations, and the methods of statistical mechanics provide a significant supplement in the direction of decreased abstraction and closer correspondence between theoretical methods and actual experience. Even in the case of simple systems of only a few degrees of freedom, where our lack of maximal knowledge is not due to difficulties arising from the complexity of the system, the methods of statistical mechanics may be applied to a system whose initial state is not completely specified.‡

He notes that, historically, statistical mechanics was developed for studying systems like gases, where (in a vague foreshadowing of the concept of computational irreducibility) "it is evident that we should be quickly lost in the complexities of our computations" if we try to trace every molecule, but where, he claims, statistical mechanics can still accurately tell us "statistically" what will happen:

> mechanics. For example, in the case of a gas, consisting, say, of a large number of simple classical particles, even if we were given at some initial time the positions and velocities of all the particles so that we could foresee the collisions that were about to take place, it is evident that we should be quickly lost in the complexities of our computations if we tried to follow the results of such collisions through any extended length of time. Nevertheless, a system such as a gas composed of many molecules is actually found to exhibit perfectly definite regularities in its behaviour, which we feel must be ultimately traceable to the laws of mechanics even though the detailed application of these laws defies our powers. For the treatment of such regularities in the behaviour of complicated systems of many degrees of freedom, the methods of statistical mechanics are adequate and especially appropriate.

But where exactly should we get the probability distributions for initial states from? Tolman says he's going to consider the kinds of mathematically defined ensembles that Gibbs discusses. And tucked away at the end of a chapter he admits that, well, yes, this setup is really all just a postulate—set up so as to make the results of statistical mechanics "merely a matter for computation":

> From the point of view adopted in the present book, the hypothesis, of equal *a priori* probabilities for equal regions in the phase space, must in any case be regarded as an essential element of statistical mechanics, which has to be introduced by postulation and which is then sufficient to provide the statistical methods used. *Without* this postulate there would be nothing to correspond to the circumstance that nature does not have any tendency to present us with systems in conditions which we regard as mechanically entirely possible but statistically improbable; and *with* the postulate the use of statistical mechanics for the determination of averages and fluctuations then becomes merely a matter for computation.

On this basis Tolman then derives Boltzmann's H theorem, and his $\overline{\overline{H}}$ "coarse-grained" generalization (where, yes, the coarse graining ultimately operates according to his postulate). For 530 pages, there's not a single mention of the Second Law. But finally, on page 558 Tolman is at least prepared to talk about an "analog of the Second Law":

> **130. The analogue of the second law of thermodynamics**
>
> In preceding sections we have discussed the statistical mechanical analogues for a considerable number of thermodynamic processes that involve the second law of thermodynamics. These processes included

And basically what Tolman argues is that his $\overline{\overline{H}}$ can reasonably be identified with thermodynamic entropy S. In the end, the argument is very similar to Boltzmann's, though Tolman seems to feel that it has achieved more:

> In concluding this chapter, it is hoped that due appreciation will be felt for the importance and significance of the great achievement of statistical mechanics in providing a fundamental, mechanical interpretation and explanation of the principles of thermodynamics.

Very different in character from Tolman's book, another widely cited book is Percy Bridgman's (1882–1961) largely philosophical 1943 *The Nature of Thermodynamics*. His chapter on the Second Law begins:

> **THE SECOND LAW OF THERMODYNAMICS**
>
> THERE have been nearly as many formulations of the second law as there have been discussions of it. Although many of these formulations are doubtless roughly equivalent, and the proof that they are equivalent has been considered to be one of the tasks of a thermodynamic analysis, I question whether any really rigorous examination has been attempted from the postulational point of view and I question whether such an examination would be of great physical interest. It does seem obvious, however, that not all these formulations can be exactly equivalent, but it is possible to distinguish stronger and weaker forms.

A decade earlier Bridgman had discussed outright violations of the Second Law, saying that he'd found that the younger generation of physicists at the time seemed to often think that "it may be possible some day to construct a machine which shall violate the Second Law on a scale large enough to be commercially profitable"—perhaps, he said, by harnessing Brownian motion:

STATISTICAL MECHANICS AND THE SECOND LAW OF THERMODYNAMICS[1]

By Dr. P. W. BRIDGMAN
HARVARD UNIVERSITY

ONE thing that has much impressed me in recent conversations with physicists, particularly those of the younger generation, is the frequency of the conviction that it may be possible some day to construct a machine which shall violate the second law of thermodynamics on a scale large enough to be commercially profitable. This constitutes a striking reversal of the attitude of the founders of thermodynamics, Kelvin and Clausius, who postulated the impossibility of perpetual motion of the second kind as a generalization from the uniformly unsuccessful attempts of

[1] The Ninth Josiah Willard Gibbs Lecture, delivered at New Orleans, December 29, 1931, under the auspices of the American Mathematical Society, at a joint meeting of the society with the American Physical Society, and Section A of the American Association for the Advancement of Science.

the entire human race to realize it. Paradoxically, one very important factor in bringing about this change in attitude is the feeling of better understanding of the second law which the present generation enjoys, and which is largely due to the universal acceptance of the explanation of the second law in statistical terms, for which Gibbs was in so large a degree responsible. Statistical mechanics reduces the second law from a law of ostensibly absolute validity to a statement about high probabilities, leaving open the possibility that once in a great while there may be important violations. Doubtless another most important factor in present scepticism as to the ultimate commercial validity of the second law is the discovery of the importance in many physical phenomena of those fluctuation effects which are demanded by sta-

At a philosophical level, a notable treatment of the Second Law appeared in Hans Reichenbach's (1891–1953) (unfinished-at-his-death) 1953 work *The Direction of Time*. Wanting to make use of the Second Law, but concerned about the reversibility objections, Reichenbach introduces the notion of "branch systems"—essentially parts of the universe that can eventually be considered isolated, but which were once connected to other parts that were responsible for determining their ("nonrandom") effective initial conditions:

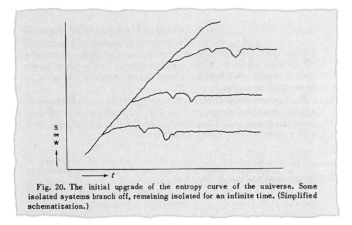

Fig. 20. The initial upgrade of the entropy curve of the universe. Some isolated systems branch off, remaining isolated for an infinite time. (Simplified schematization.)

Most textbooks that cover the Second Law use one of the formulations that we've already discussed. But there is one more formulation that also sometimes appears, usually associated with the name "Carathéodory" or the term "axiomatic thermodynamics".

Back in the first decade of the twentieth century—particularly in the circle around David Hilbert (1862–1943)—there was a lot of enthusiasm for axiomatizing things, including physics. And in 1908 the mathematician Constantin Carathéodory (1873–1950) suggested an axiomatization of thermodynamics. His essential idea—that he developed further in the 1920s—was to consider something like Gibbs's phase fluid and then roughly to assert that it gets (in some measure-theoretic sense) "so mixed up" that there aren't "experimentally doable" transformations that can unmix it. Or, in his original formulation:

> In any arbitrary neighborhood of an arbitrarily given initial point there is a state that cannot be arbitrarily approximated by adiabatic changes of state.

There wasn't much pickup of this approach—though Max Born (1882–1970) supported it, Max Planck dismissed it, and in 1939 S. Chandrasekhar (1910–1995) based his exposition of stellar structure on it. But in various forms, the approach did make it into a few textbooks. An example is Brian Pippard's (1920–2008) otherwise rather practical 1957 *The Elements of Classical Thermodynamics*:

> A third formulation, due to Carathéodory, is not so clearly related:
>
> *In the neighbourhood of any equilibrium state of a system there are states which are inaccessible by an adiathermal process.*
>
> Each formulation has its enthusiastic supporters. For the engineer or the practically minded physicist Clausius's or Kelvin's formulations are more directly meaningful, and, moreover, the derivation from them of the important consequences of the law may be made without a great deal of mathematics. On the other hand, Carathéodory's formulation is undoubtedly more economical, in that it demands the impossibility of a rather simpler type of process than that considered

> accomplished adiathermally. Carathéodory's law appears to differ in outlook from the others. The average physicist is prepared to take Clausius's and Kelvin's laws as reasonable generalizations of common experience, but Carathéodory's law (at any rate in the author's opinion) is not immediately acceptable except in the trivial cases, of which Joule's experiment is one; it is neither intuitively obvious nor supported by a mass of experimental evidence. It may be argued therefore that the further development of thermodynamics should not be made to rest on this basis, but that Carathéodory's law should be regarded, in view of the fact that it leads to the same conclusions as the others, as a statement of the minimal postulate which is needed in order to achieve the desired end. It bears somewhat the same relation to the other statements as Hamilton's principle bears to Newton's laws of motion.

Yet another (loosely related) approach is the "postulatory formulation" on which Herbert Callen's (1919–1993) 1959 textbook *Thermodynamics* is based:

> The postulatory formulation of thermodynamics features states, rather than processes, as fundamental constructs. Statements about Carnot cycles and about the impossibility of perpetual motion of various kinds do not appear in the postulates, but state functions, energy, and entropy become the fundamental concepts. An enormous simplification in the mathematics is obtained, for processes then enter simply as differentials of the state functions. The conventional method proceeds inversely from processes to state functions by the relatively difficult procedure of integration of partial differential equations.

In effect this is now "assuming the result" of the Second Law:

Postulate II. *There exists a function (called the entropy S) of the extensive parameters of any composite system, defined for all equilibrium states and having the following property. The values assumed by the extensive parameters in the absence of an internal constraint are those that maximize the entropy over the manifold of constrained equilibrium states.*

Though in an appendix he rather tautologically states:

The entropy of the system is defined as

$$S(X) = k \ln N(X) \qquad \text{(B.1)}$$

in which k is Boltzman's constant. Consequently, the macrostate (X)

observed is that corresponding to the maximum value of the entropy, consistent with the imposed constraints. The entropy of any macrostate is proportional to the logarithm of the number of microstates associated with that macrostate.

The foregoing statements, although made in reference to a highly artificial model of a system, are actually general. They are valid for any closed thermodynamic system.

So what about other textbooks? A famous set are Richard Feynman's (1918–1988) 1963 *Lectures on Physics*. Feynman starts his discussion of the Second Law quite carefully, describing it as a "hypothesis":

hot body. Now, the hypothesis of Carnot, the second law of thermodynamics, is sometimes stated as follows: heat cannot, of itself, flow from a cold to a hot object.

Feynman says he's not going to go very far into thermodynamics, though quotes (and criticizes) Clausius's statements:

The two laws of thermodynamics are often stated this way:

First law: the energy of the universe is always constant.

Second law: the entropy of the universe is always increasing.

That is not a very good statement of the second law; it does not say, for example, that in a reversible cycle the entropy stays the same, and it does not say exactly what the entropy is. It is just a clever way of remembering the two laws, but it

But then he launches into a whole chapter on "Ratchet and pawl":

46

Ratchet and pawl

46–1 How a ratchet works

In this chapter we discuss the ratchet and pawl, a very simple device which allows a shaft to turn only one way. The possibility of having something turn only one way requires some detailed and careful analysis, and there are some very interesting consequences.

The plan of the discussion came about in attempting to devise an elementary explanation, from the molecular or kinetic point of view, for the fact that there is a maximum amount of work which can be extracted from a heat engine. Of course we have seen the essence of Carnot's argument, but it would be nice to find an explanation which is elementary in the sense that we can see what is happening

46–1 How a ratchet works
46–2 The ratchet as an engine
46–3 Reversibility in mechanics
46–4 Irreversibility
46–5 Order and entropy

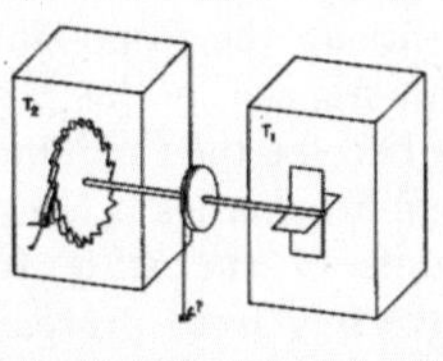

Fig. 46–1. The ratchet and pawl machine.

His goal, he explains, is to analyze a device (similar to what Marian Smoluchowski had considered in 1912) that one might think by its one-way ratchet action would be able to "harvest random heat" and violate the Second Law. But after a few pages of analysis he claims that, no, if the system is in equilibrium, thermal fluctuations will prevent systematic "one-way" mechanical work from being achieved, so that the Second Law is saved.

But now he applies this to Maxwell's demon, claiming that the same basic argument shows that the demon can't work:

> It turns out, if we build a finite-sized demon, that the demon himself gets so warm that he cannot see very well after a while. The simplest possible demon, as an example, would be a trap door held over the hole by a spring. A fast molecule comes through, because it is able to lift the trap door. The slow molecule cannot get through, and bounces back. But this thing is nothing but our ratchet and pawl in another form, and ultimately the mechanism will heat up. If we assume that the specific heat of the demon is not infinite, it must heat up. It has but a finite number of internal gears and wheels, so it cannot get rid of the extra heat that it gets from observing the molecules. Soon it is shaking from Brownian motion so much that it cannot tell whether it is coming or going, much less whether the molecules are coming or going, so it does not work.

But what about reversibility? Feynman first discusses what amounts to Boltzmann's fluctuation idea:

> Thus one possible explanation of the high degree of order in the present-day world is that it is just a question of luck. Perhaps our universe happened to have had a fluctuation of some kind in the past, in which things got somewhat separated, and now they are running back together again. This kind of theory is not un-

But then he opts instead for the argument that for some reason—then unknown—the universe started in a "low-entropy" state, and has been "running down" ever since:

> This is not to say that we understand the logic of it. For some reason, the universe at one time had a very low entropy for its energy content, and since then the entropy has increased. So that is the way toward the future. That is the origin of all irreversibility, that is what makes the processes of growth and decay, that makes us remember the past and not the future, remember the things which are closer to that moment in the history of the universe when the order was higher than now, and why we are not able to remember things where the disorder is higher than now, which we call the future. So, as we commented in an earlier chapter, the

By the beginning of the 1960s an immense number of books had appeared that discussed the Second Law. Some were based on macroscopic thermodynamics, some on kinetic theory and some on statistical mechanics. In all three of these cases there was elegant mathematical theory to be described, even if it never really addressed the ultimate origin of the Second Law.

But by the early 1960s there was something new on the scene: computer simulation. And in 1965 that formed the core of Fred Reif's (1927–2019) textbook *Statistical Physics*:

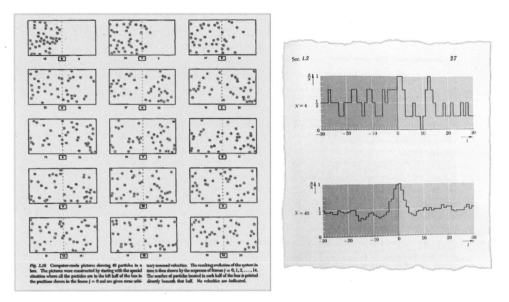

In a sense the book is an exploration of what simulated hard-sphere gases do—as analyzed using ideas from statistical mechanics. (The simulations had computational limitations, but they could go far enough to meaningfully see most of the basic phenomena of statistical mechanics.)

Even the front and back covers of the book provide a bold statement of both reversibility and the kind of randomization that's at the heart of the Second Law:

But inside the book the formal concept of entropy doesn't appear until page 147—where it's defined very concretely in terms of states one can explicitly enumerate:

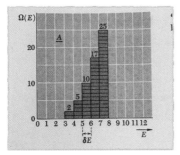

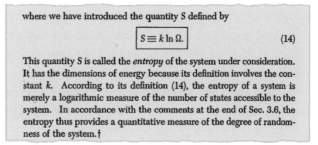

where we have introduced the quantity S defined by

$$S \equiv k \ln \Omega. \tag{14}$$

This quantity S is called the *entropy* of the system under consideration. It has the dimensions of energy because its definition involves the constant k. According to its definition (14), the entropy of a system is merely a logarithmic measure of the number of states accessible to the system. In accordance with the comments at the end of Sec. 3.6, the entropy thus provides a quantitative measure of the degree of randomness of the system.†

And finally, on page 283—after all necessary definitions have been built up—there's a rather prosaic statement of the Second Law, almost as a technical footnote:

Statement 2

We have seen that the number of states accessible to a system (or, equivalently, its entropy) is a quantity of fundamental importance in describing the macrostate of a system. In Sec. 7.2 we showed that changes in the entropy of a system can be related by (32) to the heat absorbed by this system. In Sec. 3.6 we also showed that an *isolated* system tends to approach a situation of greater probability where the number of states accessible to it (or equivalently, its entropy) is larger than initially. (As a special case, when the system is initially already in its most probable situation, it remains in equilibrium and its entropy remains unchanged.) Thus we arrive at the following statement:

Second law of thermodynamics

An equilibrium macrostate of a system can be characterized by a quantity S (called *entropy*) which has the following properties:

   (i) In any infinitesimal quasi-static process in which the system absorbs heat dQ, its entropy changes by an amount

$$dS = \frac{dQ}{T} \tag{61}$$

where T is a parameter characteristic of the macrostate of the system and is called its *absolute temperature*.

   (ii) In any process in which a thermally isolated system changes from one macrostate to another, its entropy tends to increase, i.e.,

$$\Delta S \geq 0. \tag{62}$$

Looking though many textbooks of thermodynamics and statistical mechanics it's striking how singular Reif's "show-don't-tell" computer-simulation approach is. And, as I've described in detail elsewhere, for me personally it has a particular significance, because this is the book that in 1972, at the age of 12, launched me on what has now been a 50-year journey to understand the Second Law and its origins.

When the first textbooks that described the Second Law were published nearly a century and a half ago they often (though even then not always) expressed uncertainty about the Second Law and just how it was supposed to work. But it wasn't long before the vast majority of books either just "assumed the Second Law" and got on with whatever they wanted to apply it to, or tried to suggest that the Second Law had been established from underlying principles, but that it was a sophisticated story that was "out of the scope of this book" but to be found elsewhere. And so it was that a strong sense emerged that the Second Law was something whose ultimate character and origins the typical working scientist didn't need to question—and should just believe (and protect) as part of the standard canon of science.

So Where Does This Leave the Second Law?

The Second Law is now more than 150 years old. But—at least until now—I think it's fair to say that the fundamental ideas used to discuss it haven't materially changed in more than a century. There's a lot that's been written about the Second Law. But it's always tended to follow lines of development already defined over a century ago—and mostly those from Clausius, or Boltzmann, or Gibbs.

Looking at word clouds of titles of the thousands of publications about the Second Law over the decades we see just a few trends, like the appearance of the "generalized Second Law" in the 1990s relating to black holes:

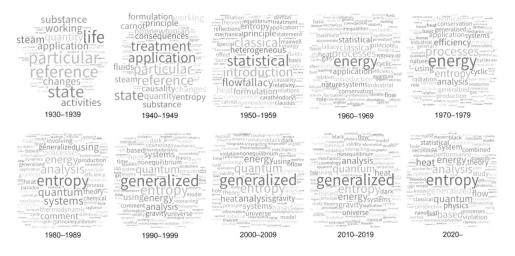

But with all this activity why hasn't more been worked out about the Second Law? How come after all this time we still don't really even understand with clarity the correspondence between the Clausius, Boltzmann and Gibbs approaches—or how their respective definitions of "entropy" are ultimately related?

In the end, I think the answer is that it needs a new paradigm—that, yes, is fundamentally based on computation and on ideas like computational irreducibility. A little more than a century ago—with people still actively arguing about what Boltzmann was saying—I don't think anyone would have been too surprised to find out that to make progress would need a new way of looking at things. (After all, just a few years earlier Boltzmann and Gibbs had needed to bring in the new idea of using probability theory.)

But as we discussed, by the beginning of the twentieth century—with other areas of physics heating up—interest in the Second Law was waning. And even with many questions unresolved people moved on. And soon several academic generations had passed. And as is typical in the history of science, by that point nobody was questioning the foundations anymore. In the particular case of the Second Law there was some sense that the uncertainties had to do with the assumption of the existence of molecules, which had by then been established. But more important, I think, was just the passage of "academic time" and the fact that what might once have been a matter of discussion had now just become a statement in the textbooks—that future academic generations should learn and didn't need to question.

One of the unusual features of the Second Law is that at the time it passed into the "standard canon of science" it was still rife with controversy. How did those different approaches relate? What about those "mathematical objections"? What about the thought experiments that seemed to suggest exceptions? It wasn't that these issues were resolved. It was just that after enough time had passed people came to assume that "somehow that must have all been worked out ages ago".

And it wasn't that there was really any pressure to investigate foundational issues. The Second Law—particularly in its implications for thermal equilibrium—seemed to work just fine in all its standard applications. And it even seemed to work in new domains like black holes. Yes, there was always a desire to extend it. But the difficulties encountered in trying to do so didn't seem in any obvious way related to issues about its foundations.

Of course, there were always a few people who kept wondering about the Second Law. And indeed I've been surprised at how much of a *Who's Who* of twentieth-century physics this seems to have included. But while many well-known physicists seem to have privately thought about the foundations of the Second Law they managed to make remarkably little progress—and as a result left very few visible records of their efforts.

But—as is so often the case—the issue, I believe, is that a fundamentally new paradigm was needed in order to make real progress. When the "standard canon" of the Second Law was formed in the latter part of the nineteenth century, calculus was the primary tool for physics—with probability theory a newfangled addition introduced specifically for studying the Second Law. And from that time it would be many decades before even the beginnings of the computational paradigm began to emerge, and nearly a century before phenomena like computational irreducibility were finally discovered. Had the sequence been different I have no doubt that what I have now been able to understand about the Second Law would have been worked out by the likes of Boltzmann, Maxwell and Kelvin.

But as it is, we've had to wait more than a century to get to this point. And having now studied the history of the Second Law—and seen the tangled manner in which it developed—I believe that we can now be confident that we have indeed successfully been able to resolve many of the core issues and mysteries that have plagued the Second Law and its foundations over the course of nearly 150 years.

Note

Almost all of what I say here is based on my reading of primary literature, assisted by modern tools and by my latest understanding of the Second Law. About some of what I discuss, there is—sometimes quite extensive—existing scholarship; some references are given in the bibliography on page 405.

Annotated Bibliography

For an extensive discussion of the history of the Second Law, see (included in this book):

S. Wolfram (2023), "How Did We Get Here? The Tangled History of the Second Law of Thermodynamics". writings.stephenwolfram.com/2023/01/how-did-we-get-here-the-tangled-history-of-the-second-law -of-thermodynamics.

For a discussion of the personal and other background history of this work, see (included in this book):

S. Wolfram (2023), "A 50-Year Quest: My Personal Journey with the Second Law of Thermodynamics". writings.stephenwolfram.com/2023/02/a-50-year-quest-my-personal-journey-with-the-second-law -of-thermodynamics.

Pointers to specific references are included as hyperlinks in the online version of this piece (wolfr.am/SW-SecondLaw).

Development of the Approach Described Here

The concept of computational irreducibility was described in (included in this book):

S. Wolfram (1985), "Undecidability and Intractability in Theoretical Physics", *Physical Review Letters* 54, 735–738. content.wolfram.com/undecidability-intractability-theoretical-physics.pdf.

An early description of the computational character of the Second Law was given in (included in this book):

S. Wolfram (1985), "Origins of Randomness in Physical Systems", *Physical Review Letters* 55, 449–452. content.wolfram.com/origins-randomness-physical-systems.pdf.

Further development was done in (included in this book):

S. Wolfram (2002), "Irreversibility and the Second Law of Thermodynamics", in *A New Kind of Science*, Wolfram Media, 441–457. wolframscience.com/nks/chap-9--fundamental-physics/#sect-9-3 --irreversibility-and-the-second-law-of-thermodynamics.

The Wolfram Physics Project is described in:

S. Wolfram (2020), "A Class of Models with the Potential to Represent Fundamental Physics". arXiv:2004.08210.

S. Wolfram (2020), *A Project to Find the Fundamental Theory of Physics*, Wolfram Media.

The "particle cellular automaton" used here was introduced in:

S. Wolfram (1986), "Minimal Cellular Automaton Approximations to Continuum Systems", presented at *Cellular Automata '86*; reprinted in *Cellular Automata and Complexity: Collected Papers* (1994), Addison-Wesley. content.wolfram.com/cellular-automaton-continuum-systems.pdf.

Other Works on the Second Law

Classic Original Sources

S. Carnot (1824), *Réflexions sur la puissance motrice du feu et sur les machines propres à développer cette puissance* (in French), Bachelier. ark:/13960/t7rn68p52. (Translated by R. Thurston (1890), as *Reflections on the Motive Power of Heat*, reprinted in *Reflections on the Motive Power of Fire* (1988), Dover. ark:/13960/t0jv2661s.)

R. Clausius (1857), "Über die Art der Bewegung die wir Wärme nennen" (in German), *Annalen der Physik* 100, 353–380. ark:/13960/t9679978p. (Translated as "On the Nature of the Motion Which We Call Heat", *Philosophical Magazine* 14, 108–127 (1857). ark:/13960/t3jw8w07q.)

J. Maxwell (1858), "Illustrations of the Dynamical Theory of Gases", *Philosophical Magazine* 19, 19–32. ark:/13960/t4sj20460.

J. Maxwell (1866), "On the Dynamical Theory of Gases", *Philosophical Transactions of the Royal Society of London* 157, 49–88. ark:/13960/t06x5c78d.

L. Boltzmann (1872), "Weitere Studien über das Wärmegleichgewicht unter Gasmolekülen" (in German), *Sitzungsberichte Akademie der Wissenschaften* 66, 275–370. ark:/13960/t4pk0sf66. (Translated as "Further Studies on the Thermal Equilibrium of Gas Molecules", in *The Kinetic Theory of Gases: An Anthology of Classic Papers with Historical Commentary* (2003), S. Brush (ed.), Imperial College Press, 262–349. doi: 10.1142/p281.)

W. Thomson (1874), "The Kinetic Theory of the Dissipation of Energy", *Proceedings of the Royal Society of Edinburgh* 8, 325–334. ark:/13960/t2b85fc5t.

L. Boltzmann (1877), "Über die Beziehung zwischen dem zweiten Hauptsatze der mechanischen Wärmetheorie und der Wahrscheinlichkeitsrechnung respective den Sätzen über das Wärmegleichgewicht" (in German), *Sitzungberichte Akademie der Wissenschaften* 76, 373–435. doi: 10.1017/CBO9781139381437.011. (Translated as "On the Relationship between the Second Fundamental Theorem of the Mechanical Theory of Heat and Probability Calculations Regarding the Conditions for Thermal Equilibrium", *Entropy* 17, 1971–2009 (2015). doi: 10.3390/e17041971.)

J. Gibbs (1902), *Elementary Principles of Statistical Mechanics*, Charles Scribner's Sons. ark:/13960/t6rz5sz8r.

Notable Collections, etc.

R. Clausius (1864), *Abhandlungen über die mechanische Wärmetheorie* (in German), Friedrich Viewweg und Sohn. ark:/13960/t53f4qn28. (Translated by W. Browne (1879), as *The Mechanical Theory of Heat*, Macmillan and Co. ark:/13960/t1wd4fr9c.)

W. Thomson (1882-1911), *Mathematical and Physical Papers* vols. 1–6, Cambridge University Press. ark:/13960/t20d2r137.

W. Niven (ed.) (1890), *The Scientific Papers of James Clerk Maxwell* vols. 1 & 2, Dover. ark:/13960/t3pv6kt6d.

A. Tuckerman (1890), "Index to the Literature of Thermodynamics", *Smithsonian Miscellaneous Collections* 34, iii–239. repository.si.edu/handle/10088/23167.

F. Hasenöhrl (ed.) (1909), *Wissenschaftliche Abhandlungen von Ludwig Boltzmann* (in German) [*Scientific Works of Ludwig Boltzmann*] vols. 1–3, Johann Ambrosius Barth. (Available on HathiTrust.)

F. Donnan and A. Haas (eds.) (1936), *A Commentary on the Scientific Writings of J. Willard Gibbs* vols. 1 & 2, Yale University Press. ark:/13960/t6sx6c24j.

J. Stachel, D. Cassidy, J. Renn and R. Schulmann (eds.) (1990), *The Collected Papers of Albert Einstein* vol. 2, Princeton University Press. einsteinpapers.press.princeton.edu/vol2-doc.

H. Ebbinghaus, C. Fraser and A. Kanamori (eds.) (2010), *Ernst Zermelo—Collected Works/Gesammelte Werke* vols. 1 & 2, Springer. doi: 10.1007/978-3-540-79384-7.

Notable Textbooks

J. Maxwell (1871), *Theory of Heat,* Longmans, Green, and Co. (Available on Wikimedia.org.) (Reprinted as *Theory of Heat* (1872), D. Aplleton and Co. ark:/13960/t26976m7w.)

L. Boltzmann (1896), *Vorlesungen über Gastheorie* (in German), Johann Ambrosius Barth. ark:/13960/t18k7bb6k. (Translated by S. Brush (1995), as *Lectures on Gas Theory*, Dover. ark:/13960/t40s8nn1q.)

M. Planck (1897), *Vorlesungen über Thermodynamik* (in German), Veit & Comp. ark:/13960/t1hh6k62v. (Translated by A. Ogg (1905), as *Treatise on Thermodynamics* 3rd edition, Dover. ark:/13960/t06x4mn0f.)

J. Jeans (1904), *The Dynamical Theory of Gases*, Cambridge University Press. ark:/13960/t8pc2w43r.

P. Ehrenfest and T. Ehrenfest (1912), "Begriffliche Grundlagen der Statistischen Auffasung in der Mechanik" (in German), *Encyklopädie der mathematischen Wissenschaften* 4, part 4, 1–90. lorentz.leidenuniv.nl/IL-publications/sources/Ehrenfest_1911b.pdf. (Translated by M. Moravcsik (1959), as *The Conceptual Foundations of the Statistical Approach in Mechanics*, Cornell University Press. ark:/13960/t8jf2nx05.)

L. Landau and E. Lifshitz (1938), *Statisticheskaia fizika* (in Russian). (Translated by E. Peierls and R. Peierls (1958), as *Statistical Physics*, Pergamon Press. ark:/13960/t07x01p71.)

R. Tolman (1938), *The Principles of Statistical Mechanics*, Oxford University Press. ark:/13960/t9w11hx7r.

A. Sommerfeld (1956), *Thermodynamics and Statistical Mechanics*, Academic Press. ark:/13960/t53g3bb8j.

H. Callen (1960), *Thermodynamics: An Introduction to the Physical Theories of Equilibrium Thermostatics and Irreversible Thermodynamics*, Wiley & Sons. ark:/13960/t77t65w24.

K. Huang (1963), *Statistical Mechanics*, Wiley & Sons. ark:/13960/t14n0bb0g.

F. Reif (1965), *Statistical Physics*, McGraw Hill Company. ark:/13960/t44q8ww6q.

D. Ruelle (1969), *Statistical Mechanics: Rigorous Results*, Imperial College Press. ark:/13960/t6rz72b6s.

L. Landau and E. Lifshitz (1981), *Physical Kinetics*, Pergamon Press. ark:/13960/t20d2fj56.

G. Gallavotti (1999), *Statistical Mechanics: A Short Treatise*, Springer. doi: 10.1007/978-3-662-03952-6.

Surveys about Foundations

P. Bridgman (1941), *The Nature of Thermodynamics*, Harvard University Press. ark:/13960/t7vm43w9f.

P. Landsberg (1956), "Foundations of Thermodynamics", *Reviews of Modern Physics* 28, 363–392. doi: 10.1103/RevModPhys.28.363.

H. Reichenbach (1956), *The Direction of Time*, University of California Press. ark:/13960/t8sc1wj45.

E. Cohen (1962), *Fundamental Problems in Statistical Mechanics*, North-Holland.

E. Jaynes (1965), "Gibbs vs. Boltzmann Entropies", *American Journal of Physics* 33, 391–398. doi: 10.1119/1.1971557.

O. Penrose (1970), *Foundations of Statistical Mechanics: A Deductive Treatment*, Dover.

P. Davies (1977), *The Physics of Time Asymmetry*, University of California Press. ark:/13960/t5n99zw8p.

C. Bennett (1987), "Demons, Engines and the Second Law", *Scientific American* 257, 108–117. doi: 10.1038/SCIENTIFICAMERICAN1187-108.

M. Mackey (1992), *Time's Arrow: The Origins of Thermodynamic Behavior*, Springer. doi: 10.1007/978-1-4613-9524-9.

J. Lebowitz (1993), "Macroscopic Laws, Microscopic Dynamics, Time's Arrow and Boltzmann's Entropy", *Physica A* 194, 1–27. doi: 10.1016/0378-4371(93)90336-3.

J. Halliwell (1996), *Physical Origins of Time Asymmetry*, Cambridge University Press.

R. Peierls (1997), "Time Reversal and the Second Law of Thermodynamics", in *Selected Scientific Papers of Sir Rudolf Peierls*, R. Dalitz and R. Peierls (eds.), World Scientific, 563–570. doi: 10.1142/9789812795779_0055.

L. Schulman (1997), *Time's Arrows and Quantum Measurement*, Cambridge University Press. doi: 10.1017/CBO9780511622878.

J. Uffink (2001), "Bluff Your Way in the Second Law of Thermodynamics", *Studies in History and Philosophy of Science Part B: Studies in History and Philosophy of Modern Physics* 32, 305–394. arXiv:cond-mat/0005327.

J. Uffink (2007), "Compendium of the Foundations of Classical Statistical Physics", in *Philosophy of Physics, Handbook of the Philosophy of Science*, J. Butterfield and J. Earman (eds.), Elsevier, 923–1074.

W. Grandy Jr. (2008), *Entropy and the Time Evolution of Macroscopic Systems*, Oxford University Press.

J. Parrondo, J. Horowitz and T. Sagawa (2015), "Thermodynamics of Information", *Nature Physics* 11, 131–139. doi: 10.1038/nphys3230.

D. Lairez (2022), "What Entropy Really Is: The Contribution of Information Theory". arXiv:2204.05747.

Historical Surveys & Collections

S. Brush (1976), *The Kind of Motion We Call Heat: A History of the Kinetic Theory of Gases in the Nineteenth Century* vols. 1 & 2, North-Holland.

J. Kestin (ed.) (1976), *The Second Law of Thermodynamics*, Cambridge University Press.

J. Kestin (ed.) (1977), *The Second Law of Thermodynamics*, Dowden, Hutchinson and Ross.

C. Truesdell (1980), *The Tragicomical History of Thermodynamics, 1822–1854*, Springer. archive.org/details/tragicomicalhist0000unse.

E. Jaynes (1984), *The Evolution of Carnot's Principle*, Springer. doi: 10.1007/978-94-009-3049-0_15.

J. von Plato (1991), "Boltzmann's Ergodic Hypothesis", *Archive for History of Exact Sciences* 42, 71–89. doi: 10.1007/BF00384333.

S. Brush (ed.) (2003), *The Kinetic Theory of Gases: An Anthology of Classic Papers with Historical Commentary*, Imperial College Press. doi: 10.1142/p281.

I. Müller (2006), *A History of Thermodynamics: The Doctrine of Energy and Entropy*, Springer.

SklogWiki (2007–). sklogwiki.org (website).

H. Leff and Andrew Rex (eds.) (2016), *Maxwell's Demon: Entropy, Information, Computing*. Princeton University Press.

O. Darrigol (2018), *Atoms, Mechanics, and Probability: Ludwig Boltzmann's Statistico-Mechanical Writings—An Exegesis*, Oxford University Press.

J. Barbour (2020), *A History of Thermodynamics*, Platonia. platonia.com/A_History_of_Thermodynamics.pdf.

Less Technical Presentations

J. Maxwell (1878), "Atom", "Attraction", "Constitution of Bodies", "Diagrams", "Ether", "Molecule", *Encyclopedia Britannica* 9th edition. ark:/13960/s22210225hh. (Reprinted in *The Scientific Papers of James Clerk Maxwell* vol. 2 (1890). ark:/13960/t56f4vm7t.)

H. Helmholtz (1885), *Popular Lectures on Scientific Subjects*, Appleton and Company. urn:oclc:record:669327403.

L. Boltzmann (1895), *Theoretical Physics and Philosophical Problems: Selected Writings*, D. Reidel Publishing Company.

E. Schrödinger (1945), *What Is Life? The Physical Aspect of the Living Cell*, Macmillan Company. ark:/13960/t4qk7tk6c.

H. Bent (1965), *The Second Law: An Introduction to Classical and Statistical Thermodynamics*, Oxford University Press. ark:/13960/t1gj7793z.

P. Atkins (1984), *The Second Law*, Freeman and Company. ark:/13960/t2h78530v.

S. Berry (2019), *Three Laws of Nature: A Little Book on Thermodynamics*, Yale University Press.

EARLIER WORKS

VOLUME 54, NUMBER 8   PHYSICAL REVIEW LETTERS   25 FEBRUARY 1985

Undecidability and Intractability in Theoretical Physics

Stephen Wolfram
The Institute for Advanced Study, Princeton, New Jersey 08540
(Received 26 October 1984)

Physical processes are viewed as computations, and the difficulty of answering questions about them is characterized in terms of the difficulty of performing the corresponding computations. Cellular automata are used to provide explicit examples of various formally undecidable and computationally intractable problems. It is suggested that such problems are common in physical models, and some other potential examples are discussed.

PACS numbers: 02.90.+p, 01.70.+w, 05.90.+m

There is a close correspondence between physical processes and computations. On one hand, theoretical models describe physical processes by computations that transform initial data according to algorithms representing physical laws. And on the other hand, computers themselves are physical systems, obeying physical laws. This paper explores some fundamental consequences of this correspondence.[1]

The behavior of a physical system may always be calculated by simulating explicitly each step in its evolution. Much of theoretical physics has, however, been concerned with devising shorter methods of calculation that reproduce the outcome without tracing each step. Such shortcuts can be made if the computations used in the calculation are more sophisticated than those that the physical system can perform. Any computations must, however, be carried out on a computer. But the computer is itself an example of a physical system. And it can determine the outcome of its own evolution only by explicitly following it through: No shortcut is possible. Such computational irreducibility occurs whenever a physical system can act as a computer. The behavior of the system can be found only by direct simulation or observation: No general predictive procedure is possible. Computational irreducibility is common among the systems investigated in mathematical and computational physics. This paper suggests that it is also common in theoretical physics. Computational irreducibility may well be the exception rather than the rule: Nature's predictive questions may be more of the kind that can require arbitrarily long computations, but can require arbitrarily long computations. A diverse set of systems can be shown to be equivalent in their computational capabilities in that particular forms of one system can emulate any other.

tine equations.[2] One expects in fact that universal computers are as powerful in their computational capabilities as any physically realizable system can be, so that they can simulate any physical system.[3] This is the case if in all physical systems there is a finite density of information, which can be transmitted only at a finite rate in a finite-dimensional space.[4] No physically implementable procedure could then short cut a computationally irreducible process.

Different physically realizable universal computers appear to require the same order of magnitude times and information storage capacities to solve particular classes of finite problems.[5] One computer may be constructed so that in a single step it carries out the equivalent of two steps on another computer. However...

Reprinted from JOURNAL OF STATISTICAL PHYSICS   Vol. 45, Nos. 3/4, November 1986
Printed in Belgium

Cellular Automaton Fluids 1: Basic Theory

Stephen Wolfram[1,2]

Received March, 1986; revision received August, 1986

Continuum equations are derived for the large-scale behavior of a class of cellular automaton models for fluids. The cellular automata are discrete analogues of molecular dynamics, in which particles with discrete velocities populate the links of a fixed array of sites. Kinetic equations for microscopic particle distributions are constructed. Hydrodynamic equations are then derived using the Chapman–Enskog expansion. Slightly modified Navier–Stokes equations are obtained in two and three dimensions with certain lattices. Viscosities and other transport coefficients are calculated using the Boltzmann transport equation approximation. Some corrections to the equations of motion for cellular automaton fluids beyond the Navier–Stokes order are given.

KEY WORDS: Cellular automata; derivation of hydrodynamics; molecular dynamics; kinetic theory; Navier–Stokes equations.

Thermodynamics and Hydrodynamics with Cellular A...

James B. Salem
Thinking Machines Corporation, 245 First Street, Cambridge, MA 02144

and

Stephen Wolfram
The Institute for Advanced Study, Princeton NJ 08540.

Simple cellular automata which seem... dynamics and hydrodynamics are discre... mata are discrete approximations to m... equilibrium. On a large scale, they be... methods for hydrodynamic simulation,...

Thermodynamics and hydrodynamics de... dent of the precise microscopic construction... hydrodynamics using simple models, which a... to mathematical analysis.

Cellular automata (CA) are discrete dy... plex physical processes [1]. This paper con... to molecular dynamics. In the simplest case,... "particle" with unit velocity in each directi... arriving at a particular site then "scatter"... well-suited to simulation on digital compute... follows simple logical rules. The rules are... simulations in this paper were performed... concurrently in each of 65536 Boolean proc...

In two dimensions, one can consider... lattice [4], the only nontrivial rule which c... of particles colliding head on scatter in th... hexagonal lattice [5], such pairs may scatter... be affected by particles in the third dire... scatter in different directions. Finally, par...

VOLUME 55, NUMBER 5   PHYSICAL REVIEW LETTERS   29 JULY 1985

Origins of Randomness in Physical Systems

Stephen Wolfram
The Institute for Advanced Study, Princeton, New Jersey 08540
(Received 4 February 1985)

Randomness and chaos in physical systems are usually ultimately attributed to external noise. But it is argued here that even without such random input, the intrinsic behavior of many nonlinear systems can be computationally so complicated as to seem random in all practical experiments. This effect is suggested as the basic origin of such phenomena as fluid turbulence.

PACS numbers: 05.45.+b, 02.90.+p, 03.40.Gc

There are many physical processes that seem random or chaotic. They appear to follow no definite rules, and to be governed merely by probabilities. But all fundamental physical laws, at least outside of quantum mechanics, are thought to be deterministic. So how, then, is apparent randomness produced?

One possibility is that its ultimate source is external noise, often from a heat bath. When the evolution of a system is unstable, so that perturbations grow, any randomness introduced through initial and boundary conditions is transmitted or amplified with time, and eventually affects many components of the system.[1] A simple example of this "homoplectic" behavior occurs in the shift mapping $x_i \to 2x_{i-1} \bmod 1$. The time sequence of bins, say, above and below $\frac{1}{2}$ visited by x_i, is a direct transcription of the binary-digit sequence of the initial real number x_0.[2] So if this digit sequence is random (as for most x_0 uniformly sampled in the unit interval) then so will the time sequence be; unpredictable behavior arises from a sensitive dependence on unknown features of initial conditions.[3] But if the initial condition is "simple," say a rational number with a periodic digit sequence, then no randomness appears.

There are, however, systems which can also generate apparent randomness internally, without external random input. Figure 1 shows an example, in which a cellular automaton evolving from a simple initial state produces a pattern so complicated that many features of it seem random. Like the shift map, this cellular automaton is homoplectic, and would yield random behavior given random input. But unlike the shift map, it can still produce random behavior even with simple input. Systems which generate randomness in this way will be called "autoplectic."

In developing a mathematical definition of autoplectic behavior, one must first discuss in what sense it is "random." Sequences are commonly considered random if no patterns can be discerned in them. But whether a pattern is found depends on how it is looked for. Different degrees of randomness can be defined in terms of the computational complexity of the procedures used.

The methods usually embodied in practical physics experiments are computationally quite simple.[4,1] They correspond to standard statistical tests for random-

ness,[6] such as relative frequencies of blocks of elements (dimensions and entropies), correlations, and power spectra. (The mathematical properties of ergodicity and mixing are related to tests of this kind.) One characteristic of these tests is that the computation time they require increases asymptotically at most like polynomial in the sequence length.[7] So if in fact no polynomial-time procedure can detect patterns in a sequence, then the sequence can be considered "effectively random" for practical purposes.

Any patterns that are identified in a sequence can be used to give a compressed specification for it. (Thus, for example, Morse coding compresses English text by exploiting the unequal frequencies of letters of the alphabet.) The length of the shortest specification measures the "information content" of a sequence with respect to a particular class of computations. (Standard Shannon information content for a stationary process[8] is associated with simple statistical computations of block frequencies.) Sequences are predictable only to the extent that they are longer than their shortest specification, and so contain information that can be recognized as "redundant" or "overdetermined."

Sequences generated by chaotic physical systems often show some redundancy or determinism under simple statistical procedures. (This happens whenever measurements extract information faster than it can be transferred from other parts of the system.[1]) But, typically, there remain compressed sequences in which no patterns are seen.

A sequence can, in general, be specified by giving an algorithm or computer program for constructing it. The length of the smallest possible program measures the "absolute" information content of the sequence.[9] For an "absolutely random" sequence the program must essentially give each element explicitly, and so be close in length to the sequence itself. But since no computation can increase the absolute information content of a closed system (except for $O(\log t)$ from input of "clock pulses"), physical processes presumably cannot generate absolute randomness.[10] However, the numbers of possible sequences and programs both increase exponentially with length, so that all but an exponentially small fraction of arbitrarily chosen sequences must be absolutely random. Nevertheless, it

    449

Undecidability and Intractability in Theoretical Physics

Stephen Wolfram

The Institute for Advanced Study, Princeton, New Jersey 08540

(Received 26 October 1984)

Physical processes are viewed as computations, and the difficulty of answering questions about them is characterized in terms of the difficulty of performing the corresponding computations. Cellular automata are used to provide explicit examples of various formally undecidable and computationally intractable problems. It is suggested that such problems are common in physical models, and some other potential examples are discussed.

PACS numbers: 02.90.+p, 01.70.+w, 05.90.+m

There is a close correspondence between physical processes and computations. On one hand, theoretical models describe physical processes by computations that transform initial data according to algorithms representing physical laws. And on the other hand, computers themselves are physical systems, obeying physical laws. This paper explores some fundamental consequences of this correspondence.[1]

The behavior of a physical system may always be calculated by simulating explicitly each step in its evolution. Much of theoretical physics has, however, been concerned with devising shorter methods of calculation that reproduce the outcome without tracing each step. Such shortcuts can be made if the computations used in the calculation are more sophisticated than those that the physical system can itself perform. Any computations must, however, be carried out on a computer. But the computer is itself an example of a physical system. And it can determine the outcome of its own evolution only by explicitly following it through: No shortcut is possible. Such computational irreducibility occurs whenever a physical system can act as a computer. The behavior of the system can be found only by direct simulation or observation: No general predictive procedure is possible. Computational irreducibility is common among the systems investigated in mathematics and computation theory.[2] This paper suggests that it is also common in theoretical physics. Computational reducibility may well be the exception rather than the rule: Most physical questions may be answerable only through irreducible amounts of computation. Those that concern idealized limits of infinite time, volume, or numerical precision can require arbitrarily long computations, and so be formally undecidable.

A diverse set of systems are known to be equivalent in their computational capabilities, in that particular forms of one system can emulate any of the others. Standard digital computers are one example of such "universal computers": With fixed intrinsic instructions, different initial states or programs can be devised to simulate different systems. Some other examples are Turing machines, string transformation systems, recursively defined functions, and Diophan-tine equations.[2] One expects in fact that universal computers are as powerful in their computational capabilities as any physically realizable system can be, so that they can simulate any physical system.[3] This is the case if in all physical systems there is a finite density of information, which can be transmitted only at a finite rate in a finite-dimensional space.[4] No physically implementable procedure could then short cut a computationally irreducible process.

Different physically realizable universal computers appear to require the same order of magnitude times and information storage capacities to solve particular classes of finite problems.[5] One computer may be constructed so that in a single step it carries out the equivalent of two steps on another computer. However, when the amount of information n specifying an instance of a problem becomes large, different computers use resources that differ only by polynomials in n. One may then distinguish several classes of problems.[6] The first, denoted P, are those such as arithmetical ones taking a time polynomial in n. The second, denoted $PSPACE$, are those that can be solved with polynomial storage capacity, but may require exponential time, and so are in practice effectively intractable. Certain problems are "complete" with respect to $PSPACE$, so that particular instances of them correspond to arbitrary $PSPACE$ problems. Solutions to these problems mimic the operation of a universal computer with bounded storage capacity: A computer that solves $PSPACE$-complete problems for any n must be universal. Many mathematical problems are $PSPACE$-complete.[6] (An example is whether one can always win from a given position in chess.) And since there is no evidence to the contrary, it is widely conjectured that $PSPACE \neq P$, so that $PSPACE$-complete problems cannot be solved in polynomial time. A final class of problems, denoted NP, consist in identifying, among an exponentially large collection of objects, those with some particular, easily testable property. An example would be to find an n-digit integer that divides a given $2n$-digit number exactly. A particular candidate divisor, guessed nondeterministically, can be tested in polynomial time, but a systematic solution may require almost all $O(2^n)$ possible candidates to be

S. Wolfram (1985), *Phys. Rev. Lett.*, 54, 735. Copyright (1985) by the American Physical Society. doi: 10.1103/PhysRevLett.54.735.

tested. A computer that could follow arbitrarily many computational paths in parallel could solve such problems in polynomial time. For actual computers that allow only boundedly many paths, it is suspected that no general polynomial time solution is possible.[5] Nevertheless, in the infinite time limit, parallel paths are irrelevant, and a computer that solves *NP*-complete problems is equivalent to other universal computers.[6]

The structure of a system need not be complicated for its behavior to be highly complex, corresponding to a complicated computation. Computational irreducibility may thus be widespread even among systems with simple construction. Cellular automata (CA)[7] provide an example. A CA consists of a lattice of sites, each with k possible values, and each updated in time steps by a deterministic rule depending on a neighborhood of R sites. CA serve as discrete approximations to partial differential equations, and provide models for a wide variety of natural systems. Figure 1 shows typical examples of their behavior. Some rules give periodic patterns, and the outcome after many steps can be predicted without following each intermediate step. Many rules, however, give complex patterns for which no predictive procedure is evident. Some CA are in fact known to be capable of universal computation, so that their evolution must be computationally irreducible. The simplest cases proved have $k = 18$ and $R = 3$ in one dimension,[8] or $k = 2$ and $R = 5$ in two dimensions.[9] It is strongly suspected that "class-4" CA are generically capable of universal computation: There are such CA with $k = 3$, $R = 3$ and $k = 2$, $R = 5$ in one dimension.[10]

Computationally, irreducibility may occur in systems that are not full universal computers. For inability to perform, specific computations need not allow all computations to be short cut. Though class-3 CA and other chaotic systems may not be universal computers, most of them are expected to be computationally irreducible, so that the solution of problems concerning their behavior requires irreducible amounts of computation.

As a first example consider finding the value of a site in a CA after t steps of evolution from a finite initial seed, as illustrated in Fig. 1. The problem is specified by giving the seed and the CA rule, together with the $\log t$ digits of t. In simple cases such as the first two shown in Fig. 1, it can be solved in the time $O(\log t)$

necessary to input this specification. However, the evolution of a universal computer CA for a polynomial in t steps can implement any computation of length t. As a consequence, its evolution is computationally irreducible, and its outcome found only by an explicit simulation with length $O(t)$: exponentially longer than for the first two in Fig. 1.

One may ask whether the pattern generated by evolution with a CA rule from a particular seed will grow forever, or will eventually die out.[11] If the evolution is computationally irreducible, then an arbitrarily long computation may be needed to answer this question. One may determine by explicit simulation whether the pattern dies out after any specified number of steps, but there is no upper bound on the time needed to find out its ultimate fate.[12] Simple criteria may be given for particular cases, but computational irreducibility implies that no shortcut is possible in general. The infinite-time limiting behavior is formally undecidable: No finite mathematical or computational process can reproduce the infinite CA evolution.

The fate of a pattern in a CA with a finite total number of sites N can always be determined in at most k^N steps. However, if the CA is a universal computer, then the problem is *PSPACE*-complete, and so presumably cannot be solved in a time polynomial in N.[13]

One may consider CA evolution not only from finite seeds, but also from initial states with all infinitely many sites chosen arbitrarily. The value $a^{(t)}$ of a site after many time steps t then in general depends on $2\lambda t \leqslant Rt$ initial site values, where λ is the rate of information transmission (essentially Lyapunov exponent) in the CA.[9] In class-1 and -2 CA, information remains localized, so that $\lambda = 0$, and $a^{(t)}$ can be found by a length $O(\log t)$ computation. For class-3 and -4 CA, however, $\lambda > 0$, and $a^{(t)}$ requires an $O(t)$ computation.[14]

The global dynamics of CA are determined by the possible states reached in their evolution. To characterize such states one may ask whether a particular string of n site values can be generated after evolution for t steps from any (length $n + 2\lambda t$) initial string. Since candidate initial strings can be tested in $O(t)$ time, this problem is in the class *NP*. When the CA is a universal computer, the problem is in general *NP*-complete, and can presumably be answered essentially only by testing all $O(k^{n + 2\lambda t})$ candidate initial

FIG. 1. Seven examples of patterns generated by repeated application of various simple cellular automaton rules. The last four are probably computationally irreducible, and can be found only by direct simulation.

VOLUME 54, NUMBER 8   PHYSICAL REVIEW LETTERS   25 FEBRUARY 1985

strings.[15] In the limit $t \rightarrow \infty$, it is in general undecidable whether particular strings can appear.[16] As a consequence, the entropy or dimension of the limiting set of CA configurations is in general not finitely computable.

Formal languages describe sets of states generated by CA.[17] The set that appears after t steps in the evolution of a one-dimensional CA forms a regular formal language: each possible state corresponds to a path through a graph with $\Xi^{(t)} < 2^{k^{Rt}}$ nodes. If, indeed, the length of computation to determine whether a string can occur increases exponentially with t for computationally irreducible CA, then the "regular language complexity" $\Xi^{(t)}$ should also increase exponentially, in agreement with empirical data on certain class-3 CA,[17] and reflecting the "irreducible computational work" achieved by their evolution.

Irreducible computations may be required not only to determine the outcome of evolution through time, but also to find possible arrangements of a system in space. For example, whether an $x \times x$ patch of site values occurs after just one step in a two-dimensional CA is in general NP-complete.[18] To determine whether there is any complete infinite configuration that satisfies a particular predicate (such as being invariant under the CA rule) is in general undecidable[18]: It is equivalent to finding the infinite-time behavior of a universal computer that lays down each row on the lattice in turn.

There are many physical systems in which it is known to be possible to construct universal computers. Apart from those modeled by CA, some examples are electric circuits, hard-sphere gases with obstructions, and networks of chemical reactions.[19] The evolution of these systems is in general computationally irreducible, and so suffers from undecidable and intractable problems. Nevertheless, the constructions used to find universal computers in these systems are arcane, and if computationally complex problems occurred only there, they would be rare. It is the thesis of this paper that such problems are in fact common.[20] Certainly there are many systems whose properties are in practice studied only by explicit simulation or exhaustive search: Few computational shortcuts (often stated in terms of invariant quantities) are known.

Many complex or chaotic dynamical systems are expected to be computationally irreducible, and their behavior effectively found only by explicit simulation. Just as it is undecidable whether a particular initial state in a CA leads to unbounded growth, to self-replication, or has some other outcome, so it may be undecidable whether a particular solution to a differential equation (studied say with symbolic dynamics) even enters a certain region of phase space, and whether, say, a certain n-body system is ultimately stable. Similarly, the existence of an attractor, say,

with a dimension above some value, may be undecidable.

Computationally complex problems can arise in finding eigenvalues or extremal states in physical systems. The minimum energy conformation for a polymer is in general NP-complete with respect to its length.[21] Finding a configuration below a specified energy in a spin-glass with particular couplings is similarly NP-complete.[22] Whenever the stationary state of a physical system such as this can be found only by lengthy computation, the dynamic physical processes that lead to it must take a correspondingly long time.[5]

Global properties of some models for physical systems may be undecidable in the infinite-size limit (like those for two-dimensional CA). An example is whether a particular generalized Ising model (or stochastic multidimensional CA[23]) exhibits a phase transition.

Quantum and statistical mechanics involve sums over possibly infinite sets of configurations in systems. To derive finite formulas one must use finite specifications for these sets. But it may be undecidable whether two finite specifications yield equivalent configurations. So, for example, it is undecidable whether two finitely specified four-manifolds or solutions to the Einstein equations are equivalent (under coordinate reparametrization).[24] A theoretical model may be considered as a finite specification of the possible behavior of a system. One may ask for example whether the consequences of two models are identical in all circumstances, so that the models are equivalent. If the models involve computations more complicated than those that can be carried out by a computer with a fixed finite number of states (regular language), this question is in general undecidable. Similarly, it is undecidable what is the simplest such model that describes a given set of empirical data.[25]

This paper has suggested that many physical systems are computationally irreducible, so that their own evolution is effectively the most efficient procedure for determining their future. As a consequence, many questions about these systems can be answered only by very lengthy or potentially infinite computations. But some questions answerable by simpler computations may still be formulated.

This work was supported in part by the U. S. Office of Naval Research under Contract No. N00014-80-C-0657. I am grateful for discussions with many people, particularly C. Bennett, G. Chaitin, R. Feynman, E. Fredkin, D. Hillis, L. Hurd, J. Milnor, N. Packard, M. Perry, R. Shaw, K. Steiglitz, W. Thurston, and L. Yaffe.

[1]For a more informal exposition see: S. Wolfram, Sci. Am. 251, 188 (1984). A fuller treatment will be given else-

where.

[2]E.g., *The Undecidable: Basic Papers on Undecidable Propositions, Unsolvable Problems, and Computable Functions,* edited by M. Davis (Raven, New York, 1965), or J. Hopcroft and J. Ullman, *Introduction to Automata Theory, Languages, and Computations* (Addison-Wesley, Reading, Mass., 1979).

[3]This is a physical form of the Church-Turing hypothesis. Mathematically conceivable systems of greater power can be obtained by including tables of answers to questions insoluble for these universal computers.

[4]Real-number parameters in classical physics allow infinite information density. Nevertheless, even in classical physics, the finiteness of experimental arrangements and measurements, implemented as coarse graining in statistical mechanics, implies finite information input and output. In relativistic quantum field theory, finite density of information (or quantum states) is evident for free fields bounded in phase space [e.g., J. Bekenstein, Phys. Rev. D **30**, 1669 (1984)]. It is less clear for interacting fields, except if space-time is ultimately discrete [but cf. B. Simon, *Functional Integration and Quantum Physics* (Academic, New York, 1979), Sec. III.9]. A finite information transmission rate is implied by relativistic causality and the manifold structure of space-time.

[5]It is just possible, however, that the parallelism of the path integral may allow quantum mechanical systems to solve any *NP* problem in polynomial time.

[6]M. Garey and D. Johnson, *Computers and Intractability: A Guide to the Theory of NP-Completeness* (Freeman, San Francisco, 1979).

[7]See S. Wolfram, Nature **311**, 419 (1984); *Cellular Automata,* edited by D. Farmer, T. Toffoli, and S. Wolfram, Physica **10D**, Nos. 1 and 2 (1984), and references therein.

[8]A. R. Smith, J. Assoc. Comput. Mach. **18**, 331 (1971).

[9]E. R. Banks, Massachusetts Institute of Technology Report No. TR-81, 1971 (unpublished). The "Game of Life," discussed in E. R. Berlekamp, J. H. Conway, and R. K. Guy, *Winning Ways for Your Mathematical Plays* (Academic, New York, 1982), is an example with $k = 2$, $R = 9$. N. Margolus, Physica (Utrecht) **10D**, 81 (1984), gives a reversible example.

[10]S. Wolfram, Physica (Utrecht) **10D**, 1 (1984), and to be published.

[11]This is analogous to the problem of whether a computer run with particular input will ever reach a "halt" state.

[12]The number of steps to check ("busy-beaver function") in general grows with the seed size faster than any finite formula can describe (Ref. 2).

[13]Cf. C. Bennett, to be published.

[14]Cf. B. Eckhardt, J. Ford, and F. Vivaldi, Physica (Utrecht) **13D**, 339 (1984).

[15]The question is a generalization of whether there exists an assignment of values to sites such that the logical expression corresponding the t-step CA mapping is true (cf. V. Sewelson, private communication).

[16]L. Hurd, to be published.

[17]S. Wolfram, Commun. Math. Phys. **96**, 15 (1984).

[18]N. Packard and S. Wolfram, to be published. The equivalent problem of covering a plane with a given set of tiles is considered in R. Robinson, Invent. Math. **12**, 177 (1971).

[19]E.g., C. Bennett, Int. J. Theor. Phys. **21**, 905 (1982); E. Fredkin and T. Toffoli, Int. J. Theor. Phys. **21**, 219 (1982); A. Vergis, K. Steiglitz, and B. Dickinson, "The Complexity of Analog Computation" (unpublished).

[20]Conventional computation theory primarily concerns possibilities, not probabilities. There are nevertheless some problems for which almost all instances are known to be of equivalent difficulty. But other problems are known to be much easier on average then in the worst case. In addition, for some *NP*-complete problems the density of candidate solutions close to the actual one is very large, so approximate solutions can easily be found [S. Kirkpatrick, C. Gelatt, and M. Vecchi, Science **220**, 671 (1983)].

[21]Compare *Time Warps, String Edites, and Macromolecules,* edited by D. Sankoff and J. Kruskal (Addison-Wesley, Reading, Mass., 1983).

[22]F. Barahona, J. Phys. A **13**, 3241 (1982).

[23]E. Domany and W. Kinzel, Phys. Rev. Lett. **53**, 311 (1984).

[24]See W. Haken, in *Word Problems,* edited by W. W. Boone, F. B. Cannonito, and R. C. Lyndon (North-Holland, Amsterdam, 1973).

[25]G. Chaitin, Sci. Am. **232**, 47 (1975), and IBM J. Res. Dev. **21**, 350 (1977); R. Shaw, to be published.

Origins of Randomness in Physical Systems

Stephen Wolfram

The Institute for Advanced Study, Princeton, New Jersey 08540
(Received 4 February 1985)

Randomness and chaos in physical systems are ususally ultimately attributed to external noise. But it is argued here that even without such random input, the intrinsic behavior of many nonlinear systems can be computationally so complicated as to seem random in all practical experiments. This effect is suggested as the basic origin of such phenomena as fluid turbulence.

PACS numbers: 05.45.+b, 02.90.+p, 03.40.Gc

There are many physical processes that seem random or chaotic. They appear to follow no definite rules, and to be governed merely by probabilities. But all fundamental physical laws, at least outside of quantum mechanics, are thought to be deterministic. So how, then, is apparent randomness produced?

One possibility is that its ultimate source is external noise, often from a heat bath. When the evolution of a system is unstable, so that perturbations grow, any randomness introduced through initial and boundary conditions is transmitted or amplified with time, and eventually affects many components of the system.[1] A simple example of this "homoplectic" behavior occurs in the shift mapping $x_t = 2x_{t-1} \bmod 1$. The time sequence of bins, say, above and below $\frac{1}{2}$ visited by x_t is a direct transcription of the binary-digit sequence of the initial real number x_0.[2] So if this digit sequence is random (as for most x_0 uniformly sampled in the unit interval) then so will the time sequence be; unpredictable behavior arises from a sensitive dependence on unknown features of initial conditions.[3] But if the initial condition is "simple," say a rational number with a periodic digit sequence, then no randomness appears.

There are, however, systems which can also generate apparent randomness internally, without external random input. Figure 1 shows an example, in which a cellular automaton evolving from a simple initial state produces a pattern so complicated that many features of it seem random. Like the shift map, this cellular automaton is homoplectic, and would yield random behavior given random input. But unlike the shift map, it can still produce random behavior even with simple input. Systems which generate randomness in this way will be called "autoplectic."

In developing a mathematical definition of autoplectic behavior, one must first discuss in what sense it is "random." Sequences are commonly considered random if no patterns can be discerned in them. But whether a pattern is found depends on how it is looked for. Different degrees of randomness can be defined in terms of the computational complexity of the procedures used.

The methods usually embodied in practical physics experiments are computationally quite simple.[4,5] They correspond to standard statistical tests for randomness,[6] such as relative frequencies of blocks of elements (dimensions and entropies), correlations, and power spectra. (The mathematical properties of ergodicity and mixing are related to tests of this kind.) One characteristic of these tests is that the computation time they require increases asymptotically at most like polynomial in the sequence length.[7] So if in fact no polynomial-time procedure can detect patterns in a sequence, then the sequence can be considered "effectively random" for practical purposes.

Any patterns that are identified in a sequence can be used to give a compressed specification for it. (Thus, for example, Morse coding compresses English text by exploiting the unequal frequencies of letters of the alphabet.) The length of the shortest specification measures the "information content" of a sequence with respect to a particular class of computations. (Standard Shannon information content for a stationary process[8] is associated with simple statistical computations of block frequencies.) Sequences are predictable only to the extent that they are longer than their shortest specification, and so contain information that can be recognized as "redundant" or "overdetermined."

Sequences generated by chaotic physical systems often show some redundancy or determinism under simple statistical procedures. (This happens whenever measurements extract information faster than it can be transferred from other parts of the system.[1]) But, typically, there remain compressed sequences in which no patterns are seen.

A sequence can, in general, be specified by giving an algorithm or computer program for constructing it. The length of the smallest possible program measures the "absolute" information content of the sequence.[9] For an "absolutely random" sequence the program must essentially give each element explicitly, and so be close in length to the sequence itself. But since no computation can increase the absolute information content of a closed system [except for $O(\log t)$ from input of "clock pulses"], physical processes presumably cannot generate absolute randomness.[10] However, the numbers of possible sequences and programs both increase exponentially with length, so that all but an exponentially small fraction of arbitrarily chosen sequences must be absolutely random. Nevertheless, it

S. Wolfram (1985), *Phys. Rev. Lett.*, 55, 449. Copyright (1985) by the American Physical Society. doi: 10.1103/PhysRevLett.55.449.

is usually undecidable what the smallest program for any particular sequence is, and thus whether the sequence is absolutely random. In general, each program of progressively greater length must be tried, and any one of them may run for an arbitrarily long time, so that the question of whether it ever generates the sequence may be formally undecidable.

Even if a sequence can ultimately be obtained from a small specification or program, and so is not absolutely random, it may nevertheless be effectively random if no feasible computation can recover the program.[11] The program can always be found by explicitly trying each possible one in turn.[12] But the total number of possible programs increases exponentially with length, and so such an exhaustive search would soon become infeasible. And if there is no better method the sequence must be effectively random.

In general, one may define the "effective information content" Θ of a sequence to be the length of the shortest specification for it that can be found by a feasible (say polynomial time) computation. A sequence can be considered "simple" if it has small Θ. Θ (often normalized by sequence length) provides a measurue of "complexity," "effective randomness," or "computational unpredictability."

Increasing Θ can be considered the defining characteristic of autoplectic behavior. Examples such as Fig. 1 suggest that Θ can increase through polynomial-time processes. The rule and initial seed have a short specification, with small Θ. But one suspects that no polynomial time computation can recover this specification from the center vertical sequence produced, or can in fact detect any pattern in it.[13] The polynomial-time process of cellular automaton evolution thus increases Θ, and generates effective randomness. It is phenomena of this kind that are the basis for cryptogra-

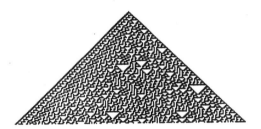

FIG. 1. Pattern generated by cellular automaton evolution from a simple initial state. Site values 0 or 1 (represented by white or black, respectively) are updated at each step according to the rule $a_i' = a_{i-1} \oplus (a_i \vee a_{i+1})$ ($\oplus$ denotes addition modulo 2, and $\vee$ Boolean disjunction). Despite the simplicity of its specification, many features of the pattern (such as the sequence of site values down the center column) appear random.

phy, in which one strives to produce effectively random sequences whose short "keys" cannot be found by any practical cryptanalysis.[14]

The simplest mathematical and physical systems (such as the shift mapping) can be decomposed into essentially uncoupled components, and cannot increase Θ. Such systems are nevertheless often homoplectic, so that they transfer information, and with random input show random behavior. But when their input is simple (low Θ), their behavior is correspondingly simple, and is typically periodic. Of course, any system with a fixed finite total number of degrees of freedom (such as a finite cellular automaton) must eventually become periodic. But the phenomena considered here occur on time scales much shorter than such exponentially long recurrences.

Another class of systems widely investigated consists of those with linear couplings between components [such as a cellular automaton in which $a_i^{(t+1)} = (a_{i-1}^{(t)} + a_{i+1}^{(t)}) \bmod 2$]. Given random input, such systems can again yield random output, and are thus homoplectic. But even with simple input, they can produce sequences which pass some statistical tests of randomness. Examples are the standard linear congruence and linear-feedback shift-register (or finite additive cellular automaton[15]) systems used for pseudorandom number generation in practical computer programs.[6,16]

Characteristic of such systems is the generation of self-similar patterns, containing sequences that are invariant under blocking or scaling transformations. These sequences are almost periodic, but may contain all possible blocks of elements with equal frequencies. They can be considered as the outputs of finite-state machines (generalized Markov processes) given the digits of the numerical positions of each element as input.[17] And although the sequences have certain statistical properties of randomness, their seeds can be found by comparatively simply polynomial-time procedures.[18] Such systems are thus not autoplectic (with respect to polynomial-time computations).

Many nonlinear mathematical systems seem, however, to be autoplectic, since they generate sequences in which no patterns have ever been found. An example is the sequence of leading digits in the fractional part of successive powers of $\frac{3}{2}$[19] (which corresponds to a vertical column in a particular $k = 6$, $r = 1$ cellular automaton with a single site seed).

Despite extensive empirical evidence, almost nothing has, however, been proved about the randomness of such sequences. It is nevertheless possible to construct sequences that are strongly expected to be effectively random.[20] An example is the lowest-order bits of $x_t = x_{t-1}^2 \bmod (pq)$, where p and q are large primes.[20] The problem of deducing the initial seed x_0, or of substantially compressing this sequence, is

VOLUME 55, NUMBER 5   PHYSICAL REVIEW LETTERS   29 JULY 1985

equivalent to the problem of factoring large integers, which is widely conjectured to require more than polynomial time.[21]

Standard statistical tests have also revealed no patterns in the digit sequences of transcendental numbers such as[22] $\sqrt{2}$, e, and π [22] (or continued-fraction expansions of π or of most cubic irrational numbers). But the polynomial-time procedure of squaring and comparing with an integer does reveal the digits of, say, $\sqrt{2}$ as nonrandom.[23] Without knowing how the sequence was generated, however, such a very special "statistical test" (or program) can probably only be found by explicit enumeration of all exponentially many possible ones. And if a sequence passes all but perhaps exponentially few polynomial-time batteries of statistical tests, it should probably be considered effectively random in practice.

Within a set of homoplectic dynamical systems (such as class 3 or 4 cellular automata) capable of transmitting information, all but the simplest seem to support sophisticated information processing, and are thus expected to be autoplectic. In some cases (quite probably including Fig. 1 [24]) the evolution of the system represents a "complete" or "universal" computation, which, with appropriate initial conditions, can mimic any other (polynomial-time) computation.[21] If short specifications for sequences generated by any one such computation could in general be found in polynomial time, it would imply that all could, which is widely conjectured to be impossible. (Such problems are called *NP*-complete.[21])

Many systems are expected to be computationally irreducible, so that the outcome of their evolution can be found essentially only by direct simulation, and no computational short cuts are possible.[25] To predict the future of these systems requires an almost complete knowledge of their current state. And it seems likely that this can be deduced from partial measurements only by essentially testing all exponentially many possibilities. The evolution of computationally irreducible systems should thus generically be autoplectic.

Autoplectic behavior is most clearly identified in discrete systems such as cellular automata. Continuous dynamical systems involve the idealization of real numbers on which infinite-precision arithmetic operations are performed. For systems such as iterated mappings of the interval there seems to be no robust notion of "simple" initial conditions. (The number of binary digits in images of, say, a dyadic rational grows like p^t, where p is the highest power of x in the map.) But in systems with many degrees of freedom, described for example by partial differential equations, autoplectism may be identified through discrete approximations.

Autoplectism is expected to be responsible for apparent randomness in many physical systems. Some features of turbulent fluid flow,[26] say in a jet ejected from a nozzle, are undoubtedly determined by details of initial or boundary conditions. But when the flow continues to appear random far from the nozzle, one suspects that other sources of effective information are present. One possibility might be thermal fluctuations or external noise, amplified by homoplectic processes.[1] But viscous damping probably allows only sufficiently large-scale perturbations to affect large-scale features of the flow. (Apparently random behavior is found to be almost exactly repeatable in some carefully controlled experiments.[27]) Thus, it seems more likely that the true origin of turbulence is an internal autoplectic process, somewhat like Fig. 1, operating on large-scale features of the flow. Numerical experiments certainly suggest that the Navier-Stokes equations can yield complicated behavior even with simple initial conditions.[28] Autoplectic processes may also be responsible for the widespread applicability of the second law of thermodynamics.

Many discussions have contributed to the material presented here; particularly those with C. Bennett, L. Blum, M. Blum, J. Crutchfield, P. Diaconis, D. Farmer, R. Feynman, U. Frisch, S. Goldwasser, D. Hillis, P. Hohenberg, E. Jen, R. Kraichnan, L. Levin, D. Lind, A. Meyer, S. Micali, J. Milnor, D. Mitchell, A. Odlyzko, N. Packard, I. Procaccia, H. Rose, and R. Shaw. This work was supported in part by the U. S. Office of Naval Research under Contract No. N00014-80-C-0657.

[1]For example, R. Shaw, Z. Naturforsch. **36A**, 80 (1981), and in *Chaos and Order in Nature*, edited by H. Haken (Springer, New York, 1981).

[2]An analogous cellular automaton [S. Wolfram, Nature (London) **311**, 419 (1984), and references therein] has evolution rule $a_i^{(t+1)} = a_{i+1}^{(t)}$, so that with time the value of a particular site is determined by the value of progressively more distant initial sites.

[3]For example, *Order in Chaos*, edited by D. Campbell and H. Rose (North-Holland, Amsterdam, 1982). Many processes analyzed in dynamical systems theory admit "Markov partitions" under which they are directly equivalent to the shift mapping. But in some measurements (say of x_t with four bins) their deterministic nature may introduce simple regularities, and "deterministic chaos" may be said to occur. (This term would in fact probably be better reserved for the autoplectic processes to be described below.)

[4]This is probably also true of at least the lower levels of human sensory processing [for example, D. Marr, *Vision* (Freeman, San Francisco, 1982); B. Julesz, Nature (London) **290**, 91 (1981)].

[5]The validity of Monte Carlo simulations tests the random sequences that they use. But most stochastic physical processes are in fact insensitive to all but the simplest equidistribution and statistical independence properties.

(Partial exceptions occur when long-range order is present.) And in general no polynomial-time simulation can reveal patterns in effectively random sequences.

[6]For example, D. Knuth, *Seminumerical Algorithms* (Addison-Wesley, Reading, Mass., 1981).

[7]Some sophisticated statistical procedures, typically involving the partitioning of high-dimensional spaces, seem to take exponential time. But most take close to linear time. It is possible that those used in practice can be characterized as needing $O(\log^p n)$ time on computers with $O(n^q)$ processors (and so be in the computational complexity class NC) [cf. N. Pippenger, in *Proceedings of the Twentieth IEEE Symposium on Foundations of Computer Science* (IEEE, New York, 1979); J. Hoover and L. Ruzzo, unpublished].

[8]For example, R. Hamming, *Coding and Information Theory* (Prentice-Hall, Englewood Cliffs, 1980).

[9]G. Chaitin, J. Assoc. Comput. Mach. **13**, 547 (1966), and **16**, 145 (1969), and Sci. Am. **232**, No. 5, 47 (1975); A. N. Kolmogorov, Problems Inform. Transmission **1**, 1 (1965); R. Solomonoff, Inform. and Control **7**, 1 (1964); L. Levin, Soviet Math. Dokl. **14**, 1413 (1973). Compare J. Ford, Phys. Today **33**, No. 4, 40 (1983). Note that the lengths of programs needed on different universal computers differ only by a constant, since each computer can simulate any other by means of a fixed "interpreter" program.

[10]Quantum mechanics suggests that processes such as radioactive decay occur purely according to probabilities, and so could perhaps give absolutely random sequences. But complete quantum mechanical measurements are an idealization, in which information on a microscopic quantum event is spread through an infinite system. In finite systems, unmeasured quantum states are like unknown classical parameters, and can presumably produce no additional randomness. Suggestions of absolute randomness probably come only when classical and quantum models are mixed, as in the claim that quantum processes near black holes may lose information to space-time regions that are causally disconnected in the classical approximation.

[11]In the cases now known, recognition of any pattern seems to involve essentially complete reconstruction of the original program, but this may not always be so (L. Levin, private communication).

[12]In some cases, such as optimization or eigenvalue problems in the complexity class NP [e.g., M. Garey and D. Johnson, *Computers and Interactability: A Guide to the Theory of NP-Completeness* (Freeman, San Francisco, 1979)], even each individual test may take exponential time.

[13]The sequence certainly passes the standard statistical tests of Ref. 6, and contains all possible subsequences up to length at least 12. It has also been proved that only at most one vertical sequence in the pattern of Fig. 1 can have a finite period [E. Jen, Los Alamos Report No. LA-UR-85-1218 (to be published)].

[14]For example, D. E. R. Denning, *Cryptography and Data Security* (Addison-Wesley, Reading, Mass., 1982). Systems like Fig. 1 can, for example, be used for "stream ciphers" by adding each bit in the sequences produced with a particular seed to a bit in a plain-text message.

[15]For example, O. Martin, A. Odlyzko, and S. Wolfram,

Commun. Math. Phys. **93**, 219 (1984).

[16]B. Jansson, *Random Number Generators* (Almqvist & Wiksells, Stockholm, 1966).

[17]They are one-symbol-deletion tag sequences [A. Cobham, Math. Systems Theory **6**, 164 (1972)], and can be represented by generating functions algebraic over $GF(k)$ [G. Christol, T. Kamae, M. Mendes France, and G. Rauzy, Bull. Soc. Math. France **108**, 401 (1980); J.-M. Deshouillers, Seminar de Theorie des Nombres, Université de Bordeaux Exposé No. 5, 1979 (unpublished); M. Dekking, M. Mendes France, and A. van der Poorten, Math. Intelligencer, **4**, 130, 173, 190 (1983)]. Their self-similarity is related to the pumping lemma for regular languages [e.g., J. Hopcroft and J. Ullman, *Introduction to Automata Theory, Languages and Computation* (Addison-Wesley, Reading Mass., 1979)]. More complicated sequences associated with context-free formal languages can also be recognized in polynomial time, but the recognition problem for context-sensitive ones is P-space complete.

[18]For example, A. M. Frieze, R. Kannan, and J. C. Lagarias, in *Twenty-Fifth IEEE Symposium on Foundations of Computer Science* (IEEE, New York, 1984). The sequences also typically fail certain statistical randomness tests, such as multidimensional spectral tests (Ref. 6). They are nevertheless probably random with respect to all NC computations [J. Reif and J. Tygar, Harvard University Computation Laboratory Report No. TR-07-84 (to be published)].

[19]For example, G. Choquet, C. R. Acad. Sci. (Paris), Ser. A **290**, 575 (1980); cf. J. Lagarias, Amer. Math. Monthly **92**, 3 (1985). (Note that with appropriate boundary conditions a finite-size version of this system is equivalent to a linear congruential pseudorandom number generator.)

[20]A. Shamir, Lecture Notes in Computer Science, **62**, 544 (1981); S. Goldwasser and S. Micali, J. Comput. Sys. Sci. **28**, 270 (1984); M. Blum and S. Micali, SIAM J. Comput. **13**, 850 (1984); A. Yao, in *Twenty-Third IEEE Symposium on Foundations of Computer Science* (IEEE, New York, 1982); L. Blum, M. Blum, and M. Shub, in *Advances in Cryptology: Proceedings of CRYPTO-82*, edited by D. Chaum, R. Rivest, and A. T. Sherman (Plenum, New York, 1983); O. Goldreich, S. Goldwasser, and S. Micali, in *Twenty-Fifth IEEE Synmposium on Foundations of Computer Science* (IEEE, New York, 1984).

[21]For example, M. Garey and D. Johnson, Ref. 12.

[22]For example, L. Kuipers and H. Niederreiter, *Uniform Distribution of Sequences* (Wiley, New York, 1974).

[23]A polynomial-time procedure is also known for recognizing solutions to more complicated algebraic or trigonometric equations (R. Kannan, A. K. Lenstra, and L. Lovasz, Carnegie-Mellon University Technical Report No. CMU-CS-84-111).

[24]Many localized structures have been found (D. Lind, private communication).

[25]S. Wolfram, Phys. Rev. Lett. **54**, 735 (1985).

[26]For example, U. Frisch, Phys. Scr. **T9**, 137 (1985).

[27]G. Ahlers and R. W. Walden, Phys. Rev. Lett. **44**, 445 (1980).

[28]For example, M. Brachet *et al.*, J. Fluid Mech. **130**, 411 (1983).

Thermodynamics and Hydrodynamics with Cellular Automata

(November 1985)

James B. Salem
Thinking Machines Corporation, 245 First Street, Cambridge, MA 02144

Stephen Wolfram
The Institute for Advanced Study, Princeton, NJ 08540

Simple cellular automata which seem to capture the essential features of thermodynamics and hydrodynamics are discussed. At a microscopic level, the cellular automata are discrete approximations to molecular dynamics, and show relaxation towards equilibrium. On a large scale, they behave like continuum fluids, and suggest efficient methods for hydrodynamic simulation.

Thermodynamics and hydrodynamics describe the overall behaviour of many systems, independent of the precise microscopic construction of each system. One can thus study thermodynamics and hydrodynamics using simple models, which are more amenable to efficient simulation, and potentially to mathematical analysis.

Cellular automata (CA) are discrete dynamical systems which give simple models for many complex physical processes [1]. This paper considers CA which can be viewed as discrete approximations to molecular dynamics. In the simplest case, each link in a regular spatial lattice carries at most one "particle" with unit velocity in each direction. At each time step, each particle moves one link; those arriving at a particular site then "scatter" according to a fixed set of rules. This discrete system is well-suited to simulation on digital computers. The state of each site is represented by a few bits, and follows simple logical rules. The rules are local, so that many sites can be updated in parallel. The simulations in this paper were performed on a Connection Machine Computer [2] which updates sites concurrently in each of 65536 Boolean processors [3].

In two dimensions, one can consider square and hexagonal (six links at 60°) lattices. On a square lattice [4], the only nontrivial rule which conserves momentum and particle number takes isolated pairs of particles colliding head on scatter in the orthogonal direction (no interaction in other cases). On a hexagonal lattice [5], such pairs may scatter in either of the other two directions, and the scattering may be affected by particles in the third direction. Four particles coming along two directions may also scatter in different directions. Finally, particles on three links separated by 120° may scatter along the other three links. At fixed boundaries, particles may either "bounce back" (yielding "no slip" on average), or reflect "specularly" through 120°.

On a microscopic scale, these rules are deterministic, reversible and discrete. But on a sufficiently large scale, a statistical description may apply, and the system may behave like a continuum fluid, with macroscopic quantities, such as hydrodynamic velocity, obtained by kinetic theory averages.

Figure 1 illustrates relaxation to "thermodynamic equilibrium". The system randomizes, and coarse-grained entropy increases. This macroscopic behaviour is robust. but microscopic details depend sensitively on initial conditions. Small perturbations (say of one particle) have microscopic effects over linearly-expanding regions [6]. Thus ensembles of "nearby" initial states usually evolve to contain widely-differing "typical" states. But in addition, individual "simply-specified" initial states can yield behaviour so complex as to seem random [7, 8], as in figure 1. The dynamics thus "encrypts" the initial data; given only coarse-grained, partial, information, the initial simplicity cannot be recovered or recognized by computationally feasible procedures [7], and the behaviour is effectively irreversible.

Microscopic instability implies that predictions of detailed behaviour are impossible without ever more extensive knowledge of initial conditions. With complete knowledge (say from a simple specification), the behaviour can always be reproduced by explicit simulation. But if effective predictions are to be made, more efficient computational procedures should be found. It is known, however, that the CA considered here can act as universal computers [9]: with appropriate initial conditions, their evolution can implement any computation. Streams of particles corresponding to "wires" can meet in logical gates implemented by fixed obstructions or other streams. As a consequence, the evolution is computationally irreducible [10]; there is no general shortcut to explicit simulation. No simpler computation can reproduce all the possible phenomena.

Some overall statistical predictions can nevertheless be made. In isolation, the CA seem to relax to an equilibrium in which links are populated effectively randomly with a particular average particle density ρ and net velocity (as in figure 1). On length scales large compared to the mean free path λ, the system then behaves like a continuum fluid. The effective fluid pressure is $p = \rho/2$, giving a speed of sound $c = 1/\sqrt{2}$. Despite the microscopic anisotropy of the lattice, circular sound wavefronts are obtained from point sources (so long as their wavelength is larger than the mean free path) [11].

Assuming local equilibrium, the large-scale behaviour of the CA can be approximated by average rules for collections of particles, with particular average densities and velocities. The rules are like finite difference approximations to partial differential equations, whose form can be found by a standard Chapman–Enskog expansion [12] of microscopic particle distributions in terms of macroscopic quantities. The results are analogous to those for systems [13] in which particles occur with an arbitrary continuous density at each point in space, but have only a finite set of possible velocities corresponding to the links of the lattice. The hexagonal lattice CA is then found to follow exactly the standard Navier–Stokes equations [5, 14]. As usual, the parameters in the Navier–Stokes equations depend on the microscopic structure of the system. Kinetic theory suggests a kinematic viscosity $\nu = \lambda/2$ [15].

Figures 2 and 3 show hydrodynamic phenomena in the large scale behaviour of the hexagonal lattice CA. An overall flow U is obtained by maintaining a difference in the numbers of left- and right-moving particles at the boundaries. Since local equilibrium is rapidly reached from almost any state, the results are insensitive to the precise arrangement used [14]. Random boundary fluxes imitate an infinite region; a regular pattern of incoming particles nevertheless also suffices, and reflecting or cyclic boundary conditions can be used on the top and bottom edges.

The hydrodynamics of the CA is much like a standard physical fluid [16]. For low Mach numbers $Ma = U/c$, the fluid is approximately incompressible, and the flows show dynamical similarity, depending only on Reynolds number $Re = UL/\nu$ ($L > \lambda$.). The patterns obtained agree qualitatively with experiment [3]. At low Re, the flows are macroscopically stable; perturbations are dissipated into microscopic "heat".

At higher Re, vortex streets are produced, and there are signs of irregular, turbulent, behaviour. Perturbations now affect details of the flow, though not its statistical properties. The macroscopic irregularity does not depend on microscopic randomness; it occurs even if microscopically simple (say spatially and temporally periodic) initial and boundary conditions are used, as illustrated in figure 2. As at the microscopic level, it seems that the evolution corresponds to a sufficiently complex computation that its results seem random [7].

The CA discussed here should serve as a basis for practical hydrodynamic simulations. They are simple to program, readily amenable to parallel processing, able to handle complex geometries easily, and presumably show no unphysical instabilities. (Generalization to three dimensions is straightforward in principle [17].)

Standard finite difference methods [18] consider discrete cells of fluid described by continuous parameters. These parameters are usually represented as digital numbers with say 64 bits of precision. Most of these bits are, however, probably irrelevant in determining observable features of flow. In the CA approach, all bits are of essentially equal importance, and the number of elementary operations performed is potentially closer to the irreducible limit.

The difficulty of computation in a particular case depends on the number of cells that must be used. Below a certain dissipation length scale a, viscosity makes physical fluids smooth. Each CA cell can thus potentially represent a packet of fluid on this length scale. In fully developed turbulence, a appears to scale approximately as $Re^{-d/4}$ in d dimensions [16]. These estimates govern cell sizes needed in finite difference schemes.

The CA discussed so far incorporate no microscopic irreversibility: macroscopic energy dissipation occurs only through degradation to microscopic "heat". But in turbulence there is continual dissipation of $\sim Re^d$ energy at small scales. To absorb this energy, the CA cells must be $Re^{1/4}(d=3)$ or $Re^{1/2}(d=2)$ times larger than a, making such CA less efficient than irreversible finite difference schemes [19]. "Turbulence models" with irreversibility in the microscopic CA rules may overcome this difficulty.

Several further extensions of the CA scheme can be considered. First, on some or all of the lattice, basic units containing say n particles, rather than single particles, can be used. The properties of these units can be specified by digital numbers with $O(\log n)$ bits, but exact conservation laws can still be maintained. This scheme comes closer to adaptive grid finite difference methods [18], and potentially avoids detailed computation in featureless parts of flows.

A second, related, extension introduces discrete internal degrees of freedom for each particle. These could represent different particle types, directions of discrete vortices [18], or internal energy (giving variable temperature [20]).

This paper has given evidence that simple cellular automata can reproduce the essential features of thermodynamic and hydrodynamic behaviour. These models make contact with results in dynamical systems theory and computation theory. They should also yield efficient practical simulations, particularly on parallel-processing computers.

Cellular automata can potentially reproduce behaviour conventionally described by partial differential equations in many other systems whose intrinsic dynamics involves many degrees of freedom with no large disparity in scales.

We thank U. Frisch, B. Hasslacher, Y. Pomeau and T. Shimomura for sharing their unpublished results with us, and we thank N. Margolus, S. Omohundro, S. Orszag, N. Packard, R. Shaw, T. Toffoli, G. Vichniac and V. Yakhot for discussions. We are also grateful to many people at Thinking Machines Corporation for their help and encouragement. The work of S.W. was supported in part by the U.S. Office of Naval Research under contract number N00014-85-K-0045.

1. See for example S. Wolfram, "Cellular Automata as Models of Complexity", *Nature* 311, 419 (1984) where applications to thermodynamics and hydrodynamics were mentioned but not explored.

2. D. Hillis, *The Connection Machine* (MIT press, 1985). This application is discussed in S. Wolfram, "Scientific Computation with the Connection Machine", Thinking Machines Corporation report (March 1985).

3. More detailed results of theory and simulation will be given in a forthcoming series of papers.

4. J. Hardy, O. de Pazzis and Y. Pomeau, "Molecular Dynamics of a Classical Lattice Gas: Transport Properties and Time Correlation Functions", *Phys. Rev.* A13, 1949 (1976).

5. U. Frisch, B. Hasslacher and Y. Pomeau, "A Lattice Gas Automaton for the Navier–Stokes Equation", Los Alamos preprint LA-UR-85-3503.

6. The expansion rate gives the Lyapunov exponent as defined in N. Packard and S. Wolfram, "Two-Dimensional Cellular Automata", *J. Stat. Phys.* 38, 901 (1985). Note that the effect involves many particles, and does not arise from instability in the motion of single particles, as in the case of hard spheres with continuous position variables (e.g. O. Penrose, "Foundations of Statistical Mechanics", *Rep. Prog. Phys.* 42, 129 (1979).)

7. S. Wolfram, "Origins of Randomness in Physical Systems", *Phys. Rev. Lett.* 55, 449 (1985); "Random Sequence Generation by Cellular Automata", *Adv. Appl. Math.* (in press).

8. Simple patterns are obtained with very simple or symmetrical initial conditions. On a hexagonal lattice, the motion of an isolated particle in a rectangular box is described by a linear congruence relation, and is ergodic when the side lengths are not commensurate.

9. N. Margolus, "Physics-Like Models of Computation", *Physica* 10D, 81 (1984).

10. S. Wolfram, "Undecidability and Intractability in Theoretical Physics", *Phys. Rev. Lett.* 54, 735 (1985).

11. *cf* T. Toffoli, "CAM: A High-Performance Cellular Automaton Machine", *Physica* 10D, 195 (1984).

12. e.g. S. Harris, *The Boltzmann Equation*, (Holt, Reinhart and Winston, 1971).

13. R. Gatignol, *Theorie cinetique des gaz a repartition discrete de vitesse*, (Springer, 1975); J. Broadwell, "Shock Structure in a Simple Discrete Velocity Gas", *Phys. Fluids* 7, 1243 (1964).

14. On a square lattice, the total momentum in each row is separately conserved, and so cannot be convected by velocity in the orthogonal direction. Symmetric three particle collisions on a hexagonal lattice remove this spurious conservation law.

15. The rank three and four tensor coefficients of the nonlinear and viscous terms in the Navier–Stokes equations are isotropic for a hexagonal but not a square lattice. Higher order coefficients are anisotropic in both cases. In two dimensions, there can be logarithmic corrections to the Newtonian fluid approximation: these can apparently be ignored on the length scales considered, but yield a formal divergence in the viscosity (*cf* [3]).

16. e.g. D. J. Tritton, *Physical fluid dynamics*, (Van Nostrand, 1977).

17. Icosahedral symmetry yields isotropic fluid behaviour, and can be achieved to an arbitrary degree of approximation with a suitable periodic lattice (*cf* D. Levine *et al.*, "Elasticity and Dislocations in Pentagonal and Icosahedral Quasicrystals", *Phys. Rev. Lett.* 54, 1520 (1985); P. Bak, "Symmetry, Stability, and Elastic Properties of Icosahedral Incommensurate Crystals", *Phys. Rev.* B32, 5764 (1985)).

18. e.g. P. Roache, *Computational Fluid Dynamics*, (Hermosa, Albuquerque, 1976).

19. S. Orszag and V. Yakhot, "Reynolds Number Scaling of Cellular Automaton Hydrodynamics", Princeton University Applied and Computational Math. report (November 1985).

20. In simple cases the resulting model is analogous to a deterministic microcanonical spin system (M. Creutz, "Deterministic Ising Dynamics", *Ann. Phys.*, to be published.)

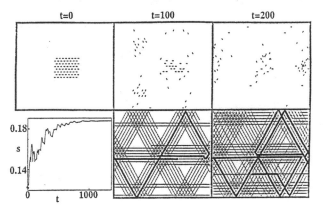

Figure 1. Relaxation to "thermodynamic equilibrium" in the hexagonal lattice cellular automaton (CA) described in the text. Discrete particles are initially in a simple array in the centre of a 32×32 site square box. The upper sequence shows the randomization of this pattern with time; the lower sequence shows the cells visited in the discrete phase space (one particle track is drawn thicker). The graph illustrates the resulting increase of coarse-grained entropy $\sum p_i \log_2 p_i$ calculated from particle densities in 32×32 regions of a 256×256 box.

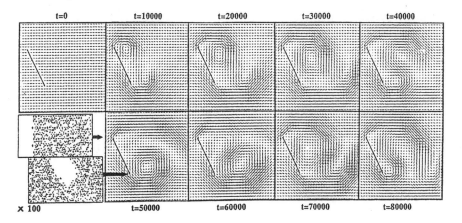

Figure 2. Time evolution of hydrodynamic flow around an inclined plate in the CA of figure 1 on a 2048×2048 site lattice. Hydrodynamic velocities are obtained as indicated by averaging over 64×64 site regions. There is an average density of 0.3 particles per link (giving a total of 7×10^6 particles). An overall velocity $U = 0.1$ is maintained by introducing an excess of particles (here in a regular pattern) on the left hand boundary.

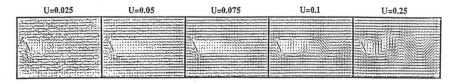

Figure 3. Hydrodynamic flows obtained after 10^5 time steps in the CA of figure 2. for various overall velocities U.

Cellular Automaton Fluids 1: Basic Theory

(March 1986; final version August 1986)

Stephen Wolfram

The Institute for Advanced Study, Princeton, New Jersey

Thinking Machines Corporation, 245 First Street, Cambridge, Massachusetts 02142

Continuum equations are derived for the large-scale behavior of a class of cellular automaton models for fluids. The cellular automata are discrete analogues of molecular dynamics, in which particles with discrete velocities populate the links of a fixed array of sites. Kinetic equations for microscopic particle distributions are constructed. Hydrodynamic equations are then derived using the Chapman–Enskog expansion. Slightly modified Navier–Stokes equations are obtained in two and three dimensions with certain lattices. Viscosities and other transport coefficients are calculated using the Boltzmann transport equation approximation. Some corrections to the equations of motion for cellular automaton fluids beyond the Navier–Stokes order are given.

1. Introduction

Cellular automata (e.g., Refs. 1 and 2) are arrays of discrete cells with discrete values. Yet sufficiently large cellular automata often show seemingly continuous macroscopic behavior (e.g., Refs. 1 and 3). They can thus potentially serve as models for continuum systems, such as fluids. Their underlying discreteness, however, makes them particularly suitable for digital computer simulation and for certain forms of mathematical analysis.

On a microscopic level, physical fluids also consist of discrete particles. But on a large scale, they, too, seem continuous, and can be described by the partial differential equations of hydrodynamics (e.g., Ref. 4). The form of these equations is in fact quite insensitive to microscopic details. Changes in molecular interaction laws can affect parameters such as viscosity, but do not alter the basic form of the macroscopic equations. As a result, the overall behavior of fluids can be found without accurately reproducing the details of microscopic molecular dynamics.

This paper is the first in a series which considers models of fluids based on cellular automata whose microscopic rules give discrete approximations to molecular dynamics.[1] The paper uses methods from kinetic theory to show that the macroscopic behavior of certain cellular automata corresponds to the standard Navier–Stokes equations for fluid flow. The next paper in the series [16] describes computer experiments on such cellular automata, including simulations of hydrodynamic phenomena.

Figure 1 shows an example of the structure of a cellular automaton fluid model. Cells in an array are connected by links carrying a bounded number of discrete "particles." The particles move in steps and "scatter" according to a fixed set of deterministic rules. In most cases, the rules are chosen so

"Cellular Automaton Fluids 1: Basic Theory" was originally published in the *Journal of Statistical Physics*, volume 45, pages 471–526 (November 1986). doi: 10.1007/BF01021083.

that quantities such as particle number and momentum are conserved in each collision. Macroscopic variations of such conserved quantities can then be described by continuum equations.

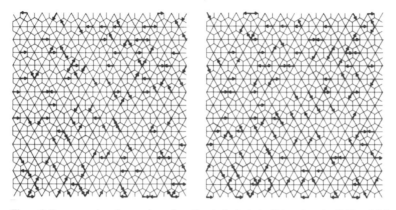

Figure 1. Two successive microscopic configurations in the typical cellular automaton fluid model discussed in Section 2. Each arrow represents a discrete "particle" on a link of the hexagonal grid. Continuum behavior is obtained from averages over large numbers of particles.

Particle configurations on a microscopic scale are rapidly randomized by collisions, so that a local equilibrium is attained, described by a few statistical average quantities. (The details of this process will be discussed in a later paper.) A master equation can then be constructed to describe the evolution of average particle densities as a result of motion and collisions. Assuming slow variations with position and time, one can then write these particle densities as an expansion in terms of macroscopic quantities such as momentum density. The evolution of these quantities is determined by the original master equation. To the appropriate order in the expansion, certain cellular automaton models yield exactly the usual Navier–Stokes equations for hydrodynamics.

The form of such macroscopic equations is in fact largely determined simply by symmetry properties of the underlying cellular automaton. Thus, for example, the structure of the nonlinear and viscous terms in the Navier–Stokes equations depends on the possible rank three and four tensors allowed by the symmetry of the cellular automaton array. In two dimensions, a square lattice of particle velocities gives anisotropic forms for these terms.[6] A hexagonal lattice, however, has sufficient symmetry to ensure isotropy.[7] In three dimensions, icosahedral symmetry would guarantee isotropy, but no crystallographic lattice with such a high degree of symmetry exists. Various structures involving links beyond nearest neighbors on the lattice can instead be used.

Although the overall form of the macroscopic equations can be established by quite general arguments, the specific coefficients which appear in them depend on details of the underlying model. In most cases, such transport coefficients are found from explicit simulations. But, by using a Boltzmann approximation to the master equation, it is possible to obtain some exact results for such coefficients, potentially valid in the low-density limit.

This paper is organized as follows. Section 2 describes the derivation of kinetic and hydrodynamic equations for a particular sample cellular automaton fluid model. Section 3 generalizes these results and discusses the basic symmetry conditions necessary to obtain standard hydrodynamic behavior. Section 4 then uses the Boltzmann equation approximation to investigate microscopic behavior and obtain results for transport coefficients. Section 5 discusses a few extensions of the model. The Appendix gives an SMP program[17] used to find macroscopic equations for cellular automaton fluids.

2. Macroscopic Equations for a Sample Model

2.1 Structure of the Model

The model[7] is based on a regular two-dimensional lattice of hexagonal cells, as illustrated in Fig. 1. The site at the center of each cell is connected to its six neighbors by links corresponding to the unit vectors $\mathbf{e}_1$ through $\mathbf{e}_6$ given by

$$\mathbf{e}_a = (\cos(2\pi a/6), \sin(2\pi a/6)) \tag{2.1.1}$$

At each time step, zero or one particles lie on each directed link. Assuming unit time steps and unit particle masses, the velocity and momentum of each particle is given simply by its link vector $\mathbf{e}_a$. In this model, therefore, all particles have equal kinetic energy, and have zero potential energy.

The configuration of particles evolves in a sequence of discrete time steps. At each step, every particle first moves by a displacement equal to its velocity $\mathbf{e}_a$. Then the particles on the six links at each site are rearranged according to a definite set of rules. The rules are chosen to conserve the number and total momentum of the particles. In a typical case, pairs of particles meeting head on might scatter through 60°, as would triples of particles 120° apart. The rules may also rearrange other configurations, such as triples of particles meeting asymmetrically. Such features are important in determining parameters such as viscosity, but do not affect the form of the macro-scopic equations derived in this section.

To imitate standard physical processes, the collision rules are usually chosen to be microscopically reversible. There is therefore a unique predecessor, as well as a unique successor, for each microscopic particle configuration. The rules for collisions in each cell thus correspond to a simple permutation of the possible particle arrangements. Often the rules are self-inverse. But in any case, the evolution of a complete particle configuration can be reversed by applying inverse collision rules at each site.

The discrete nature of the cellular automaton model makes such precise reversal in principle possible. But the rapid randomization of microscopic particle configurations implies that very complete knowledge of the current configuration is needed. With only partial information, the evolution may be effectively irreversible.[8, 19]

2.2 Basis for Kinetic Theory

Cellular automaton rules specify the precise deterministic evolution of microscopic configurations. But if continuum behavior is seen, an approximate macroscopic description must also be possible. Such a description will typically be a statistical one, specifying not, for example, the exact config-uration of particles, but merely the probabilities with which different configurations appear.

A common approach is to consider ensembles in which each possible microscopic configuration occurs with a particular probability (e.g., Ref. 18). The reversibility of the microscopic dynamics ensures that the total probability for all configurations in the ensemble must remain constant with time. The probabilities for individual configurations may, however, change, as described formally by the Liouville equation.

An ensemble is in "equilibrium" if the probabilities for configurations in it do not change with time. This is the case for an ensemble in which all possible configurations occur with equal probability. For cellular automata with collision rules that conserve momentum and particle number, the subsets of this ensemble that contain only those configurations with particular total values of the conserved quantities also correspond to equilibrium ensembles.

If the collision rules effectively conserved absolutely no other quantities, then momentum and particle number would uniquely specify an equilibrium ensemble. This would be the case if the system were ergodic, so that starting from any initial configuration, the system would eventually visit all other microscopic configurations with the same values of the conserved quantities. The time required would, however, inevitably be exponentially long, making this largely irrelevant for practical purposes.

A more useful criterion is that starting from a wide range of initial ensembles, the system evolves rapidly to ensembles whose statistical properties are determined solely from the values of conserved quantities. In this case, one could assume for statistical purposes that the ensemble reached contains all configurations with these values of the conserved quantities, and that the configurations occur with equal probabilities. This assumption then allows for the immediate construction of kinetic equations that give the average rates for processes in the cellular automaton.

The actual evolution of a cellular automaton does not involve an ensemble of configurations, but rather a single, specific configuration. Statistical results may nevertheless be applicable if the behavior of this single configuration is in some sense "typical" of the ensemble.

This phenomenon is in fact the basis for statistical mechanics in many different systems. One assumes that appropriate space or time averages of an individual configuration agree with averages obtained from an ensemble of different configurations. This assumption has never been firmly established in most practical cases; cellular automata may in fact be some of the clearest systems in which to investigate it.

The assumption relies on the rapid randomization of microscopic configurations, and is closely related to the second law of thermodynamics. At least when statistical or coarse-grained measurements are made, configurations must seem progressively more random, and must, for example, show increasing entropies. Initially ordered configurations must evolve to apparent disorder.

The reversibility of the microscopic dynamics nevertheless implies that ordered initial configurations can always in principle be reconstructed from a complete knowledge of these apparently disordered states. But just as in pseudorandom sequence generators or cryptographic systems, the evolution may correspond to a sufficiently complex transformation that any regularities in the initial conditions cannot readily be discerned. One suspects in fact that no feasibly simple computation can discover such regularities from typical coarse-grained measurements.[19, 20] As a result, the configurations of the system seem random, at least with respect to standard statistical procedures.

While most configurations may show progressive randomization, some special configurations may evolve quite differently. Configurations obtained by computing time-reversed evolution from ordered states will, for example, evolve back to ordered states. Nevertheless, one suspects that the systematic construction of such "antithermodynamic" states must again require detailed computations of a complexity beyond that corresponding to standard macroscopic experimental arrangements.

Randomization requires that no additional conserved quantities are present. For some simple choices of collision rules, spurious conservation laws can nevertheless be present, as discussed in Section 4.5. For most of the collision rules considered in this paper, however, rapid microscopic randomization does seem to occur.

As a result, one may use a statistical ensemble description. Equilibrium ensembles in which no statistical correlations are present should provide adequate approximations for many macroscopic properties. At a microscopic level, however, the deterministic dynamics does lead to correlations in the detailed configurations of particles.[2] Such correlations are crucial in determining local properties of the system. Different levels of approximation to macroscopic behavior are obtained by ignoring correlations of different orders.

Transport and hydrodynamic phenomena involve systems whose properties are not uniform in space and time. The uniform equilibrium ensembles discussed above cannot provide exact descriptions of such systems. Nevertheless, so long as macroscopic properties vary slowly enough, collisions should maintain approximate local equilibrium, and should make approximations based on such ensembles accurate.

2.3 Kinetic Equations

An ensemble of microscopic particle configurations can be described by a phase space distribution function which gives the probability for each complete configuration. In studying macroscopic phenomena, it is, however, convenient to consider reduced distribution functions, in which an average has been taken over most degrees of freedom in the system. Thus, for example, the one-particle distribution function $f_a(\mathbf{x}, t)$ gives the probability of finding a particle with velocity $\mathbf{e}_a$ at position $\mathbf{x}$ and time t, averaged over all other features of the configuration (e.g., Ref. 23).

Two processes lead to changes in f_a with time: motion of particles from one cell to another, and interactions between particles in a given cell. A master equation can be constructed to describe these processes.

In the absence of collisions, the cellular automaton rules imply that all particles in a cell at position $\mathbf{X}$ with velocity $\mathbf{e}_a$ move at the next time step to the adjacent cell at position $\mathbf{X} + \mathbf{e}_a$. As a result, the distribution function evolves according to

$$f_a(\mathbf{X} + \mathbf{e}_a, T + 1) = f_a(\mathbf{X}, T) \tag{2.3.1}$$

For large lattices and long time intervals, position and time may be approximated by continuous variables. One may define, for example, scaled variables $\mathbf{x} = \delta_x \mathbf{X}$ and $t = \delta_t T$, where $\delta_x, \delta_t \ll 1$. In terms of these scaled variables, the difference equation (2.3.1) becomes

$$f_a(\mathbf{x} + \mathbf{e}_a \delta_x, t + \delta_t) - f_a(\mathbf{x}, t) = 0 \tag{2.3.2}$$

In deriving macroscopic transport equations, this must be converted to a differential equation. Carrying out a Taylor expansion, one obtains[24]

$$\delta_t \, \partial_t f_a + \delta_x \, \mathbf{e}_a \cdot \nabla f_a + \frac{1}{2} \delta_t^2 \, \partial_{tt} f_a + \delta_x \delta_t (\mathbf{e}_a \cdot \nabla) \, \partial_t f_a + \frac{1}{2} \delta_x^2 (\mathbf{e}_a \cdot \nabla)^2 f_a + O(\delta^3) = 0 \tag{2.3.3}$$

If all variations in the f_a are assumed small, and certainly less than $O\left(1/\delta_x, 1/\delta_t\right)$, it suffices to keep only first-order terms in δ_x, δ_t. In this way one obtains the basic transport equation

$$\partial_t f_a(\mathbf{x}, t) + \mathbf{e}_a \cdot \nabla f_a(\mathbf{x}, t) = 0 \qquad (2.3.4)$$

This has the form of a collisionless Boltzmann transport equation for f_a (e.g., Ref. 25). It implies, as expected, that f_a is unaffected by particle motion in a spatially uniform system.

Collisions can, however, change f_a even in a uniform system, and their effect can be complicated. Consider, for example, collisions that cause particles in directions $\mathbf{e}_1$ and $\mathbf{e}_4$ to scatter in directions $\mathbf{e}_2$ and $\mathbf{e}_5$. The rate for such collisions is determined by the probability that particles in directions $\mathbf{e}_1$ and $\mathbf{e}_4$ occur together in a particular cell. This probability is defined as the joint two-particle distribution function $\tilde{F}^{(2)}_{14}$. The collisions deplete the population of particles in direction $\mathbf{e}_1$ at a rate $\tilde{F}^{(2)}_{14}$. Microscopic reversibility guarantees the existence of an inverse process, which increases the population of particles in direction $\mathbf{e}_1$ at a rate given in this case by $\tilde{F}^{(2)}_{25}$. Notice that in a model where there can be at most one particle on each link, the scattering to directions $\mathbf{e}_2$ and $\mathbf{e}_5$ in a particular cell can occur only if no particles are already present on these links. The distribution function $\tilde{F}$ is constructed to include this effect, which is mathematically analogous to the Pauli exclusion principle for fermions.

The details of collisions are, however, irrelevant to the derivation of macroscopic equations given in this section. As a result, the complete change due to collisions in a one-particle distribution function f_a will for now be summarized by a simple "collision term" Ω_a, which in general depends on two-particle and higher order distribution functions. (In the models considered here, Ω_a is always entirely local, and cannot depend directly on, for example, derivatives of distribution functions.) In terms of Ω_a, the kinetic equation (2.3.3) extended to include collisions becomes

$$\partial_t f_a + \mathbf{e}_a \cdot \nabla f_a = \Omega_a \qquad (2.3.5)$$

With the appropriate form for Ω_a, this is an exact statistical equation for f_a (at least to first order in δ).

But the equation is not in general sufficient to determine f_a. It gives the time evolution of f_a in terms of the two-particle and higher order distribution functions that appear in Ω_a. The two-particle distribution function then in turn satisfies an equation involving three-particle and higher order distribution functions, and so on. The result is the exact BBGKY hierarchy of equations,[23] of which Eq. (2.3.5) is the first level.

The Boltzmann transport equation approximates (2.3.5) by assuming that Ω_a depends only on one-particle distribution functions. In particular, one may make a "molecular chaos" assumption that all sets of particles are statistically uncorrelated before each collision, so that multiple-particle distribution functions can be written as products of one-particle ones. The distribution function $\tilde{F}^{(2)}_{14}$ is thus approximated as $f_1 f_4 (1 - f_2)(1 - f_3)(1 - f_5)(1 - f_6)$. The resulting Boltzmann equations will be used in Section 4. In this section, only the general form (2.3.5) is needed.

The derivation of Eq. (2.3.5) has been discussed here in the context of a cellular automaton model in which particles are constrained to lie on the links of a fixed array. In this case, the maintenance of terms in (2.3.3) only to first order in δ_x, δ_t is an approximation, and corrections can arise, as discussed in Section 2.5.[24] Equation (2.3.5) is, however, exact for a slightly different class of models, in which particles have a discrete set of possible velocities, but follow continuous trajectories with arbitrary spatial positions. Such "discrete velocity gases" have often been considered, particularly in studies of highly rarefied fluids, in which the mean distance between collisions is comparable to the overall system size.[11, 14]

2.4 Conservation Laws

The one-particle distribution functions typically determine macroscopic average quantities. In particular, the total particle density n is given by

$$\sum_a f_a = n \tag{2.4.1}$$

while the momentum density $n\mathbf{u}$, where $\mathbf{u}$ is the average fluid velocity, is given by

$$\sum_a \mathbf{e}_a f_a = n\mathbf{u} \tag{2.4.2}$$

The conservation of these quantities places important constraints on the behavior of the f_a.

In a uniform system $\nabla f_a = 0$, so that Eq. (2.3.5) becomes

$$\partial_t f_a = \Omega_a \tag{2.4.3}$$

and Eqs. (2.4.1) and (2.4.2) imply

$$\sum_a \Omega_a = 0 \tag{2.4.4}$$

$$\sum_a \mathbf{e}_a \Omega_a = 0 \tag{2.4.5}$$

Using the kinetic equation (2.3.5), Eq. (2.4.4) implies

$$\partial_t \sum_a f_a + \sum_a \mathbf{e}_a \cdot \nabla f_a = 0 \tag{2.4.6}$$

With the second term in the form $\nabla \cdot (\sum \mathbf{e}_a f_a)$, Eq. (2.4.6) can be written exactly in terms of macroscopic quantities as

$$\partial_t n + \nabla \cdot (n\mathbf{u}) = 0 \tag{2.4.7}$$

This is the usual continuity equation, which expresses the conservation of fluid. It is a first example of a macroscopic equation for the average behavior of a cellular automaton fluid.

Momentum conservation yields the slightly more complicated equation

$$\partial_t \sum_a \mathbf{e}_a f_a + \sum_a \mathbf{e}_a (\mathbf{e}_a \cdot \nabla f_a) = 0 \tag{2.4.8}$$

Defining the momentum flux density tensor

$$\Pi_{ij} = \sum_a (\mathbf{e}_a)_i \, (\mathbf{e}_a)_j f_a \tag{2.4.9}$$

Eq. (2.4.8) becomes

$$\partial_t(nu_i) + \partial_j \Pi_{ij} = 0 \tag{2.4.10}$$

No simple macroscopic result for Π_{ij} can, however, be obtained directly from the definitions (2.4.1) and (2.4.2).

Equations (2.4.7) and (2.4.10) have been derived here from the basic transport equation (2.3.5). However, as discussed in Section 2.3, this transport equation is only an approximation, valid to first order in the lattice scale parameters δ_x, δ_t.[24] Higher order versions of (2.4.7) and (2.4.10) may be derived from the original Taylor expansion (2.3.3), and in some cases, correction terms are obtained.[24]

Assuming $\delta_x = \delta_t = \delta$, Eq. (2.4.6) to second order becomes

$$\sum_a [(\partial_t + \mathbf{e}_a \cdot \nabla) + \frac{1}{2} \delta (\partial_t + \mathbf{e}_a \cdot \nabla)^2] = 0 \tag{2.4.11}$$

Writing the $O(\delta)$ term in the form

$$\partial_t \sum_a (\partial_t + \mathbf{e}_a \cdot \nabla) f_a + \nabla \cdot \sum_a (\partial_t + \mathbf{e}_a \cdot \nabla) \mathbf{e}_a f_a \tag{2.4.12}$$

this term is seen to vanish for any f_a which satisfy the first-order equations (2.4.7) and (2.4.10). Lattice discretization effects thus do not affect the continuity equation (2.4.7), at least to second order.

Corrections do, however, appear at this order in the momentum equation (2.4.10). To second order, Eq. (2.4.8) can be written as

$$\sum_a (\partial_t + \mathbf{e}_a \cdot \nabla) \mathbf{e}_a f_a + \frac{1}{2} \delta \partial_t \sum_a (\partial_t + \mathbf{e}_a \cdot \nabla) \mathbf{e}_a f_a + \frac{1}{2} \delta \sum_a [\mathbf{e}_a \cdot \nabla \partial_t + (\mathbf{e}_a \cdot \nabla)^2] \mathbf{e}_a f_a = 0 \tag{2.4.13}$$

The second term vanishes if f_a satisfies the first-order equation (2.4.8). The third term, however, contains a piece trilinear in the $\mathbf{e}_a$, which gives a correction to the momentum equation (2.4.10).[24]

2.5 Chapman–Enskog Expansion

If there is local equilibrium, as discussed in Section 2.2, then the microscopic distribution functions $f_a(\mathbf{x}, t)$ should depend, on average, only on the macroscopic parameters $\mathbf{u}(\mathbf{x}, t)$ and $n(\mathbf{x}, t)$ and their derivatives. In general, this dependence may be very complicated. But in hydrodynamic processes, $\mathbf{u}$ and n vary only slowly with position and time. In addition, in the subsonic limit, $|\mathbf{u}| \ll 1$.

With these assumptions, one may approximate the f_a by a series or Chapman–Enskog expansion in the macroscopic variables. To the order required for standard hydrodynamic phenomena, the possible terms are

$$f_a = f \{1 + c^{(1)} \mathbf{e}_a \cdot \mathbf{u} + c^{(2)} [(\mathbf{e}_a \cdot \mathbf{u})^2 - \frac{1}{2}|\mathbf{u}|^2] + c_\nabla^{(2)} [(\mathbf{e}_a \cdot \nabla)(\mathbf{e}_a \cdot \mathbf{u}) - \frac{1}{2} \nabla \cdot \mathbf{u}] + \cdots\} \tag{2.5.1}$$

where the $c^{(i)}$ are undetermined coefficients. The first three terms here represent the change in microscopic particle densities as a consequence of changes in macroscopic fluid velocity; the fourth term accounts for first-order dependence of the particle densities on macroscopic spatial variations in the fluid velocity. The structures of these terms can be deduced merely from the need to form scalar quantities f_a from the vectors $\mathbf{e}_a$, $\mathbf{u}$, and ∇.

The relation

$$\sum_a (\mathbf{e}_a)_i \, (\mathbf{e}_a)_j = \frac{M}{d} \, \delta_{ij} \tag{2.5.2}$$

where here $M = 6$ and $d = 2$, and i and j are space indices, has been used in Eq. (2.5.1) to choose the forms of the $|\mathbf{u}|^2$ and ∇u terms so as to satisfy the constraints (2.4.1) and (2.4.2), independent of the values of the coefficients $c^{(2)}$ and $c_\nabla^{(2)}$. In terms of (2.5.1), Eq. (2.4.1) yields immediately

$$f = n/6 \tag{2.5.3}$$

while (2.4.2) gives

$$c^{(1)} = 2 \tag{2.5.4}$$

The specific values of $c^{(2)}$ and $c_\nabla^{(2)}$ can be determined only by explicit solution of the kinetic equation (2.3.5) including collision terms. (Some approximate results for these coefficients based on the Boltzmann transport equation will be given in Section 4.) Nevertheless, the structure of macroscopic equations can be derived from (2.5.1) without knowledge of the exact values of these parameters.

For a uniform equilibrium system with $\mathbf{u} = 0$, all the f_a are given by

$$f_a = f = n/6 \tag{2.5.5}$$

In this case, the momentum flux tensor (2.4.9) is equal to the pressure tensor, given, as in the standard kinetic theory of gases, by

$$P_{ij} = \sum_a (\mathbf{e}_a)_i \, (\mathbf{e}_a)_j f = \frac{1}{2} \, n \delta_{ij} \tag{2.5.6}$$

where the second equality follows from Eq. (2.5.2). Note that this form is spatially isotropic, despite the underlying anisotropy of the cellular automaton lattice. This result can be deduced from general symmetry considerations, as discussed in Section 3. Equation (2.5.6) gives the equation of state relating the scalar pressure to the number density of the cellular automaton fluid:

$$p = n/2 \tag{2.5.7}$$

When $\mathbf{u} \neq 0$, Π_{ij} can be evaluated in the approximation (2.5.1) using the relations

$$\sum_a (\mathbf{e}_a)_i \, (\mathbf{e}_a)_j \, (\mathbf{e}_a)_k = 0 \tag{2.5.8}$$

and

$$\sum_a (\mathbf{e}_a)_i \, (\mathbf{e}_a)_j \, (\mathbf{e}_a)_k \, (\mathbf{e}_a)_l = \frac{M}{d(d+2)} \, (\delta_{ij} \, \delta_{kl} + \delta_{ik} \, \delta_{jl} + \delta_{il} \, \delta_{jk}) \tag{2.5.9}$$

The result is

$$\Pi_{ij} = \frac{n}{2} \, \delta_{ij} + \frac{n}{4} \, c^{(2)} \, [u_i \, u_j - \frac{1}{2} |\mathbf{u}|^2 \delta_{ij}] + \frac{n}{4} \, c_\nabla^{(2)} \, [\partial_i u_j - \frac{1}{2} \, \nabla \cdot \mathbf{u}] \tag{2.5.10}$$

Substituting the result into Eq. (2.4.10), one obtains the final macroscopic equation

$$\partial_t(n\mathbf{u}) + \frac{1}{4} \, nc^{(2)} \, \{(\mathbf{u} \cdot \nabla) \, \mathbf{u} + [\mathbf{u}(\nabla \cdot \mathbf{u}) - \frac{1}{2} \, \nabla |\mathbf{u}|^2]\} = -\frac{1}{2} \, \nabla n - \frac{1}{8} \, nc_\nabla^{(2)} \, \nabla^2 \mathbf{u} - \frac{1}{4} \, \Xi \tag{2.5.11}$$

where

$$\Xi = \mathbf{u}(\mathbf{u} \cdot \nabla) \, (nc^{(2)}) - \frac{1}{2} |\mathbf{u}|^2 \nabla(nc^{(2)}) + (\mathbf{u} \cdot \nabla) \, (nc_\nabla^{(2)}) - \frac{1}{2} \, (\nabla \cdot \mathbf{u}) \, \nabla(nc_\nabla^{(2)}) \tag{2.5.12}$$

The form (2.5.10) for Π_{ij} follows exactly from the Chapman–Enskog expansion (2.5.1). But to obtain Eq. (2.5.11), one must use the momentum equation (2.4.10). Equation (2.4.13) gives corrections to this equation that arise at second order in the lattice size parameter δ. These corrections must be compared with other effects included in Eq. (2.5.11). The rescaling $\mathbf{x} = \delta_x \, \mathbf{X}$ implies that spatial gradient terms in the Chapman–Enskog expansion can be of the same order as the $O(\delta_x)$ correction terms in Eq. (2.4.13). When the $\mathbf{e}_a \cdot \mathbf{u}$ term in the Chapman–Enskog expansion (2.5.1) for the f_a is substituted into the last term of Eq. (2.4.13), it gives a contribution[24]

$$\Psi = -\frac{1}{16} \, nc^{(1)} \, \nabla^2 \mathbf{u} = -\frac{1}{8} \, n\nabla^2 \mathbf{u} \tag{2.5.13}$$

to the right-hand side of Eq. (2.5.11). Note that Ψ depends solely on the choice of $\mathbf{e}_a$, and must, for example, vary purely linearly with the particle density f.

2.6 Navier–Stokes Equation

The standard Navier–Stokes equation for a continuum fluid in d dimensions can be written in the form

$$\partial_t(n\mathbf{u}) + \mu n(\mathbf{u} \cdot \nabla) \, \mathbf{u} = -\nabla p + \eta \, \nabla^2 \mathbf{u} + (\zeta + \frac{1}{d} \, \eta) \, \nabla(\nabla \cdot \mathbf{u}) \tag{2.6.1}$$

where p is pressure, and η and ζ are, respectively, shear and bulk viscosities (e.g., Ref. 27). The coefficient μ of the convective term is usually constrained to have value 1 by Galilean invariance. Note that the coefficient of the last term in Eq. (2.6.1) is determined by the requirement that the term in Π_{ij} proportional to η be traceless.[27,57]

The macroscopic equation (2.5.11) for the cellular automaton fluid is close to the Navier–Stokes form (2.6.1). The convective and viscous terms are present, and have the usual structure. The pressure term appears according to the equation of state (2.5.7). There are, however, a few additional terms.

Terms proportional to $\mathbf{u} \, \nabla n$ must be discounted, since they depend on features of the microscopic distribution functions beyond those included in the Chapman–Enskog expansion (2.5.1). The continuity equation (2.4.7) shows that terms proportional to $\mathbf{u}(\nabla \cdot \mathbf{u})$ must also be neglected.

The term proportional to $\nabla|\mathbf{u}|^2$ remains, but can be combined with the ∇n term to yield an effective pressure term which includes fluid kinetic energy contributions.

The form of the viscous terms in (2.5.11) implies that for a cellular automaton fluid, considered here, bulk viscosity is given by

$$\zeta = 0 \tag{2.6.2}$$

The value of η is determined by the coefficient $c_\nabla^{(2)}$ that appears in the microscopic distribution function (2.5.1), according to

$$\eta = n\nu = -\frac{1}{8} n c_\nabla^{(2)} \tag{2.6.3}$$

where ν is the kinematic viscosity. An approximate method of evaluating $c_\nabla^{(2)}$ is discussed in Section 4.6.

The convective term in Eq. (2.5.11) has the same structure as in the Navier–Stokes equation (2.6.1), but includes a coefficient

$$\mu = \frac{1}{4} c^{(2)} \tag{2.6.4}$$

which is not in general equal to 1. In continuum fluids, the covariant derivative usually has the form $D_t = \partial_t + \mathbf{u} \cdot \nabla$ implied by Galilean invariance. The cellular automaton fluid acts, however, as a mixture of components, each with velocities $\mathbf{e}_a$, and these components can contribute with different weights to the covariant derivatives of different quantities, leading to convective terms with different coefficients.

The usual coefficient of the convective term can be recovered in Eq. (2.6.1) and thus Eq. (2.5.11) by a simple rescaling in velocity: setting

$$\tilde{\mathbf{u}} = \mu \mathbf{u} \tag{2.6.5}$$

the equation for $\tilde{\mathbf{u}}$ has coefficient 1 for the $(\tilde{\mathbf{u}} \cdot \nabla)\, \tilde{\mathbf{u}}$ term.

Small perturbations from a uniform state may be represented by a linearized approximation to Eqs. (2.4.7) and (2.5.11), which has the standard sound wave equation form, with a sound speed obtained from the equation of state (2.5.7) as

$$c = 1/\sqrt{2} \tag{2.6.6}$$

The form of the Navier–Stokes equation (2.6.1) is usually obtained by simple physical arguments. Detailed derivations suggest, however, that more elaborate equations may be necessary, particularly in two dimensions (e.g., Ref. 28). The Boltzmann approximation used in Section 4 yields definite values for $c^{(2)}$ and $c_\nabla^{(2)}$. Correlation function methods indicate, however, that additional effects yield logarithmically divergent contributions to $c_\nabla^{(2)}$ in two dimensions (e.g., Ref. 29). The full viscous term in this case may in fact be of the rough form $\nabla^2 \log(\nabla^2)\, \mathbf{u}$.

2.7 Higher Order Corrections

The derivation of the Navier–Stokes form (2.5.11) neglects all terms in the Chapman–Enskog expansion beyond those given explicitly in Eq. (2.5.1). This approximation is expected to be adequate only when $|\mathbf{u}| \ll c$. Higher order corrections may be particularly significant for supersonic flows involving shocks (e.g., Ref. 30).

Since the dynamics of shocks are largely determined just by conservation laws (e.g., Ref. 27), they are expected to be closely analogous in cellular automaton fluids and in standard continuum fluids. For $|\mathbf{u}|/c \gtrsim 2$, however, shocks become so strong and thin that continuum descriptions of physical fluids can no longer be applied in detail (e.g., Ref. 14). The structure of shocks in such cases can apparently be found only through consideration of explicit particle dynamics.[11,14]

In the transonic flow regime $|\mathbf{u}| \approx c$, however, continuum equations may be used, but corrections to the Navier–Stokes form may be significant. A class of such corrections can potentially be found by maintaining terms $O(u^3)$ and higher in the Chapman–Enskog expansion (2.5.1).

In the homogeneous fluid approximation $\nabla \mathbf{u} = 0$, one may take

$$\begin{aligned}
f_a = \; & f\{1 + c^{(1)}\, \mathbf{e}_a \cdot \mathbf{u} + c^{(2)}[(\mathbf{e}_a \cdot \mathbf{u})^2 + \sigma_2|\mathbf{u}|^2] \\
& + c^{(3)}[(\mathbf{e}_a \cdot \mathbf{u})^3 + \sigma_3|\mathbf{u}|^2(\mathbf{e}_a \cdot \mathbf{u})] \\
& + c^{(4)}[(\mathbf{e}_a \cdot \mathbf{u})^4 + \sigma_{4,1}|\mathbf{u}|^2(\mathbf{e}_a \cdot \mathbf{u})^2 + \sigma_{4,2}|\mathbf{u}|^4] + \cdots\}
\end{aligned} \tag{2.7.1}$$

The constraints (2.4.1) and (2.4.2) imply

$$c^{(1)} = d \tag{2.7.2}$$

$$\sigma_2 = -\frac{1}{d} \tag{2.7.3}$$

$$\sigma_3 = -\frac{3}{d+2} \tag{2.7.4}$$

$$\frac{3}{d(d+2)} + \frac{1}{d}\,\sigma_{4,1} + \sigma_{4,2} = 0 \tag{2.7.5}$$

where d is the space dimension, equal to two for the model of this section.

Corrections to (2.5.11) can be found by substituting (2.7.1) in the kinetic equation (2.4.8). For the hexagonal lattice model, one obtains, for example,

$$\begin{aligned}
\partial_t(nu_x) + \; & \frac{1}{4} nc^{(2)}(u_x\,\partial_x u_x + u_x\,\partial_y u_y + u_y\,\partial_y u_x - u_y\,\partial_x u_y) \\
& + \frac{1}{8} nc^{(4)}\{[(5 + 4\sigma_{4,1})\,u_x^3 - 3u_x u_y^2]\,\partial_x u_x \\
& + [(3 + 2\sigma_{4,1})\,u_y^3 + (3 + 6\sigma_{4,1})\,u_x^2\,u_y]\,\partial_y u_x \\
& - [(3 + 4\sigma_{4,1})\,u_y^3 + 3u_x^2\,u_y]\,\partial_x u_y \\
& + [(1 + 2\sigma_{4,1})\,u_x^3 + (9 + 6\sigma_{4,1})\,u_x u_y^2]\,\partial_y u_y\} = 0
\end{aligned} \tag{2.7.6}$$

The $O(u^2)$ term in Eq. (2.7.6) has the isotropic form given in Eq. (2.5.11). The $O(u^4)$ term is, however, anisotropic.

To obtain an isotropic $O(u^4)$ term, one must generalize the model, as discussed in Section 3. One possibility is to allow vectors $\mathbf{e}_a$ corresponding to corners of an M-sided polygon with $M > 6$. In this case, the continuum equation deduced from the Chapman–Enskog expansion (2.7.1) becomes

$$
\begin{aligned}
\partial_t (n\mathbf{u}) \;+\; & \frac{1}{4}\, nc^{(2)} [(\mathbf{u}\cdot\nabla)\,\mathbf{u} + \mathbf{u}(\nabla\cdot\mathbf{u}) - \frac{1}{2}\nabla|\mathbf{u}|^2] \\
+\; & \frac{1}{4}\, nc^{(4)}(1+\sigma_{4,\,1})\,\{|\mathbf{u}|^2[(\mathbf{u}\cdot\nabla)\,\mathbf{u} + \mathbf{u}(\nabla\cdot\mathbf{u}) - \nabla|\mathbf{u}|^2] \\
+\; & \mathbf{u}(\mathbf{u}\cdot\nabla)|\mathbf{u}|^2\} = 0
\end{aligned}
\tag{2.7.7}
$$

This gives a definite form for the next-order corrections to the convective part of the Navier–Stokes equation.

Corrections to the viscous part can be found by including terms proportional to $\nabla\mathbf{u}$ in the Chapman–Enskog expansion (2.7.1). The possible fourth-order terms are given by contractions of $u_i\, u_j\, \partial_k u_l$ with products of $(\mathbf{e}_a)_m$ or δ_{mn}. They yield a piece in the Chapman–Enskog expansion of the form

$$
\begin{aligned}
c_\nabla^{(4)}[\tau_1\,(\mathbf{e}_a\cdot\mathbf{u})^2\,(\mathbf{e}_a\cdot\nabla)\,(\mathbf{e}_a\cdot\mathbf{u}) + \tau_2|\mathbf{u}|^2(\mathbf{e}_a\cdot\nabla)\,(\mathbf{e}_a\cdot\mathbf{u}) \\
+ \tau_3(\mathbf{e}_a\cdot\mathbf{u})\,(\mathbf{u}\cdot\nabla)\,(\mathbf{e}_a\cdot\mathbf{u}) + \tau_4(\mathbf{e}_a\cdot\mathbf{u})^2\,(\nabla\cdot\mathbf{u}) + \tau_5|\mathbf{u}|^2(\nabla\cdot\mathbf{u})]
\end{aligned}
\tag{2.7.8}
$$

where Eq. (2.4.1) implies the constraints (for $d = 2$)

$$
\tau_1 + 2\,\tau_3 = 0
\tag{2.7.9}
$$

$$
\tau_1 + 4\,\tau_2 + 4\,\tau_4 + 8\,\tau_5 = 0
\tag{2.7.10}
$$

The resulting continuum equations may be written in terms of vectors formed by contractions of $u_i\, u_j\, \partial_k\, \partial_l u_m$ and $u_i\, \partial_j u_k\, \partial_l u_m$. The complete result is

$$
\begin{aligned}
\partial_t(n\mathbf{u}) \;+\; & \frac{1}{4}\, nc^{(2)}[(\mathbf{u}\cdot\nabla)\,\mathbf{u} + \mathbf{u}(\nabla\cdot\mathbf{u}) - \frac{1}{2}\nabla|\mathbf{u}|^2] \\
+\; & \frac{1}{4}\, nc^{(4)}(1+\sigma_{4,\,1})\,\{|\mathbf{u}|^2[(\mathbf{u}\cdot\nabla)\,\mathbf{u} + \mathbf{u}(\nabla\cdot\mathbf{u}) - \nabla|\mathbf{u}|^2] + \mathbf{u}(\mathbf{u}\cdot\nabla)|\mathbf{u}|^2\} \\
=\; & -\frac{1}{8}\, nc_\nabla^{(2)}\,\nabla^2\mathbf{u} \\
& -\frac{1}{32}\, nc_\nabla^{(4)}[((\tau_1 - 4\tau_2 + 12\tau_4)\,\mathbf{u}(\nabla\cdot\mathbf{u})^2 - (\tau_1 - 4\tau_2 + 4\tau_4) \\
& \times\; \mathbf{u}\,\{\nabla[(\mathbf{u}\cdot\nabla)\,\mathbf{u}] - (\mathbf{u}\cdot\nabla)\,(\nabla\cdot\mathbf{u})\} + 8\tau_4\,\{[(\mathbf{u}\cdot\nabla)\,\mathbf{u}]\cdot\nabla\}\,\mathbf{u} \\
& +\; \frac{1}{2}(\tau_1 + 4\tau_2)\,[(\nabla|\mathbf{u}|^2)\cdot\nabla]\,\mathbf{u} \\
& +\; 2\tau_1\,\mathbf{u}\,[\frac{1}{2}\nabla\cdot(\nabla|\mathbf{u}|^2) - \mathbf{u}\cdot(\nabla^2\mathbf{u})] - 4\tau_4(\nabla\cdot\mathbf{u})\,\nabla|\mathbf{u}|^2) \\
& +\; \{8\tau_4[\mathbf{u}(\mathbf{u}\cdot\nabla)\,(\nabla\cdot\mathbf{u}) - \frac{1}{2}|\mathbf{u}|^2\nabla(\nabla\cdot\mathbf{u})] \\
& +\; 2\tau_1\,\mathbf{u}[\mathbf{u}\cdot(\nabla^2\mathbf{u})] + 4\tau_2|\mathbf{u}|^2\nabla^2\mathbf{u}\}]
\end{aligned}
\tag{2.7.11}
$$

where, on the right-hand side, the first group of terms are all $O((\nabla \mathbf{u})^2)$, while the second group are $O(\nabla \nabla \mathbf{u})$. Further corrections involve higher derivative terms, such as $u_i \, \partial_j \, \partial_k \, \partial_l u_m$.

For a channel flow with $u_x = ax^2$, $u_y = 0$, the time-independent terms in Eq. (2.7.11) have an x component

$$\frac{1}{4} \, ac_\nabla^{(2)} + \frac{5}{8} \, a^3 \, x^4 \, c_\nabla^{(4)} (\tau_1 + 2\tau_2 + 2\tau_4) + \frac{1}{2} \, a_2 \, x^3 \, c^{(2)} + a^4 \, x^7 \, c^{(4)} (1 + \sigma_{4,1}) \tag{2.7.12}$$

and zero y component.

3. Symmetry Considerations

3.1 Tensor Structure

The form of the macroscopic equations (2.4.7) and (2.5.11) depends on few specific properties of the hexagonal lattice cellular automaton model. The most important properties relate to the symmetries of the tensors

$$\mathbf{E}^{(n)}_{i_1 i_2 \cdots i_n} = \sum_a (\mathbf{e}_a)_{i_1} \cdots (\mathbf{e}_a)_{i_n} \tag{3.1.1}$$

These tensors are determined in any cellular automaton fluid model simply from the choice of the basic particle directions $\mathbf{e}_a$. The momentum flux tensor (2.4.9) is given in terms of them by

$$\begin{aligned}
\Pi_{ij} = & \; f(\mathbf{E}^{(2)}_{ij} + c^{(1)} \, \mathbf{E}^{(3)}_{ijk} \, u_k + c^{(2)} [\mathbf{E}^{(4)}_{ijkl} \, u_k \, u_l + \sigma \mathbf{E}^{(2)}_{ij} \, u_k u_k] \\
& + c_\nabla^{(2)} [\mathbf{E}^{(4)}_{ijkl} \, \partial_k u_l + \sigma \mathbf{E}^{(2)}_{ij} \, \partial_k u_k])
\end{aligned} \tag{3.1.2}$$

where repeated indices are summed, and to satisfy the conditions (2.4.1) and (2.4.2)

$$\sigma = - \mathbf{E}^{(4)}_{ijkk} / \mathbf{E}^{(2)}_{ij} \tag{3.1.3}$$

The basic condition for standard hydrodynamic behavior is that the tensors $\mathbf{E}^{(n)}$ for $n \leq 4$ which appear in (3.1.2) should be isotropic. From the definition (3.1.1), the tensors must always be invariant under the discrete symmetry group of the underlying cellular automaton array. What is needed is that they should in addition be invariant under the full continuous rotation group.

The definition (3.1.1) implies that the $\mathbf{E}^{(n)}$ must be totally symmetric in their space indices. With no further conditions, the $\mathbf{E}^{(n)}$ could have $\binom{n+d-1}{n}$ independent components in d space dimensions. Symmetries in the underlying cellular automaton array provide constraints which can reduce the number of independent components.

Tensors that are invariant under all rotations and reflections (or inversions) can have only one independent component. Such invariance is obtained with a continuous set of vectors $\mathbf{e}_a$ uniformly distributed on the unit sphere. Invariance up to finite n can also be obtained with certain finite sets of vectors $\mathbf{e}_a$.

Isotropic tensors $\mathbf{E}^{(n)}$ obtained with sets of M vectors $\mathbf{e}_a$ in d space dimensions must take the form

$$\mathbf{E}^{(2n+1)} = 0 \tag{3.1.4}$$

$$\mathbf{E}^{(2n)} = \frac{M}{d(d+2)\cdots(d+2n-2)}\Delta^{(2n)} \tag{3.1.5}$$

where

$$\Delta_{ij}^{(2)} = \delta_{ij} \tag{3.1.6}$$

$$\Delta_{ijkl}^{(4)} = \delta_{ij}\,\delta_{kl} + \delta_{ik}\,\delta_{jl} + \delta_{il}\,\delta_{jk} \tag{3.1.7}$$

and in general $\Delta^{(2n)}$ consists of a sum of all the $(2n-1)!!$ possible products of Kronecker delta symbols of pairs of indices, given by the recursion relation

$$\Delta_{i_1 i_2 \cdots i_{2n}}^{(2n)} = \sum_{j=2}^{2n} \delta_{i_1 i_j} \Delta_{i_2 \cdots i_{j-1} i_{j+1} \cdots i_{2n}}^{(2n-2)} \tag{3.1.8}$$

The form of the $\Delta^{(2n)}$ can also be specified by giving their upper simplicial components (whose indices form a nonincreasing sequence). Thus, in two dimensions,

$$\Delta^{(4)} = [3, 0, 1, 0, 3] \tag{3.1.9}$$

where the 1111, 2111, 2211, 2221, and 2222 components are given. In three dimensions,

$$\Delta^{(4)} = [3, 0, 1, 0, 3, 0, 0, 0, 0, 1, 0, 1, 0, 0, 3] \tag{3.1.10}$$

Similarly,

$$\Delta^{(6)} = [5, 0, 1, 0, 1, 0, 5] \tag{3.1.11}$$

and

$$\Delta^{(6)} = [15, 0, 3, 0, 3, 0, 15, 0, 0, 0, 0, 0, 0, 3, 0, 1, 0, 3, 0, 0, 0, 0, 3, 0, 3, 0, 0, 15] \tag{3.1.12}$$

in two and three dimensions, respectively.

For isotropic sets of vectors $\mathbf{e}_a$, one finds from (3.1.5)

$$\frac{1}{M}\sum_a (\mathbf{e}_a \cdot \mathbf{v})^{2n} = Q_{2n}|\mathbf{v}|^{2n} = \frac{(2n-1)!!}{d(d+2)\cdots(d+2n-2)}|\mathbf{v}|^{2n} \tag{3.1.13}$$

so that for $d=2$

$$Q_2 = \frac{1}{2}, \quad Q_4 = \frac{3}{8}, \quad Q_6 = \frac{5}{16}, \quad Q_8 = \frac{35}{128} \tag{3.1.14}$$

while for $d=3$

$$Q_{2n} = \frac{1}{2n+1} \tag{3.1.15}$$

Similarly,

$$\frac{1}{M}\sum_a (\mathbf{e}_a \cdot \mathbf{v})^{2n}\, \mathbf{e}_a \cdot \mathbf{v} = Q_{2n}|\mathbf{v}|^{2n}\mathbf{v} \tag{3.1.16}$$

In the model of Section 2, all the particle velocities $\mathbf{e}_a$ are fundamentally equivalent, and so are added with equal weight in the tensor (3.1.1). In some cellular automaton fluid models, however, one may, for example, allow particle velocities $\mathbf{e}_a$ with unequal magnitudes (e.g., Ref. 31). The relevant tensors in such cases are

$$\mathbf{E}^{(n)}_{i_1 i_2 \cdots i_n} = \sum_a w(|\mathbf{e}_a|^2)\, (\mathbf{e}_a)_{i_1} \cdots (\mathbf{e}_a)_{i_n} \tag{3.1.17}$$

where the weights $w(|\mathbf{e}_a|^2)$ are typically determined from coefficients in the Chapman–Enskog expansion.

3.2 Polygons

As a first example, consider a set of unit vectors $\mathbf{e}_a$ corresponding to the vertices of a regular M-sided polygon:

$$\mathbf{e}_a = (\cos\frac{2\pi a}{M},\ \sin\frac{2\pi a}{M}) \tag{3.2.1}$$

For sufficiently large M, any tensor $\mathbf{E}^{(n)}$ constructed from these $\mathbf{e}_a$ must be isotropic. Table 1 gives the conditions on M necessary to obtain isotropic $\mathbf{E}^{(n)}$. In general, it can be shown that $\mathbf{E}^{(n)}$ is isotropic if and only if M does not divide any of integers n, $n-2$, $n-4$, ...[32] Thus, for example, $\mathbf{E}^{(n)}$ must be isotropic whenever $n > M$.

$\mathbf{E}^{(2)}$	$M > 2$
$\mathbf{E}^{(3)}$	$M \geq 2, M \neq 3$
$\mathbf{E}^{(4)}$	$M > 2, M \neq 4$
$\mathbf{E}^{(5)}$	$M \geq 2, M \neq 3, 5$
$\mathbf{E}^{(6)}$	$M > 4, M \neq 6$
$\mathbf{E}^{(7)}$	$M \geq 2, M \neq 3, 5, 7$

Table 1. Conditions for the tensors $\mathbf{E}^{(n)}$ of Eq. (3.1.1) to be isotropic with the lattice vectors $\mathbf{e}_a$ chosen to correspond to the vertices of regular M-sided polygons.

In the case $M = 6$, corresponding to the hexagonal lattice considered in Section 2, the $\mathbf{E}^{(n)}$ are isotropic up to $n = 5$. The macroscopic equations obtained in this case thus have the usual hydrodynamic form. However, a square lattice, with $M = 4$, yields an anisotropic $\mathbf{E}^{(4)}$, given by

$$\mathbf{E}^{(4)}|_{M=4} = 2\delta^{(4)} \tag{3.2.2}$$

where $\delta^{(n)}$ is the Kronecker delta symbol with n indices. The macroscopic equation obtained in this case is

$$\partial_t(nu_x) \; + \; \frac{1}{2} \, nc^{(2)}(u_x \, \partial_x u_x - u_y \, \partial_x u_y)$$

$$= - \frac{1}{2} \, \partial_x n - \frac{1}{8} \, nc_\nabla^{(2)}(\partial_{xx} u_x - \partial_{xy} u_y) - \frac{1}{4} \, (u_x^2 - u_y^2) \, \partial_x(nc^{(2)}) \tag{3.2.3}$$

$$- \frac{1}{8} \, (\partial_x u_x - \partial_y u_y) \, \partial_x(nc_\nabla^{(2)})$$

which does not have the standard Navier–Stokes form.[6], [3]

On a hexagonal lattice, $\mathbf{E}^{(4)}$ is isotropic, but $\mathbf{E}^{(6)}$ has the component form

$$\mathbf{E}^{(6)}|_{M=6} = \frac{1}{16} \, [33, 0, 3, 0, 9, 0, 27] \tag{3.2.4}$$

which differs from the isotropic result (3.1.11). The corrections (2.7.6) to the Navier–Stokes equation are therefore anisotropic in this case.

3.3 Polyhedra

As three-dimensional examples, one can consider vectors $\mathbf{e}_a$ corresponding to the vertices of regular polyhedra. Only for the five Platonic solids are all the $|\mathbf{e}_a|^2$ equal. Table 2 gives results for the isotropy of the $\mathbf{E}^{(n)}$ in these cases. Only for the icosahedron and dodecahedron is $\mathbf{E}^{(4)}$ found to be isotropic, so that the usual hydrodynamic equations are obtained. As in two dimensions, the $\mathbf{E}^{(2n)}$ for the cube are all proportional to a single Kronecker delta symbol over all indices.

In five and higher dimensions, the only regular polytopes are the simplex, and the hypercube and its dual.[34] These give isotropic $\mathbf{E}^{(n)}$ only for $n < 3$, and for $n < 4$ and $n < 4$, respectively.

In four dimensions, there are three additional regular polytopes,[34] specified by Schlafli symbols $\{3, 4, 3\}$, $\{3, 3, 5\}$, and $\{5, 3, 3\}$. (The elements of these lists give the number of edges around each vertex, face, and 3-cell, respectively.) The $\{3, 4, 3\}$ polytope has 24 vertices with coordinates corresponding to permutations of $(\pm 1, \pm 1, 0, 0)$. It yields $\mathbf{E}^{(n)}$ that are isotropic up to $n = 4$. The $\{3, 3, 5\}$ polytope has 120 vertices corresponding to $(\pm 1, \pm 1, \pm 1, \pm 1)$, all permutations of $(\pm 2, 0, 0, 0)$, and even-signature permutations of $(\pm \phi, \pm 1, \phi^{-1}, 0)$, where $\phi = (1 + \sqrt{5})/2$. The $\{5, 3, 3\}$ polytope is the dual of $\{3, 3, 5\}$. Both yield $\mathbf{E}^{(n)}$ that are isotropic up to $n = 8$.

	$\mathbf{e}_a$	M	$\mathbf{E}^{(2)}$	$\mathbf{E}^{(3)}$	$\mathbf{E}^{(4)}$	$\mathbf{E}^{(5)}$	$\mathbf{E}^{(6)}$
Tetrahedron	$(1, 1, 1)$, cyc: $(1, -1, -1)$	4	Y	N	N	N	N
Cube	$(\pm 1, \pm 1, \pm 1)$	8	Y	Y	N	Y	N
Octahedron	cyc: $(\pm 1, 0, 0)$	6	Y	Y	N	Y	N
Dodecahedron	$(\pm 1, \pm 1, \pm 1)$, cyc: $(0, \pm \phi^{-1}, \pm \phi)$	20	Y	Y	Y	Y	N
Icosahedron	cyc: $(0, \pm \phi, \pm 1)$	12	Y	Y	Y	Y	N

Table 2. Isotropy of the tensors $\mathbf{E}^{(n)}$ with $\mathbf{e}_a$ chosen as the M vertices of regular polyhedra. In the forms for $\mathbf{e}_a$ (which are given without normalization), the notation "cyc:" indicates all cyclic permutations. (All possible combinations of signs are chosen in all cases.) ϕ is the golden ratio $(1 + \sqrt{5})/2 \approx 1.618$.

3.4 Group Theory

The structure of the $\mathbf{E}^{(n)}$ was found above by explicit calculations based on particular choices for the $\mathbf{e}_a$. The general form of the results is, however, determined solely by the symmetries of the set of $\mathbf{e}_a$. A finite group $\mathbf{G}$ of transformations leaves the $\mathbf{e}_a$ invariant. (For the hexagonal lattice model of Section 2, it is the hexagonal group $\mathbf{S}_6$.) In general $\mathbf{G}$ is a finite subgroup of the d-dimensional rotation group $O(d)$.

The $\mathbf{e}_a$ form the basis for a representation of $\mathbf{G}$, as do their products $\mathbf{E}^{(n)}$. If the representation $\mathbf{R}^{(n)}$ carried by the $\mathbf{E}^{(n)}$ is irreducible, then the $\mathbf{E}^{(n)}$ can have only one independent component, and must be rotationally invariant. But $\mathbf{R}^{(n)}$ is in general reducible. The number of irreducible representations that it contains gives the number of independent components of $\mathbf{E}^{(n)}$ allowed by invariance under $\mathbf{G}$.

This number can be found using the method of characters (e.g., Refs. 35 and 36). Each class of elements of $\mathbf{G}$ in a particular representation $\mathbf{R}$ has a character that receives a fixed contribution from each irreducible component of $\mathbf{R}$. Characters for the representation $\mathbf{R}^{(n)}$ of $\mathbf{G}$ can be found by first evaluating them for arbitrary rotations, and then specializing to the particular sets of rotations (typically through angles of the form $2\pi/k$) that appear in $\mathbf{G}$. To find characters for arbitrary rotations, one writes the $\mathbf{E}^{(n)}$ as sums of completely traceless tensors $\mathbf{U}^{(n)}$ which form irreducible representations of $O(d)$ (e.g., Ref. 37):

$$\mathbf{E}^{(n)} = \mathbf{U}^{(n)} + \mathbf{U}^{(n-2)} + \cdots + \mathbf{U}^{(0)} \tag{3.4.1}$$

The characters of the $\mathbf{E}^{(n)}$ are then sums of the characters $\chi^{(m)}$ for the irreducible tensors $\mathbf{U}^{(m)}$. For proper rotations through an angle ϕ, the $\chi^{(m)}$ are given by (e.g., Ref. 37)

$$\chi^{(m)}(\phi) = e^{2\pi i m \phi} \quad (d=2)$$

$$\chi^{(m)}(\phi) = \frac{\sin[(2m+1)\phi/2]}{\sin \phi/2} \quad (d=3) \tag{3.4.2}$$

The resulting characters for the representations $\mathbf{R}^{(n)}$ formed by the $\mathbf{E}^{(n)}$ are given in Table 3.

Dimension	Rank	Character
2	2	$4c^2 - 1$
2	4	$(4c^2 + 2c - 1)(4c^2 - 2c - 1)$
3	2	$4c^2 + 2c$
3	4	$(2c+1)(2c-1)(4c^2 + 2c - 1)$

Table 3. Characters of transformations of totally symmetric rank n tensors $\mathbf{E}^{(n)}$ in d dimensions. $c = \cos(\phi)$, where ϕ is the rotation angle. For improper rotations in three dimensions, $\pi - \phi$ must be used.

The number of irreducible representations in $\mathbf{R}^{(n)}$ can be found as usual by evaluating the characters for each class in $\mathbf{R}^{(n)}$ (e.g., Ref. 35). Consider as an example the case of $\mathbf{R}^{(4)}$ with $\mathbf{G}$ the octahedral group $\mathbf{O}$. This group has classes $E, 8C_3, 9C_2, 6C_4$, where E represents the identity, and C_k represents a proper rotation by $\phi = 2\pi/k$ about a k-fold symmetry axis. The characters for these classes in the representation $\mathbf{R}^{(4)}$ can be found from Table 3. Adding the results, and dividing by the total number of classes in $\mathbf{G}$, one finds that $\mathbf{R}^{(4)}$ contains exactly two irreducible representations of $\mathbf{O}$. Rank 4 symmetric tensors can thus have up to two independent components while still being invariant under the octahedral group.[38]

In general, one may consider sets of vectors $\mathbf{e}_a$ that are invariant under any point symmetry group. Typically, the larger the group is, the smaller the number of independent components in the $\mathbf{E}^{(n)}$ can be. In two dimensions, there are an infinite number of point groups, corresponding to transformations of regular polygons. There are only a finite of nontrivial additional point groups in three dimensions. The largest is the group $\mathbf{Y}$ of symmetries of the icosahedron (or dodecahedron). Second largest is the cubic group $\mathbf{E}$. As seen in Table 2, only $\mathbf{Y}$ guarantees isotropy of all tensors $\mathbf{E}^{(n)}$ up to $n = 4$ (compare Ref. 39).

It should be noted, however, that such group-theoretic considerations can only give upper bounds on the number of independent components in the $\mathbf{E}^{(n)}$. The actual number of independent components depends on the particular choice of the $\mathbf{e}_a$, and potentially on the values of weights such as those in Eq. (3.1.16).

3.5 Regular Lattices

If the vectors $\mathbf{e}_a$ correspond to particle velocities, then the possible displacements of particles at each time step must be of the form $\sum_a k_a \mathbf{e}_a$. In discrete velocity gases, particle positions are not constrained. But in a cellular automaton model, they are usually taken to correspond to the sites of a regular lattice.

Only a finite number of such "crystallographic" lattices can be constructed in any space dimension (e.g., Refs. 40 and 41). As a result, the point symmetry groups that can occur are highly constrained. In two dimensions, the most symmetrical lattices are square and hexagonal ones. In three dimensions, the most symmetrical are hexagonal and cubic. The group-theoretic arguments of Section 3.4 suffice to show that in two dimensions, hexagonal lattices must give tensors $\mathbf{E}^{(n)}$ that are isotropic up to $n = 4$, and so yield standard hydrodynamic equations (2.5.11). In three dimensions, group-theoretic arguments alone fail to establish the isotropy of $\mathbf{E}^{(4)}$ for hexagonal and cubic lattices. A system with icosahedral point symmetry would be guaranteed to yield an isotropic $\mathbf{E}^{(4)}$, but since it is not possible to tesselate three-dimensional space with regular icosahedra, no regular lattice with such a large point symmetry group can exist.

Crystallographic lattices are classified not only by point symmetries, but also by the spatial arrangement of their sites. The lattices consist of "unit cells" containing a definite arrangement of sites, which can be repeated to form a regular tesselation. In two dimensions, five distinct such Bravais lattice structures exist; in three dimensions, there are 14 (e.g., Refs. 40 and 41).

Sites in these lattices can correspond directly to the sites in a cellular automaton. The links which carry particles in cellular automaton fluid models are obtained by joining pairs of sites, usually in a regular arrangement. The link vectors give the velocities $\mathbf{e}_a$ of the particles.

In the simplest cases, the links join each site to its nearest neighbors. The regularity of the lattice implies that in such cases, all the $\mathbf{e}_a$ are of equal length, so that all particles have the same speed.

For two-dimensional square and hexagonal lattices, the $\mathbf{e}_a$ with this nearest neighbor arrangement have the form (3.2.1). The results of Section 3.2 then show that with hexagonal lattices, such $\mathbf{e}_a$ give $\mathbf{E}^{(n)}$ that are isotropic up to $n = 4$, and so yield the standard hydrodynamic continuum equations (2.6.1).

Table 4 gives the forms of $\mathbf{E}^{(n)}$ for the most symmetrical three-dimensional lattices with nearest neighbor choices for the $\mathbf{e}_a$. None yield isotropic $\mathbf{E}^{(4)}$ (compare Ref. 38).

	$\mathbf{e}_a$	M	$\mathbf{E}^{(2)}$	$\mathbf{E}^{(4)}$	$\mathbf{E}^{(6)}$
Primitive cubic	cyc: $(\pm1, 0, 0)$	6	$2\delta^{(2)}$	$2\delta^{(4)}$	$2\delta^{(6)}$
Body-centered cubic	$(\pm1, \pm1, \pm1)$	8	$8\delta^{(2)}$	$8(\Delta^{(4)} - 2\delta^{(4)})$	$8(\Delta^{(6)} - 2\Delta^{(4,2)} + 16\delta^{(6)})$
Face-centered cubic	cyc: $(\pm1, \pm1, 0)$	12	$8\delta^{(2)}$	$4(\Delta^{(4)} - \delta^{(4)})$	$4(\Delta^{(4,2)} - 13\delta^{(6)})$

Table 4. Forms of the tensors $\mathbf{E}^{(n)}$ for the most symmetrical three-dimensional Bravais lattices. The basic vectors $\mathbf{e}_a$ (used here without normalization) are taken to join each site with its M nearest neighbors. $\delta^{(n)}$ represents the Kronecker delta symbol of n indices; $\Delta^{(n)}$ represents the rotationally invariant tensor defined in Eqs. (3.1.6)–(3.1.8). $\Delta^{(n,m)}$ is the sum of all possible products of pairs of Kronecker delta symbols with n and m indices, respectively.

The hexagonal and face-centered cubic lattices, which have the largest point symmetry groups in two and three dimensions, respectively, are also the lattices that give the densest packings of circles and spheres (e.g., Ref. 42). One suspects that in more than three dimensions (compare Ref. 43) the lattices with the largest point symmetry continue to be those with the densest sphere packing. The spheres are placed on lattice sites; the positions of their nearest neighbors are defined by a Voronoi polyhedron or Wigner–Seitz cell. The densest sphere packing is obtained when this cell, and thus the nearest neighbor vectors $\mathbf{e}_a$, are closest to forming a sphere. In dimensions $d \leq 8$, it has been found that the optimal lattices for sphere packing are those based on the sets of root vectors for a sequence of simple Lie groups (e.g., Ref. 44). Results on the isotropy of the tensors $\mathbf{E}^{(n)}$ for these lattices are given in Table 5.

d	Group		M	n_{max}
1	A_1	$SU(2)$	2	
2	A_2	$SU(3)$	6	4
3	A_3	$SU(4)$	12	2
4	D_4	$SO(8)$	24	4
5	D_5	$SO(10)$	40	2
6	E_6		72	0

Table 5. Sequence of simple Lie groups whose sets of root vectors yield optimal lattices for sphere packing in d dimensions. These lattices may also yield maximal isotropy for the tensors $\mathbf{E}^{(n)}$. Results are given for the maximum even n at which the $\mathbf{E}^{(n)}$ are found to be isotropic. The root vectors are given in Ref. 45.

More isotropic sets of $\mathbf{e}_a$ can be obtained by allowing links to join sites on the lattice beyond nearest neighbors.[31] On a square lattice, one may, for example, include diagonal links, yielding a set of vectors

$$\mathbf{e}_a = (0, \pm1), (\pm1, 0), (\pm1, \pm1) \tag{3.5.1}$$

Including weights $w(|\mathbf{e}_a|^2)$ as in Eq. (3.1.16), this choice of $\mathbf{e}_a$ yields

$$\mathbf{E}^{(2)} = 2[w(1) + 2w(2)]\,\delta^{(2)} \tag{3.5.2}$$

$$\mathbf{E}^{(4)} = 4w(2)\,\Delta^{(4)} + 2[w(1) - 4w(2)]\,\delta^{(4)} \tag{3.5.3}$$

If the ratio of particles on diagonal and orthogonal links can be maintained so that

$$w(1) = 4w(2) \tag{3.5.4}$$

then Eq. (3.5.3) shows that $\mathbf{E}^{(4)}$ will be isotropic. This choice effectively weights the individual vectors $(0, \pm1)$ and $(\pm1, 0)$ with a factor $\sqrt{2}$. As a result, the vectors (3.5.1) are effectively those for a regular octagon, given by Eq. (3.2.1) with $M = 8$.

Including all 24 $\mathbf{e}_a$ with components $|(\mathbf{e}_a)_i| \leq 2$ on a square lattice, one obtains

$$\mathbf{E}^{(2)} = 2[w(1) + 2w(2) + 4w(4) + 10w(5) + 8w(8)]\,\delta^{(2)} \tag{3.5.5}$$

$$\mathbf{E}^{(4)} = 4[w(2) + 8w(5) + 16w(8)]\,\Delta^{(4)} + 2[w(1) - 4w(2) + 16w(4) - 14w(5) - 64w(8)]\,\delta^{(4)} \tag{3.5.6}$$

$$\mathbf{E}^{(6)} = \frac{4}{3}[w(2) + 20w(5) + 64w(8)]\,\Delta^{(6)} + 2[w(1) - 8w(2) + 64w(4) - 70w(5) - 512w(8)]\,\delta^{(6)} \tag{3.5.7}$$

With $w(5) = w(8) = 0$, $\mathbf{E}^{(4)}$ and $\mathbf{E}^{(6)}$ are isotropic if

$$\frac{w(2)}{w(1)} = \frac{3}{8}, \quad \frac{w(4)}{w(1)} = \frac{1}{32} \tag{3.5.8}$$

They cannot both be isotropic if $w(4)$ also vanishes.

In three dimensions, one may consider a cubic lattice with sites at distances 1, $\sqrt{2}$, and $\sqrt{3}$ joined. The $\mathbf{e}_a$ in this case contain all those for primitive, face-centered, and body-centered cubic lattices, as given in Table 4. The $\mathbf{E}^{(n)}$ can then be deduced from the results of Table 4, and are given by

$$\mathbf{E}^{(2)} = 2[w(1) + 4w(2) + 4w(3)]\,\delta^{(2)} \tag{3.5.9}$$

$$\mathbf{E}^{(4)} = 4[w(2) + 2w(3)]\,\Delta^{(4)} + 2[w(1) - 2w(2) - 8w(3)]\,\delta^{(4)} \tag{3.5.10}$$

$$\mathbf{E}^{(6)} = 8w(2)\,\Delta^{(6)} + 4[w(2) - 4w(3)]\,\Delta^{(4,2)} + 2[w(1) - 26w(2) + 64w(3)]\,\delta^{(6)} \tag{3.5.11}$$

Isotropy of $\mathbf{E}^{(4)}$ is obtained when

$$w(1) = 2w(2) + 8w(3) \tag{3.5.12}$$

and of $\mathbf{E}^{(6)}$ when

$$w(1) = 10w(2) = 40w(3) \tag{3.5.13}$$

Notice that (3.5.12) and (3.5.13) cannot simultaneously be satisfied by any nonzero choice of weights. Nevertheless, so long as (3.5.12) holds, isotropic hydrodynamic behavior is obtained in this three-dimensional cellular automaton fluid. Isotropic $\mathbf{E}^{(6)}$ can be obtained by including in addition vectors $\mathbf{e}_a$ of the form $(\pm 2, 0, 0)$ (and permutations), and choosing

$$w(2) = \frac{1}{2}\,w(1), \ w(3) = \frac{1}{8}\,w(1), \ w(4) = \frac{1}{16}\,w(1) \tag{3.5.14}$$

The weights in Eq. (3.1.17) give the probabilities for particles with different speeds to occur. These probabilities are determined by microscopic equilibrium conditions. They can potentially be controlled by using different collision rules on different time steps (as discussed in Section 4.9). Each set of collision rules can, for example, be arranged to yield each particle speed with a certain probability. Then the frequency with which different collision rules are used can determine the densities of particles with different speeds.

3.6 Irregular Lattices

The general structure of cellular automaton fluid models considered here requires that particles can occur only at definite positions and with definite discrete velocities. But the possible particle positions need not necessarily correspond with the sites of a regular lattice. The directions of particle velocities should be taken from the directions of links. But the particle speeds may consistently be taken independent of the lengths of links.

As a result, one may consider constructing cellular automaton fluids on quasilattices (e.g., Ref. 46), such as that illustrated in Fig. 2. Particle velocities are taken to follow the directions of the links, but to have unit magnitude, independent of the spatial lengths of the links. Almost all intersections involve just two links, and so can support only two-particle interactions. These intersections occur at a seemingly irregular set of points, perhaps providing a more realistic model of collisions in continuum fluids.

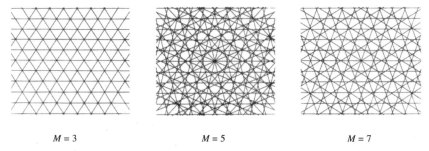

$M = 3$	$M = 5$	$M = 7$

Figure 2. Lattices and quasilattices constructed from grids oriented in the directions of the vertices of regular M-sided polygons. An appropriate dual of the $M = 5$ pattern is the Penrose aperiodic tiling.

The possible $\mathbf{e}_a$ on regular lattices are highly constrained, as discussed in Section 3.5. But it is possible to construct quasilattices which yield any set of $\mathbf{e}_a$. Given a set of generator vectors $\mathbf{g}_a$, one constructs a grid of equally spaced lines orthogonal to each of them.[47] The directions of these lines correspond to the $\mathbf{e}_a$.

If the tangent of the angles between the $\mathbf{g}_a$ are rational, then these lines must eventually form a periodic pattern, corresponding to a regular lattice. But if, for example, the $\mathbf{g}_a$ correspond to the vertices of a pentagon, then the pattern never becomes exactly periodic, and only a quasilattice is obtained. A suitable dual of the quasilattice gives in fact the standard Penrose aperiodic tiling.[48]

In three dimensions, one may form grids of planes orthogonal to generator vectors $\mathbf{g}_a$. Possible particle positions and velocities are obtained from the lines in which these planes intersect.

Continuum equations may be derived for cellular automaton fluids on quasilattices by the same methods as were used for regular lattices above. But by appropriate choices of generator vectors, three-dimensional quasilattices with effective icosahedral point symmetry may be obtained, so that isotropic fluid behavior can be obtained even with a single particle speed.

Quasilattices yield an irregular array of particle positions, but allow only a limited number of possible particle velocities. An entirely random lattice would also allow arbitrary particle velocities. Momentum conservation cannot be obtained exactly with discrete collision rules on such a lattice, but may be arranged to hold on average.

4. Evaluation of Transport Coefficients

4.1 Introduction

Section 2 gave a derivation of the general form of the hydrodynamic equations for a sample cellular automaton fluid model. This section considers the evaluation of the specific transport coefficients that appear in these equations. While these coefficients may readily be found by explicit simulation, as discussed in the second paper in this series, no exact mathematical procedure is known for calculating them. This section considers primarily an approximation method based on the Boltzmann transport equation. The results obtained are expected to be accurate for certain transport coefficients at low particle densities.[4]

4.2 Basis for Boltzmann Transport Equation

The kinetic equation (2.3.5) gives an exact result for the evolution of the one-particle distribution function f_a. But the collision term Ω_a in this equation depends on two-particle distribution functions, which in turn depend on higher order distribution functions, forming the BBGKY hierarchy of kinetic equations. To obtain explicit results for the f_a one must close or truncate this hierarchy.

The simplest assumption is that there are no statistical correlations between the particles participating in any collision. In this case, the multiparticle distribution functions that appear in Ω_a can be replaced by products of one-particle distribution functions f_a, yielding an equation of the standard Boltzmann transport form, which can in principle be solved explicitly for the f_a.

Even if particles were uncorrelated before a collision, they must necessarily show correlations after the collision. As a result, the factorization of multiparticle distribution functions used to obtain the Boltzmann transport equation cannot formally remain consistent. At low densities, it may nevertheless in some cases provide an adequate approximation.

Correlations produced by a particular collision are typically important only if the particles involved collide again before losing their correlations. At low densities, particles usually travel large distances between collisions, so that most collisions involve different sets of particles. The particles involved in one collision will typically suffer many other collisions before meeting again, so that they are unlikely to maintain correlations. At high densities, however, the same particles often undergo many successive collisions, so that correlations can instead be amplified.

In the Boltzmann transport equation approximation, correlations and deviations from equilibrium decay exponentially with time. Microscopic perturbations may, however, lead to collective, hydrodynamic, effects, which decay only as a power of time.[29] Such effects may lead to transport coefficients that are nonanalytic functions of density and other parameters, as mentioned in Section 2.6.

4.3 Construction of Boltzmann Transport Equation

This subsection describes the formulation of the Boltzmann transport equation for the sample cellular automaton fluid model discussed in Section 2.

The possible classes of particle collisions in this model are illustrated in Fig. 3. The rules for different collisions within each class are related by lattice symmetries. But, as illustrated in Fig. 3, several choices of overall rules for each class are often allowed by conservation laws.

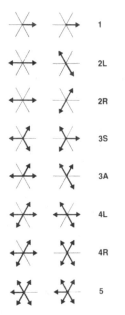

Figure 3. Possible types of initial and final states for collisions in the cellular automaton fluid model of Section 2.

In the simplest case, the same rule is chosen for a particular class of collisions at every site. But it is often convenient to allow different choices of rules at different sites. Thus, for example, there could be a checkerboard arrangement of sites on which two-body collisions lead alternately to scattering to the left and to the right. In general, one may apply a set of rules denoted by k at some fraction γ_k of the sites in a cellular automaton. (A similar procedure was mentioned in Section 3.5 as a means for obtaining isotropic behavior on three-dimensional cubic lattices.) The randomness of microscopic particle configurations suggests that the γ_k should serve merely to change the overall probabilities for different types of collisions.

The term Ω_a in the kinetic equation (2.3.5) for f_a is a sum of terms representing possible collisions involving particles of type a. Each term gives the change in the number of type a particles due to a particular type of collision, multiplied by the probability for the arrangement of particles involved in the collision to occur. In the Boltzmann equation approximation, the probability for a particular particle arrangement is taken to be a simple product of the densities f_b for particles that should be present, multiplied by factors $(1-f_c)$ for particles that should be absent.

The complete Boltzmann transport equation for the model of Section 2 thus becomes

$$\partial_t f_a + \mathbf{e}_a \cdot \nabla f_a = \Omega_a \qquad (4.3.1)$$

where

$$\begin{aligned}
\Omega = \ & [\gamma_{2L}\,\Lambda(1,4) + (\gamma_2 - \gamma_{2L})\,\Lambda(2,5)] - \gamma_2\,\Lambda(0,3) \\
& + \gamma_{3S}[\Lambda(1,3,5) - \Lambda(0,2,4)] \\
& + \gamma_{3A}[\Lambda(2,4,5) + \Lambda(1,2,5) - \Lambda(0,3,5) - \Lambda(0,2,3) \\
& + \Lambda(1,4,5) + \Lambda(1,2,4) - \Lambda(0,3,4) - \Lambda(0,1,3)] \\
& + [\gamma_4\,\Lambda(1,2,4,5) - \gamma_{4L}\,\Lambda(0,2,3,5) - (\gamma_4 - \gamma_{4L})\,\Lambda(0,1,3,4)]
\end{aligned} \tag{4.3.2}$$

Here

$$\Lambda_a(i_1, i_2, \ldots, i_k) = \frac{f_{a+i_1}}{1 - f_{a+i_1}}\,\frac{f_{a+i_2}}{1 - f_{a+i_2}}\,\cdots\,\frac{f_{a+i_k}}{1 - f_{a+i_k}}\,\prod_{j=1}^{M}(1 - f_{a+j}) \tag{4.3.3}$$

where all indices on the f_b are evaluated modulo M, and in this case $M = 6$. Note that in Eq. (4.3.2), the index a has been dropped on both Ω and Λ.

The Boltzmann transport equations for any cellular automaton fluid model have the overall form of Eqs. (4.3.1) and (4.3.2). In a more general case, the simple addition of constants i_j to the indices a in the definition of Λ can be replaced by transformations with appropriate lattice symmetry group operations.

Independent of the values of the γ_k, Ω_a is seen to satisfy the momentum and particle number constraints (2.4.4) and (2.4.5).

In the following calculations it is often convenient to maintain arbitrary values for the γ_k so as to trace the contributions of different classes of collisions. But to obtain a form for Ω_a that is invariant under the complete lattice symmetry group, one must take

$$\gamma_{2L} = \gamma_{2R} = \frac{1}{2}\,\gamma_2 \tag{4.3.4}$$

$$\gamma_{4L} = \gamma_{4R} = \frac{1}{2}\,\gamma_4 \tag{4.3.5}$$

4.4 Linear Approximation to Boltzmann Transport Equation

In studying macroscopic behavior, one assumes that the distribution functions f_a differ only slightly from their equilibrium values, as in the Chapman–Enskog expansion (2.5.1). The f_a may thus be approximated as

$$f_a = f(1 + \phi_a) \quad (|\phi_a| \ll 1) \tag{4.4.1}$$

With this approximation, the collision term Ω_a in the Boltzmann transport equation may be approximated by a power series expansion in the ϕ_a:

$$\Omega_a = \sum_b \omega_{ab}^{(1)}\,\phi_b + \sum_{b,c} \omega_{abc}^{(2)}\,\phi_b\,\phi_c + \cdots \tag{4.4.2}$$

The matrix $\omega^{(1)}$ here is analogous to the usual linearized collision operator (e.g., Ref. 26). Notice that for a cellular automaton fluid model with collisions involving at most K particles, the expansion (4.4.2) terminates at $O(\phi^K)$.

Microscopic reversibility immediately implies that the tensors $\omega^{(n)}$ are all completely symmetric in their indices. The conservation laws (2.4.4) and (2.4.5) yield conditions on all the $\omega^{(n)}$ of the form

$$\sum_{abc..} \omega^{(n)}_{abc..} = 0 \qquad (4.4.3)$$

$$\sum_{abc..} \mathbf{e}_a \, \omega^{(n)}_{abc..} = 0 \qquad (4.4.4)$$

In the particular case of $\omega^{(1)}$, the more stringent conditions

$$\sum_b \omega^{(1)}_{ab} = 0 \qquad (4.4.5)$$

and

$$\sum_b \mathbf{e}_a \, \omega^{(1)}_{ab} = 0 \qquad (4.4.6)$$

also apply.

In the model of Section 2, all particle types a are equivalent up to lattice symmetry transformations. As a result, $\omega^{(n)}_{(a+1)\,bc..}$ is always given simply by a cyclic shift of $\omega^{(n)}_{abc..}$, so that the complete form of $\omega^{(n)}$ can be determined from the first row $\omega^{(n)}_{1bc..}$. The $\omega^{(n)}$ are thus circulant tensors (e.g., Ref. 50), and the values of their components depend only on numerical differences between their indices, evaluated modulo M.

Expansion of (4.3.2) now yields

$$\begin{aligned}
\omega^{(1)}_{ab} = \ & f^2(1-f)\ \text{circ}\ \{-[\gamma_2 \bar{f}^2 + (\gamma_{3S} + 4\,\gamma_{3A})\bar{f}f + \gamma_4 f^2], \\
& \gamma_{2L}\bar{f}^2 + (\gamma_{3S} + 2\,\gamma_{3A})\bar{f}f + \gamma_{4L}f^2, \\
& (1 - \gamma_{2L})\bar{f}^2 + (-\gamma_{3S} + 2\,\gamma_{3A})\bar{f}f + (\gamma_4 - \gamma_{4L})f^2, \\
& -[\gamma_2 \bar{f}^2 + (-\gamma_{3S} + 4\,\gamma_{3A})\bar{f}f + \gamma_4 f^2], \\
& \gamma_{2L}\bar{f}^2 + (-\gamma_{3S} + 2\,\gamma_{3A})\bar{f}f + \gamma_{4L}f^2, \\
& (1 - \gamma_{2L})\bar{f}^2 + (\gamma_{3S} + 2\,\gamma_{3A})\bar{f}f + (\gamma_4 - \gamma_{4L})f^2\}
\end{aligned} \qquad (4.4.7)$$

where $\bar{f} = (1-f)$. Taking for simplicity $\gamma_2 = 1$, $\gamma_{2L} = \frac{1}{2}$, $\gamma_{3S} = 1$, $\gamma_{3A} = \gamma_{4i} = 0$, one finds

$$\omega^{(1)}_{ab} = f^2(1-f)^2\ \text{circ}[-1, \tfrac{1}{2}(1+f), \tfrac{1}{2}(1-3f), 2f-1, \tfrac{1}{2}(1-3f), \tfrac{1}{2}(1+f)] \qquad (4.4.8)$$

$$\omega_{abc}^{(2)} = \frac{1}{2} f^2 (1-f) \, \text{circ}[$$

$$
\begin{array}{cccccc}
0 & -f(f-1) & f(3f-1) & -2(f-1)(2f-1) & f(3f-1) & f(f-1) \\
-f(f-1) & 0 & 2f(f-1) & -f(5f-3) & (f-1)(2f-1) & -2f^2 \\
f(3f-1) & 2f(f-1) & 0 & -f(f-1) & 2f(3f-2) & (f-1)(2f-1) \\
-2(f-1)(2f-1) & -f(5f-3) & -f(f-1) & 0 & -f(f-1) & -f(5f-3) \\
f(3f-1) & (f-1)(2f-1) & 2f(3f-2) & -f(f-1) & 0 & 2f(f-1) \\
-f(f-1) & -2f^2 & (f-1)(2f-1) & -f(5f-3) & 2f(f-1) & 0
\end{array}
]
$$

(4.4.9)

4.5 Approach to Equilibrium

In a spatially uniform system close to equilibrium, one may use a linear approximation to the Boltzmann equation (4.3.1):

$$\partial_t (f \, \phi_a) = \sum_b \omega_{ab}^{(1)} \, \phi_b \tag{4.5.1}$$

This equation can be solved in terms of the eigenvalues and eigenvectors of the matrix $\omega_{ab}^{(1)}$. The circulant property of $\omega_{ab}^{(1)}$ considerably simplifies the computations required.

An $M \times M$ circulant matrix U_{ab} can in general be written in the form[50]

$$U_{ab} = \mathbf{u}[(a-b+1) \bmod M] = U_{11} I + U_{12} \Pi + \cdots + U_{1M} \Pi^{(M-1)} \tag{4.5.2}$$

where

$$\Pi = \text{circ}[0, 1, 0, 0, \ldots, 0] \tag{4.5.3}$$

is an $M \times M$ cyclic permutation matrix, and I is the $M \times M$ identity matrix. From this representation, it follows that all $M \times M$ circulants have the same set of right eigenvectors $\mathbf{v}_c$, with components given by

$$(\mathbf{v}_c)_a = \frac{1}{\sqrt{M}} \exp \frac{2\pi i (c-1)(a-1)}{M} \tag{4.5.4}$$

Writing

$$\Gamma(z) = \sum_{a=1}^{M} U_{1a} z^{a-1} \tag{4.5.5}$$

the corresponding eigenvalues are found to be

$$\lambda_c = \Gamma(\exp[\frac{2\pi i (c-1)}{M}]) \tag{4.5.6}$$

Using these results, the eigenvectors of $\omega_{ab}^{(1)}$ for the model of Section 2 are found to be

$$\mathbf{v}_1 = \frac{1}{\sqrt{6}}(1,1,1,1,1,1)$$

$$\mathbf{v}_2 = \frac{1}{\sqrt{6}}(1,\sigma,-\sigma^*,-1,-\sigma,\sigma^*) = (\mathbf{v}_6)^*$$

$$\mathbf{v}_3 = \frac{1}{\sqrt{6}}(1,-\sigma^*,-\sigma,1,-\sigma^*,-\sigma) = (\mathbf{v}_5)^*$$

$$\mathbf{v}_4 = \frac{1}{\sqrt{6}}(1,-1,1,-1,1,-1)$$

$$\frac{1}{2}(\mathbf{v}_2+\mathbf{v}_6) = \frac{1}{2\sqrt{6}}(2,1,-1,-2,-1,1)$$

$$\frac{1}{2i}(\mathbf{v}_2-\mathbf{v}_6) = \frac{\sqrt{3}}{2\sqrt{6}}(0,1,1,0,-1,-1)$$

(4.5.7)

where

$$\sigma = \exp(i\pi/3) = \frac{1}{2}(1+i\sqrt{3})$$

and the corresponding eigenvalues are

$$\lambda_1 = 0$$

$$\lambda_2 = 0$$

$$\lambda_3 = -3f^2(1-f)\{[\gamma_2(1-f)^2+4\gamma_{3A}f(1-f)+\gamma_4 f^2]$$
$$-\frac{4i}{\sqrt{3}}[(1-f)^2(\frac{\gamma_2}{2}-\gamma_{2L})+f^2(\frac{\gamma_4}{2}-\gamma_{4L})]\}$$

(4.5.8)

$$\lambda_4 = -6\gamma_{3S}f^3(1-f)^2$$

$$\lambda_5 = (\lambda_3)^*$$

$$\lambda_6 = 0$$

Combinations of the ϕ_a corresponding to eigenvectors with zero eigenvalue are conserved with time according to Eq. (4.5.1). Three such combinations are associated with the conservation laws (2.4.1) and (2.4.2). $\mathbf{v}_1$ corresponds to $\sum_a\phi_a$, which is the total particle number density. $(\mathbf{v}_2+\mathbf{v}_6)/2$ and $(\mathbf{v}_2-\mathbf{v}_6)/2i$ correspond, respectively, to the x and y components of the momentum density $\sum_a\mathbf{e}_a\phi_a$.

The ϕ_a may always be written as sums of pieces proportional to each of the orthogonal eigenvectors $\mathbf{v}_c$ of Eq. (4.5.7):

$$\phi_a = \sum_c \psi_c(\mathbf{v}_c)_a$$

(4.5.9)

The coefficients ψ_1, $(\psi_2 + \psi_6)/2$, and $(\psi_2 - \psi_6)/2i$ give the values of the conserved particle and momentum densities in this representation, and remain fixed with time.

The general solution of Eq. (4.5.1) is given in terms of Eq. (4.5.9) by

$$\psi_c(t) = \psi_c(0)\, e^{\lambda_c t} \tag{4.5.10}$$

Equation (4.5.8) shows that for any positive choices of the γ_k, all nonzero λ_c have negative real parts. As a result, the associated ψ_c must decay exponentially with time. Only the combinations of ϕ_a associated with conserved quantities survive at large times.

This result supports the local equilibrium assumption used for the derivation of hydrodynamic equations in Section 2. It implies that regardless of the initial average densities ϕ_a, collisions bring the system to an equilibrium that depends only on the values of the macroscopic conserved quantities (2.4.1) and (2.4.2). One may thus expect to be able to describe the local state of the cellular automaton fluid on time scales large compared to $|\lambda_c|^{-1}$ ($\lambda_c \neq 0$) solely in terms of these macroscopic conserved quantities. [Section 4.2 nevertheless mentioned some effects not accounted for by the Boltzmann equation (4.3.1) that can slow the approach to equilibrium.]

One notable feature of the results (4.5.8) is that they imply that the final equilibrium values of the ϕ_a are not affected by the choice of the parameters γ_{2L} and γ_{4L}, which determine the mixtures of two- and four-particle collisions with different chiralities. When the rate for collisions with different chiralities are unequal, however, λ_3 and λ_5 acquire imaginary parts, which lead to damped oscillations in the ϕ_a as a function of time.

When all the types of collisions illustrated in Fig. 3 can occur, Eq. (4.5.8) implies that momentum and particle number are indeed the only conserved quantities. If, however, only two-particle collisions are allowed, then there are additional conserved quantities. In fact, whenever symmetric three-particle collisions are absent, so that $\gamma_{3S} = 0$, Eq. (4.5.8) implies that the quantities

$$Q_i = \sum_{a=i}^{M/2+i} f_a \tag{4.5.11}$$

where the index a is evaluated modulo $M = 6$, is conserved. Thus, independent of the value of γ_{2L}, the total momenta on the two sides of any line (not along a lattice direction) through the cellular automaton must independently be conserved.

If three-particle symmetric collisions are absent, the cellular automaton thus exhibits a spurious additional conservation law, which prevents the attainment of standard local equilibrium, and modifies the hydrodynamic behavior discussed in Section 2. Section 4.8 considers some general conditions which avoid such spurious conservation laws.

4.6 Equilibrium Conditions and Transport Coefficients

Section 4.5 discussed the solution of the Boltzmann transport equation for uniform cellular automaton fluids. This section considers nonuniform fluids, and gives some approximate results for transport coefficients.

The Chapman–Enskog expansion (2.5.1) gives the general form for approximations to the microscopic distribution functions f_a. The coefficients $c^{(2)}$ and $c_\nabla^{(2)}$ that appear in this expansion can be estimated using the Boltzmann transport equation (4.3.1) from the microscopic equilibrium condition

$$\partial_t f_a = 0 \tag{4.6.1}$$

In estimating $c^{(2)}$, one must maintain terms in Ω_a to the second order in ϕ_b, but one can neglect spatial variation in the ϕ_a. As a result, the Boltzmann equation (4.3.1) becomes

$$\sum_b \omega_{ab}^{(1)} \phi_b + \sum_{b,c} \omega_{abc}^{(2)} \phi_b \phi_c = 0 \tag{4.6.2}$$

Substituting forms for the ϕ_a from the Chapman–Enskog expansion (2.5.1), one obtains

$$c^{(2)} \sum_b \omega_{ab}^{(1)} (\mathbf{u} \cdot \mathbf{e}_b)^2 + [c^{(1)}]^2 \sum_{b,c} \omega_{abc}^{(2)} (\mathbf{u} \cdot \mathbf{e}_b)(\mathbf{u} \cdot \mathbf{e}_c) = 0 \tag{4.6.3}$$

where $c^{(1)} = 2$ according to Eq. (2.5.4). Using the forms for $\omega^{(1)}$ and $\omega^{(2)}$ determined by the expansion of Eq. (4.3.2), one finds that the two terms in (4.6.3) show exactly the same dependence on the γ_k. The final result for $c^{(2)}$ is thus independent of the γ_k, and is given by

$$c^{(2)} = 2\,(1 - 2f)/(1 - f) \tag{4.6.4}$$

In the Boltzmann equation approximation, this implies that the coefficient μ of the $n(\mathbf{u} \cdot \nabla)\,\mathbf{u}$ term in the hydrodynamic equation (2.6.1) is $(1 - 2f)/[2\,(1 - f)]$. Notice that, as discussed in Section 2.6, this coefficient is not in general equal to 1.

The value of the coefficient $c_\nabla^{(2)}$ can be found by a slightly simpler calculation, which depends only on the linear part $\omega_{ab}^{(1)}$ of the expansion of the collision term Ω_a. Keeping now first-order spatial derivatives of the ϕ_a, one can determine $c_\nabla^{(2)}$ from the equilibrium condition

$$\sum_b \omega_{ab}^{(1)} \phi_b = f \mathbf{e}_a \cdot \nabla \phi_b \tag{4.6.5}$$

which yields

$$\sum_b c_\nabla^{(2)} \omega_{ab}^{(1)} (\mathbf{e}_b \cdot \nabla)(\mathbf{e}_b \cdot \mathbf{u}) = c^{(1)} f (\mathbf{e}_a \cdot \nabla)(\mathbf{e}_a \cdot \mathbf{u}) \tag{4.6.6}$$

With the approximations used, Eq. (2.4.7) implies that $\nabla \cdot \mathbf{u} = 0$. Then Eq. (4.6.6) gives the result

$$c_\nabla^{(2)} = -2\,\{12 f (1 - f)[\gamma_2 (1 - f)^2 + 4\,\gamma_{3A} f (1 - f) + \gamma_4 f^2]\}^{-1} \tag{4.6.7}$$

Using Eq. (2.6.3), this gives the kinematic viscosity of the cellular automaton fluid in the Boltzmann equation approximation as

$$\nu = \{12 f (1 - f)[\gamma_2 (1 - f)^2 + 4\,\gamma_{3A} f (1 - f) + \gamma_4 f^2]\}^{-1} \tag{4.6.8}$$

Some particular values are

$$\begin{aligned}
\nu &= [12 f (1 - f)^3]^{-1} \quad (\gamma_2 = 1, \gamma_{3A} = \gamma_4 = 0) \\
\nu &= [12 f (1 - f)(1 + 2f - 2f^2)]^{-1} \quad (\gamma_2 = \gamma_{3A} = \gamma_4 = 1)
\end{aligned} \tag{4.6.9}$$

For $f = 1/6$ one obtains in these cases $\nu \approx 0.86$ and $\nu \approx 0.47$, respectively, while for $f = 1/3$, $\nu \approx 0.84$ and $\nu \approx 0.26$.

4.7 A General Nonlinear Approximation

At least for homogeneous systems, Boltzmann's H theorem (e.g., Ref. 51) yields a general form for the equilibrium solution of the full nonlinear Boltzmann equation (4.3.1). The H function can be defined as

$$H = \sum_a \tilde{f}_a \log(\tilde{f}_a) \tag{4.7.1}$$

where

$$\tilde{f}_a = \frac{f_a}{1 - f_a} \tag{4.7.2}$$

This definition is analogous to that used for Fermi–Dirac particles (e.g., Refs. 51, 52): the factors $(1 - f_a)$ account for the exclusion of more than one particle on each link, as in Eq. (4.3.1). The microscopic reversibility of (4.3.1) implies that when the equilibrium condition $\partial_t H = 0$ holds, all products $\tilde{f}_{a_1} \tilde{f}_{a_2} \cdots$ must be equal for all initial and final sets of particles $\{a_1, a_2, \ldots\}$ that can participate in collisions. As a result, the $\log(\tilde{f}_a)$ must be simple linear combinations of the quantities conserved in the collisions. If only particle number and momentum are conserved, and there are no spurious conserved quantities such as (4.5.11), the $\tilde{f}_a$ can always be written in the form[7, 49, 54]

$$\tilde{f}_a = \exp(-\alpha - \beta \mathbf{u} \cdot \mathbf{e}_a) \tag{4.7.3}$$

The one-particle distribution functions thus have the usual Fermi–Dirac form

$$f_a = [1 + \exp(\alpha + \beta \mathbf{u} \cdot \mathbf{e}_a)]^{-1} \tag{4.7.4}$$

where α and β are in general functions of the conserved quantities n and $|\mathbf{u}|^2$.

For small $|\mathbf{u}|^2$, one may write

$$\alpha = \alpha_0 + \alpha_1 |\mathbf{u}|^2 + \cdots, \quad \beta = \beta_0 + \beta_1 |\mathbf{u}|^2 + \cdots \tag{4.7.5}$$

These expansions can be substituted into Eq. (4.7.4), and the results compared with the Chapman–Enskog expansion (2.7.1).

For $\mathbf{u} = 0$, one finds immediately the "fugacity relation"

$$\exp(-\alpha_0) = \frac{f}{1 - f} \tag{4.7.6}$$

Then, from the expansion (related to that for generating Euler polynomials)

$$(1+\xi e^x)^{-1} = \frac{1}{1+\xi} - \frac{\xi}{(1+\xi)^2}x - \frac{\xi(1-\xi)}{2(1+\xi)^3}x^2$$
$$- \frac{\xi(1-4\xi+\xi^2)}{6(1+\xi)^4}x^3 - \frac{\xi(1-\xi)(1-10\xi+\xi^2)}{24(1+\xi)^5}x^4 + \cdots \tag{4.7.7}$$

together with the constraints (2.7.3)–(2.7.5) one obtains (for $d=2$)

$$\beta_0 = -\frac{2}{1-f}$$

$$\alpha_1 = \frac{1-2f}{(1-f)^2}$$

$$\beta_1 = \frac{1-2f+2f^2}{(1-f)^3} \tag{4.7.8}$$

$$\alpha_2 = -\frac{(1-2f)(3-4f+4f^2)}{16(1-f)^4}$$

where it has been assumed that the $\mathbf{e}_a$ form an isotropic set of unit vectors, satisfying Eq. (3.1.5). The complete Chapman–Enskog expansion (2.7.1) then becomes

$$f_a = f\{1 + d\mathbf{u}\cdot\mathbf{e}_a + \frac{d^2}{2}\frac{1-2f}{1-f}[(\mathbf{u}\cdot\mathbf{e}_a)^2 - \frac{1}{d}|\mathbf{u}|^2]$$

$$+ \frac{d^3}{6}\frac{1-6f+6f^2}{(1-f)^2}[(\mathbf{u}\cdot\mathbf{e}_a)^3 - \frac{3}{d+2}|\mathbf{u}|^2(\mathbf{u}\cdot\mathbf{e}_a)]$$

$$+ \frac{1}{48}\frac{1-2f}{(1-f)^3}[32(1-12f+12f^2)(\mathbf{u}\cdot\mathbf{e}_a)^4$$

$$+ 384f(1-f)|\mathbf{u}|^2(\mathbf{u}\cdot\mathbf{e}_a)^2 + 3(11-36f+36f^2)|\mathbf{u}|^4] + \cdots\} \tag{4.7.9}$$

where for the last term it has been assumed that $d=2$.

The result (4.6.4) for $c^{(2)}$ follows immediately from this expansion. For cellular automaton fluid models with $\mathbf{E}^{(6)}$ isotropic, the continuum equation (2.7.7) holds. The results for the coefficients that appear in this equation can be obtained from the approximation (4.7.9), and have the simple forms

$$c^{(2)} = \frac{d^2(1-2f)}{2(1-f)} \tag{4.7.10}$$

$$c^{(4)}(1+\sigma_{4,1}) = \frac{2(1-2f)}{3(1-f)^3} \quad (d=2) \tag{4.7.11}$$

These results allow an estimate of the importance of the next-order corrections to the Navier–Stokes equations included in Eq. (2.7.7). They suggest that the corrections may be important whenever $|\mathbf{u}/(1-f)^2|^2$ is not small compared to 1. The corrections can thus potentially be important both at high average velocities and high particle densities.

The hexagonal lattice model of Section 2 yields a continuum equation of the form (2.7.6), with an anisotropic $O(\mathbf{u}^2 \nabla \mathbf{u})$ term. Equation (4.7.9) gives in this case

$$
\begin{aligned}
\partial_t(6f\,u_x) \;+\;& \frac{3f(1-2f)}{1-f}\,(u_x\,\partial_x u_x + u_x\,\partial_y u_y + u_y\,\partial_y u_x - u_y\,\partial_x u_y) \\[4pt]
+\;& \frac{f(1-2f)}{4(1-f)^3}\,\{[(55-84f+84f^2)\,u_x^2 \\[4pt]
+\;& 3(13+4f-4f^2)\,u_y^2]\,u_x\,\partial_x u_x \\[4pt]
+\;& 2[(1+12f-12f^2)\,u_x^2 + 9\,(1-2f)^2\,u_y^2]\,u_x\,\partial_y u_y \\[4pt]
+\;& 6[(1+12f-12f^2)\,u_x^2 + (1-2f)^2]\,u_y\,\partial_y u_x \\[4pt]
+\;& 3[(13+4f-4f^2)\,u_x^2 + (13-28f+28f^2)\,u_y^2]\,u_y\,\partial_x u_y\} = 0
\end{aligned} \tag{4.7.12}
$$

$$
\begin{aligned}
\partial_t(6f\,u_y) \;+\;& \frac{3f(1-2f)}{1-f}\,(-u_x\,\partial_x u_x + u_x\,\partial_x u_y + u_y\,\partial_x u_x - u_y\,\partial_y u_y) \\[4pt]
+\;& \frac{f(1-2f)}{4(1-f)^3}\,\{[(35-36f+36f^2)\,u_x^2 \\[4pt]
+\;& 3(17-44f+44f^2)\,u_y^2]\,u_x\,\partial_y u_x \\[4pt]
+\;& 2[(1+12f-12f^2)\,u_x^2 + 9\,(1-2f)^2\,u_y^2]\,u_x\,\partial_x u_y \\[4pt]
+\;& 6[(1+12f-12f^2)\,u_x^2 + (1-2f)^2]\,u_y\,\partial_x u_x \\[4pt]
+\;& 3[(17-44f+44f^2)\,u_x^2 + (17-12f+12f^2)\,u_y^2]\,u_y\,\partial_y u_y\} = 0
\end{aligned} \tag{4.7.13}
$$

The $O(\mathbf{u}\,\nabla\mathbf{u})$ term is as given in Eq. (2.5.11). The $O(\mathbf{u}^3 \nabla\mathbf{u})$ terms are anisotropic, and are not even invariant under exchange of x and y coordinates ($\pi/2$ rotation). For small densities f, Eqs. (4.7.12) and (4.7.13) become

$$
\begin{aligned}
\partial_t(u_x) \;+\;& \frac{1}{2}\,(u_x\,\partial_x u_x + u_x\,\partial_y u_y + u_y\,\partial_y u_x - u_y\,\partial_x u_y) \\[4pt]
+\;& \frac{1}{24}\,[(55\,u_x^2 + 39\,u_y^2)\,u_x\,\partial_x u_x + 2\,(u_x^2 + 9\,u_y^2)\,u_x\,\partial_y u_y \\[4pt]
+\;& 6\,(u_x^2 + u_y^2)\,u_y\,\partial_y u_x + 39\,(u_x^2 + u_y^2)\,u_y\,\partial_x u_y] = 0
\end{aligned} \tag{4.7.14}
$$

$$\partial_t(u_y) + \frac{1}{2}(-u_x\,\partial_x u_x + u_x\,\partial_x u_y + u_y\,\partial_x u_x - u_y\,\partial_y u_y)$$

$$+ \frac{1}{24}[(35\,u_x^2 + 51\,u_y^2)\,u_x\,\partial_y u_x + 2\,(u_x^2 + 9\,u_y^2)\,u_x\,\partial_x u_y$$

$$+ 6(u_x^2 + u_y^2)\,u_y\,\partial_x u_x + 51\,(u_x^2 + u_y^2)\,u_y\,\partial_y u_y] = 0 \tag{4.7.15}$$

The results (4.7.10) and (4.7.11) follow from the Fermi–Dirac particle distribution (4.7.4). If instead an arbitrary number of particles were allowed at each site, the equilibrium particle distribution (4.7.4) would take on the Maxwell–Boltzmann form

$$f_a = \exp(-\alpha - \beta\mathbf{u}\cdot\mathbf{e}_a) \tag{4.7.16}$$

With this simpler form, more complete results for f_a as a function of n and u can be found. Results which are isotropic to all orders in $\mathbf{u}$ can be obtained only for an infinite set of possible particle directions, parametrized, say, by a continuous angle θ. In this case, the number and momentum densities (2.4.1) and (2.4.2) may be obtained as integrals

$$\frac{1}{2\pi}\int_0^{2\pi} f(\theta)\,d\theta = f \tag{4.7.17}$$

$$\frac{1}{2\pi}\int_0^{2\pi} \mathbf{e}\,(\theta)\,f(\theta)\,d\theta = f\mathbf{u} \tag{4.7.18}$$

With the distribution (4.7.16), these integrals become

$$\frac{1}{2\pi}\int_0^{2\pi} \exp(-\alpha - \beta u\cos\theta)\,d\theta = e^{-\alpha}\,I_0(\beta u) = f \tag{4.7.19}$$

$$\frac{1}{2\pi}\int_0^{2\pi} \exp(-\alpha - \beta u\cos\theta)\cos\theta\,\mathbf{u}/u\,d\theta = -e^{-\alpha}\,I_1(\beta u)\,\mathbf{u}/u = f\mathbf{u} \tag{4.7.20}$$

where $u = |\mathbf{u}|$, and the $I_\nu\,(z)$ are modified Bessel functions (e.g., Ref. 53)

$$I_\nu(z) = \sum_{n=0}^{\infty} \frac{(z/2)^{\nu+2n}}{n!\,(\nu+n)!}, \quad I_0(0) = 1, \quad I_\nu(0) = 0 \ \ (\nu > 0) \tag{4.7.21}$$

$$\int_{-1}^{1}(1-x^2)^{\nu-1/2}\,e^{-zx}\,dx = \pi z^{-\nu}(2\,\nu-1)!!\,I_\nu(z) \tag{4.7.22}$$

$$I_{\nu-1}(z) - I_{\nu+1}(z) = (2\,\nu/z)\,I_\nu(z) \tag{4.7.23}$$

The rapid convergence of the series (4.7.21) means that Eqs. (4.7.19) and (4.7.20) provide highly accurate approximations even for a small number of discrete directions $\mathbf{e}_a$. [For example, with $M = 6$, $\alpha = 0$, $\beta = 1$, and $\mathbf{u} = (1, 1)$, the error in Eq. (4.7.19) is less than 10^{-9}.]

For the simple distribution (4.7.16) the momentum flux density tensor (2.4.9) may be evaluated in direct analogy with Eqs. (4.7.19) and (4.7.20) as

$$\Pi_{ij} = e^{-\alpha}[\frac{1}{2}\,I_0(\beta u)\,\delta_{ij} + I_2(\beta u)\,(\frac{u_i\,u_j}{u^2} - \frac{1}{2}\,\delta_{ij})] \tag{4.7.24}$$

Using the recurrence relation (4.7.23), and substituting the results (4.7.19) and (4.7.20), this may be rewritten in the form

$$\Pi_{ij} = f[\frac{1}{2} I_0(\beta u) \delta_{ij} + (1+\frac{2}{\beta})(\frac{u_i u_j}{u^2} - \frac{1}{2}\delta_{ij})] \tag{4.7.25}$$

Combining Eqs. (4.7.19) and (4.7.20), one finds that the function $\beta(f, u)$ is independent of f, and can be determined from the implicit equation

$$\frac{I_1(\beta u)}{I_0(\beta u)} = \frac{I_0'(\beta u)}{I_0(\beta u)} = -u \tag{4.7.26}$$

Expanding in powers of u^2, as in Eq. (4.7.5), yields

$$\beta(u) = \sum_{n=0} \beta_n u^{2n}, \ \beta_0 = -2, \ \beta_1 = -1, \ \beta_2 = -\frac{5}{6}, \ \beta_3 = -\frac{19}{24}, \ \beta_4 = -\frac{143}{180} \tag{4.7.27}$$

Equation (4.7.19) then gives

$$\alpha(u, f) = \sum_{n=0} \alpha_n u^{2n}, \ \exp(-\alpha_0) = f, \ \alpha_1 = 1, \ \alpha_2 = \frac{3}{4}, \ \alpha_3 = \frac{25}{36}, \ \alpha_4 = \frac{133}{192} \tag{4.7.28}$$

In the limit $u \to 1, \beta \to -\infty$.

The above results immediately yield values for the transport coefficients $c^{(n)}$ in the Chapman–Enskog expansion:

$$c^{(2)} = 2 \tag{4.7.29}$$

$$c^{(4)}(1+\sigma_{4,1}) = \frac{2}{3} \tag{4.7.30}$$

independent of density. Equation (4.7.29) implies that the coefficient μ of the convective term in the Navier–Stokes equation (2.6.1) is equal to 1/2. The deviation from the Galilean invariant result 1 is associated with the constraint of fixed speed particles.

Figure 4 shows the exact result for $\beta(u)$ obtained from Eq. (4.7.26), compared with series expansions to various orders. Significant deviations from the $O(u^2)$ "Navier–Stokes" approximation are seen for $u \gtrsim 0.4$.

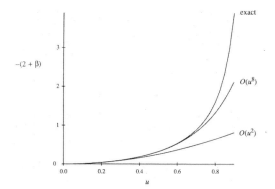

Figure 4. Dependence of $\beta(u)$ from Eq. (4.7.16) on the magnitude u of the macroscopic velocity. The results are for Maxwell–Boltzmann particles with unit speeds and arbitrary directions in two dimensions. The function $\beta(u)$ appears both in the microscopic distribution function (4.7.16) and in the macroscopic momentum flux tensor (4.7.25). The result for $\beta(u)$ from an exact solution of the implicit equation (4.7.26) is given, together with results from the series expansion (4.7.27). The $O(u^2)$ result corresponds to the Navier–Stokes approximation. Deviation from the exact result is seen for $u \gtrsim 0.4$.

For Fermi–Dirac distributions of the form (4.7.4), the integrals (4.7.19) and (4.7.20) can only be expressed as infinite sums of Bessel functions.

4.8 Other Models

The results obtained so far can be generalized directly to a large class of cellular automaton fluid models.

In the main case considered in Section 3, particles have velocities corresponding to a set of M unit vectors $\mathbf{e}_a$. If this set is invariant under inversion, then both $\mathbf{e}_a$ and $-\mathbf{e}_a$ always occur. As a result, two particles colliding head on with velocities $\mathbf{e}_a$ and $-\mathbf{e}_a$ can always scatter in any directions $\mathbf{e}_b$ and $-\mathbf{e}_b$ with $b \neq a$. One simple possibility is to choose the rules at different sites so that each scattering direction occurs with equal probability. If only such two-particle collisions are possible (as in a low-density approximation), and only one particle is allowed on each link, then the Boltzmann transport equation becomes

$$\partial_t f_a + \mathbf{e}_a \cdot \nabla f_a = -f_a f_{\bar{a}} \prod_{c \neq a, \bar{a}} (1 - f_c) + \frac{1}{M-2} \sum_{b \neq a, \bar{a}} f_b f_{\bar{b}} \prod_{c \neq b, \bar{b}} (1 - f_c) \tag{4.8.1}$$

where $f_{\bar{a}}$ is the distribution function for particles with direction $-\mathbf{e}_a$. To second order in the expansion (4.4.2) this gives

$$\begin{aligned}
\partial_t (f \phi_a) + \mathbf{e}_a \cdot \nabla (f \phi_a) &= f^2 (1-f)^{M-3} [-(\phi_a + \phi_{\bar{a}}) + \frac{1}{M-2} \sum_{b \neq a, \bar{a}} (\phi_b + \phi_{\bar{b}})] \\
&+ f^2 (1-f)^{M-4} (1-2f) [-\phi_a \phi_{\bar{a}} + \frac{1}{M-2} \sum'_{b, \bar{b} \neq a, \bar{a}} \phi_b \phi_{\bar{b}}] \\
&+ f^3 (1-f)^{M-4} [- \sum'_{b, c \neq a, \bar{a}; c \neq \bar{b}} \phi_b \phi_c + \frac{1}{M-2} \sum_{b \neq a} (\phi_a \phi_b + \phi_{\bar{a}} \phi_{\bar{b}})]
\end{aligned} \tag{4.8.2}$$

where $\sum'$ denotes summation over the triangular region in which the indices form a strictly increasing sequence.

The form of the ϕ_a for a homogeneous system can be obtained from the general equilibrium conditions of Section 4.7. The coefficients $c^{(n)}$ in the Chapman–Enskog expansion are then given by Equations (4.7.10) and (4.7.11). The convective transport coefficient μ in the Navier–Stokes equation (2.6.1) is thus given by

$$\mu = \frac{d^2 (1 - 2f)}{8 (1 - f)} \tag{4.8.3}$$

The $c_\nu^{(n)}$ cannot be obtained by the methods of Section 4.7. But from Eq. (4.8.2) one may deduce immediately the linearized collision term

$$\omega_{ab}^{(1)} = f^2(1-f)^{M-3} \operatorname{circ}[\chi_a] \cdot$$

$$\chi_1 = \chi_{1+M/2} = -1, \quad \chi_a = \frac{2}{M-2} \tag{4.8.4}$$

Then, in analogy with Eq. (4.6.8), the kinematic viscosity for the cellular automaton fluid is found to be

$$\nu = \frac{M-2}{2d(d+2)\, Mf^2(1-f)^{M-3}} \tag{4.8.5}$$

For an icosahedral set of $\mathbf{e}_a$, with $d=3$ and $M=12$, this yields

$$\nu = [36f^2(1-f)^9]^{-1} \tag{4.8.6}$$

Several generalizations may now be considered. First, one may allow not just one, but, say, up to κ particles on each link of the cellular automaton array. In the limit $\kappa \to \infty$ an arbitrary density of particles is thus allowed in each cell. The Boltzmann equation for this case is the same as (4.8.1), but with all $(1-f_c)$ factors omitted. The resulting transport coefficients are

$$\mu = \frac{1}{2} \tag{4.8.7}$$

$$\nu = \frac{M-2}{2d(d+2)\, Mf^2} \tag{4.8.8}$$

Another generalization is to allow collisions that involve more than two particles. The simplest such collisions are "composite" ones, formed by superposing collisions involving two or less particles. The presence of such collisions changes the values of transport coefficients, but cannot affect the basic properties of the model. The four-particle and asymmetric three-particle collisions in the hexagonal lattice model of Section 2 are examples of composite collisions. They increase the total collision rate, and thus, for example, decrease the viscosity, but do not change the overall macroscopic behavior of the model.

In general, collisions involving k particles can occur if the possible $\mathbf{e}_a$ are such that

$$\sum_{i=1}^{k} \mathbf{e}_{a_i} = \sum_{i=1}^{k} \mathbf{e}_{b_i} \tag{4.8.9}$$

for some sets of incoming and outgoing particles a_i and b_i. Cases in which all the a_i and b_i are distinct may be considered "elementary" collisions. In the hexagonal lattice model of Section 2, only two-particle and symmetric three-particle collisions are elementary.

No elementary three-particle collisions are possible on primitive and body-centered cubic three-dimensional lattices, or with $\mathbf{e}_a$ corresponding to the vertices of icosahedra or dodecahedra. For a face-centered cubic lattice, however, eight distinct triples of $\mathbf{e}_a$ sum to zero [an example is $(1, -1, 0) + (0, 1, -1) + (-1, 0, 1)$], so that elementary three-particle collisions are possible.

One feature of the hexagonal lattice model discussed in Section 2 is the existence of the conservation law (4.5.11) when elementary three-body symmetric collisions are absent. Such spurious conservation laws exist in any cellular automaton fluid model in which all particles have the same speed, and only two-particle collisions can occur. Elementary three-particle collisions provide one mechanism for avoiding these conservation laws and allowing the equilibrium of Section 4.7 to be attained.

4.9 Multiple Speed Models

A further generalization is to allow particles with velocities $\mathbf{e}_a$ of different magnitudes. This generalization is significant not only in allowing two-particle collisions alone to avoid the spurious conservation laws of Section 4.5, but also in making it possible to obtain isotropic hydrodynamic behavior on cubic lattices, as discussed in Section 3.5.

One may define a kinetic energy $1/2|\mathbf{e}_a|^2$ that differs for particles of different speeds. In studies of processes such as heat conduction, one must account for the conservation of total kinetic energy. In many cases, however, one considers systems in contact with a heat bath, so that energy need not be conserved in individual collisions.

In a typical case, one may then take pairs of particles with speed s_i colliding head on to give pairs of particles with some other speed s_j. In general, different collision rules may be used on different sites, typically following some regular pattern, as discussed in Section 4.3. Thus, for example, collisions between speed s_i particles may yield speed s_j particles at a fraction $\gamma_{i \to j}$ of the sites.

The number m_i of possible particles with speed $s_i = (|\mathbf{e}_a|^2)^{1/2}$ that can occur at each site is determined by the structure of the lattice. The collision rules at different sites may be arranged, as in Section 4.8, to yield particles of a particular speed s_i with equal probabilities in each of the m_i possible directions.

In a homogenous system, the probability f_i for a link with speed s_i to be populated should satisfy the master equation

$$\partial_t \tilde{f}_i = \sum_j \Gamma_{ij} \tilde{f}_j^2 = \sum_{j \neq i} (-\gamma_{i \to j} \, m_i \tilde{f}_i^2 + \gamma_{j \to i} \, m_j \tilde{f}_j^2) \tag{4.9.1}$$

where it assumed that

$$\sum_j \gamma_{i \to j} = \sum_j \gamma_{j \to i} = 1 \tag{4.9.2}$$

and $\tilde{f}$ is the reduced particle density given by Eq. (4.7.2). With two speeds, Γ_{ij} becomes

$$\Gamma_{ij} = \begin{pmatrix} -\gamma_{1 \to 2} \, m_1 & \gamma_{2 \to 1} \, m_2 \\ \gamma_{1 \to 2} \, m_1 & -\gamma_{2 \to 1} \, m_2 \end{pmatrix} \tag{4.9.3}$$

The solutions of Eq. (4.9.1) can be found in terms of the eigenvalues and corresponding eigenvectors of this matrix:

$$\lambda = 0: \qquad\qquad (\gamma_{2\to1}\, m_2, \gamma_{1\to2}\, m_1) \qquad\qquad\qquad (4.9.4)$$

$$\lambda = -(\gamma_{1\to2}\, m_1 + \gamma_{2\to1}\, m_2): \qquad (-1, 1) \qquad\qquad\qquad (4.9.5)$$

In the large-time limit, only the equilibrium eigenvector (4.9.4) should survive, giving a ratio of reduced particle densities

$$\frac{\tilde{f}_2^2}{\tilde{f}_1^2} = \frac{\gamma_{1\to2} m_1}{\gamma_{2\to1} m_2} \qquad\qquad\qquad (4.9.6)$$

For three particle speeds, one finds the equilibrium conditions

$$\frac{\tilde{f}_2^2}{\tilde{f}_1^2} = \frac{m_1}{m_2}\, \frac{\gamma_{1\to2}\gamma_{3\to1} + \gamma_{1\to2}\gamma_{3\to2} + \gamma_{1\to3}\gamma_{3\to2}}{\gamma_{2\to1}\gamma_{3\to1} + \gamma_{2\to1}\gamma_{3\to2} + \gamma_{2\to3}\gamma_{3\to1}}$$

$$\frac{\tilde{f}_3^2}{\tilde{f}_1^2} = \frac{m_1}{m_3}\, \frac{\gamma_{1\to2}\gamma_{2\to3} + \gamma_{1\to3}\gamma_{2\to1} + \gamma_{1\to3}\gamma_{2\to3}}{\gamma_{2\to1}\gamma_{3\to1} + \gamma_{2\to1}\gamma_{3\to2} + \gamma_{2\to3}\gamma_{3\to1}} \qquad (4.9.7)$$

Different choices for the γ_k yield different equilibrium speed distributions. The probabilities f_i give the weights $w(s_i^2)$ that appear in Eq. (3.1.16). Equation (4.9.6) shows that by choosing

$$\gamma_{2\to1} \approx 2\gamma_{1\to2} \qquad\qquad\qquad (4.9.8)$$

one obtains a ratio of weights for the model of Eq. (3.5.1) that satisfy the condition (3.5.5) for the isotropy of $\mathbf{E}^{(4)}$. [There is a small correction to equality in Eq. (4.9.8) associated with the difference between f and $\tilde{f}$.]

On a cubic lattice, one may similarly satisfy the condition (3.5.12) for the isotropy of $\mathbf{E}^{(4)}$ simply by taking $f_3 = 0$, and

$$\gamma_{2\to1} \approx 2\gamma_{1\to2} \qquad\qquad\qquad (4.9.9)$$

In this way, one may obtain approximate isotropic hydrodynamic behavior on a three-dimensional cubic lattice.

4.10 Tagged Particle Dynamics

In the discussion above, all the particles in the cellular automaton fluid were assumed indistinguishable. This section considers the behavior of a small concentration of special "tagged" particles.

The density g_a of tagged particles with direction $\mathbf{e}_a$ satisfies an equation of the Fokker–Planck type (e.g., Ref. 26):

$$\partial_t g_a + (\mathbf{e}_a \cdot \nabla)\, g_a = \Theta_a \qquad\qquad\qquad (4.10.1)$$

Assuming as in the Boltzmann equation approximation that there are no correlations between particles at different sites, the collision term of Eq. (4.10.1) may be written in the form

$$\Theta_a = \sum_b \theta_{ab}\, g_b \qquad (4.10.2)$$

where θ_{ab} gives the probability that a particle that arrives at a particular site from direction e_b leaves in direction e_a with $a \neq b$. The probability is averaged over different arrangements of ordinary particles. Various deterministic rules may be chosen for collisions between ordinary and tagged particles. The simplest assumption is that on average the tagged particles take the place of any of the outgoing particles with equal probability.

Conservation of the total number of tagged particles implies

$$\sum_a g_a = gM \qquad (4.10.3)$$

The total momentum of tagged particles is not conserved; the background of ordinary particles acts like a "heat bath" which can exchange momentum with the tagged particles through the noise term Θ_a. Assuming a uniform background fluid, one may make an expansion for the g_a of the form

$$g_a = (g + d^{(1)}\, e_a \cdot \nabla g + \cdots) \qquad (4.10.4)$$

The total number of tagged particles then satisfies the equation

$$\partial_t g + d^{(1)}\, \frac{1}{M}\, E_{ij}^{(2)}\, \partial_i \partial_j g = 0 \qquad (4.10.5)$$

where the collision term disappears as a result of Eq. (4.10.3). With the e_a chosen so that $E_{ij}^{(2)}$ is isotropic, Eq. (4.10.5) becomes the standard equation for self-diffusion,

$$\partial_t g = D \nabla^2 g \qquad (4.10.6)$$

with the diffusion coefficient D given by

$$D = -\frac{1}{d}\, d^{(1)} \qquad (4.10.7)$$

The value of $d^{(1)}$ must be found by solving Eq. (4.10.1) for g_a using the approximation (4.10.4). The equilibrium condition for Eq. (4.10.1) in this case becomes

$$(e_a \cdot \nabla)\, g = d^{(1)} \sum_b \theta_{ab}\, e_b \cdot \nabla g \qquad (4.10.8)$$

Thus $-d^{(1)}$ is given in this approximation by the mean free path λ for particle scattering, so that the diffusion coefficient is given by the standard kinetic theory formula

$$D = \frac{1}{d}\, \lambda \qquad (4.10.9)$$

For the hexagonal lattice model of Section 2,

$$D = \{2 f^2 (1-f)^2 [(1-f)^2 + (\gamma_{3S} + 4\,\gamma_{3A}) f(1-f) + \gamma_4 f^2]\}^{-1} \qquad (4.10.10)$$

5. Some Extensions

The simple physical basis for cellular automaton fluid models makes it comparatively straightforward for them to include many of the physical effects that occur in actual fluid experiments.

Boundaries can be represented by special sites in the cellular automaton array. Collisions with boundaries conserve particle number, but not particle momentum. One possibility is to choose boundary collision rules that exactly reverse the velocities of all particles, so that particles in a layer close to the boundary have zero average momentum. This choice yields macroscopic "no slip" boundary conditions, appropriate for many solid surfaces (e.g., Ref. 27). For boundaries that consist of flat segments aligned along lattice directions, an alternative is to take particles to undergo "specular" reflection, yielding a zero average only for the transverse component of particle momentum, and giving "free slip" macroscopic boundary conditions. The roughness of surfaces may be modeled explicitly by including various combinations of these microscopic boundary conditions (corresponding, say, to different coefficients of accommodation).

Arbitrarily complex solid boundaries may be modeled by appropriate arrangements of boundary cells. To model, for example, a porous medium one can, for example, use a random array of "boundary" cells with appropriate statistical properties.

A net flux of fluid can be maintained by continually inserting particles on one edge with an appropriate average momentum and extracting particles on an opposite edge. The precise arrangement of the inserted particles should not affect the macroscopic properties of the system, since microscopic processes should rapidly establish a microscopically random state of local equilibrium. Large-scale inhomogeneities, perhaps representing "free stream turbulence" (e.g., Ref. 4), can be included explicitly.

External pressure and density constraints, whether static or time-dependent, can be modeled by randomly inserting or extracting particles so that local average particle densities correspond to the macroscopic distribution required.

External forces can be modeled by randomly changing velocities of individual particles so as to impart momentum to the fluid at the required average rate. Moving boundaries can then be modeled by explicit motion of the special boundary cells, together with the inclusion of an appropriate average momentum change for particles striking the boundary. Gravitational and other force fields can also be represented in a "quantized approximation" by explicit local changes in particle velocities.

Many other physical effects depend on the existence of surfaces that separate different phases of a fluid or distinct immiscible fluids. The existence of such surfaces requires collective ordering effects within the system. For some choices of parameters, no such ordering can typically occur. But as the parameters change, phase transitions may occur, allowing large correlated regions to form. Such phenomena will be studied elsewhere. (Surface tension effects have been observed in other two-dimensional cellular automata.[3])

6. Discussion

Partial differential equations have conventionally formed the basis for mathematical models of continuum systems such as fluids. But only in rather simple circumstances can exact mathematical solutions to such equations be found. Most actual studies of fluid dynamics must thus be based on digital computer simulations, which use discrete approximations to the original partial differential equations (e.g., Ref. 55).

Cellular automata provide an alternative approach to modeling fluids and other continuum systems. Their basic constituent cells are discrete, and ideally suited to simulation by digital computers. Yet collections of large numbers of these cells can show overall continuum behavior. This paper has given theoretical arguments that with appropriate rules for the individual cells, the overall behavior obtained should follow that described by partial differential equations for fluids.

The cellular automata considered give simple idealized models for the motion and collision of microscopic particles in a fluid. As expected from the Second Law of thermodynamics, precise particle configurations are rapidly randomized, and may be considered to come to some form of equilibrium. In this equilibrium, it should be adequate to describe configurations merely in terms of probabilities that depend on a few macroscopic quantities, such as momentum and particle number, that are conserved in the microscopic particle interactions. Such averaged macroscopic quantities change only slowly relative to the rate of particle interactions. Partial differential equations for their behavior can be found from the transport equations for the average microscopic particle dynamics.

So long as the underlying lattice is sufficiently isotropic, many cellular automata yield in the appropriate approximation the standard Navier–Stokes equations for continuum fluids. The essential features necessary for the derivation of these equations are the conservation of a few macroscopic quantities, and the randomization of all other quantities, by microscopic particle interactions. The Navier–Stokes equations follow with approximations of low fluid velocities and velocity gradients. The simplicity of the cellular automaton model in fact makes it possible to derive in addition next order corrections to these equations.

The derivation of hydrodynamic behavior from microscopic dynamics has never been entirely rigorous. Cellular automata can be considered as providing a simple example in which the necessary assumptions and approximations can be studied in detail. But strong support for the conclusions comes from explicit simulations of cellular automaton fluid models and the comparison of results with those from actual experiments. The next paper in this series will present many such simulations.

The cellular automaton method of this paper can potentially be applied to a wide variety of processes conventionally described by partial differential equations.

One example is diffusion. At a microscopic level, diffusion arises from random particle motions. The cellular automata used above can potentially reproduce diffusion phenomena, as discussed in Section 4.10. But much simpler cellular automaton rules should suffice. The derivation of the diffusion equation requires that the number of particles be conserved. But it is not necessary for total particle momentum to be conserved. Instead, particle directions should be randomized at each site. Such randomization can potentially be achieved by very simple cellular automaton rules, such as that of Ref. 20. Thus, one may devise cellular automaton methods for the solution of the diffusion equation,[56] which in turn gives a relaxation method for solving Laplace, Poisson, and related equations.

Whenever the physical basis for partial differential equations involves large numbers of particles or other components with local interactions, one can expect to derive an effective cellular automaton model. For systems such as electromagnetic or gravitational fields, such models can perhaps be obtained as analogues of lattice gauge theories.

Appendix: SMP Programs

This appendix contains a sample SMP[17] computation of the macroscopic equations for the hexagonal lattice cellular automaton fluid model of Section 2.

The SMP definitions are as follows:

```
/* two-dimensional case */
d:2

/* define position and velocity vectors */
r:{x,y}
u:{ux,uy}

/* generate polygonal set of lattice vectors */
<XTrig
polygon[$n] :: (e:Ar[$n,{Cos[2Pi $/$n],Sin[2Pi $/$n]}])

/* calculate terms in number density, momentum vector and stress tensor */
suma[$x] :: Ex[Sum[$x,{a,1,Len[e]}]]
nterm[$f] :: suma[$f[a]]
uterm[$f] :: suma[e[a] $f[a]]
piterm[$f] :: suma[e[a]**e[a] $f[a]]

/* define vector analysis operators */
egrad[$x,$a] :: Sum[e[$a][i] Dt[$x,r[i]],{i,1,d}]
div[$x] :: Sum[Dt[$x[i],r[i]],{i,1,d}]

/* terms in Chapman-Enskog expansion */
n : f Len[e]
ce0[$a] : f
ce1[$a] : f e[$a].u
ce2[$a] : f ((e[$a].u)^2 - u.u/2)
ce2d[$a] : f (egrad[e[$a].u,$a] - div[u]/2)
celist : {ce0,ce1,ce2,ce2d}

/* specify commutativity of second derivatives */
Dt[$f,$1,{$2_=(Ord[$2,$1]>0),1}] :: Dt[$f,$2,$1]

/* define printing of derivatives */
_Dt[Pr][[$1,$2]]::Fmt[{{0,0},{1,-1},{2,0}},D,$2,$1]
_Dt[Pr][[$1,$2{$3,1}]]::Fmt[{{0,0},{1,-1},{2,-1},{3,0}},D,$2,$3,$1]
```

The following is a transcript of an interactive SMP session:

```
#I[1]:: <"cafluid.smp"    /* load definitions */

#I[2]:: polygon[6]        /* set up for hexagonal lattice */

            1/2         1/2              1/2          1/2
           3           3              - 3           -3
#0[2]:  {{1/2,----},{-1/2,----},{-1,0},{-1/2,------},{1/2,------},{1,0}}
            2           2                2            2


#I[3]:: Map[nterm,celist]     /* find contributions to number density from
                                 terms in Chapman-Enskog expansion */

#0[3]:  {6f,0,0,0}

#I[4]:: Map[uterm,celist]  /* find contributions to momentum vector */

#0[4]:  {{0,0},{3f ux,3f uy},{0,0},{0,0}}

#I[5]:: Map[piterm,celist]  /* stress tensor */

#0[5]:* {{{3f,0},{0,3f}},{{0,0},{0,0}},
```

$$\left\{\left\{\frac{3f\ ux^2}{4} - \frac{3f\ uy^2}{4}, \frac{3f\ ux\ uy}{2}\right\}, \left\{\frac{3f\ ux\ uy}{2}, \frac{-3f\ ux^2}{4} + \frac{3f\ uy^2}{4}\right\}\right\},$$

$$\left\{\left\{\frac{3f\ D_x ux}{4} - \frac{3f\ D_y uy}{4}, \frac{3f\ D_y ux}{4} + \frac{3f\ D_x uy}{4}\right\},\right.$$

$$\left.\left\{\frac{3f\ D_y ux}{4} + \frac{3f\ D_x uy}{4}, \frac{-3f\ D_x ux}{4} + \frac{3f\ D_y uy}{4}\right\}\right\}\right\}$$

```
#I[6]:: Dt[f,$$]:0 ;  /* make incompressibility approximation */

#I[7]:: Fac[Map[div,@5]]  /* contributions to momentum equation */
```

$$\text{#0[7]:* } \{\{0,0\},\{0,0\},\{\frac{3f\ (ux\ D_x ux + ux\ D_y uy + uy\ D_y ux - uy\ D_x uy)}{2},$$

$$\frac{-3f\ (ux\ D_y ux - ux\ D_x uy - uy\ D_x ux - uy\ D_y uy)}{2}\},$$

$$\{\frac{3f\ (D_{xx} ux + D_{yy} ux)}{4}, \frac{3f\ (D_{xx} uy + D_{yy} uy)}{4}\}\}$$

Acknowledgments

Many people have contributed in various ways to the material presented here. For general discussions I thank: Uriel Frisch, Brosl Hasslacher, David Levermore, Steve Orszag, Yves Pomeau, and Victor Yakhot. For specific suggestions I thank Roger Dashen, Dominique d'Humieres, Leo Kadanoff, Paul Martin, John Milnor, Steve Omohundro, Paul Steinhardt, and Larry Yaffe. Most of the calculations described here were made possible by using the SMP general-purpose computer mathematics system.[17] I thank Thinking Machines Corporation for much encouragement and partial support of this work.

References

1. S. Wolfram ed., *Theory and Applications of Cellular Automata* (World Scientific, 1986).

2. S. Wolfram, "Cellular Automata as Models of Complexity", *Nature* 311, 419 (1984).

3. N. Packard and S. Wolfram, "Two-Dimensional Cellular Automata", *J. Stat. Phys.* 38, 901 (1985).

4. D. J. Tritton, *Physical Fluid Dynamics* (Van Nostrand, 1977).

5. W. W. Wood, "Computer Studies on Fluid Systems of Hard-Core Particles", in *Fundamental Problems in Statistical Mechanics 3*, E. D. G. Cohen, ed. (North-Holland, 1975).

6. J. Hardy, Y. Pomeau, and O. de Pazzis, "Time Evolution of a Two-Dimensional Model System. I. Invariant States and Time Correlation Functions", *J. Math. Phys.* 14, 1746 (1973); J. Hardy, O. de Pazzis, and Y. Pomeau, "Molecular Dynamics of a Classical Lattice Gas: Transport Properties and Time Correlation Functions", *Phys. Rev. A* 13, 1949 (1976).

7. U. Frisch, B. Hasslacher, and Y. Pomeau, "Lattice Gas Automata for the Navier–Stokes Equation", *Phys. Rev. Lett.* 56, 1505 (1986).

8. J. Salem and S. Wolfram, "Thermodynamics and Hydrodynamics with Cellular Automata", in *Theory and Applications of Cellular Automata*, S. Wolfram, ed. (World Scientific, 1986).

9. D. d'Humieres, P. Lallemand, and T. Shimomura, "An Experimental Study of Lattice Gas Hydrodynamics", Los Alamos preprint LA-UR-85-4051; D. d'Humieres, Y. Pomeau, and P. Lallemand, "Simulation d'allees de Von Karman Bidimensionnelles a l'aide d'un gaz sur reseau", *C. R. Acad. Sci. Paris II* 301, 1391 (1985).

10. J. Broadwell, "Shock Structure in a Simple Discrete Velocity Gas", *Phys. Fluids* 7, 1243 (1964).

11. H. Cabannes, "The Discrete Boltzmann Equation", Lecture Notes, Berkeley (1980).

12. R. Gatignol, *Theorie cinetique des gaz a repartition discrete de vitesse* (Springer, 1975).

13. J. Hardy and Y. Pomeau, "Thermodynamics and Hydrodynamics for a Modeled Fluid", *J. Math. Phys.* 13, 1042 (1972).

14. S. Harris, *The Boltzmann Equation* (Holt, Rinehart and Winston, 1971).

15. R. Caflisch and G. Papanicolaou, "The Fluid-Dynamical Limit of a Nonlinear Model Boltzmann Equation", *Commun. Pure Appl. Math.* 32, 589 (1979).

16. B. Nemnich and S. Wolfram, "Cellular Automaton Fluids 2: Basic Phenomenology", in preparation.

17. S. Wolfram, *SMP Reference Manual* (Inference Corporation, Los Angeles, 1983); S. Wolfram, "Symbolic Mathematical Computation", *Commun. ACM* 28, 390 (1985).

18. A. Sommerfeld, *Thermodynamics and Statistical Mechanics* (Academic Press, 1955).

19. S. Wolfram, "Origins of Randomness in Physical Systems", *Phys. Rev. Lett.* 55, 449 (1985).

20. S. Wolfram, "Random Sequence Generation by Cellular Automata", *Adv. Appl. Math.* 7, 123 (1986).

21. J. P. Boon and S. Yip, *Molecular Hydrodynamics* (McGraw–Hill, 1980).

22. E. M. Lifshitz and L. P. Pitaevskii, *Statistical Mechanics, Part 2* (Pergamon, 1980), Chapter 9.

23. R. Liboff, *The Theory of Kinetic Equations* (Wiley, 1969).

24. D. Levermore, "Discretization Effects in the Macroscopic Properties of Cellular Automaton Fluids", in preparation.

25. E. M. Lifshitz and L. P. Pitaevskii, *Physical Kinetics* (Pergamon, 1981).

26. P. Resibois and M. De Leener, *Classical Kinetic Theory of Fluids* (Wiley, 1977).

27. L. D. Landau and E. M. Lifshitz, *Fluid Mechanics* (Pergamon, 1959).

28. M. H. Ernst, B. Cichocki, J. R. Dorfman, J. Sharma, and H. van Beijeren, "Kinetic Theory of Nonlinear Viscous Flow in Two and Three Dimensions", *J. Stat. Phys.* 18, 237 (1978).

29. J. R. Dorfman, "Kinetic and Hydrodynamic Theory of Time Correlation Functions", in *Fundamental Problems in Statistical Mechanics 3*, E. D. G. Cohen, ed. (North-Holland, 1975).

30. R. Courant and K. O. Friedrichs, *Supersonic Flows and Shock Waves* (Interscience, 1948).

31. D. Levermore, private communication.

32. J. Milnor, private communication.

33. V. Yakhot, B. Bayley, and S. Orszag, "Analogy between Hyperscale Transport and Cellular Automaton Fluid Dynamics", Princeton University preprint (February 1986).

34. H. S. M. Coxeter, *Regular Polytopes* (Macmillan, 1963).

35. M. Hammermesh, *Group Theory* (Addison–Wesley, 1962), Chapter 9.

36. L. D. Landau and E. M. Lifshitz, *Quantum Mechanics* (Pergamon, 1977), Chapter 12.

37. H. Boerner, *Representations of Groups* (North–Holland, 1970), Chapter 7.

38. L. D. Landau and E. M. Lifshitz, *Theory of Elasticity* (Pergamon, 1975), Section 10.

39. D. Levine *et al.*, "Elasticity and Dislocations in Pentagonal and Icosahedral Quasicrystals", *Phys. Rev. Lett.* 14, 1520 (1985).

40. L. D. Landau and E. M. Lifshitz, *Statistical Physics* (Pergamon, 1978), Chapter 13.

41. B. K. Vainshtein, *Modern Crystallography*, (Springer, 1981), Chapter 2.

42. J. H. Conway and N. J. A. Sloane, to be published.

43. R. L. E. Schwarzenberger, *N-Dimensional Crystallography* (Pitman, 1980).

44. J. Milnor, Hilbert's problem 18: "On Crystallographic Groups, Fundamental Domains, and on Sphere Packing", *Proc. Symp. Pure Math.* 28, 491 (1976).

45. B. G. Wybourne, *Classical Groups for Physicists* (Wiley, 1974), p. 78; R. Slansky, "Group Theory for Unified Model Building", *Phys. Rep.* 79, 1 (1981).

46. B. Grunbaum and G. C. Shephard, *Tilings and Patterns* (Freeman, in press); D. Levine and P. Steinhardt, "Quasicrystals I: Definition and Structure", Univ. of Pennsylvania preprint.

47. N. G. de Bruijn, "Algebraic Theory of Penrose's Non-periodic Tilings of the Plane", *Nedl. Akad. Wetensch. Indag. Math.* 43, 39 (1981); J. Socolar, P. Steinhardt, and D. Levine, "Quasicrystals with Arbitrary Orientational Symmetry", *Phys. Rev. B* 32, 5547 (1985).

48. R. Penrose, "Pentaplexity: A Class of Nonperiodic Tilings of the Plane", *Math. Intelligencer* 2, 32 (1979).

49. J. P. Rivet and U. Frisch, Automates sur gaz de reseau dans l'approximation de Boltzmann, *C. R. Acad. Sci. Paris II* 302, 267 (1986).

50. P. J. Davis, *Circulant Matrices* (Wiley, 1979).

51. L. D. Landau and E. M. Lifshitz, *Statistical Physics* (Pergamon, 1978), Chapter 5.

52. E. Kolb and S. Wolfram, "Baryon Number Generation in the Early Universe", *Nucl. Phys. B* 172, 224 (1980), Appendix A.

53. I. S. Gradestyn and I. M. Ryzhik, *Table of Integrals, Series and Products* (Academic Press, 1965).

54. U. Frisch, private communication.

55. P. Roache, *Computational Fluid Mechanics* (Hermosa, Albuquerque, 1976).

56. S. Omohundro and S. Wolfram, unpublished (July 1985).

57. D. d'Humieres, private communication.

Notes

1. This work has many precursors. A discrete model of exactly the kind considered here was discussed in Ref. 6. A version on a hexagonal lattice was introduced in Ref. 7, and further studied in Refs. 8, 9. Related models in which particles have a discrete set of possible velocities, but can have continuously variable positions and densities, were considered much earlier.[10–14] Detailed derivations of hydrodynamic behavior do not, however, appear to have been given even in these cases (see, however, e.g., Ref. 15).

2. The kinetic theory approach used in this paper concentrates on average particle distribution functions. An alternative but essentially equivalent approach concentrates on microscopic correlation functions (e.g., Refs. 21, 22).

3. Note that even the linearized equation for sound waves is anisotropic on a square lattice. The waves propagate isotropically, but are damped with an effective viscosity that varies with direction, and can be negative.[33]

4. Some similar results have been obtained by a slightly different method in Ref. 49.

Excerpts from

A NEW KIND
OF SCIENCE (2002)

STEPHEN
WOLFRAM
A NEW
KIND OF
SCIENCE

Contents

7.2 Three Mechanisms for Randomness

In nature one of the single most common things one sees is apparent randomness. And indeed, there are a great many different kinds of systems that all exhibit randomness. And it could be that in each case the cause of randomness is different. But from my investigations of simple programs I have come to the conclusion that one can in fact identify just three basic mechanisms for randomness, as illustrated in the pictures below.

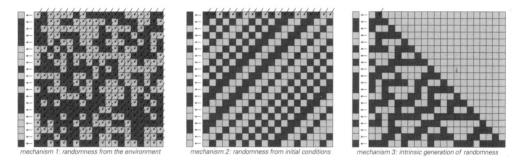

mechanism 1: randomness from the environment   mechanism 2: randomness from initial conditions   mechanism 3: intrinsic generation of randomness

Three possible mechanisms that can be responsible for randomness. The diagonal arrows represent external input. In the first case, there is random input from the environment at every step. In the second case, there is random input only in the initial conditions. And in the third case, there is effectively no random input at all. Yet despite their different underlying structure, each of these mechanisms leads to randomness in the column shown at the left. The first mechanism corresponds to randomness produced by external noise, as captured in so-called stochastic models. The second mechanism is essentially the one suggested by chaos theory. The third mechanism is new, and is suggested by the results on the behavior of simple programs in this book. I will give evidence that this third mechanism is the most common one in nature.

In the first mechanism, randomness is explicitly introduced into the underlying rules for the system, so that a random color is chosen for every cell at each step.

This mechanism is the one most commonly considered in the traditional sciences. It corresponds essentially to assuming that there is a random external environment which continually affects the system one is looking at, and continually injects randomness into it.

In the second mechanism shown above, there is no such interaction with the environment. The initial conditions for the system are chosen randomly, but then the subsequent evolution of the system is assumed to follow definite rules that involve no randomness.

A crucial feature of these rules, however, is that they make the system behave in a way that depends sensitively on the details of its initial conditions. In the particular case shown, the rules are simply set up to shift every color one position to the left at each step.

And what this does is to make the sequence of colors taken on by any particular cell depend on the colors of cells progressively further and further to the right in the initial conditions. Insofar as the initial conditions are random, therefore, so also will the sequence of colors of any particular cell be correspondingly random.

In general, the rules can be more complicated than those shown in the example on the previous page. But the basic idea of this mechanism for randomness is that the randomness one sees arises from some kind of transcription of randomness that is present in the initial conditions.

The two mechanisms for randomness just discussed have one important feature in common: they both assume that the randomness one sees in any particular system must ultimately come from outside of that system. In a sense, therefore, neither of these mechanisms takes any real responsibility for explaining the origins of randomness: they both in the end just say that randomness comes from outside whatever system one happens to be looking at.

Yet for quite a few years, this rather unsatisfactory type of statement has been the best that one could make. But the discoveries about simple programs in this book finally allow new progress to be made.

The crucial point that we first saw on page 27 is that simple programs can produce apparently random behavior even when they are given no random input whatsoever. And what this means is that there is a third possible mechanism for randomness, which this time does not rely in any way on randomness already being present outside the system one is looking at.

If we had found only a few examples of programs that could generate randomness in this way, then we might think that this third mechanism was a rare and special one. But in fact over the past few chapters we have seen that practically every kind of simple program that we can construct is capable of generating such randomness.

Note: page numbers in the text refer to the original work *A New Kind of Science* (2002).

And as a result, it is reasonable to expect that this same mechanism should also occur in many systems in nature. Indeed, as I will discuss in this chapter and the chapters that follow, I believe that this mechanism is in fact ultimately responsible for a large fraction, if not essentially all, of the randomness that we see in the natural world.

But that is not to say that the other two mechanisms are never relevant in practice. For even though they may not be able to explain how randomness is produced at the lowest level, they can still be useful in describing observations about randomness in particular systems.

And in the next few sections, I will discuss various kinds of systems where the randomness that is seen can be best described by each of the three mechanisms for randomness identified here.

7.5 **The Intrinsic Generation of Randomness**

In the past two sections, we have studied two possible mechanisms that can lead to observed randomness. But as we have discussed, neither of these in any real sense themselves generate randomness. Instead, what they essentially do is just to take random input that comes from outside, and transfer it to whatever system one is looking at.

One of the important results of this book, however, is that there is also a third possible mechanism for randomness, in which no random input from outside is needed, and in which randomness is instead generated intrinsically inside the systems one is looking at.

The picture below shows the rule 30 cellular automaton in which I first identified this mechanism for randomness. The basic rule for the system is very simple. And the initial condition is also very simple.

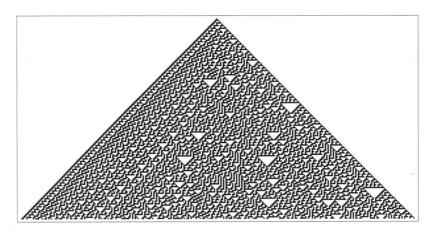

The rule 30 cellular automaton from page 27 that was the first example I found of intrinsic randomness generation. There is no random input to this system, yet its behavior seems in many respects random. I suspect that this is how much of the randomness that we see in nature arises.

Yet despite the lack of anything that can reasonably be considered random input, the evolution of the system nevertheless intrinsically yields behavior which seems in many respects random.

As we have discussed before, traditional intuition makes it hard to believe that such complexity could arise from such a simple

Note: page numbers in the text refer to the original work *A New Kind of Science* (2002).

underlying process. But the past several chapters have demonstrated that this is not only possible, but actually quite common.

Yet looking at the cellular automaton on the previous page there are clearly at least some regularities in the pattern it produces—like the diagonal stripes on the left. But if, say, one specifically picks out the color of the center cell on successive steps, then what one gets seems like a completely random sequence.

But just how random is this sequence really?

For our purposes here the most relevant point is that so far as one can tell the sequence is at least as random as sequences one gets from any of the phenomena in nature that we typically consider random.

When one says that something seems random, what one usually means in practice is that one cannot see any regularities in it. So when we say that a particular phenomenon in nature seems random, what we mean is that none of our standard methods of analysis have succeeded in finding regularities in it. To assess the randomness of a sequence produced by something like a cellular automaton, therefore, what we must do is to apply to it the same methods of analysis as we do to natural systems.

As I will discuss in Chapter 10, some of these methods have been well codified in standard mathematics and statistics, while others are effectively implicit in our processes of visual and other perception. But the remarkable fact is that none of these methods seem to reveal any real regularities whatsoever in the rule 30 cellular automaton sequence. And thus, so far as one can tell, this sequence is at least as random as anything we see in nature.

But is it truly random?

Over the past century or so, a variety of definitions of true randomness have been proposed. And according to most of these definitions, the sequence is indeed truly random. But there are a certain class of definitions which do not consider it truly random.

For these definitions are based on the notion of classifying as truly random only sequences which can never be generated by any simple procedure whatsoever. Yet starting with a simple initial condition and then applying a simple cellular automaton rule constitutes a simple procedure. And as a result, the center column of rule 30 cannot be considered truly random according to such definitions.

But while definitions of this type have a certain conceptual appeal, they are not likely to be useful in discussions of randomness in nature. For as we will see later in this book, it is almost certainly impossible for any natural process ever to generate a sequence which is guaranteed to be truly random according to such definitions.

For our purposes more useful definitions tend to concentrate not so much on whether there exists in principle a simple way to generate a particular sequence, but rather on whether such a way can realistically be recognized by applying various kinds of analysis to the sequence. And as discussed above, there is good evidence that the center column of rule 30 is indeed random according to all reasonable definitions of this kind.

So whether or not one chooses to say that the sequence is truly random, it is, as far as one can tell, at least random for all practical purposes. And in fact sequences closely related to it have been used very successfully as sources of randomness in practical computing.

For many years, most kinds of computer systems and languages have had facilities for generating what they usually call random numbers. And in *Mathematica*—ever since it was first released—Random[Integer] has generated 0's and 1's using exactly the rule 30 cellular automaton.

The way this works is that every time Random[Integer] is called, another step in the cellular automaton evolution is performed, and the value of the cell in the center is returned. But one difference from the picture two pages ago is that for practical reasons the pattern is not allowed to grow wider and wider forever. Instead, it is wrapped around in a region that is a few hundred cells wide.

One consequence of this, as discussed on page 259, is that the sequence of 0's and 1's that is generated must then eventually repeat. But even with the fastest foreseeable computers, the actual period of repetition will typically be more than a billion billion times the age of the universe.

Another issue is that if one always ran the cellular automaton from page 315 with the particular initial condition shown there, then one would always get exactly the same sequence of 0's and 1's. But by using different initial conditions one can get completely different

Note: page numbers in the text refer to the original work *A New Kind of Science* (2002).

sequences. And in practice if the initial conditions are not explicitly specified, what *Mathematica* does, for example, is to use as an initial condition a representation of various features of the exact state of the computer system at the time when Random was first called.

The rule 30 cellular automaton provides a particularly clear and good example of intrinsic randomness generation. But in previous chapters we have seen many other examples of systems that also intrinsically produce apparent randomness. And it turns out that one of these is related to the method used since the late 1940s for generating random numbers in almost all practical computer systems.

The pictures on the next page show what happens if one successively multiplies a number by various constant factors, and then looks at the digit sequences of the numbers that result. As we first saw on page 119, the patterns of digits obtained in this way seem quite random. And the idea of so-called linear congruential random number generators is precisely to make use of this randomness.

For practical reasons, such generators typically keep only, say, the rightmost 31 digits in the numbers at each step. Yet even with this restriction, the sequences generated are random enough that at least until recently they were almost universally what was used as a source of randomness in practical computing.

So in a sense linear congruential generators are another example of the general phenomenon of intrinsic randomness generation. But it turns out that in some respects they are rather unusual and misleading.

Keeping only a limited number of digits at each step makes it inevitable that the sequences produced will eventually repeat. And one of the reasons for the popularity of linear congruential generators is that with fairly straightforward mathematical analysis it is possible to tell exactly what multiplication factors will maximize this repetition period.

It has then often been assumed that having maximal repetition period will somehow imply maximum randomness in all aspects of the sequence one gets. But in practice over the years, one after another linear congruential generator that has been constructed to have maximal repetition period has turned out to exhibit very substantial deviations from perfect randomness.

Note: page numbers in the text refer to the original work *A New Kind of Science* (2002).

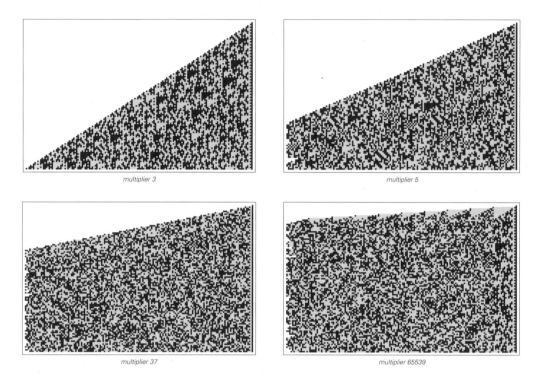

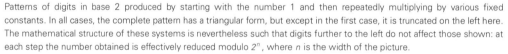

Patterns of digits in base 2 produced by starting with the number 1 and then repeatedly multiplying by various fixed constants. In all cases, the complete pattern has a triangular form, but except in the first case, it is truncated on the left here. The mathematical structure of these systems is nevertheless such that digits further to the left do not affect those shown: at each step the number obtained is effectively reduced modulo 2^n, where n is the width of the picture.

A typical kind of failure, illustrated in the pictures on the next page, is that points with coordinates determined by successive numbers from the generator turn out to be distributed in an embarrassingly regular way. At first, such failures might suggest that more complicated schemes must be needed if one is to get good randomness. And indeed with this thought in mind all sorts of elaborate combinations of linear congruential and other generators have been proposed. But although some aspects of the behavior of such systems can be made quite random, deviations from perfect randomness are still often found.

And seeing this one might conclude that it must be essentially impossible to produce good randomness with any kind of system that has reasonably simple rules. But the rule 30 cellular automaton that we discussed above demonstrates that in fact this is absolutely not the case.

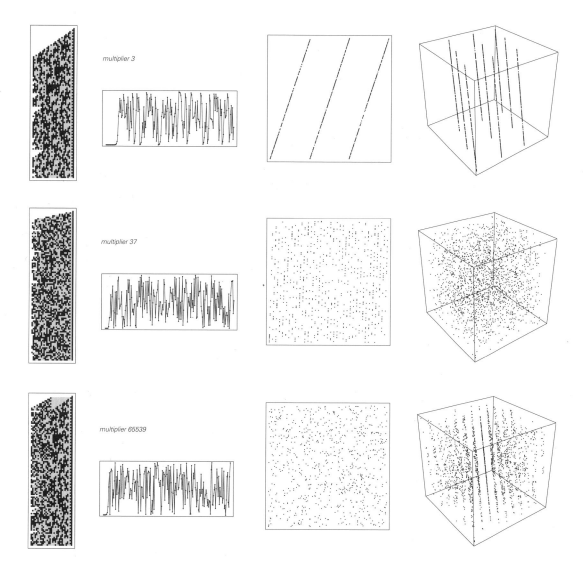

multiplier 3

multiplier 37

multiplier 65539

Examples of three so-called linear congruential random number generators. In each case they start with the number 1, then successively multiply by the specified multiplier, keeping only the rightmost 31 digits in the base 2 representation of the number obtained at each step. A version of the case with multiplier 3 was already shown on page 120. Multiplier 65539 was used as the random number generator on many computer systems, starting with mainframes in the 1960s. The last two pictures in each row above give the distribution of points whose coordinates in two and three dimensions are obtained by taking successive numbers from the linear congruential generator. If the output from the generator was perfectly random, then in each case these points would be uniformly distributed. But as the pictures demonstrate, stripes are visible in either two or three dimensions, or both.

Note: page numbers in the text refer to the original work *A New Kind of Science* (2002).

Indeed, the rules for this cellular automaton are in some respects much simpler than for even a rather basic linear congruential generator. Yet the sequences it produces seem perfectly random, and do not suffer from any of the problems that are typically found in linear congruential generators.

So why do linear congruential generators not produce better randomness? Ironically, the basic reason is also the reason for their popularity. The point is that unlike the rule 30 cellular automaton that we discussed above, linear congruential generators are readily amenable to detailed mathematical analysis. And as a result, it is possible for example to guarantee that a particular generator will indeed have a maximal repetition period.

Almost inevitably, however, having such a maximal period implies a certain regularity. And in fact, as we shall see later in this book, the very possibility of any detailed mathematical analysis tends to imply the presence of at least some deviations from perfect randomness.

But if one is not constrained by the need for such analysis, then as we saw in the cellular automaton example above, remarkably simple rules can successfully generate highly random behavior.

And indeed the existence of such simple rules is crucial in making it plausible that the general mechanism of intrinsic randomness generations can be widespread in nature. For if the only way for intrinsic randomness generation to occur was through very complicated sets of rules, then one would expect that this mechanism would be seen in practice only in a few very special cases.

But the fact that simple cellular automaton rules are sufficient to give rise to intrinsic randomness generation suggests that in reality it is rather easy for this mechanism to occur. And as a result, one can expect that the mechanism will be found often in nature.

So how does the occurrence of this mechanism compare to the previous two mechanisms for randomness that we have discussed?

The basic answer, I believe, is that whenever a large amount of randomness is produced in a short time, intrinsic randomness generation is overwhelmingly likely to be the mechanism responsible.

We saw in the previous section that random details of the initial conditions for a system can lead to a certain amount of randomness in the behavior of a system. But as we discussed, there is in most practical situations a limit on the lengths of sequences whose randomness can realistically be attributed to such a mechanism. With intrinsic randomness generation, however, there is no such limit: in the cellular automaton above, for example, all one need do to get a longer random sequence is to run the cellular automaton for more steps.

But it is also possible to get long random sequences by continual interaction with a random external environment, as in the first mechanism for randomness discussed in this chapter.

The issue with this mechanism, however, is that it can take a long time to get a given amount of good-quality randomness from it. And the point is that in most cases, intrinsic randomness generation can produce similar randomness in a much shorter time.

Indeed, in general, intrinsic randomness generation tends to be much more efficient than getting randomness from the environment. The basic reason is that intrinsic randomness generation in a sense puts all the components in a system to work in producing new randomness, while getting randomness from the environment does not.

Thus, for example, in the rule 30 cellular automaton discussed above, every cell in effect actively contributes to the randomness we see. But in a system that just amplifies randomness from the environment, none of the components inside the system itself ever contribute any new randomness at all. Indeed, ironically enough, the more components that are involved in the process of amplification, the slower it will typically be to get each new piece of random output. For as we discussed two sections ago, each component in a sense adds what one can consider to be more inertia to the amplification process.

But with a larger number of components it becomes progressively easier for randomness to be generated through intrinsic randomness generation. And indeed unless the underlying rules for the system somehow explicitly prevent it, it turns out in the end that intrinsic randomness generation will almost inevitably occur—often producing so much randomness that it completely swamps any randomness that might be produced from either of the other two mechanisms.

Yet having said this, one can ask how one can tell in an actual experiment on some particular system in nature to what extent intrinsic randomness generation is really the mechanism responsible for whatever seemingly random behavior one observed.

The clearest sign is a somewhat unexpected phenomenon: that details of the random behavior can be repeatable from one run of the experiment to another. It is not surprising that general features of the behavior will be the same. But what is remarkable is that if intrinsic randomness generation is the mechanism at work, then the precise details of the behavior can also be repeatable.

In the mechanism where randomness comes from continual interaction with the environment, no repeatability can be expected. For every time the experiment is run, the state of the environment will be different, and so the behavior one sees will also be correspondingly different. And similarly, in the mechanism where randomness comes from the details of initial conditions, there will again be little, if any, repeatability. For the details of the initial conditions are typically affected by the environment of the system, and cannot realistically be kept the same from one run to another.

But the point is that with the mechanism of intrinsic randomness generation, there is no dependence on the environment. And as a result, so long as the setup of the system one is looking at remains the same, the behavior it produces will be exactly the same. Thus for example, however many times one runs a rule 30 cellular automaton, starting with a single black cell, the behavior one gets will always be exactly the same. And so for example the sequence of colors of the center cell, while seemingly random, will also be exactly the same.

But how easy is it to disturb this sequence? If one makes a fairly drastic perturbation, such as changing the colors of cells all the way from white to black, then the sequence will indeed often change, as illustrated in the pictures at the top of the next page.

But with less drastic perturbations, the sequence can be quite robust. As an example, one can consider allowing each cell to be not just black or white, but any shade of gray, as in the continuous cellular automata we discussed on page 155. And in such systems, one can

Note: page numbers in the text refer to the original work *A New Kind of Science* (2002).

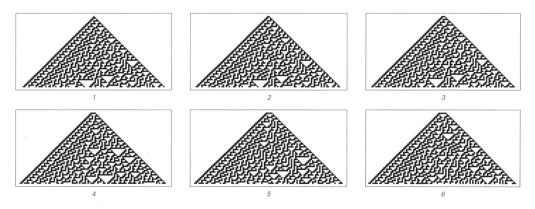

The effect of changing the number of initial black cells in the rule 30 cellular automaton shown above. With only 2 or 3 black cells, the sequence in the center of the pattern does not change. But as soon as more black cells are added, it does change.

investigate what happens if at every step one randomly perturbs the gray level of each cell by a small amount.

The pictures on the next page show results for perturbations of various sizes. What one sees is that when the perturbations are sufficiently large, the sequence of colors of the center cell does indeed change. But the crucial point is that for perturbations below a certain critical size, the sequence always remains essentially unchanged.

Even though small perturbations are continually being made, the evolution of the system causes these perturbations to be damped out, and produces behavior that is in practice indistinguishable from what would be seen if there were no perturbations.

The question of what size of perturbations can be tolerated without significant effect depends on the details of the underlying rules. And as the pictures suggest, rules which yield more complex behavior tend to be able to tolerate only smaller sizes of perturbations. But the crucial point is that even when the behavior involves intrinsic randomness generation, perturbations of at least some size can still be tolerated.

And the reason this is important is that in any real experiment, there are inevitably perturbations on the system one is looking at.

With more care in setting up the experiment, a higher degree of isolation from the environment can usually be achieved. But it is never possible to eliminate absolutely all interaction with the environment.

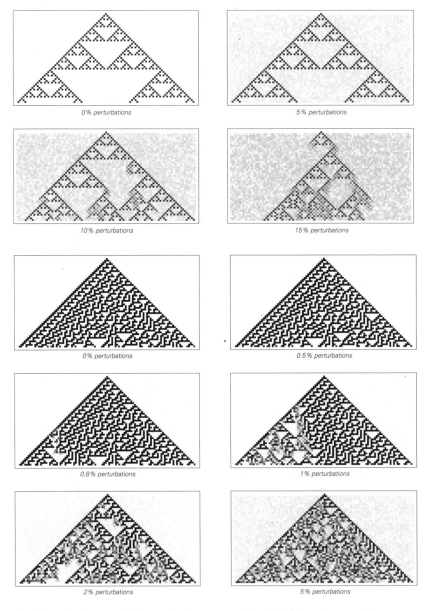

The effects of various levels of external randomness on the behavior of continuous cellular automata with generalizations of rules 90 and 30. The value of each cell can be any gray level between 0 and 1. For the generalization of rule 90, the values of the left and right cells are added together, and the value of the cell on the next step is then found by applying the continuous generalization of the modulo 2 function shown at the right. For the generalization of rule 30, a similar scheme based on an algebraic representation of the rule is used. In both cases, every value at each step is also perturbed by a random amount up to the percentage indicated for each picture.

And as a result, the system one is looking at will be subjected to at least some level of random perturbations from the environment.

But what the pictures on the previous page demonstrate is that when such perturbations are small enough, they will have essentially no effect. And what this means is that when intrinsic randomness generation is the dominant mechanism it is indeed realistic to expect at least some level of repeatability in the random behavior one sees in real experiments.

So has such repeatability actually been seen in practice?

Unfortunately there is so far very little good information on this point, since without the idea of intrinsic randomness generation there was never any reason to look for such repeatability when behavior that seemed random was observed in an experiment.

But scattered around the scientific literature—in various corners of physics, chemistry, biology and elsewhere—I have managed to find at least some cases where multiple runs of the same carefully controlled experiment are reported, and in which there are clear hints of repeatability even in behavior that looks quite random.

If one goes beyond pure numerical data of the kind traditionally collected in scientific experiments, and instead looks for example at the visual appearance of systems, then sometimes the phenomenon of repeatability becomes more obvious. Indeed, for example, as I will discuss in Chapter 8, different members of the same biological species often have many detailed visual similarities—even in features that on their own seem complex and apparently quite random.

And when there are, for example, two symmetrical sides to a particular system, it is often possible to compare the visual patterns produced on each side, and see what similarities exist. And as various examples in Chapter 8 demonstrate, across a whole range of physical, biological and other systems there can indeed be remarkable similarities.

So in all of these cases the randomness one sees cannot reasonably be attributed to randomness that is introduced from the environment—either continually or through initial conditions. And instead, there is no choice but to conclude that the randomness must in fact come from the mechanism of intrinsic randomness generation that I have discovered in simple programs, and discussed in this section.

NOTES FOR CHAPTER 7 OF *A NEW KIND OF SCIENCE*

Mechanisms in Programs and Nature

Universality of Behavior

■ **History.** That very different natural and artificial systems can show similar forms has been noted for many centuries. Informal studies have been done by a whole sequence of architects interested both in codifying possible forms and in finding ways to make structures fit in with nature and with our perception of it. Beginning in the Renaissance the point has also been noted by representational and decorative artists, most often in the context of developing a theory of the types of forms to be studied by students of art. The growth of comparative anatomy in the 1800s led to attempts at more scientific treatments, with analogies between biological and physical systems being emphasized particularly by D'Arcy Thompson in 1917. Yet despite all this, the phenomenon of similarity between forms remained largely a curiosity, discussed mainly in illustrated books with no clear basis in either art or science. In a few cases (such as work by Peter Stevens in 1974) general themes were however suggested. These included for example symmetry, the golden ratio, spirals, vortices, minimal surfaces, branching patterns, and—since the 1980s—fractals. The suggestion is also sometimes made that we perceive a kind of harmony in nature because we see only a limited number of types of forms in it. And particularly in classical architecture the idea is almost universally used that structures will seem more comfortable to us if they repeat in ornament or otherwise forms with which we have become familiar from nature. Whenever a scientific model has the same character for different systems this means that the systems will tend to show similar forms. And as models like cellular automata capable of dealing with complexity have become more widespread it has been increasingly popular to show that they can capture similar forms seen in very different systems.

Three Mechanisms for Randomness

■ **Page 299 · Definition.** How randomness can be defined is discussed at length on page 552.

■ **History.** In antiquity, it was often assumed that all events must be governed by deterministic fate—with any apparent randomness being the result of arbitrariness on the part of the gods. Around 330 BC Aristotle mentioned that instead randomness might just be associated with coincidences outside whatever system one is looking at, while around 300 BC Epicurus suggested that there might be randomness continually injected into the motion of all atoms. The rise of emphasis on human free will (see page 1135) eroded belief in determinism, but did not especially address issues of randomness. By the 1700s the success of Newtonian physics seemed again to establish a form of determinism, and led to the assumption that whatever randomness was actually seen must reflect lack of knowledge on the part of the observer— or particularly in astronomy some form of error of measurement. The presence of apparent randomness in digit sequences of square roots, logarithms, numbers like π, and other mathematical constructs was presumably noticed by the 1600s (see page 911), and by the late 1800s it was being taken for granted. But the significance of this for randomness in nature was never recognized. In the late 1800s and early 1900s attempts to justify both statistical mechanics and probability theory led to ideas that perfect microscopic randomness might somehow be a fundamental feature of the physical world. And particularly with the rise of quantum mechanics it came to be thought that meaningful calculations could be done only on probabilities, not on individual random sequences. Indeed, in almost every area where quantitative methods were used, if randomness was observed, then either a different system was studied, or efforts were made to remove the randomness by averaging or some other statistical method. One case where there was occasional discussion of origins of randomness from at least

Note: page numbers in the text refer to the original work *A New Kind of Science* (2002).

the early 1900s was fluid turbulence (see page 997). Early theories tended to concentrate on superpositions of repetitive motions, but by the 1970s ideas of chaos theory began to dominate. And in fact the widespread assumption emerged that between randomness in the environment, quantum randomness and chaos theory almost any observed randomness in nature could be accounted for. Traditional mathematical models of natural systems are often expressed in terms of probabilities, but do not normally involve anything one can explicitly consider as randomness. Models used in computer simulations, however, do very often use explicit randomness. For not knowing about the phenomenon of intrinsic randomness generation, it has normally been assumed that with the kinds of discrete elements and fairly simple rules common in such models, realistically complicated behavior can only ever be obtained if explicit randomness is continually introduced.

■ **Applications of randomness.** See page 1192.

■ **Sources of randomness.** Two simple mechanical methods for generating randomness seem to have been used in almost every civilization throughout recorded history. One is to toss an object and see which way up or where it lands; the other is to select an object from a collection mixed by shaking. The first method has been common in games of chance, with polyhedral dice already existing in 2750 BC. The second—often called drawing lots—has normally been used when there is more at stake. It is mentioned several times in the Bible, and even today remains the most common method for large lotteries. (See page 969.) Variants include methods such as drawing straws. In antiquity fortune-telling from randomness often involved looking say at growth patterns of goat entrails or sheep shoulder blades; today configurations of tea leaves are sometimes considered. In early modern times the matching of fracture patterns in broken tally sticks was used to identify counterparties in financial contracts. Horse races and other events used as a basis for gambling can be viewed as randomness sources. Children's games like musical chairs in effect generate randomness by picking arbitrary stopping times. Games of chance based on wheels seem to have existed in Roman times; roulette developed in the 1700s. Card shuffling (see page 974) has been used as a source of randomness since at least the 1300s. Pegboards (as on page 312) were used to demonstrate effects of randomness in the late 1800s. An explicit table of 40,000 random digits was created in 1927 by Leonard Tippett from details of census data. And in 1938 further tables were generated by Ronald Fisher from digits of logarithms. Several tables based on physical processes were produced, with the RAND Corporation in 1955 publishing a table of a million random

digits obtained from an electronic roulette wheel. Beginning in the 1950s, however, it became increasingly common to use pseudorandom generators whenever long sequences were needed—with linear feedback shift registers being most popular in standalone electronic devices, and linear congruential generators in programs (see page 974). There nevertheless continued to be occasional work done on mechanical sources of randomness for toys and games, and on physical electronic sources for cryptography systems (see page 969).

The Intrinsic Generation of Randomness

■ **Autoplectic processes.** In the 1985 paper where I introduced intrinsic randomness generation I called processes that show this autoplectic, while I called processes that transcribe randomness from outside homoplectic.

■ **Page 316 · Algorithmic randomness.** The idea of there being no simple procedure that can generate a particular sequence can be stated more precisely by saying that there is no program shorter than the sequence itself which can be used to generate the sequence, as discussed in more detail on page 1067.

■ **Page 317 · Randomness in *Mathematica*.** SeedRandom[n] is the function that sets up the initial conditions for the cellular automaton. The idea of using this kind of system in general and this system in particular as a source of randomness was described in my 1987 U.S. patent number 4,691,291.

■ **Page 321 · Cellular automata.** From the discussion here it should not be thought that in general there is necessarily anything better about generating randomness with cellular automata than with systems based on numbers. But the point is that the specific method used for making practical linear congruential generators does not yield particularly good randomness and has led to some incorrect intuition about the generation of randomness. If one goes beyond the specifics of linear congruential generators, then one can find many features of systems based on numbers that seem to be

Note: page numbers in the text refer to the original work *A New Kind of Science* (2002).

perfectly random, as discussed in Chapter 4. In addition, one should recognize that while the complete evolution of the cellular automaton may effectively generate perfect randomness, there may be deviations from randomness introduced when one constructs a practical random number generator with a limited number of cells. Nevertheless, no such deviations have so far been found except when one looks at sequences whose lengths are close to the repetition period. (See however page 603.)

■ **Page 321 · Card shuffling.** Another rather poor example of intrinsic randomness generation is perfect card shuffling. In a typical case, one splits the deck of cards in two, then carefully riffles the cards so as to make alternate cards come from each part of the deck. Surprisingly enough, this simple procedure, which can be represented by the function

$s[list\_] := Flatten[$
$\qquad Transpose[Reverse[Partition[list, Length[list]/2]]]]$

with or without the *Reverse*, is able to produce orderings which at least in some respects seem quite random. But by doing *Nest[s, Range[52], 26]* one ends up with a simple reversal of the original deck, as in the pictures below.

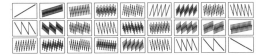

■ **Random number generators.** A fairly small number of different types of random number generators have been used in practice, so it is possible to describe all the major ones here.

Linear congruential generators. The original suggestion made by Derrick Lehmer in 1948 was to take a number n and at each step to replace it by *Mod[a n, m]*. Lehmer used $a = 23$ and $m = 10^8 + 1$. Most subsequent implementations have used $m = 2^j$, often with $j = 31$. Such choices are particularly convenient on computers where machine integers are represented by 32 binary digits. The behavior of the linear congruential generator depends greatly on the exact choice of a. Starting with the so-called RANDU generator used on mainframe computers in the 1960s, a common choice made was $a = 65539$. But as shown in the main text, this choice leads to embarrassingly obvious regularities. Starting in the mid-1970s, another common choice was $a = 69069$. This was also found to lead to regularities, but only in six or more dimensions. (Small values of a also lead to an excess of runs of identical digits, as mentioned on page 903.)

The repetition period for a generator with rule $n \to Mod[a\,n, m]$ is given (for a and m relatively prime) by *MultiplicativeOrder[a, m]*. If m is of the form 2^j, this implies a

maximum period for any a of $m/4$, achieved when *MemberQ[{3, 5}, Mod[a, 8]]*. In general the maximum period is *CarmichaelLambda[m]*, where the value $m - 1$ can be achieved for prime m.

As illustrated in the main text, when $m = 2^j$ the right-hand base 2 digits in numbers produced by linear congruential generators repeat with short periods; a digit k positions from the right will typically repeat with period no more than 2^k. When $m = 2^j - 1$ is prime, however, even the rightmost digit repeats only with period $m - 1$ for many values of a.

More general linear congruential generators use the basic rule $n \to Mod[a\,n + b, m]$, and in this case, $n = 0$ is no longer special, and a repetition period of exactly m can be achieved with appropriate choices of a, b and m. Note that if the period is equal to its absolute maximum of m, then every possible n is always visited, whatever n one starts from. Page 962 showed diagrams that represent the evolution for all possible starting values of n.

Each point in the 2D plots in the main text has coordinates of the form $\{n[i], n[i + 1]\}$ where $n[i + 1] = Mod[a\,n[i], m]$. If one could ignore the *Mod*, then the coordinates would simply be $\{n[i], a\,n[i]\}$, so the points would lie on a single straight line with slope a. But the presence of the *Mod* takes the points off this line whenever $a\,n[i] \geq m$. Nevertheless, if a is small, there are long runs of $n[i]$ for which the *Mod* is never important. And that is why in the case $a = 3$ the points in the plot fall on obvious lines.

In the case $a = 65539$, the points lie on planes in 3D. The reason for this is that

$n[i + 2] == Mod[65539^2\,n[i], 2^{31}] ==$
$\qquad Mod[6\,n[i + 1] - 9\,n[i], 2^{31}]$

so that in computing $n[i + 2]$ from $n[i + 1]$ and $n[i]$ only small coefficients are involved.

It is a general result related to finding short vectors in lattices that for some d the quantity $n[i + d]$ can always be written in terms of the $n[i + k]$; $k < d$ using only small coefficients. And as a consequence, the points produced by any linear congruential generator must lie on regular hyperplanes in some number of dimensions.

(For cryptanalysis of linear congruential generators see page 1089.)

Linear feedback shift registers. Used since the 1950s, particularly in special-purpose electronic devices, these systems are effectively based on running additive cellular automata such as rule 60 in registers with a limited number

Note: page numbers in the text refer to the original work *A New Kind of Science* (2002).

494

of cells and with a certain type of spiral boundary conditions. In a typical case, each cell is updated using

    LFSRStep[list_] :=
        Append[Rest[list], Mod[list[[1]] + list[[2]], 2]]

with a step of cellular automaton evolution corresponding to the result of updating all cells in the register. As with additive cellular automata, the behavior obtained depends greatly on the length n of the register. The maximal repetition period of $2^n - 1$ can be achieved only if $Factor[1 + x + x^n, Modulus \rightarrow 2]$ finds no factors. (For $n < 512$, this is true when $n = 1, 2, 3, 4, 6, 7, 9, 15, 22, 28, 30, 46, 60, 63, 127, 153, 172, 303$ or 471. Maximal period is assured when in addition $PrimeQ[2^n - 1]$.) The pictures below show the evolution obtained for $n = 30$ with

    NestList[Nest[LFSRStep, #, n] &,
        Append[Table[0, {n - 1}], 1], t]

Like additive cellular automata as discussed on page 951, states in a linear feedback shift register can be represented by a polynomial $FromDigits[list, x]$. Starting from a single 1, the state after t steps is then given by

    PolynomialMod[x^t, {1 + x + x^n, 2}]

This result illustrates the analogy with linear congruential generators. And if the distribution of points generated is studied with the Cantor set geometry, the same kind of problems occur as in the linear congruential case (compare page 1094).

In general, linear feedback shift registers can have "taps" at any list of positions on the register, so that their evolution is given by

    LFSRStep[taps_List, list_] :=
        Append[Rest[list], Mod[Apply[Plus, list[[taps]]], 2]]

(With taps specified by the positions of 1's in a vector of 0's, the inside of the *Mod* can be replaced by *vec . list* as on page 1087.) For a register of size n the maximal period of $2^n - 1$ is obtained whenever $x^n + Apply[Plus, x^{taps-1}]$ is one of the $EulerPhi[2^n - 1]/n$ primitive polynomials that appear in $Factor[Cyclotomic[2^n - 1, x], Modulus \rightarrow 2]$. (See pages 963 and 1084.)

One can also consider nonlinear feedback shift registers, as discussed on page 1088.

Generalized Fibonacci generators. It was suggested in the late 1950s that the Fibonacci sequence $f[n_] := f[n-1] + f[n-2]$

modulo 2^k might be used with different choices of $f[0]$ and $f[1]$ as a random number generator (see page 891). This particular idea did not work well, but generalizations based on the recurrence $f[n_] := Mod[f[n-p] + f[n-q], 2^k]$ have been studied extensively, for example with $p = 24$, $q = 55$. Such generators are directly related to linear feedback shift registers, since with a list of length q, each step is simply

    Append[Rest[list], Mod[list[[1]] + list[[q - p + 1]], 2^k]]

Cryptographic generators. As discussed on page 598, so-called stream cipher cryptographic systems work essentially by generating a repeatable random sequence. Practical stream cipher systems can thus be used as random number generators. Starting in the 1980s, the most common example has been the Data Encryption Standard (DES) introduced by the U.S. government (see page 1085). Unless special-purpose hardware is used, however, this method has not usually been efficient enough for practical random number generation applications.

Quadratic congruential generators. Several generalizations of linear congruential generators have been considered in which nonlinear functions of n are used at each step. In fact, the first known generator for digital computers was John von Neumann's "middle square method"

    n → FromDigits[Take[IntegerDigits[n^2, 10, 20], {5, 15}], 10]

In practice this generator has too short a repetition period to be useful. But in the early 1980s studies of public key cryptographic systems based on number theoretical problems led to some reinvestigation of quadratic congruential generators. The simplest example uses the rule

    n → Mod[n^2, m]

It was shown that for $m = pq$ with p and q prime the sequence $Mod[n, 2]$ was in a sense as difficult to predict as the number m is to factor (see page 1090). But in practice, the period of the generator in such cases is usually too short to be useful. In addition, there has been the practical problem that if n is stored on a computer as a 32-bit number, then n^2 can be 64 bits long, and so cannot be stored in the same way. In general, the period divides $CarmichaelLambda[CarmichaelLambda[m]]$. When m is a prime, this implies that the period can then be as long as $(m - 3)/2$. The largest m less than 2^{16} for which this is true is 65063, and the sequence generated in this case appears to be fairly random.

Cellular automaton generators. I invented the rule 30 cellular automaton random number generator in 1985. Since that time the generator has become quite widely used for a variety of applications. Essentially all the other generators discussed here have certain linearity properties which

Note: page numbers in the text refer to the original work *A New Kind of Science* (2002).

495

allow for fairly complete analysis using traditional mathematical methods. Rule 30 has no such properties. Empirical studies, however, suggest that the repetition period, for example, is about $2^{0.63n}$, where n is the number of cells (see page 260). Note that rule 45 can be used as an alternative to rule 30. It has a somewhat longer period, but does not mix up nearby initial conditions as quickly as rule 30. (See also page 603.)

■ **Unequal probabilities.** Given a sequence *a* of *n* equally probable 0's and 1's, the following generates a single 0 or 1 with probabilities approximating *{1 - p, p}* to *n* digits:

> *Fold[{BitAnd, BitOr}[[1 + First[#2]]][[#1, Last[#2]] &, 0,*
> *Reverse[Transpose[{First[RealDigits[p, 2, n, -1]], a}]]]*

This can be generalized to allow a whole sequence to be generated with as little as an average of two input digits being used for each output digit.

■ **Page 323 · Sources of repeatable randomness.** In using repeatability to test for intrinsic randomness generation, one must avoid systems in which there is essentially some kind of static randomness in the environment. Sources of this include the profile of a rough solid surface, or the detailed patterns of grains inside a solid.

■ **Page 324 · Probabilistic rules.** There appears to be a discrete transition as a function of the size of the perturbations, similar to phase transitions seen in the phenomenon of directed percolation. Note that if one just uses the original cellular automata rules, then with any nonzero probability of reversing the colors of cells, the patterns will be essentially destroyed. With more complicated cellular automaton rules, one can get behavior closer to the continuous cellular automata shown here. (See also page 591.)

■ **Page 325 · Noisy cellular automata.** In correspondence with electronics, the continuous cellular automata used here can be thought of as analog models for digital cellular automata. The specific form of the continuous generalization of the modulo 2 function used is

> $\lambda[x\_] := Exp[-10(x-1)^2] + Exp[-10(x-3)^2]$

Each cell in the system is then updated according to $\lambda[a+c]$ for rule 90, and $\lambda[a+b+c+bc]$ for rule 30. Perturbations of size δ are then added using $v + Sign[v - 1/2] Random[]\delta$.

Note that the basic approach used here can be extended to allow discrete cellular automata to be approximated by partial differential equations where not only color but also space and time are continuous. (Compare page 464.)

■ **Page 326 · Repeatably random experiments.** Over the years, I have asked many experimental scientists about repeatability in seemingly random data, and in almost all cases they have told me that they have never looked for such a thing. But in a

few cases they say that in fact on thinking about it they remember various forms of repeatability.

Examples where I have seen evidence of repeatable randomness as a function of time in published experimental data include temperature differences in thermal convection in closed cells of liquid helium, reaction rates in oxidation of carbon monoxide on catalytic surfaces, and output voltages from firings of excited single nerve cells. Typically there are quite long periods of time where the behavior is rather accurately repeatable—even though it may wiggle tens or hundreds in a seemingly random way—interspersed with jumps of some kind. In most cases the only credible models seem to be ones based on intrinsic randomness generation. But insofar as there is any definite model, it is inevitable that looking in sufficient detail at sufficiently many components of the system will reveal regularities associated with the underlying mechanism.

Note: page numbers in the text refer to the original work *A New Kind of Science* (2002).

496

10.3 Defining the Notation of Randomness

Many times in this book I have said that the behavior of some system or another seems random. But so far I have given no precise definition of what I mean by randomness. And what we will discover in this section is that to come up with an appropriate definition one has no choice but to consider issues of perception and analysis.

One might have thought that from traditional mathematics and statistics there would long ago have emerged some standard definition of randomness. But despite occasional claims for particular definitions, the concept of randomness has in fact remained quite obscure. And indeed I believe that it is only with the discoveries in this book that one is finally now in a position to develop a real understanding of what randomness is.

At the level of everyday language, when we say that something seems random what we usually mean is that there are no significant regularities in it that we can discern—at least with whatever methods of perception and analysis we use.

We would not usually say, therefore, that either of the first two pictures at the top of the next page seem random, since we can readily recognize highly regular repetitive and nested patterns in them. But the third picture we would probably say does seem random, since at least at the level of ordinary visual perception we cannot recognize any significant regularities in it.

So given this everyday notion of randomness, how can we build on it to develop more precise definitions? The first step is to clarify what it means not to be able to recognize regularities in something. Following the discussion in the previous section, we know that whenever we find regularities, it implies that redundancy is present, and this in turn means that a shorter description can be given. So when we say that we cannot recognize any regularities, this is equivalent to saying that we cannot find a shorter description.

The three pictures on the next page can always be described by explicitly giving a list of the colors of each of the 6561 cells that they contain. But by using the regularities that we can see in the first two

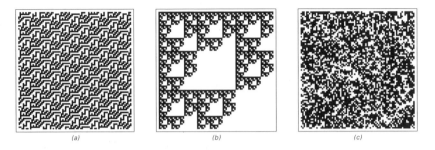

Pictures exhibiting different degrees of apparent randomness. Pictures (a) and (b) have obvious regularities, and would never be considered particularly random. But picture (c) has almost no obvious regularities, and would typically be considered quite random. As it turns out, picture (c), like (a) and (b), can actually be generated by a quite simple process. But the point is that the simplicity of this process does not affect the fact that with our standard methods of perception and analysis picture (c) is for practical purposes random.

pictures, we can readily construct much shorter—yet still complete—descriptions of these pictures.

The repetitive structure of picture (a) implies that to reproduce this picture all we need do is to specify the colors in a 49×2 block, and then say that this block should be repeated an appropriate number of times. Similarly, the nested structure of picture (b) implies that to reproduce this picture, all we need do is to specify the colors in a 3×3 block, and then say that as in a two-dimensional substitution system each black cell should repeatedly be replaced by this block.

But what about picture (c)? Is there any short description that can be given of this picture? Or do we have no choice but just to specify explicitly the color of every one of the cells it contains?

Our powers of visual perception certainly do not reveal any significant regularities that would allow us to construct a shorter description. And neither, it turns out, do any standard methods of mathematical or statistical analysis. And so for practical purposes we have little choice but just to specify explicitly the color of each cell.

But the fact that no short description can be found by our usual processes of perception and analysis does not in any sense mean that no such description exists at all. And indeed, as it happens, picture (c) in fact allows a very short description. For it can be generated just by starting with a single black cell and then applying a simple two-dimensional cellular automaton rule 250 times.

But does the existence of this short description mean that picture (c) should not be considered random? From a practical point of view the fact that a short description may exist is presumably not too relevant if we can never find this description by any of the methods of perception and analysis that are available to us. But from a conceptual point of view it may seem unsatisfactory to have a definition of randomness that depends on our methods of perception and analysis, and is not somehow absolute.

So one possibility is to define randomness so that something is considered random only if no short description whatsoever exists of it. And before the discoveries in this book such a definition might have seemed not far from our everyday notion of randomness. For we would probably have assumed that anything generated from a sufficiently short description would necessarily look fairly simple. But what we have discovered in this book is that this is absolutely not the case, and that in fact even from rules with very short descriptions it is easy to generate behavior in which our standard methods of perception and analysis recognize no significant regularities.

So to say that something is random only if no short description whatsoever exists of it turns out to be a highly restrictive definition of randomness. And in fact, as I mentioned in Chapter 7, it essentially implies that no process based on definite rules can ever manage to generate randomness when there is no randomness before. For since the rules themselves have a short description, anything generated by following them will also have a correspondingly short description, and will therefore not be considered random according to this definition.

And even if one is not concerned about where randomness might come from, there is still a further problem: it turns out in general to be impossible to determine in any finite way whether any particular thing can ever be generated from a short description. One might imagine that one could always just try running all programs with progressively longer descriptions, and see whether any of them ever generate what one wants. But the problem is that one can never in general tell in advance how many steps of evolution one will need to look at in order to be sure that any particular piece of behavior will not occur. And as a result, no finite process can in general be used to guarantee that there is no short description that exists of a particular thing.

By setting up various restrictions, say on the number of steps of evolution that will be allowed, it is possible to obtain slightly more tractable definitions of randomness. But even in such cases the amount of computational work required to determine whether something should be considered random is typically astronomically large. And more important, while such definitions may perhaps be of some conceptual interest, they correspond very poorly with our intuitive notion of randomness. In fact, if one followed such a definition most of the pictures in this book that I have said look random—including for example picture (c) on page 553—would be considered not random. And following the discussion of Chapter 7, so would at least many of the phenomena in nature that we normally think of as random.

Indeed, what I suspect is that ultimately no useful definition of randomness can be based solely on the issue of what short descriptions of something may in principle exist. Rather, any useful definition must, I believe, make at least some reference to how such short descriptions are supposed to be found.

Over the years, a variety of definitions of randomness have been proposed that are based on the absence of certain specific regularities. Often these definitions are presented as somehow being fundamental. But in fact they typically correspond just to seeing whether some particular process—and usually a rather simple one—succeeds in recognizing regularities and thus in generating a shorter description.

A common example—to be discussed further two sections from now—involves taking, say, a sequence of black and white cells, and then counting the frequency with which each color and each block of colors occurs. Any deviation from equality among these frequencies represents a regularity in the sequence and reveals nonrandomness. But despite some confusion in the past it is certainly not true that just checking equality of frequencies of blocks of colors—even arbitrarily long ones—is sufficient to ensure that no regularities at all exist. This

Note: page numbers in the text refer to the original work *A New Kind of Science* (2002).

procedure can indeed be used to check that no purely repetitive pattern exists, but as we will see later in this chapter, it does not successfully detect the presence of even certain highly regular nested patterns.

So how then can we develop a useful yet precise definition of randomness? What we need is essentially just a precise version of the statement at the beginning of this section: that something should be considered random if none of our standard methods of perception and analysis succeed in detecting any regularities in it. But how can we ever expect to find any kind of precise general characterization of what all our various standard methods of perception and analysis do?

The key point that will emerge in this chapter is that in the end essentially all these methods can be viewed as being based on rather simple programs. So this suggests a definition that can be given of randomness: something should be considered to be random whenever there is essentially no simple program that can succeed in detecting regularities in it.

Usually if what one is studying was itself created by a simple program then there will be a few closely related programs that always succeed in detecting regularities. But if something can reasonably be considered random, then the point is that the vast majority of simple programs should not be able to detect any regularities in it.

So does one really need to try essentially all sufficiently simple programs in order to determine this? In my experience, the answer tends to be no. For once a few simple programs corresponding to a few standard methods of perception and analysis have failed to detect regularities, it is extremely rare for any other simple program to succeed in detecting them.

So this means that the everyday definition of randomness that we discussed at the very beginning of this section is in the end already quite unambiguous. For it typically will not matter much which of the standard methods of perception and analysis we use: after trying a few of them we will almost always be in a position to come to a quite definite conclusion about whether or not something should be considered random.

NOTES FOR CHAPTER 10 OF *A NEW KIND OF SCIENCE*

Processes of Perception and Analysis

Defining the Notion of Randomness

■ **Page 554 · Algorithmic information theory.** A description of a piece of data can always be thought of as some kind of program for reproducing the data. So if one could find the shortest program that works then this must correspond to the shortest possible description of the data—and in algorithmic information theory if this is no shorter than the data itself then the data is considered to be algorithmically random.

How long the shortest program is for a given piece of data will in general depend on what system is supposed to run the program. But in a sense the program will on the whole be as short as possible if the system is universal (see page 642). And between any two universal systems programs can differ in length by at most a constant: for one can always just add a fixed interpreter program to the programs for one system in order to make them run on the other system.

As mentioned in the main text, any data generated by a simple program can by definition never be algorithmically random. And so even though algorithmic randomness is often considered in theoretical discussions (see note below) it cannot be directly relevant to the kind of randomness we see in so many systems in this book—or, I believe, in nature.

If one considers all 2^n possible sequences (say of 0's and 1's) of length n then it is straightforward to see that most of them must be more or less algorithmically random. For in order to have enough programs to generate all 2^n sequences most of the programs one uses must themselves be close to length n. (In practice there are subtleties associated with the encoding of programs that make this hold only for sufficiently large n.) But even though one knows that almost all long sequences must be algorithmically random, it turns out to be undecidable in general whether any particular sequence is algorithmically random. For in general one can give no upper limit to how much computational effort one might have to expend in order to find out whether any given short

program—after any number of steps—will generate the sequence one wants.

But even though one can never expect to construct them explicitly, one can still give formal descriptions of sequences that are algorithmically random. An example due to Gregory Chaitin is the digits of the fraction Ω of initial conditions for which a universal system halts (essentially a compressed version—with various subtleties about limits—of the sequence from page 1127 giving the outcome for each initial condition). As emphasized by Chaitin, it is possible to ask questions purely in arithmetic (say about sequences of values of a parameter that yield infinite numbers of solutions to an integer equation) whose answers would correspond to algorithmically random sequences. (See page 786.)

As a reduced analog of algorithmic information theory one can for example ask what the simplest cellular automaton rule is that will generate a given sequence if started from a single black cell. Page 1186 gives some results, and suggests that sequences which require more complicated cellular automaton rules do tend to look to us more complicated and more random.

■ **History.** Randomness and unpredictability were discussed as general notions in antiquity in connection both with questions of free will (see page 1135) and games of chance. When probability theory emerged in the mid-1600s it implicitly assumed sequences random in the sense of having limiting frequencies following its predictions. By the 1800s there was extensive debate about this, but in the early 1900s with the advent of statistical mechanics and measure theory the use of ensembles (see page 1020) turned discussions of probability away from issues of randomness in individual sequences. With the development of statistical hypothesis testing in the early 1900s various tests for randomness were proposed (see page 1084). Sometimes these were claimed to have some kind of general significance, but mostly they were just viewed as simple practical methods. In many fields

Note: page numbers in the text refer to the original work *A New Kind of Science* (2002).

outside of statistics, however, the idea persisted even to the 1990s that block frequencies (or flat frequency spectra) were somehow the only ultimate tests for randomness. In 1909 Emile Borel had formulated the notion of normal numbers (see page 912) whose infinite digit sequences contain all blocks with equal frequency. And in the 1920s Richard von Mises—attempting to capture the observed lack of systematically successful gambling schemes—suggested that randomness for individual infinite sequences could be defined in general by requiring that "collectives" consisting of elements appearing at positions specified by any procedure should show equal frequencies. To disallow procedures say specially set up to pick out all the infinite number of 1's in a sequence Alonzo Church in, 1940 suggested that only procedures corresponding to finite computations be considered. (Compare page 1021 on coarse-graining in thermodynamics.) Starting in the late 1940s the development of information theory began to suggest connections between randomness and inability to compress data, but emphasis on $p\,Log[p]$ measures of information content (see page 1071) reinforced the idea that block frequencies are the only real criterion for randomness. In the early 1960s, however, the notion of algorithmic randomness (see note above) was introduced by Gregory Chaitin, Andrei Kolmogorov and Ray Solomonoff. And unlike earlier proposals the consequences of this definition seemed to show remarkable consistency (in 1966 for example Per Martin-Löf proved that in effect it covered all possible statistical tests)—so that by the early 1990s it had become generally accepted as the appropriate ultimate definition of randomness. In the 1980s, however, work on cryptography had led to the study of some slightly weaker definitions of randomness based on inability to do cryptanalysis or make predictions with polynomial-time computations (see page 1089). But quite what the relationship of any of these definitions might be to natural science or everyday experience was never much discussed. Note that definitions of randomness given in dictionaries tend to emphasize lack of aim or purpose, in effect following the common legal approach of looking at underlying intentions (or say at physical construction of dice) rather than trying to tell if things are random from their observed behavior.

■ **Inevitable regularities and Ramsey theory.** One might have thought that there could be no meaningful type of regularity that would be present in all possible data of a given kind. But through the development since the late 1920s of Ramsey theory it has become clear that this is not the case. As one example, consider looking for runs of m equally spaced squares of the same color embedded in sequences of black and white squares of length n. The pictures below show results with $m = 3$ for various n. For $n < 9$ there are always some sequences in which no runs of length 3 exist. But it turns out that for $n \geq 9$ every single possible sequence contains at least one run of length 3. For any m the same is true for sufficiently large n; it is known that $m = 4$ requires $n \geq 35$ and $m = 5$ requires $n \geq 178$. (In problems like this the analog of n often grows extremely rapidly with m.) If one has a sufficiently long sequence, therefore, just knowing that a run of equally spaced identical elements exists in it does not narrow down at all what the sequence actually is, and can so cannot ultimately be considered a useful regularity.

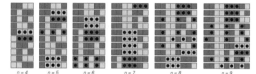

$n = 4$ $n = 5$ $n = 6$ $n = 7$ $n = 8$ $n = 9$

(Compare pattern-avoiding sequences on page 944.)

Note: page numbers in the text refer to the original work *A New Kind of Science* (2002).

9.2 The Notion of Reversibility

At any particular step in the evolution of a system like a cellular automaton the underlying rule for the system tells one how to proceed to the next step. But what if one wants to go backwards? Can one deduce from the arrangement of black and white cells at a particular step what the arrangement of cells must have been on previous steps?

All current evidence suggests that the underlying laws of physics have this kind of reversibility. So this means that given a sufficiently precise knowledge of the state of a physical system at the present time, it is therefore possible to deduce not only what the system will do in the future, but also what it did in the past.

In the first cellular automaton shown below it is also straightforward to do this. For any cell that has one color at a particular step must always have had the opposite color on the step before.

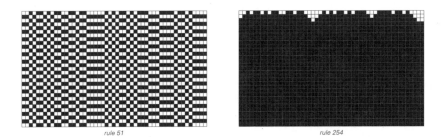

rule 51   *rule 254*

Examples of cellular automata that are and are not reversible. Rule 51 is reversible, so that it preserves enough information to allow one to go backwards from any particular step as well as forwards. Rule 254 is not reversible, since it always evolves to uniform black and preserves no information about the arrangement of cells on earlier steps.

But the second cellular automaton works differently, and does not allow one to go backwards. For after just a few steps, it makes every cell black, regardless of what it was before—with the result that there is no way to tell what color might have occurred on previous steps.

There are many examples of systems in nature which seem to organize themselves a little like the second case above. And indeed the conflict between this and the known reversibility of underlying laws of physics is related to the subject of the next section in this chapter.

But my purpose here is to explore what kinds of systems can be reversible. And of the 256 elementary cellular automata with two colors and nearest-neighbor rules, only the six shown below turn out to be reversible. And as the pictures demonstrate, all of these exhibit fairly trivial behavior, in which only rather simple transformations are ever made to the initial configuration of cells.

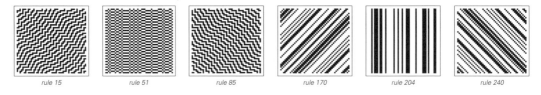

| rule 15 | rule 51 | rule 85 | rule 170 | rule 204 | rule 240 |

Examples of the behavior of the six elementary cellular automata that are reversible. In all cases the transformations made to the initial conditions are simple enough that it is straightforward to go backwards as well as forwards in the evolution.

So is it possible to get more complex behavior while maintaining reversibility? There are a total of 7,625,597,484,987 cellular automata with three colors and nearest-neighbor rules, and searching through these one finds just 1800 that are reversible. Of these 1800, many again exhibit simple behavior, much like the pictures above. But some exhibit more complex behavior, as in the pictures below.

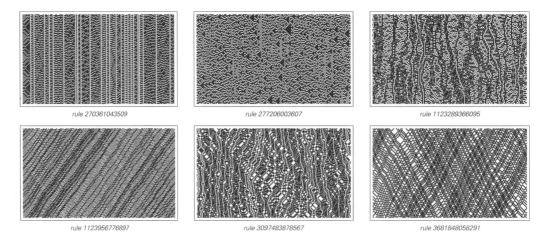

| rule 270361043509 | rule 277206003607 | rule 1123289366095 |
| rule 1123956776897 | rule 3097483878567 | rule 3681848058291 |

Examples of some of the 1800 reversible cellular automata with three colors and nearest-neighbor rules. Even though these systems exhibit complex behavior that scrambles the initial conditions, all of them are still reversible, so that starting from the configuration of cells at the bottom of each picture, it is always possible to deduce the configurations on all previous steps.

How can one now tell that such systems are reversible? It is no longer true that their evolution leads only to simple transformations of the initial conditions. But one can still check that starting with the specific configuration of cells at the bottom of each picture, one can evolve backwards to get to the top of the picture. And given a particular rule it turns out to be fairly straightforward to do a detailed analysis that allows one to prove or disprove its reversibility.

But in trying to understand the range of behavior that can occur in reversible systems it is often convenient to consider classes of cellular automata with rules that are specifically constructed to be reversible. One such class is illustrated below. The idea is to have rules that explicitly remain the same even if they are turned upside-down, thereby interchanging the roles of past and future.

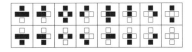

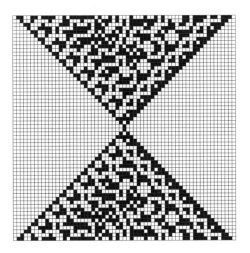

An example of a cellular automaton that is explicitly set up to be reversible. The rule for the system remains unchanged if all its elements are turned upside-down—effectively interchanging the roles of past and future. Patterns produced by the rule must exhibit the same time reversal symmetry, as shown on the left. The specific rule used here is based on taking elementary rule 214, then adding the specification that the new color of a cell should be inverted whenever the cell was black two steps back. Note that by allowing a total of four rather than two colors, a version of the rule that depends only on the immediately preceding step can be constructed.

Such rules can be constructed by taking ordinary cellular automata and adding dependence on colors two steps back.

The resulting rules can be run both forwards and backwards. In each case they require knowledge of the colors of cells on not one but two successive steps. Given this knowledge, however, the rules can be used to determine the configuration of cells on either future or past steps.

The next two pages show examples of the behavior of such cellular automata with both random and simple initial conditions.

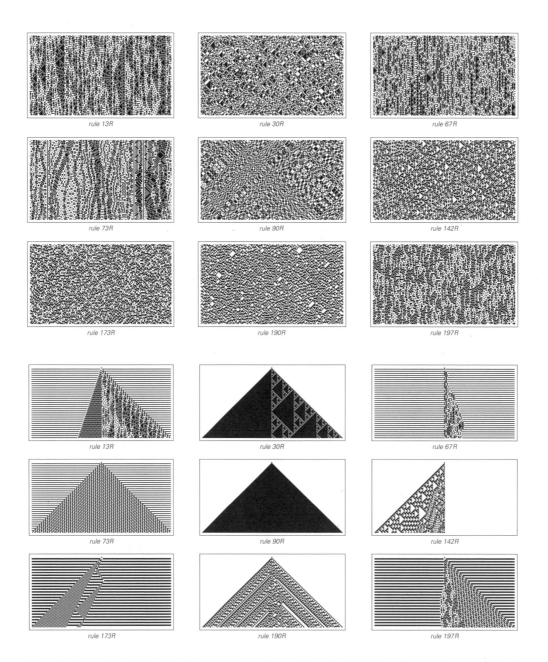

Examples of reversible cellular automata starting from random and from simple initial conditions. In the upper block of pictures, every cell is chosen to be black or white with equal probability on the two successive first steps. In the lower block of pictures, only the center cell is taken to be black on these steps.

rule 150R

rule 154R

rule 214R

The evolution of three reversible cellular automata for 300 steps. In the first case, a regular nested pattern is obtained. In the other cases, the patterns show many features of randomness.

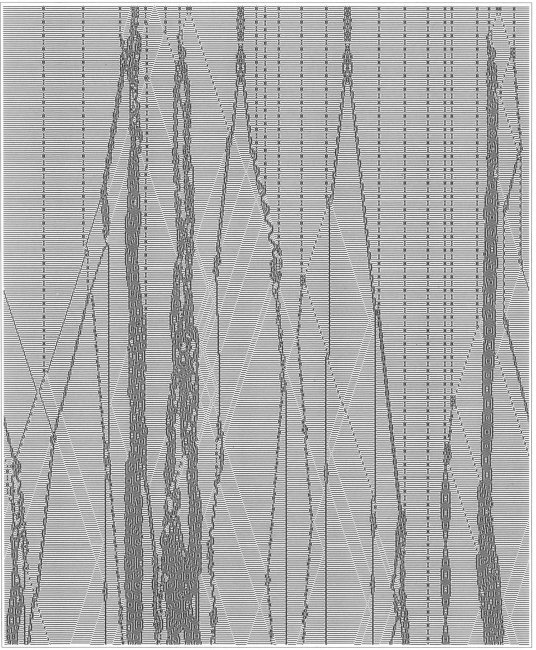

rule 37R

An example of a reversible cellular automaton whose evolution supports localized structures. Because of the reversibility of the underlying rule, every collision must be able to occur equally well when its initial and final states are interchanged.

In some cases, the behavior is fairly simple, and the patterns obtained have simple repetitive or nested structures. But in many cases, even with simple initial conditions, the patterns produced are highly complex, and seem in many respects random.

The reversibility of the underlying rules has some obvious consequences, such as the presence of triangles pointing sideways but not down. But despite their reversibility, the rules still manage to produce the kinds of complex behavior that we have seen in cellular automata and many other systems throughout this book.

So what about localized structures?

The picture on the facing page demonstrates that these can also occur in reversible systems. There are some constraints on the details of the kinds of collisions that are possible, but reversible rules typically tend to work very much like ordinary ones.

So in the end it seems that even though only a very small fraction of possible systems have the property of being reversible, such systems can still exhibit behavior just as complex as one sees anywhere else.

9.3 Irreversibility and the Second Law of Thermodynamics

All the evidence we have from particle physics and elsewhere suggests that at a fundamental level the laws of physics are precisely reversible. Yet our everyday experience is full of examples of seemingly irreversible phenomena. Most often, what happens is that a system which starts in a fairly regular or organized state becomes progressively more and more random and disorganized. And it turns out that this phenomenon can already be seen in many simple programs.

The picture at the top of the next page shows an example based on a reversible cellular automaton of the type discussed in the previous section. The black cells in this system act a little like particles which bounce around inside a box and interact with each other when they collide.

At the beginning the particles are placed in a simple arrangement at the center of the box. But over the course of time the picture shows that the arrangement of particles becomes progressively more random.

A reversible cellular automaton that exhibits seemingly irreversible behavior. Starting from an initial condition in which all black cells or particles lie at the center of a box, the distribution becomes progressively more random. Such behavior appears to be the central phenomenon responsible for the Second Law of Thermodynamics. The specific cellular automaton used here is rule 122R. The system is restricted to a region of size 100 cells.

Typical intuition from traditional science makes it difficult to understand how such randomness could possibly arise. But the discovery in this book that a wide range of systems can generate randomness even with very simple initial conditions makes it seem considerably less surprising.

But what about reversibility? The underlying rules for the cellular automaton used in the picture above are precisely reversible. Yet the picture itself does not at first appear to be at all reversible. For there appears to be an irreversible increase in randomness as one goes down successive panels on the page.

The resolution of this apparent conflict is however fairly straightforward. For as the picture on the facing page demonstrates, if the

An extended version of the picture on the facing page, in which the reversibility of the underlying cellular automaton is more clearly manifest. An initial condition is carefully constructed so that halfway through the evolution shown a simple arrangement of particles will be produced. If one starts with this arrangement, then the randomness of the system will effectively increase whether one goes forwards or backwards in time from that point.

simple arrangement of particles occurs in the middle of the evolution, then one can readily see that randomness increases in exactly the same way—whether one goes forwards or backwards from that point.

Yet there is still something of a mystery. For our everyday experience is full of examples in which randomness increases much as in the second half of the picture above. But we essentially never see the kind of systematic decrease in randomness that occurs in the first half.

By setting up the precise initial conditions that exist at the beginning of the whole picture it would certainly in principle be possible to get such behavior. But somehow it seems that initial conditions like these essentially never actually occur in practice.

There has in the past been considerable confusion about why this might be the case. But the key to understanding what is going on is simply to realize that one has to think not only about the systems one is studying, but also about the types of experiments and observations that one uses in the process of studying them.

The crucial point then turns out to be that practical experiments almost inevitably end up involving only initial conditions that are fairly simple for us to describe and construct. And with these types of initial conditions, systems like the one on the previous page always tend to exhibit increasing randomness.

But what exactly is it that determines the types of initial conditions that one can use in an experiment? It seems reasonable to suppose that in any meaningful experiment the process of setting up the experiment should somehow be simpler than the process that the experiment is intended to observe.

But how can one compare such processes? The answer that I will develop in considerable detail later in this book is to view all such processes as computations. The conclusion is then that the computation involved in setting up an experiment should be simpler than the computation involved in the evolution of the system that is to be studied by the experiment.

It is clear that by starting with a simple state and then tracing backwards through the actual evolution of a reversible system one can find initial conditions that will lead to decreasing randomness. But if one looks for example at the pictures on the last couple of pages the complexity of the behavior seems to preclude any less arduous way of finding such initial conditions. And indeed I will argue in Chapter 12 that the Principle of Computational Equivalence suggests that in general no such reduced procedure should exist.

The consequence of this is that no reasonable experiment can ever involve setting up the kind of initial conditions that will lead to decreases in randomness, and that therefore all practical experiments will tend to show only increases in randomness.

It is this basic argument that I believe explains the observed validity of what in physics is known as the Second Law of Thermodynamics. The law was first formulated more than a century ago, but despite many related technical results, the basic reasons for its validity have until now remained rather mysterious.

The field of thermodynamics is generally concerned with issues of heat and energy in physical systems. A fundamental fact known since the mid-1800s is that heat is a form of energy associated with the random microscopic motions of large numbers of atoms or other particles.

One formulation of the Second Law then states that any energy associated with organized motions of such particles tends to degrade irreversibly into heat. And the pictures at the beginning of this section show essentially just such a phenomenon. Initially there are particles which move in a fairly regular and organized way. But as time goes on, the motion that occurs becomes progressively more random.

There are several details of the cellular automaton used above that differ from actual physical systems of the kind usually studied in thermodynamics. But at the cost of some additional technical complication, it is fairly straightforward to set up a more realistic system.

The pictures on the next two pages show a particular two-dimensional cellular automaton in which black squares representing particles move around and collide with each other, essentially like particles in an ideal gas. This cellular automaton shares with the cellular automaton at the beginning of the section the property of being reversible. But it also has the additional feature that in every collision the total number of particles in it remains unchanged. And since each particle can be thought of as having a certain energy, it follows that the total energy of the system is therefore conserved.

In the first case shown, the particles are taken to bounce around in an empty square box. And it turns out that in this particular case only very simple repetitive behavior is ever obtained. But almost any change destroys this simplicity.

And in the second case, for example, the presence of a small fixed obstacle leads to rapid randomization in the arrangement of particles—very much like the randomization we saw in the one-dimensional cellular automaton that we discussed earlier in this section.

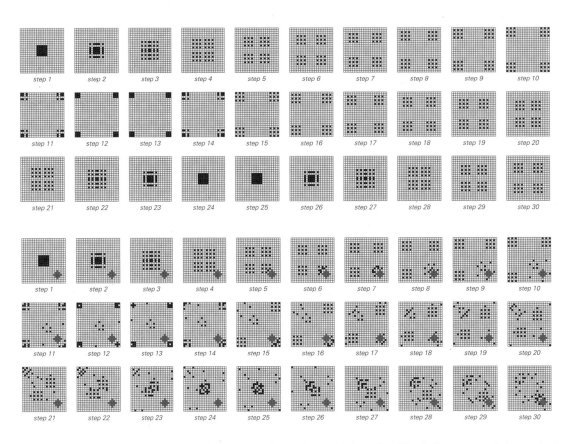

The behavior of a simple two-dimensional cellular automaton that emulates an ideal gas of particles. In the top group of pictures, the particles bounce around in an empty square box. In the bottom group of pictures, the box contains a small fixed obstacle. In the top group of pictures, the arrangement of particles shows simple repetitive behavior. In the bottom group, however, it becomes progressively more random with time. The underlying rules for the cellular automaton used here are reversible, and conserve the total number of particles. The specific rules are based on 2 × 2 blocks—a two-dimensional generalization of the block cellular automata to be discussed in the next section. For each 2 × 2 block the configuration of particles is taken to remain the same at a particular step unless there are exactly two particles arranged diagonally within the block, in which case the particles move to the opposite diagonal.

So even though the total of the energy of all particles remains the same, the distribution of this energy becomes progressively more random, just as the usual Second Law implies.

An important practical consequence of this is that it becomes increasingly difficult to extract energy from the system in the form of systematic mechanical work. At an idealized level one might imagine trying to do this by inserting into the system some kind of paddle which would experience force as a result of impacts from particles.

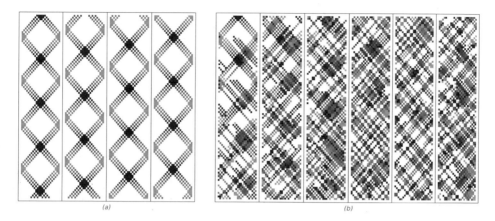

Time histories of the cellular automata from the facing page. In each case a slice is taken through the midline of the box. Black cells that are further from the midline are shown in progressively lighter shades of gray. Case (a) corresponds to an empty square box, and shows simple repetitive behavior. Case (b) corresponds to a box containing a fixed obstacle, and in this case rapid randomization is seen. Each panel corresponds to 100 steps in the evolution of the system; the box is 24 cells across.

The pictures below show how such force might vary with time in cases (a) and (b) above. In case (a), where no randomization occurs, the force can readily be predicted, and it is easy to imagine harnessing it to produce systematic mechanical work. But in case (b), the force quickly randomizes, and there is no obvious way to obtain systematic mechanical work from it.

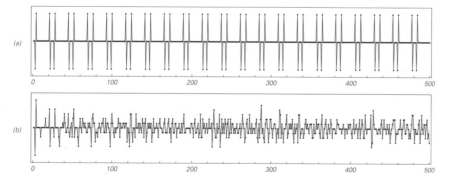

The force on an idealized paddle placed on the midline of the systems shown above. The force reflects an imbalance in the number of particles at each step arriving at the midline from above and below. In case (a) this imbalance is readily predictable. In case (b), however, it rapidly becomes for most practical purposes random. This randomness is essentially what makes it impossible to build a physical perpetual motion machine which continually turns heat into mechanical work.

One might nevertheless imagine that it would be possible to devise a complicated machine, perhaps with an elaborate arrangement of paddles, that would still be able to extract systematic mechanical work even from an apparently random distribution of particles. But it turns out that in order to do this the machine would effectively have to be able to predict where every particle would be at every step in time.

And as we shall discuss in Chapter 12, this would mean that the machine would have to perform computations that are as sophisticated as those that correspond to the actual evolution of the system itself. The result is that in practice it is never possible to build perpetual motion machines that continually take energy in the form of heat—or randomized particle motions—and convert it into useful mechanical work.

The impossibility of such perpetual motion machines is one common statement of the Second Law of Thermodynamics. Another is that a quantity known as entropy tends to increase with time.

Entropy is defined as the amount of information about a system that is still unknown after one has made a certain set of measurements on the system. The specific value of the entropy will depend on what measurements one makes, but the content of the Second Law is that if one repeats the same measurements at different times, then the entropy deduced from them will tend to increase with time.

If one managed to find the positions and properties of all the particles in the system, then no information about the system would remain unknown, and the entropy of the system would just be zero. But in a practical experiment, one cannot expect to be able to make anything like such complete measurements.

And more realistically, the measurements one makes might for example give the total numbers of particles in certain regions inside the box. There are then a large number of possible detailed arrangements of particles that are all consistent with the results of such measurements. The entropy is defined as the amount of additional information that would be needed in order to pick out the specific arrangement that actually occurs.

We will discuss in more detail in Chapter 10 the notion of amount of information. But here we can imagine numbering all the possible arrangements of particles that are consistent with the results of our measurements, so that the amount of information needed to pick out a single arrangement is essentially the length in digits of one such number.

The pictures below show the behavior of the entropy calculated in this way for systems like the one discussed above. And what we see is that the entropy does indeed tend to increase, just as the Second Law implies.

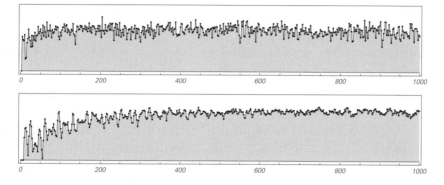

The entropy as a function of time for systems of the type shown in case (b) from page 447. The top plot is exactly for case (b); the bottom one is for a system three times larger in size. The entropy is found in each case by working out how many possible configurations of particles are consistent with measurements of the total numbers of particles in a 6 × 6 grid of regions within the system. Just as the Second Law of Thermodynamics suggests, the entropy tends to increase with time. Note that the plots above would be exactly symmetrical if they were continued to the left: the entropy would increase in the same way going both forwards and backwards from the simple initial conditions used.

In effect what is going on is that the measurements we make represent an attempt to determine the state of the system. But as the arrangement of particles in the system becomes more random, this attempt becomes less and less successful.

One might imagine that there could be a more elaborate set of measurements that would somehow avoid these problems, and would not lead to increasing entropy. But as we shall discuss in Chapter 12, it again turns out that setting up such measurements would have to involve the same level of computational effort as the actual evolution of the system itself. And as a result, one concludes that the entropy associated with measurements done in practical experiments will always tend to increase, as the Second Law suggests.

Note: page numbers in the text refer to the original work *A New Kind of Science* (2002).

In Chapter 12 we will discuss in more detail some of the key ideas involved in coming to this conclusion. But the basic point is that the phenomenon of entropy increase implied by the Second Law is a more or less direct consequence of the phenomenon discovered in this book that even with simple initial conditions many systems can produce complex and seemingly random behavior.

One aspect of the generation of randomness that we have noted several times in earlier chapters is that once significant randomness has been produced in a system, the overall properties of that system tend to become largely independent of the details of its initial conditions.

In any system that is reversible it must always be the case that different initial conditions lead to at least slightly different states—otherwise there would be no unique way of going backwards. But the point is that even though the outcomes from different initial conditions differ in detail, their overall properties can still be very much the same.

The pictures on the facing page show an example of what can happen. Every individual picture has different initial conditions. But whenever randomness is produced the overall patterns that are obtained look in the end almost indistinguishable.

The reversibility of the underlying rules implies that at some level it must be possible to recognize outcomes from different kinds of initial conditions. But the point is that to do so would require a computation far more sophisticated than any that could meaningfully be done as part of a practical measurement process.

So this means that if a system generates sufficient randomness, one can think of it as evolving towards a unique equilibrium whose properties are for practical purposes independent of its initial conditions.

This fact turns out in a sense to be implicit in many everyday applications of physics. For it is what allows us to characterize all sorts of physical systems by just specifying a few parameters such as temperature and chemical composition—and avoids us always having to know the details of the initial conditions and history of each system.

The existence of a unique equilibrium to which any particular system tends to evolve is also a common statement of the Second Law of

The approach to equilibrium in a reversible cellular automaton with a variety of different initial conditions. Apart from exceptional cases where no randomization occurs, the behavior obtained with different initial conditions is eventually quite indistinguishable in its overall properties. Because the underlying rule is reversible, however, the details with different initial conditions are always at least slightly different—otherwise it would not be possible to go backwards in a unique way. The rule used here is 122R. Successive pairs of pictures have initial conditions that differ only in the color of a single cell at the center.

Thermodynamics. And once again, therefore, we find that the Second Law is associated with basic phenomena that we already saw early in this book.

But just how general is the Second Law? And does it really apply to all of the various kinds of systems that we see in nature?

Starting nearly a century ago it came to be widely believed that the Second Law is an almost universal principle. But in reality there is surprisingly little evidence for this.

Indeed, almost all of the detailed applications ever made of the full Second Law have been concerned with just one specific area: the behavior of gases. By now there is therefore good evidence that gases obey the Second Law—just as the idealized model earlier in this section suggests. But what about other kinds of systems?

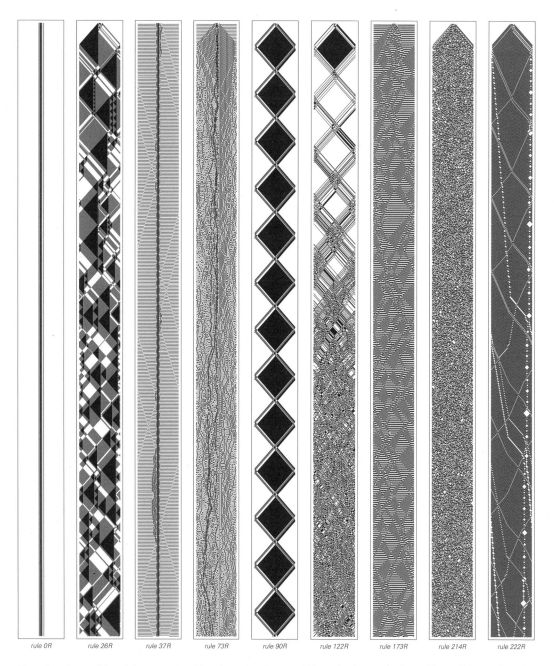

rule 0R   rule 26R   rule 37R   rule 73R   rule 90R   rule 122R   rule 173R   rule 214R   rule 222R

Examples of reversible cellular automata with various rules. Some quickly randomize, as the Second Law of Thermodynamics would suggest. But others do not—and thus in effect do not obey the Second Law of Thermodynamics.

The pictures on the facing page show examples of various reversible cellular automata. And what we see immediately from these pictures is that while some systems exhibit exactly the kind of randomization implied by the Second Law, others do not.

The most obvious exceptions are cases like rule 0R and rule 90R, where the behavior that is produced has only a very simple fixed or repetitive form. And existing mathematical studies have indeed identified these simple exceptions to the Second Law. But they have somehow implicitly assumed that no other kinds of exceptions can exist.

The picture on the next page, however, shows the behavior of rule 37R over the course of many steps. And in looking at this picture, we see a remarkable phenomenon: there is neither a systematic trend towards increasing randomness, nor any form of simple predictable behavior. Indeed, it seems that the system just never settles down, but rather continues to fluctuate forever, sometimes becoming less orderly, and sometimes more so.

So how can such behavior be understood in the context of the Second Law? There is, I believe, no choice but to conclude that for practical purposes rule 37R simply does not obey the Second Law.

And as it turns out, what happens in rule 37R is not so different from what seems to happen in many systems in nature. If the Second Law was always obeyed, then one might expect that by now every part of our universe would have evolved to completely random equilibrium.

Yet it is quite obvious that this has not happened. And indeed there are many kinds of systems, notably biological ones, that seem to show, at least temporarily, a trend towards increasing order rather than increasing randomness.

How do such systems work? A common feature appears to be the presence of some kind of partitioning: the systems effectively break up into parts that evolve at least somewhat independently for long periods of time.

The picture on page 456 shows what happens if one starts rule 37R with a single small region of randomness. And for a while what one sees is that the randomness that has been inserted persists. But eventually the system instead seems to organize itself to yield just a small number of simple repetitive structures.

Note: page numbers in the text refer to the original work *A New Kind of Science* (2002).

| steps 0-3000 | steps 5000-8000 | steps 10000-13000 | steps 20000-23000 | steps 100000-103000 | steps 200000-203000 |

More steps in the evolution of the reversible cellular automaton with rule 37R. This system is an example of one that does not in any meaningful way obey the Second Law of Thermodynamics. Instead of exhibiting progressively more random behavior, it appears to fluctuate between quite ordered and quite disordered states.

This kind of self-organization is quite opposite to what one would expect from the Second Law. And at first it also seems inconsistent with the reversibility of the system. For if all that is left at the end are a few simple structures, how can there be enough information to go backwards and reconstruct the initial conditions?

The answer is that one has to consider not only the stationary structures that stay in the middle of the system, but also all various small structures that were emitted in the course of the evolution. To go backwards one would need to set things up so that one absorbs exactly the sequence of structures that were emitted going forwards.

If, however, one just lets the emitted structures escape, and never absorbs any other structures, then one is effectively losing information. The result is that the evolution one sees can be intrinsically not reversible, so that all of the various forms of self-organization that we saw earlier in this book in cellular automata that do not have reversible rules can potentially occur.

If we look at the universe on a large scale, then it turns out that in a certain sense there is more radiation emitted than absorbed. Indeed, this is related to the fact that the night sky appears dark, rather than having bright starlight coming from every direction. But ultimately the asymmetry between emission and absorption is a consequence of the fact that the universe is expanding, rather than contracting, with time.

The result is that it is possible for regions of the universe to become progressively more organized, despite the Second Law, and despite the reversibility of their underlying rules. And this is a large part of the reason that organized galaxies, stars and planets can form.

Allowing information to escape is a rather straightforward way to evade the Second Law. But what the pictures on the facing page demonstrate is that even in a completely closed system, where no information at all is allowed to escape, a system like rule 37R still does not follow the uniform trend towards increasing randomness that is suggested by the Second Law.

What instead happens is that kinds of membranes form between different regions of the system, and within each region orderly behavior can then occur, at least while the membrane survives.

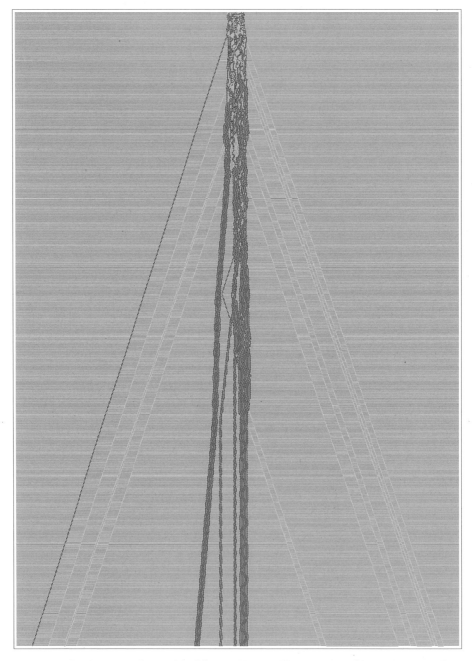

An example of evolution according to rule 37R from an initial condition containing a fairly random region. Even though the system is reversible, this region tends to organize itself so as to take on a much simpler form. Information on the initial conditions ends up being carried by localized structures which radiate outwards.

This basic mechanism may well be the main one at work in many biological systems: each cell or each organism becomes separated from others, and while it survives, it can exhibit organized behavior.

But looking at the pictures of rule 37R on page 454 one may ask whether perhaps the effects we see are just transients, and that if we waited long enough something different would happen.

It is an inevitable feature of having a closed system of limited size that in the end the behavior one gets must repeat itself. And in rules like 0R and 90R shown on page 452 the period of repetition is always very short. But for rule 37R it usually turns out to be rather long. Indeed, for the specific example shown on page 454, the period is 293,216,266.

In general, however, the maximum possible period for a system containing a certain number of cells can be achieved only if the evolution of the system from any initial condition eventually visits all the possible states of the system, as discussed on page 258. And if this in fact happens, then at least eventually the system will inevitably spend most of its time in states that seem quite random.

But in rule 37R there is no such ergodicity. And instead, starting from any particular initial condition, the system will only ever visit a tiny fraction of all possible states. Yet since the total number of states is astronomically large—about 10^{60} for size 100—the number of states visited by rule 37R, and therefore the repetition period, can still be extremely long.

There are various subtleties involved in making a formal study of the limiting behavior of rule 37R after a very long time. But irrespective of these subtleties, the basic fact remains that so far as I can tell, rule 37R simply does not follow the predictions of the Second Law.

And indeed I strongly suspect that there are many systems in nature which behave in more or less the same way. The Second Law is an important and quite general principle—but it is not universally valid. And by thinking in terms of simple programs we have thus been able in this section not only to understand why the Second Law is often true, but also to see some of its limitations.

Note: page numbers in the text refer to the original work *A New Kind of Science* (2002).

9.4 Conserved Quantities and Continuum Phenomena

Reversibility is one general feature that appears to exist in the basic laws of physics. Another is conservation of various quantities—so that for example in the evolution of any closed physical system, total values of quantities like energy and electric charge appear always to stay the same.

With most rules, systems like cellular automata do not usually exhibit such conservation laws. But just as with reversibility, it turns out to be possible to find rules that for example conserve the total number of black cells appearing on each step.

Among elementary cellular automata with just two colors and nearest-neighbor rules, the only types of examples are the fairly trivial ones shown in the pictures below.

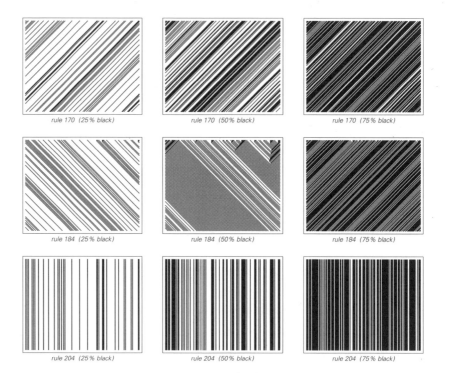

rule 170 (25% black)   rule 170 (50% black)   rule 170 (75% black)

rule 184 (25% black)   rule 184 (50% black)   rule 184 (75% black)

rule 204 (25% black)   rule 204 (50% black)   rule 204 (75% black)

Elementary cellular automata whose evolution conserves the total number of black cells. The behavior of the rules shown here is simple enough that in each case it is fairly obvious how the number of black cells manages to stay the same on every step.

But with next-nearest-neighbor rules, more complicated examples become possible, as the pictures below demonstrate.

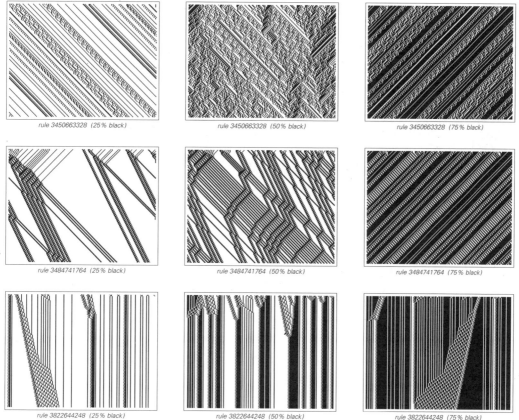

Examples of cellular automata with next-nearest-neighbor rules whose evolution conserves the total number of black cells. Even though it is not immediately obvious by eye, the total number of black cells stays exactly the same on each successive step in each picture. Among the 4,294,967,296 possible next-neighbor rules, only 428 exhibit the kind of conservation property shown here.

One straightforward way to generate collections of systems that will inevitably exhibit conserved quantities is to work not with ordinary cellular automata but instead with block cellular automata. The basic idea of a block cellular automaton is illustrated at the top of the next page. At each step what happens is that blocks of adjacent cells are replaced by other blocks of the same size according to some definite rule. And then on successive steps the alignment of these blocks shifts by one cell.

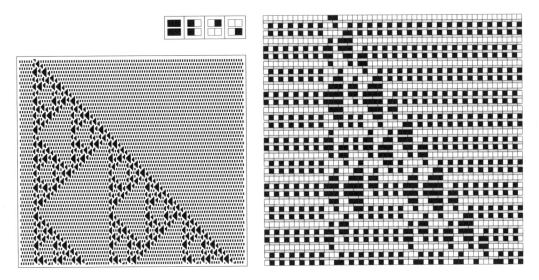

An example of a block cellular automaton. The system works by partitioning the sequence of cells that exists at each step into pairs, then replacing these pairs by other pairs according to the rule shown. The choice of whether to pair a cell with its left or right neighbor alternates on successive steps. Like many block cellular automata, the system shown is reversible, since in the rule each pair has a unique predecessor. It does not, however, conserve the total number of black cells.

And with this setup, if the underlying rules replace each block by one that contains the same number of black cells, it is inevitable that the system as a whole will conserve the total number of black cells.

With two possible colors and blocks of size two the only kinds of block cellular automata that conserve the total number of black cells are the ones shown below—and all of these exhibit rather trivial behavior.

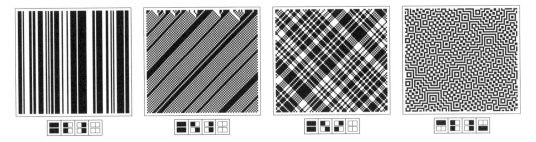

Block cellular automata with two possible colors and blocks of size two that conserve the total number of black cells (the last example has this property only on alternate steps). It so happens that all but the second of the rules shown here not only conserve the total number of black cells but also turn out to be reversible.

But if one allows three possible colors, and requires, say, that the total number of black and gray cells together be conserved, then more complicated behavior can occur, as in the pictures below.

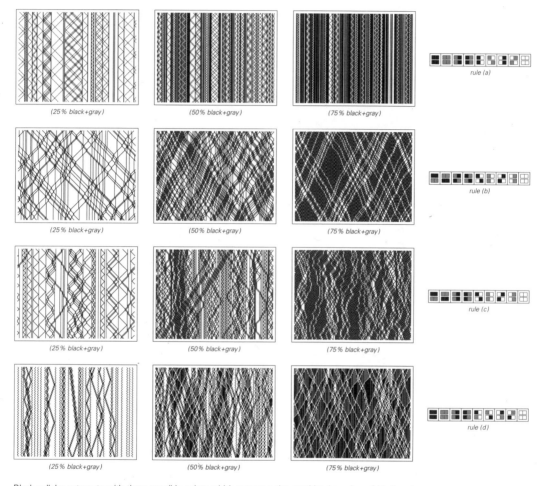

Block cellular automata with three possible colors which conserve the combined number of black and gray cells. In rule (a), black and gray cells remain in localized regions. In rule (b), they move in fairly simple ways, and in rules (c) and (d), they move in a seemingly somewhat random way. The rules shown here are reversible, although their behavior is similar to that of non-reversible rules, at least after a few steps.

Indeed, as the pictures on the next page demonstrate, such systems can produce considerable randomness even when starting from very simple initial conditions.

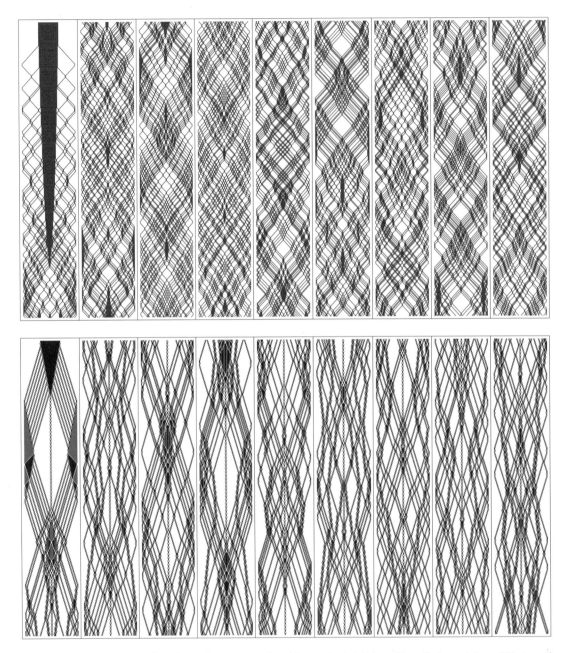

The behavior of rules (c) and (d) from the previous page, starting with very simple initial conditions. Each panel shows 500 steps of evolution, and rapid randomization is evident. The black and gray cells behave much like physical particles: their total number is conserved, and with the particular rules used here, their interactions are reversible. Note that the presence of boundaries is crucial; for without them there would in a sense be no collisions between particles, and the behavior of both systems would be rather trivial.

But there is still an important constraint on the behavior: even though black and gray cells may in effect move around randomly, their total number must always be conserved. And this means that if one looks at the total average density of colored cells throughout the system, it must always remain the same. But local densities in different parts of the system need not—and in general they will change as colored cells flow in and out.

The pictures below show what happens with four different rules, starting with higher density in the middle and lower density on the sides. With rules (a) and (b), each different region effectively remains separated forever. But with rules (c) and (d) the regions gradually mix.

As in many kinds of systems, the details of the initial arrangement of cells will normally have an effect on the details of the behavior that occurs. But what the pictures below suggest is that if one looks only at the overall distribution of density, then these details will become largely irrelevant—so that a given initial distribution of density will always tend to evolve in the same overall way, regardless of what particular arrangement of cells happened to make up that distribution.

rule (a)   rule (b)   rule (c)   rule (d)

The block cellular automata from previous pages started from initial conditions containing regions of different density. In rules (a) and (b) the regions remain separated forever, but in rules (c) and (d) they gradually diffuse into each other.

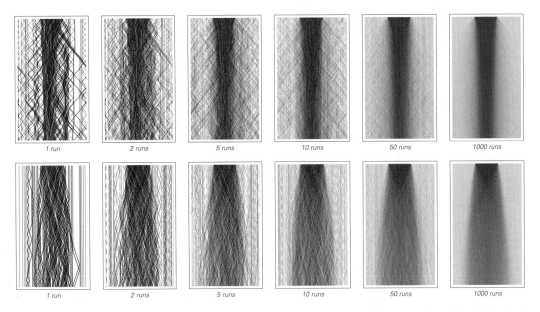

The evolution of overall density for block cellular automata (c) and (d) from the previous page. Even though at an underlying level these systems consist of discrete cells, their overall behavior seems smooth and continuous. The results shown here are obtained by averaging over progressively larger numbers of runs with initial conditions that differ in detail, but have the same overall density distribution. In the limit of an infinite number of runs (or infinite number of cells), the behavior in the second case approaches the form implied by the continuum diffusion equation. (In the first case correlations in effect last too long to yield exactly such behavior.)

The pictures above then show how the average density evolves in systems (c) and (d). And what is striking is that even though at the lowest level both of these systems consist of discrete cells, the overall distribution of density that emerges in both cases shows smooth continuous behavior.

And much as in physical systems like fluids, what ultimately leads to this is the presence of small-scale apparent randomness that washes out details of individual cells or molecules—as well as of conserved quantities that force certain overall features not to change too quickly. And in fact, given just these properties it turns out that essentially the same overall continuum behavior always tends to be obtained.

One might have thought that continuum behavior would somehow rely on special features of actual systems in physics. But in fact what we have seen here is that once again the fundamental mechanisms responsible already occur in a much more minimal way in programs that have some remarkably simple underlying rules.

NOTES FOR CHAPTER 9 OF *A NEW KIND OF SCIENCE*

Fundamental Physics

The Notion of Reversibility

■ **Page 437 · Testing for reversibility.** To show that a cellular automaton is reversible it is sufficient to check that all configurations consisting of repetitions of different blocks have different successors. This can be done for blocks up to length n in a 1D cellular automaton with k colors using

> ReversibleQ[rule_, k_, n_] := Catch[Do[
>   If[Length[Union[Table[CAStep[rule, IntegerDigits[i, k, m]],
>   {i, 0, k^m - 1}]]] ≠ k^m, Throw[False]], {m, n}]; True]

For $k = 2$, $r = 1$ it turns out that it suffices to test only up to $n = 4$ (128 out of the 256 rules fail at $n = 1$, 64 at $n = 2$, 44 at $n = 3$ and 14 at $n = 4$); for $k = 2$, $r = 2$ it suffices to test up to $n = 15$, and for $k = 3$, $r = 1$, up to $n = 9$. But although these results suggest that in general it should suffice to test only up to $n = k^{2r}$, all that has so far been rigorously proved is that $n = k^{2r} (k^{2r} - 1) + 2r + 1$ (or $n = 15$ for $k = 2$, $r = 1$) is sufficient.

For 2D cellular automata an analogous procedure can in principle be used, though there is no upper limit on the size of blocks that need to be tested, and in fact the question of whether a particular rule is reversible is directly equivalent to the tiling problem discussed on page 213 (compare page 942), and is thus formally undecidable.

■ **Numbers of reversible rules.** For $k = 2$, $r = 1$, there are 6 reversible rules, as shown on page 436. For $k = 2$, $r = 2$ there are 62 reversible rules, in 20 families inequivalent under symmetries, out of a total of 2^{32} or about 4 billion possible rules. For $k = 3$, $r = 1$ there are 1800 reversible rules, in 172 families. For $k = 4$, $r = 1$, some of the reversible rules can be constructed from the second-order cellular automata below. Note that for any k and r, no non-trivial totalistic rule can ever be reversible.

■ **Inverse rules.** Some reversible rules are self-inverse, so that applying the same rule twice yields the identity. Other rules come in distinct pairs. Most often a rule that involves r neighbors has an inverse that also involves at most r neighbors. But for both $k = 2$, $r = 2$ and $k = 3$, $r = 1$ there turn out to be reversible rules whose inverses involve larger

numbers of neighbors. For any given rule one can define the neighborhood size s to be the largest block of cells that is ever needed to determine the color of a single new cell. In general $s ≤ 2r + 1$, and for a simple identity or shift rule, $s = 1$. For $k = 2$, $r = 1$, it then turns out that all the reversible rules and their inverses have $s = 1$. For $k = 2$, $r = 2$, the reversible rules have values of s from 1 to 5, but their inverses have values $\bar{s}$ from 1 to 6. There are only 8 rules (the inequivalent ones being 16740555 and 3327051468) where $\bar{s} > s$, and in each case $\bar{s} = 6$ while $s = 5$. For $k = 3$, $r = 1$, there are a total of 936 rules with this property: 576, 216 and 144 with $\bar{s} = 4$, 5 and 6, and in all cases $s = 3$. Examples with $\bar{s} = 3$, 4, 5 and 6 are shown below. For arbitrary k and r, it is not clear what the maximum $\bar{s}$ can be; the only bound rigorously established so far is $\bar{s} ≤ r + 1/2 k^{2r+1} (k^{2r} - 1)$.

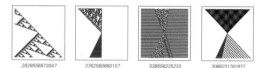

2828556973047   3762560660157   538556225233   3066231781977

■ **Surjectivity and injectivity.** See page 959.

■ **Directional reversibility.** Even if successive time steps in the evolution of a cellular automaton do not correspond to an injective map, it is still possible to get an injective map by looking at successive lines at some angle in the spacetime evolution of the system. Examples where this works include the surjective rules 30 and 90.

■ **Page 437 · Second-order cellular automata.** Second-order elementary rules can be implemented using

> CA2EvolveList[rule_List, {a_List, b_List}, t_Integer] :=
>   Map[First, NestList[CA2Step[rule, #] &, {a, b}, t]]
> CA2Step[rule_List, {a_, b_}] := {b, Mod[a + rule[[
>   8 - (RotateLeft[b] + 2 (b + 2 RotateRight[b]))]], 2]}

where *rule* is obtained from the rule number using *IntegerDigits[n, 2, 8]*.

Note: page numbers in the text refer to the original work *A New Kind of Science* (2002).

535

The combination *Drop[list, -1] + 2 Drop[list, 1]* of the result from *CA2EvolveList* corresponds to evolution according to a first-order $k = 4$, $r = 1$ rule.

▪ **History.** The concept of getting reversibility in a cellular automaton by having a second-order rule was apparently first suggested by Edward Fredkin around 1970 in the context of 2D systems—on the basis of an analogy with second-order differential equations in physics. Similar ideas had appeared in numerical analysis in the 1960s in connection with so-called symmetric or self-adjoint discrete approximations to differential equations.

▪ **Page 438 · Properties.** The pattern from rule 67R with simple initial conditions grows irregularly, at an average rate of about 1 cell every 5 steps. The right-hand side of the pattern from rule 173R consists three triangles that repeat progressively larger at steps of the form $2(9^s - 1)$. Rule 90R has the property that of the diamond of cells at relative positions *{{-n, 0}, {0, -n}, {n, 0}, {0, n}}* it is always true for any n that an even number are black.

▪ **Page 439 · Properties.** The initial conditions used here have a single black cell on two successive initial steps. For rule 150R, however, there is no black cell on the first initial step. The pattern generated by rule 150R has fractal dimension $Log[2, 3 + \sqrt{17}] - 1$ or about 1.83. In rule 154R, each diagonal stripe is followed by at least one 0; otherwise, the positions of the stripes appear to be quite random, with a density around 0.44.

▪ **Generalized additive rules.** Additive cellular automata of the kind discussed on page 952 can be generalized by allowing the new value of each cell to be obtained from combinations of cells on s previous steps. For rule 90 the combination c can be specified as *{{1, 0, 1}}*, while for rule 150R it can be specified as *{{0, 1, 0}, {1, 1, 1}}*. All generalized additive rules ultimately yield nested patterns. Starting with a list of the initial conditions for s steps, the configurations for the next s steps are given by

```
Append[Rest[list],
  Map[Mod[Apply[Plus, Flatten[c #]], 2] &, Transpose[
    Table[RotateLeft[list, {0, i}], {i, -r, r}], {3, 2, 1}]]]
```

where *r = (Length[First[c]] - 1)/2*.

Just as for ordinary additive rules on page 1091, an algebraic analysis for generalized additive rules can be given. The objects that appear are solutions to linear recurrences of order s, and in general involve s^{th} roots. For rule 150R, the configuration at step t as shown in the picture on page 439 is given by $(u^t - v^t)/Sqrt[4 + h^2]$, where $\{u, v\} = z$ /. $Solve[z^2 == h z + 1]$ and $h = 1/x + 1 + x$. (See also page 1078.)

▪ **Page 440 · Rule 37R.** Complicated structures are fairly easy to get with this rule. The initial condition *{1, 0, 1}* with all cells 0 on the previous step yields a structure that repeats but only every 666 steps. The initial condition *{{0, 1, 1}, {1, 0, 0}}* yields a pattern that grows sporadically for 3774 steps, then breaks into two repetitive structures. The typical background repeats every 3 steps.

▪ **Classification of reversible rules.** In a reversible system it is possible with suitable initial conditions to get absolutely any arrangement of cells to appear at any step. Despite this, however, the overall spacetime pattern of cells is not arbitrary, but is instead determined by the underlying rules. If one starts with completely random initial conditions then class 2 and class 3 behavior are often seen. Class 1 behavior can never occur in a reversible system. Class 4 behavior can occur, as in rule 37R, but is typically obvious only if one starts say with a low density of black cells.

For arbitrary rules, difference patterns of the kind shown on page 250 can get both larger and smaller. In a reversible rule, such patterns can grow and shrink, but can never die out completely.

▪ **Emergence of reversibility.** Once on an attractor, any system—even if it does not have reversible underlying rules—must in some sense show approximate reversibility. (Compare page 959.)

▪ **Other reversible systems.** Reversible examples can be found of essentially all the types of systems discussed in this book. Reversible mobile automata can for instance be constructed using

```
Table[(IntegerDigits[i, 2, 3] → If[First[#] == 0, {#, -1},
    {Reverse[#], 1}] &)[IntegerDigits[perm[[i]], 2, 3]], {i, 8}]
```

where *perm* is an element of *Permutations[Range[8]]*. An example that exhibits complex behavior is:

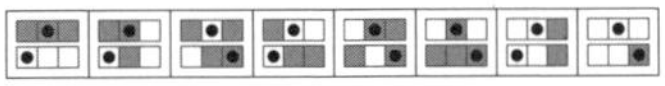

Systems based on numbers are typically reversible whenever the mathematical operations they involve are invertible. Thus, for example, the system on page 121 based on successive multiplication by 3/2 is reversible by using division by 3/2. Page 905 gives another example of a reversible system based on numbers.

Multiway systems are reversible whenever both $a \to b$ and $b \to a$ are present as rules, so that the system corresponds mathematically to a semigroup. (See page 938.)

▪ **Reversible computation.** Typical practical computers—and computer languages—are not even close to reversible: many inputs can lead to the same output, and there is no unique

Note: page numbers in the text refer to the original work *A New Kind of Science* (2002).

way to undo the steps of a computation. But despite early confusion (see page 1020), it has been known since at least the 1970s that there is nothing in principle which prevents computation from being reversible. And indeed—just like with the cellular automata in this section—most of the systems in Chapter 11 that exhibit universal computation can readily be made reversible with only slight overhead.

Irreversibility and the Second Law of Thermodynamics

■ **Time reversal invariance.** The reversibility of the laws of physics implies that given the state of a physical system at a particular time, it is always possibly to work out uniquely both its future and its past. Time reversal invariance would further imply that the rules for going in each direction should be identical. To a very good approximation this appears to be true, but it turns out that in certain esoteric particle physics processes small deviations have been found. In particular, it was discovered in 1964 that the decay of the K^0 particle violated time reversal invariance at the level of about one part in a thousand. In current theories, this effect is not attributed any particularly fundamental origin, and is just assumed to be associated with the arbitrary setting of certain parameters. K^0 decay was for a long time the only example of time reversal violation that had explicitly been seen, although recently examples in B particle decays have probably also been seen. It also turns out that the only current viable theories of the apparent preponderance of matter over antimatter in the universe are based on the idea that a small amount of time reversal violation occurred in the decays of certain very massive particles in the very early universe.

The basic formalism used for particle physics assumes not only reversibility, but also so-called CPT invariance. This means that same rules should apply if one not only reverses the direction of time (T), but also simultaneously inverts all spatial coordinates (P) and conjugates all charges (C), replacing particles by antiparticles. In a certain mathematical sense, CPT invariance can be viewed as a generalization of relativistic invariance: with a speed faster than light, something close to an ordinary relativistic transformation is a CPT transformation.

Originally it was assumed that C, P and T would all separately be invariances, as they are in classical mechanics. But in 1957 it was discovered that in radioactive beta decay, C and P are in a sense each maximally violated: among other things, the correlation between spin and motion direction is exactly opposite for neutrinos and for antineutrinos that are emitted. Despite this, it was still assumed that CP and T

would be true invariances. But in 1964 these too were found to be violated. Starting with a pure beam of K^0 particles, it turns out that quantum mechanical mixing processes lead after about 10^{-8} seconds to a certain mixture of $\overline{K}^0$ particles— the antiparticles of the K^0. And what effectively happens is that the amount of mixing differs by about 0.1% in the positive and negative time directions. (What is actually observed is a small probability for the long-lived component of a K^0 beam to decay into two rather than three pions. Some analysis is required to connect this with T violation.) Particle physics experiments so far support exact CPT invariance. Simple models of gravity potentially suggest CPT violation (as a consequence of deviations from pure special relativistic invariance), but such effects tend to disappear when the models are refined.

■ **History of thermodynamics.** Basic physical notions of heat and temperature were established in the 1600s, and scientists of the time appear to have thought correctly that heat is associated with the motion of microscopic constituents of matter. But in the 1700s it became widely believed that heat was instead a separate fluid-like substance. Experiments by James Joule and others in the 1840s put this in doubt, and finally in the 1850s it became accepted that heat is in fact a form of energy. The relation between heat and energy was important for the development of steam engines, and in 1824 Sadi Carnot had captured some of the ideas of thermodynamics in his discussion of the efficiency of an idealized engine. Around 1850 Rudolf Clausius and William Thomson (Kelvin) stated both the First Law—that total energy is conserved—and the Second Law of Thermodynamics. The Second Law was originally formulated in terms of the fact that heat does not spontaneously flow from a colder body to a hotter. Other formulations followed quickly, and Kelvin in particular understood some of the law's general implications. The idea that gases consist of molecules in motion had been discussed in some detail by Daniel Bernoulli in 1738, but had fallen out of favor, and was revived by Clausius in 1857. Following this, James Clerk Maxwell in 1860 derived from the mechanics of individual molecular collisions the expected distribution of molecular speeds in a gas. Over the next several years the kinetic theory of gases developed rapidly, and many macroscopic properties of gases in equilibrium were computed. In 1872 Ludwig Boltzmann constructed an equation that he thought could describe the detailed time development of a gas, whether in equilibrium or not. In the 1860s Clausius had introduced entropy as a ratio of heat to temperature, and had stated the Second Law in terms of the increase of this quantity. Boltzmann then showed that his

Note: page numbers in the text refer to the original work *A New Kind of Science* (2002).

537

equation implied the so-called H Theorem, which states that a quantity equal to entropy in equilibrium must always increase with time. At first, it seemed that Boltzmann had successfully proved the Second Law. But then it was noticed that since molecular collisions were assumed reversible, his derivation could be run in reverse, and would then imply the opposite of the Second Law. Much later it was realized that Boltzmann's original equation implicitly assumed that molecules are uncorrelated before each collision, but not afterwards, thereby introducing a fundamental asymmetry in time. Early in the 1870s Maxwell and Kelvin appear to have already understood that the Second Law could not formally be derived from microscopic physics, but must somehow be a consequence of human inability to track large numbers of molecules. In responding to objections concerning reversibility Boltzmann realized around 1876 that in a gas there are many more states that seem random than seem orderly. This realization led him to argue that entropy must be proportional to the logarithm of the number of possible states of a system, and to formulate ideas about ergodicity. The statistical mechanics of systems of particles was put in a more general context by Willard Gibbs, beginning around 1900. Gibbs introduced the notion of an ensemble—a collection of many possible states of a system, each assigned a certain probability. He argued that if the time evolution of a single state were to visit all other states in the ensemble—the so-called ergodic hypothesis—then averaged over a sufficiently long time a single state would behave in a way that was typical of the ensemble. Gibbs also gave qualitative arguments that entropy would increase if it were measured in a "coarse-grained" way in which nearby states were not distinguished. In the early 1900s the development of thermodynamics was largely overshadowed by quantum theory and little fundamental work was done on it. Nevertheless, by the 1930s, the Second Law had somehow come to be generally regarded as a principle of physics whose foundations should be questioned only as a curiosity. Despite neglect in physics, however, ergodic theory became an active area of pure mathematics, and from the 1920s to the 1960s properties related to ergodicity were established for many kinds of simple systems. When electronic computers became available in the 1950s, Enrico Fermi and others began to investigate the ergodic properties of nonlinear systems of springs. But they ended up concentrating on recurrence phenomena related to solitons, and not looking at general questions related to the Second Law. Much the same happened in the 1960s, when the first simulations of hard sphere gases were led to concentrate on the specific phenomenon of long-time tails. And by the 1970s, computer experiments were mostly oriented towards ordinary differential equations and strange attractors, rather than towards systems with large numbers of components, to which the Second Law might apply. Starting in the 1950s, it was recognized that entropy is simply the negative of the information quantity introduced in the 1940s by Claude Shannon. Following statements by John von Neumann, it was thought that any computational process must necessarily increase entropy, but by the early 1970s, notably with work by Charles Bennett, it became accepted that this is not so (see page 1018), laying some early groundwork for relating computational and thermodynamic ideas.

■ **Current thinking on the Second Law.** The vast majority of current physics textbooks imply that the Second Law is well established, though with surprising regularity they say that detailed arguments for it are beyond their scope. More specialized articles tend to admit that the origins of the Second Law remain mysterious. Most ultimately attribute its validity to unknown constraints on initial conditions or measurements, though some appeal to external perturbations, to cosmology or to unknown features of quantum mechanics.

An argument for the Second Law from around 1900, still reproduced in many textbooks, is that if a system is ergodic then it will visit all its possible states, and the vast majority of these will look random. But only very special kinds of systems are in fact ergodic, and even in such systems, the time necessary to visit a significant fraction of all possible states is astronomically long. Another argument for the Second Law, arising from work in the 1930s and 1940s, particularly on systems of hard spheres, is based on the notion of instability with respect to small changes in initial conditions. The argument suffers however from the same difficulties as the ones for chaos theory discussed in Chapter 6 and does not in the end explain in any real way the origins of randomness, or the observed validity of the Second Law.

With the Second Law accepted as a general principle, there is confusion about why systems in nature have not all dissipated into complete randomness. And often the rather absurd claim is made that all the order we see in the universe must just be a fluctuation—leaving little explanatory power for principles such as the Second Law.

■ **My explanation of the Second Law.** What I say in this book is not incompatible with much of what has been said about the Second Law before; it is simply that I make more definite some key points that have been left vague before. In particular, I use notions of computation to specify what kinds of initial conditions can reasonably be prepared, and what kinds of measurements can reasonably be made. In a sense

Note: page numbers in the text refer to the original work *A New Kind of Science* (2002).

538

what I do is just to require that the operation of coarse graining correspond to a computation that is less sophisticated than the actual evolution of the system being studied. (See also Chapters 10 and 12.)

■ **Biological systems and Maxwell's demon.** Unlike most physical systems, biological systems typically seem capable of spontaneously organizing themselves. And as a result, even the original statements of the Second Law talked only about "inanimate systems". In the mid-1860s James Clerk Maxwell then suggested that a demon operating at a microscopic level could reduce the randomness of a system such as a gas by intelligently controlling the motion of molecules. For many years there was considerable confusion about Maxwell's demon. There were arguments that the demon must use a flashlight that generates entropy. And there were extensive demonstrations that actual biological systems reduce their internal entropy only at the cost of increases in the entropy of their environment. But in fact the main point is that if the evolution of the whole system is to be reversible, then the demon must store enough information to reverse its own actions, and this limits how much the demon can do, preventing it, for example, from unscrambling a large system of gas molecules.

■ **Self-gravitating systems.** The observed existence of structures such as galaxies might lead one to think that any large number of objects subject to mutual gravitational attraction might not follow the Second Law and become randomized, but might instead always form orderly clumps. It is difficult to know, however, what an idealized self-gravitating system would do. For in practice, issues such as the limited size of a galaxy, its overall rotation, and the details of stellar collisions all seem to have major effects on the results obtained. (And it is presumably not feasible to do a small-scale experiment, say in Earth orbit.) There are known to be various instabilities that lead in the direction of clumping and core collapse, but how these weigh against effects such as the transfer of energy into tight binding of small groups of stars is not clear. Small galaxies such as globular clusters that contain less than a million stars seem to exhibit a certain uniformity which suggests a kind of equilibrium. Larger galaxies such as our own that contain perhaps 100 billion stars often have intricate spiral or other structure, whose origin may be associated with gravitational effects, or may be a consequence of detailed processes of star formation and explosion. (There is some evidence that older galaxies of a given size tend to develop more regularities in their structure.) Current theories of the early universe tend to assume that galaxies originally began to form as a result of density fluctuations of non-gravitational origin (and reflected in the cosmic microwave background). But there is evidence that a widespread fractal structure develops—with a correlation function of the form $r^{-1.8}$—in the distribution of stars in our galaxy, galaxies in clusters and clusters in superclusters, perhaps suggesting the existence of general overall laws for self-gravitating systems. (See also page 973.)

As mentioned on page 880, it so happens that my original interest in cellular automata around 1981 developed in part from trying to model the growth of structure in self-gravitating systems. At first I attempted to merge and generalize ideas from traditional areas of mathematical physics, such as kinetic theory, statistical mechanics and field theory. But then, particularly as I began to think about doing explicit computer simulations, I decided to take a different tack and instead to look for the most idealized possible models. And in doing this I quickly came up with cellular automata. But when I started to investigate cellular automata, I discovered some very remarkable phenomena, and I was soon led away from self-gravitating systems, and into the process of developing the much more general science in this book. Over the years, I have occasionally come back to the problem of self-gravitating systems, but I have never succeeded in developing what I consider to be a satisfactory approach to them.

■ **Cosmology and the Second Law.** In the standard big bang model it is assumed that all matter in the universe was initially in completely random thermal equilibrium. But such equilibrium implies uniformity, and from this it follows that the initial conditions for the gravitational forces in the universe must have been highly regular, resulting in simple overall expansion, rather than random expansion in some places and contraction in others. As I discuss on page 1026 I suspect that in fact the universe as a whole probably had what were ultimately very simple initial conditions, and it is just that the effective rules for the evolution of matter led to rapid randomization, whereas those for gravity did not.

■ **Alignment of time in the universe.** Evidence from astronomy clearly suggests that the direction of irreversible processes is the same throughout the universe. The reason for this is presumably that all parts of the universe are expanding—with the local consequence that radiation is more often emitted than absorbed, as evidenced by the fact that the night sky is dark. Olbers' paradox asks why one does not see a bright star in every direction in the night sky. The answer is that locally stars are clumped, and light from stars further away is progressively red-shifted to lower energy. Focusing a larger and larger distance away, the light one sees was emitted longer and longer ago. And eventually one sees light emitted when the universe was filled with hot opaque

Note: page numbers in the text refer to the original work *A New Kind of Science* (2002).

539

gas—now red-shifted to become the 2.7K cosmic microwave background.

■ **Poincaré recurrence.** Systems of limited size that contain only discrete elements inevitably repeat their evolution after a sufficiently long time (see page 258). In 1890 Henri Poincaré established the somewhat less obvious fact that even continuous systems also always eventually get at least arbitrarily close to repeating themselves. This discovery led to some confusion in early interpretations of the Second Law, but the huge length of time involved in a Poincaré recurrence makes it completely irrelevant in practice.

■ **Page 446 · Billiards.** The discrete system I consider here is analogous to continuous so-called billiard systems consisting of circular balls in the plane. The simplest case involves one ball bouncing around in a region of a definite shape. In a rectangular region, the position is given by $Mod[a\,t, \{w, h\}]$ and every point will be visited if the parameters have irrational ratios. In a region that contains fixed circular obstructions, the motion can become sensitively dependent on initial conditions. (This setup is similar to a so-called Lorentz gas.) For a system of balls in a region with cyclic boundaries, a complicated proof due to Yakov Sinai from the 1960s purports to show that every ball eventually visits every point in the region, and that certain simple statistical properties of trajectories are consistent with randomness. (See also page 971.)

■ **Page 449 · Entropy of particles in a box.** The number of possible states of a region of m cells containing q particles is $Binomial[m, q]$. In the large size limit, the logarithm of this can be approximated by $q\,Log[m/q]/m$.

■ **Page 457 · Periods in rule 37R.** With a system of size n, the maximum possible repetition period is 2^{2n}. In actuality, however, the periods are considerably shorter. With all cells 0 on one step, and a block of nonzero cells on the next step, the periods are for example: $\{1\}$: 21; $\{1, 1\}$: $3n-8$; $\{1, 0, 1\}$: 666; $\{1, 1, 1\}$: $3n-8$; $\{1, 0, 0, 1\}$: irregular ($<24n$; peaks at $6j+1$); $\{1, 0, 0, 1, 0, 1\}$: irregular ($\lesssim 2^n$; 857727 for $n=26$; 13705406 for $n=100$). With completely random initial conditions, there are great fluctuations, but a typical period is around $2^{n/3}$.

Conserved Quantities and Continuum Phenomena

■ **Physics.** The quantities in physics that so far seem to be exactly conserved are: energy, momentum, angular momentum, electric charge, color charge, lepton number (as well as electron number, muon number and τ lepton number) and baryon number.

■ **Implementation.** Whether a k-color cellular automaton with range r conserves total cell value can be determined from

```
Catch[Do[
  (If[Apply[Plus, CAStep[rule, #]]-#] ≠ 0, Throw[False]] &)[
    IntegerDigits[i, k, m]], {m, w}, {i, 0, k^m - 1}]; True]
```

where w can be taken to be k^{2r}, and perhaps smaller. Among the 256 elementary cellular automata just 5 conserve total cell value. Among the 2^{32} $k=2$, $r=2$ rules 428 do, and of these 2 are symmetric, and 6 are reversible, and all these are just shift and identity rules.

■ **More general conserved quantities.** Some rules conserve not total numbers of cells with given colors, but rather total numbers of blocks of cells with given forms—or combinations of these. The pictures below show the simplest quantities of these kinds that end up being conserved by various elementary rules.

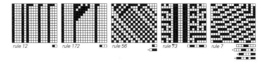

Among the 256 elementary rules, the total numbers that have conserved quantities involving at most blocks of lengths 1 through 10 are $\{5, 38, 66, 88, 102, 108, 108, 114, 118, 118\}$.

Rules that show complicated behavior usually do not seem to have conserved quantities, and this is true for example of rules 30, 90 and 110, at least up to blocks of length 10.

One can count the number of occurrences of each of the k^b possible blocks of length b in a given state using

```
BC[list_] :=
  With[{z = Map[FromDigits[#, k] &, Partition[list, b, 1, 1]]},
    Map[Count[z, #] &, Range[0, k^b - 1]]]
```

Conserved quantities of the kind discussed here are then of the form $q . BC[a]$ where q is some fixed list. A way to find candidates for q is to compute

```
NullSpace[Table[With[{u = Table[Random[Integer,
  {0, k - 1}], {m}]], BC[CAStep[u]] - BC[u]], {s}]]]
```

for progressively larger m and s, and to see what lists continue to appear. For block size b, k^{b-1} lists will always appear as a result of trivial conserved quantities. (With $k = 2$, for $b = 1$, $\{1, 1\}$ represents conservation of the total number of cells, regardless of color, while for $b = 2$, $\{1, 1, 1, 1\}$ represents the same thing, while $\{0, 1, -1, 0\}$ represents the fact that in going along in any state the number of black-to-white transitions must equal the number of white-to-black ones.) If more than k^{b-1} lists appear, however, then some must correspond to genuine non-trivial conserved quantities. To identify any such quantity with certainty, it turns out to be enough to look at the k^{b+2r-1} states where no block of length

Note: page numbers in the text refer to the original work *A New Kind of Science* (2002).

540

$b + 2r - 1$ appears more than once (and perhaps even just some fairly small subset of these).

(See also page 981.)

■ **Other conserved quantities.** The conserved quantities discussed so far can all be thought of as taking values assigned to blocks of different kinds in a given state and then just adding them up as ordinary numbers. But one can also imagine using other operations to combine such values. Addition modulo n can be handled by inserting $Modulus \rightarrow n$ in *NullSpace* in the previous note. And doing this shows for example that rule 150 conserves the total number of black cells modulo 2. But in general not many additional conserved quantities are found in this way. One can also consider combining values of blocks by the multiplication operation in a group—and seeing whether the conjugacy class of the result is conserved.

■ **PDEs.** In the early 1960s it was discovered that certain nonlinear PDEs support an infinite number of distinct conserved quantities, associated with so-called integrability and the presence of solitons. Systematic methods now exist to find conserved quantities that are given by integrals of polynomials of any given degree in the dependent variables and their derivatives. Most randomly chosen PDEs appear, however, to have no such conserved quantities.

■ **Local conservation laws.** Whenever a system like a cellular automaton (or PDE) has a global conserved quantity there must always be a local conservation law which expresses the fact that every point in the system the total flux of the conserved quantity into a particular region must equal the rate of increase of the quantity inside it. (If the conserved quantity is thought of like charge, the flux is then current.) In any 1D $k = 2$, $r = 1$ cellular automaton, it follows from the basic structure of the rule that one can tell what the difference in values of a particular cell on two successive steps will be just by looking at the cell and its immediate neighbor on each side. But if the number of black cells is conserved, then one can compute this difference instead by defining a suitable flux, and subtracting its values on the left and right of the cell. What the flux should be depends on the rule. For rule 184, it can be taken to be 1 for each ■□ block, and to be 0 otherwise. For rule 170, it is 1 for both □□ and ■□. For rule 150, it is 1 for □□ and ■■, with all computations done modulo 2. In general, if the global conserved quantity involves blocks of size b, the flux can be computed by looking at blocks of size $b + 2r - 1$. What the values for these blocks should be can be found by solving a system of linear equations; that a solution must exist can be seen by looking at the de Bruijn network (see page 941), with nodes labelled by size $b + 2r - 1$ blocks,

and connections by value differences between size b blocks at the center of the possible size $b + 2r$ blocks. (Note that the same basic kind of setup works in any number of dimensions.)

■ **Block cellular automata.** With a rule of the form $\{\{1, 1\} \rightarrow \{1, 1\}, \{1, 0\} \rightarrow \{1, 0\}, \{0, 1\} \rightarrow \{0, 0\}, \{0, 0\} \rightarrow \{0, 1\}\}$ the evolution of a block cellular automaton with blocks of size n can be implemented using

```
BCAEvolveList[{n_Integer, rule_}, init_, t_] :=
  FoldList[BCAStep[{n, rule}, #1, #2] &, init, Range[t]] /;
    Mod[Length[init], n] == 0
BCAStep[{n_, rule_}, a_, d_] := RotateRight[
  Flatten[Partition[RotateLeft[a, d], n] /. rule], d]
```

Starting with a single black cell, none of the $k = 2$, $n = 2$ block cellular automata generate anything beyond simple nested patterns. In general, there are $k^{n k^n}$ possible rules for block cellular automata with k colors and blocks of size n. Of these, $k^n!$ are reversible. For $k = 2$, the number of rules that conserve the total number of black cells can be computed from $q = Binomial[n, Range[0, n]]$ as $Apply[Times, q^q]$. The number of these rules that are also reversible is $Apply[Times, q!]$. In general, a block cellular automaton is reversible only if its rule simply permutes the k^n possible blocks.

Compressing each block into a single cell, and n steps into one, any block cellular automaton with k colors and block size n can be translated directly into an ordinary cellular automaton with k^n colors and range $r = n/2$.

■ **Page 461 · Block rules.** These pictures show the behavior of rule (c) starting from some special initial conditions.

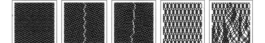

The repetition period with a total of n cells can be 3^n steps. With random initial conditions, the period is typically up to about $3^{n/2}$. Starting with a block of q black cells, the period can get close to this. For $n = 20$, $q = 17$, for example, it is 31,300.

Note that even in rule (b) wraparound phenomena can lead to repetition periods that increase rapidly with n (e.g. 4820 for $n = 20$, $q = 15$), but presumably not exponentially.

In rule (d), the repetition periods can typically be larger than in rule (c): e.g. 803,780 for $n = 20$, $q = 13$.

■ **Page 464 · Limiting procedures.** Several different limiting procedures all appear to yield the same continuum behavior for the cellular automata shown here. In the pictures on this

Note: page numbers in the text refer to the original work *A New Kind of Science* (2002).

page a large ensemble of different initial conditions is considered, and the density of each individual cell averaged over this ensemble is computed. In a more direct analogy to actual physical systems, one would consider instead a very large number of cells, then compute the density in a single state of the system by averaging over regions that contain many cells but are nevertheless small compared to the size of the whole system.

■ **PDE approximations.** Cellular automaton (d) in the main text can be viewed as minimal discrete approximations to the diffusion equation. The evolution of densities in the ensemble average is analogous to a traditional finite difference method with a real number at each site. The cellular automaton itself uses in effect a distributed representation of the density.

■ **Diffusion equation.** In an appropriate limit the density distribution for cellular automaton (d) appears to satisfy the usual diffusion equation $\partial_t f[x, t] == c\, \partial_{xx} f[x, t]$ discussed on page 163. The solution to this equation with an impulse initial condition is $Exp[-x^2/t]$, and with a block from $-a$ to a it is $(Erf[(a-x)/\sqrt{t}] + Erf[(a+x)/\sqrt{t}])/a$.

■ **Derivation of the diffusion equation.** With some appropriate assumptions, it is fairly straightforward to derive the usual diffusion equation from a cellular automaton. Let the density of black cells at position x and time t be $f[x, t]$, where this density can conveniently be computed by averaging over many instances of the system. If we assume that the density varies slowly with position and time, then we can make series expansions such as

$$f[x + dx, t] == f[x, t] + \partial_x f[x, t]\, dx + 1/2\, \partial_{xx} f[x, t]\, dx^2 + \ldots$$

where the coordinates are scaled so that adjacent cells are at positions $x - dx$, x, $x + dx$, etc. If we then assume perfect underlying randomness, the density at a particular position must be given in terms of the densities at neighboring positions on the previous step by

$$f[x, t + dt] == p_1\, f[x - dx, t] + p_2\, f[x, t] + p_3\, f[x + dx, t]$$

Density conservation implies that $p_1 + p_2 + p_3 == 1$, while left-right symmetry implies $p_1 == p_3$. And from this it follows that

$$f[x, t + dt] == c\, (f[x - dx, t] + f[x + dx, t]) + (1 - 2\,c)\, f[x, t]$$

Performing a series expansion then yields

$$f[x, t] + dt\, \partial_t f[x, t] == f[x, t] + c\, dx^2\, \partial_{xx} f[x, t]$$

which in turn gives exactly the usual 1D diffusion equation $\partial_t f[x, t] == \xi\, \partial_{xx} f[x, t]$, where ξ is the diffusion coefficient for the system. I first gave this derivation in 1986, together with extensive generalizations.

■ **Page 464 · Non-standard diffusion.** To get ordinary diffusion behavior of the kind that occurs in gases—and is described by the diffusion equation—it is in effect necessary to have perfect uncorrelated randomness, with no structure that persists too long. But for example in the rule (a) picture on page 463 there is in effect a block of solid that persists in the middle—so that no ordinary diffusion behavior is seen. In rule (c) there is considerable apparent randomness, but it turns out that there are also fluctuations that last too long to yield ordinary diffusion. And thus for example whenever there is a structure containing s identical cells (as on page 462), this typically takes about s^2 steps to decay away. The result is that on page 464 the limiting form of the average behavior does not end up being an ordinary Gaussian.

■ **Conservation of vector quantities.** Conservation of the total number of colored cells is analogous to conservation of a scalar quantity such as energy or particle number. One can also consider conservation of a vector quantity such as momentum which has not only a magnitude but also a direction. Direction makes little sense in 1D, but is meaningful in 2D. The 2D cellular automaton used as a model of an idealized gas on page 446 provides an example of a system that can be viewed as conserving a vector quantity. In the absence of fixed scatterers, the total fluxes of particles in the horizontal and the vertical directions are conserved. But in a sense there is too much conservation in this system, and there is no interaction between horizontal and vertical motions. This can be achieved by having more complicated underlying rules. One possibility is to use a hexagonal rather than square grid, thereby allowing six particle directions rather than four. On such a grid it is possible to randomize microscopic particle motions, but nevertheless conserve overall momenta. This is essentially the model used in my discussion of fluids on page 378.

Note: page numbers in the text refer to the original work *A New Kind of Science* (2002).

12.6 Computational Irreducibility

When viewed in computational terms most of the great historical triumphs of theoretical science turn out to be remarkably similar in their basic character. For at some level almost all of them are based on finding ways to reduce the amount of computational work that has to be done in order to predict how some particular system will behave.

Most of the time the idea is to derive a mathematical formula that allows one to determine what the outcome of the evolution of the system will be without explicitly having to trace its steps.

And thus, for example, an early triumph of theoretical science was the derivation of a formula for the position of a single idealized planet orbiting a star. For given this formula one can just plug in numbers to work out where the planet will be at any point in the future, without ever explicitly having to trace the steps in its motion.

But part of what started my whole effort to develop the new kind of science in this book was the realization that there are many common systems for which no traditional mathematical formulas have ever been found that readily describe their overall behavior.

At first one might have thought this must be some kind of temporary issue, that could be overcome with sufficient cleverness. But from the discoveries in this book I have come to the conclusion that in fact it is not, and that instead it is one of the consequences of a very fundamental phenomenon that follows from the Principle of Computational Equivalence and that I call computational irreducibility.

If one views the evolution of a system as a computation, then each step in this evolution can be thought of as taking a certain amount of computational effort on the part of the system. But what traditional theoretical science in a sense implicitly relies on is that much of this effort is somehow unnecessary—and that in fact it should be possible to find the outcome of the evolution with much less effort.

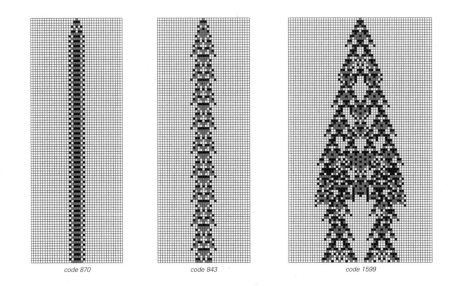

code 870   code 843   code 1599

Examples of computational reducibility and irreducibility in the evolution of cellular automata. The first two rules yield simple repetitive computationally reducible behavior in which the outcome after many steps can readily be deduced without tracing each step. The third rule yields behavior that appears to be computationally irreducible, so that its outcome can effectively be found only by explicitly tracing each step. The cellular automata shown here all have 3-color totalistic rules.

And certainly in the first two examples above this is the case. For just as with the orbit of an idealized planet there is in effect a straightforward formula that gives the state of each system after any number of steps. So even though the systems themselves generate their behavior by going through a whole sequence of steps, we can readily shortcut this process and find the outcome with much less effort.

But what about the third example shown above? What does it take to find the outcome in this case? It is always possible to do an experiment and explicitly run the system for a certain number of steps and see how it behaves. But to have any kind of traditional theory one must find a shortcut that involves much less computation.

Yet from the picture on the previous page it is certainly not obvious how one might do this. And looking at the pictures on the next page it begins to seem quite implausible that there could ever in fact be any way to find a significant shortcut in the evolution of this system.

So while the behavior of the first two systems on the previous page is readily seen to be computationally reducible, the behavior of the third system appears instead to be computationally irreducible.

In traditional science it has usually been assumed that if one can succeed in finding definite underlying rules for a system then this means that ultimately there will always be a fairly easy way to predict how the system will behave.

Several decades ago chaos theory pointed out that to have enough information to make complete predictions one must in general know not only the rules for a system but also its complete initial conditions.

But now computational irreducibility leads to a much more fundamental problem with prediction. For it implies that even if in principle one has all the information one needs to work out how some particular system will behave, it can still take an irreducible amount of computational work actually to do this.

Indeed, whenever computational irreducibility exists in a system it means that in effect there can be no way to predict how the system will behave except by going through almost as many steps of computation as the evolution of the system itself.

In traditional science it has rarely even been recognized that there is a need to consider how systems that are used to make predictions actually operate. But what leads to the phenomenon of computational irreducibility is that there is in fact always a fundamental competition

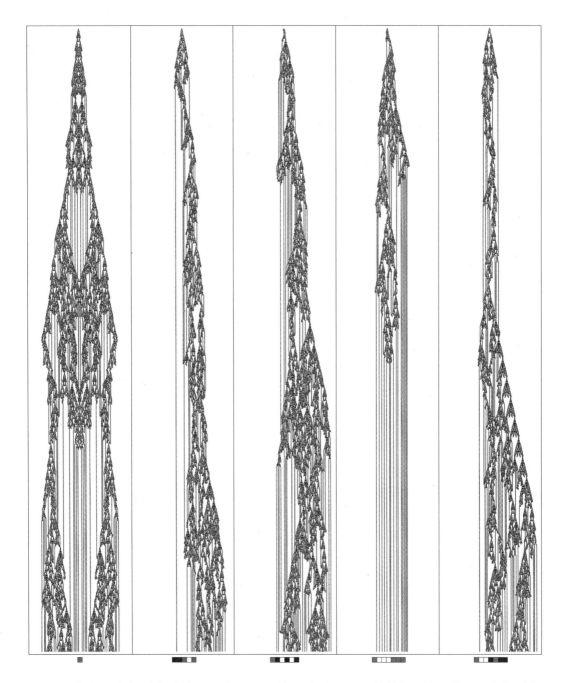

5000 steps in the evolution of the third system from page 738, starting from several initial conditions. The complexity of the behavior makes it seem inconceivable that there could ever be a procedure that would always immediately find its outcome.

Note: page numbers in the text refer to the original work *A New Kind of Science* (2002).

between systems used to make predictions and systems whose behavior one tries to predict.

For if meaningful general predictions are to be possible, it must at some level be the case that the system making the predictions be able to outrun the system it is trying to predict. But for this to happen the system making the predictions must be able to perform more sophisticated computations than the system it is trying to predict.

In traditional science there has never seemed to be much problem with this. For it has normally been implicitly assumed that with our powers of mathematics and general thinking the computations we use to make predictions must be almost infinitely more sophisticated than those that occur in most systems in nature and elsewhere whose behavior we try to predict.

But the remarkable assertion that the Principle of Computational Equivalence makes is that this assumption is not correct, and that in fact almost any system whose behavior is not obviously simple performs computations that are in the end exactly equivalent in their sophistication.

So what this means is that systems one uses to make predictions cannot be expected to do computations that are any more sophisticated than the computations that occur in all sorts of systems whose behavior we might try to predict. And from this it follows that for many systems no systematic prediction can be done, so that there is no general way to shortcut their process of evolution, and as a result their behavior must be considered computationally irreducible.

If the behavior of a system is obviously simple—and is say either repetitive or nested—then it will always be computationally reducible. But it follows from the Principle of Computational Equivalence that in practically all other cases it will be computationally irreducible.

And this, I believe, is the fundamental reason that traditional theoretical science has never managed to get far in studying most types of systems whose behavior is not ultimately quite simple.

For the point is that at an underlying level this kind of science has always tried to rely on computational reducibility. And for example its whole idea of using mathematical formulas to describe behavior makes sense only when the behavior is computationally reducible.

So when computational irreducibility is present it is inevitable that the usual methods of traditional theoretical science will not work. And indeed I suspect the only reason that their failure has not been more obvious in the past is that theoretical science has typically tended to define its domain specifically in order to avoid phenomena that do not happen to be simple enough to be computationally reducible.

But one of the major features of the new kind of science that I have developed is that it does not have to make any such restriction. And indeed many of the systems that I study in this book are no doubt computationally irreducible. And that is why—unlike most traditional works of theoretical science—this book has very few mathematical formulas but a great many explicit pictures of the evolution of systems.

It has in the past couple of decades become increasingly common in practice to study systems by doing explicit computer simulations of their behavior. But normally it has been assumed that such simulations are ultimately just a convenient way to do what could otherwise be done with mathematical formulas.

But what my discoveries about computational irreducibility now imply is that this is not in fact the case, and that instead there are many common systems whose behavior cannot in the end be determined at all except by something like an explicit simulation.

Knowing that universal systems exist already tells one that this must be true at least in some situations. For consider trying to outrun the evolution of a universal system. Since such a system can emulate any system, it can in particular emulate any system that is trying to outrun it. And from this it follows that nothing can systematically outrun the universal system. For any system that could would in effect also have to be able to outrun itself.

But before the discoveries in this book one might have thought that this could be of little practical relevance. For it was believed that except among specially constructed systems universality was rare. And it was also assumed that even when universality was present, very special initial conditions would be needed if one was ever going to perform computations at anything like the level of sophistication involved in most methods of prediction.

But the Principle of Computational Equivalence asserts that this is not the case, and that in fact almost any system whose behavior is not obviously simple will exhibit universality and will perform sophisticated computations even with typical simple initial conditions.

So the result is that computational irreducibility can in the end be expected to be common, so that it should indeed be effectively impossible to outrun the evolution of all sorts of systems.

One slightly subtle issue in thinking about computational irreducibility is that given absolutely any system one can always at least nominally imagine speeding up its evolution by setting up a rule that for example just executes several steps of evolution at once.

But insofar as such a rule is itself more complicated it may in the end achieve no real reduction in computational effort. And what is more important, it turns out that when there is true computational reducibility its effect is usually much more dramatic.

The pictures on the next page show typical examples based on cellular automata that exhibit repetitive and nested behavior. In the patterns on the left the color of each cell at any given step is in effect found by tracing the explicit evolution of the cellular automaton up to that step. But in the pictures on the right the results for particular cells are instead found by procedures that take much less computational effort.

These procedures are again based on cellular automata. But now what the cellular automata do is to take specifications of positions of cells, and then in effect compute directly from these the colors of cells.

The way things are set up the initial conditions for these cellular automata consist of digit sequences of numbers that give positions. The color of a particular cell is then found by evolving for a number of steps equal to the length of these input digit sequences.

And this means for example that the outcome of a million steps of evolution for either of the cellular automata on the left is now determined by just 20 steps of evolution, where 20 is the length of the base 2 digit sequence of the number 1,000,000.

And this turns out to be quite similar to what happens with typical mathematical formulas in traditional theoretical science. For the point of such formulas is usually to allow one to give a number as

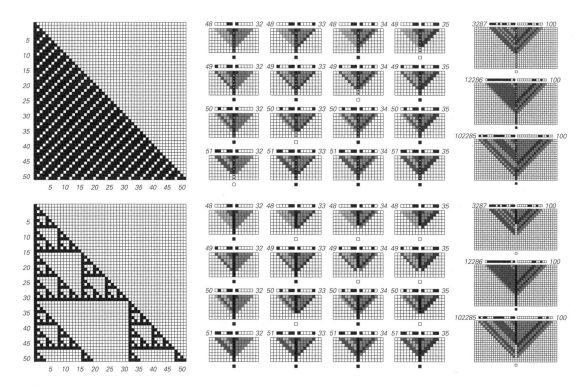

Examples of computational reducibility in action. The pictures on the left show patterns produced by the ordinary evolution of cellular automata with elementary rules 188 and 60. The pictures on the right show how colors of particular cells in these patterns can be found with much less computational effort. In each case the position of a cell is specified by a pair of numbers given as base 2 digit sequences in the initial conditions for a cellular automaton. The evolution of the cellular automaton then quickly determines what the color of the cell at that position in the pattern on the left will be. For rule 188 the cellular automaton that does this involves 12 colors; for rule 60 it involves 6. In general, to find the color of a cell after *t* steps of rule 188 or rule 60 evolution takes about *Log[2, t]* steps. Compare page 608.

input, and then to compute directly something that corresponds, say, to the outcome of that number of steps in the evolution of a system.

In traditional mathematics it is normally assumed that once one has an explicit formula involving standard mathematical functions then one can in effect always evaluate this formula immediately.

But evaluating a formula—like anything else—is a computational process. And unless some digits effectively never matter, this process cannot normally take less steps than there are digits in its input.

Indeed, it could in principle be that the process could take a number of steps proportional to the numerical value of its input. But if this were so, then it would mean that evaluating the formula would

Note: page numbers in the text refer to the original work *A New Kind of Science* (2002).

require as much effort as just tracing each step in the original process whose outcome the formula was supposed to give.

And the crucial point that turns out to be the basis for much of the success of traditional theoretical science is that in fact most standard mathematical functions can be evaluated in a number of steps that is far smaller than the numerical value of their input, and that instead normally grows only slowly with the length of the digit sequence of their input.

So the result of this is that if there is a traditional mathematical formula for the outcome of a process then almost always this means that the process must show great computational reducibility.

In practice, however, the vast majority of cases for which traditional mathematical formulas are known involve behavior that is ultimately either uniform or repetitive. And indeed, as we saw in Chapter 10, if one uses just standard mathematical functions then it is rather difficult even to reproduce many simple examples of nesting.

But as the pictures on the facing page and in Chapter 10 illustrate, if one allows more general kinds of underlying rules then it becomes quite straightforward to set up procedures that with very little computational effort can find the color of any element in any nested pattern.

So what about more complex patterns, like the rule 30 cellular automaton pattern at the bottom of the page?

When I first generated such patterns I spent a huge amount of time trying to analyze them and trying to find a procedure that would allow me to compute directly the color of each cell. And indeed it was the fact that I was never able to make much progress in doing this that first led me to consider the possibility that there could be a phenomenon like computational irreducibility.

And now, what the Principle of Computational Equivalence implies is that in fact almost any system whose behavior is not obviously simple will tend to exhibit computational irreducibility.

But particularly when the underlying rules are simple there is often still some superficial computational reducibility. And so, for example, in the rule 30 pattern on the right one can tell whether a cell at a given position has any chance of not being white just by doing a

An example of a pattern where it is difficult to compute directly the color of a particular cell.

very short computation that tests whether that position lies outside the center triangular region of the pattern. And in a class 4 cellular automaton such as rule 110 one can readily shortcut the process of evolution for at least a limited number of steps in places where there happen to be only a few well-separated localized structures present.

And indeed in general almost any regularities that we manage to recognize in the behavior of a system will tend to reflect some kind of computational reducibility in this behavior.

If one views the pattern of behavior as a piece of data, then as we discussed in Chapter 10 regularities in it allow a compressed description to be found. But the existence of a compressed description does not on its own imply computational reducibility. For any system that has simple rules and simple initial conditions—including for example rule 30—will always have such a description.

But what makes there be computational reducibility is when only a short computation is needed to find from the compressed description any feature of the actual behavior.

And it turns out that the kinds of compressed descriptions that can be obtained by the methods of perception and analysis that we use in practice and that we discussed in Chapter 10 all essentially have this property. So this is why regularities that we recognize by these methods do indeed reflect the presence of computational reducibility.

But as we saw in Chapter 10, in almost any case where there is not just repetitive or nested behavior, our normal powers of perception and analysis recognize very few regularities—even though at some level the behavior we see may still be generated by extremely simple rules.

And this supports the assertion that beyond perhaps some small superficial amount of computational reducibility a great many systems are in the end computationally irreducible. And indeed this assertion explains, at least in part, why our methods of perception and analysis cannot be expected to go further in recognizing regularities.

But if behavior that we see looks complex to us, does this necessarily mean that it can exhibit no computational reducibility? One way to try to get an idea about this is just to construct patterns

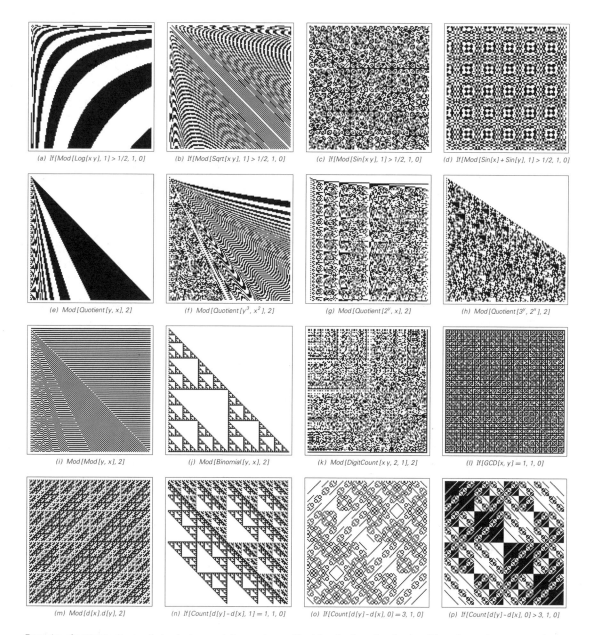

(a) If[Mod[Log[x y], 1] > 1/2, 1, 0]

(b) If[Mod[Sqrt[x y], 1] > 1/2, 1, 0]

(c) If[Mod[Sin[x y], 1] > 1/2, 1, 0]

(d) If[Mod[Sin[x] + Sin[y], 1] > 1/2, 1, 0]

(e) Mod[Quotient[y, x], 2]

(f) Mod[Quotient[y^3, x^2], 2]

(g) Mod[Quotient[2^y, x], 2]

(h) Mod[Quotient[3^y, 2^x], 2]

(i) Mod[Mod[y, x], 2]

(j) Mod[Binomial[y, x], 2]

(k) Mod[DigitCount[x y, 2, 1], 2]

(l) If[GCD[x, y] = 1, 1, 0]

(m) Mod[d[x].d[y], 2]

(n) If[Count[d[y] - d[x], 1] = 1, 1, 0]

(o) If[Count[d[y] - d[x], 0] = 3, 1, 0]

(p) If[Count[d[y] - d[x], 0] > 3, 1, 0]

Examples of patterns set up so that a short computation can be used to determine the color of each cell from the numbers representing its position. Most such patterns look to us quite simple, but the examples shown here were specifically chosen to be ones that look more complicated. In most of them fairly standard mathematical functions are used, but in unusual combinations. In every picture both x and y run from 1 to 127. *d[n]* stands for *IntegerDigits[n, 2, 7]*. (h) is equivalent to digit sequences of powers of 3 in base 2 (see page 120). (j) is essentially Pascal's triangle (see page 611). (l) was discussed on page 613. (m) is a nested pattern seen on page 583. The only pattern that is known to be obtainable by evolving down the page according to a simple local rule is (j), which corresponds to the rule 60 elementary cellular automaton.

Note: page numbers in the text refer to the original work *A New Kind of Science* (2002).

where we explicitly set up the color of each cell to be determined by some short computation from the numbers that represent its position.

When we look at such patterns most of them appear to us quite simple. But as the pictures on the previous page demonstrate, it turns out to be possible to find examples where this is not so, and where instead the patterns appear to us at least somewhat complex.

But for such patterns to yield meaningful examples of computational reducibility it must also be possible to produce them by some process of evolution—say by repeated application of a cellular automaton rule. Yet for the majority of cases shown here there is at least no obvious way to do this.

I have however found one class of systems—already mentioned in Chapter 10—whose behavior does not appear simple, but nevertheless turns out to be computationally reducible, as in the pictures on the facing page. However, I strongly suspect that systems like this are very rare, and that in the vast majority of cases where the behavior that we see in nature and elsewhere appears to us complex it is in the end indeed associated with computational irreducibility.

So what does this mean for science?

In the past it has normally been assumed that there is no ultimate limit on what science can be expected to do. And certainly the progress of science in recent centuries has been so impressive that it has become common to think that eventually it should yield an easy theory—perhaps a mathematical formula—for almost anything.

But the discovery of computational irreducibility now implies that this can fundamentally never happen, and that in fact there can be no easy theory for almost any behavior that seems to us complex.

It is not that one cannot find underlying rules for such behavior. Indeed, as I have argued in this book, particularly when they are formulated in terms of programs I suspect that such rules are often extremely simple. But the point is that to deduce the consequences of these rules can require irreducible amounts of computational effort.

One can always in effect do an experiment, and just watch the actual behavior of whatever system one wants to study. But what one

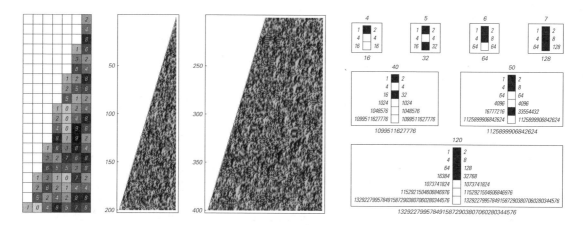

A system whose behavior looks complex but still turns out to be computationally reducible. The system is a cellular automaton with 10 possible colors for each cell. But it can also be viewed as a system based on numbers, in which successive rows are the base 10 digit sequences of successive powers of 2. And it turns out that there is a fast way to compute row *n* just from the base 2 digit sequence of *n*, as the pictures on the right illustrate. This procedure is based on the standard repeated squaring method of finding 2^n by starting from 2, and then successively squaring the numbers one gets, multiplying by 2 if the corresponding base 2 digit in *n* is 1. Using this procedure one can certainly compute the color of any cell on row *n* by doing about $n \, Log[n]^3$ operations—instead of the n^2 needed if one carried out the cellular automaton evolution explicitly.

cannot in general do is to find an easy theory that will tell one without much effort what every aspect of this behavior will be.

So given this, can theoretical science still be useful at all?

The answer is definitely yes. For even in its most traditional form it can often deal quite well with those aspects of behavior that happen to be simple enough to be computationally reducible. And since one can never know in advance how far computational reducibility will go in a particular system it is always worthwhile at least to try applying the traditional methods of theoretical science.

But ultimately if computational irreducibility is present then these methods will fail. Yet there are still often many reasons to want to use abstract theoretical models rather than just doing experiments on actual systems in nature and elsewhere. And as the results in this book suggest, by using the right kinds of models much can be achieved.

Any accurate model for a system that exhibits computational irreducibility must at some level inevitably involve computations that are as sophisticated as those in the system itself. But as I have shown in this book even systems with very simple underlying rules can still perform computations that are as sophisticated as in any system.

And what this means is that to capture the essential features even of systems with very complex behavior it can be sufficient to use models that have an extremely simple basic structure. Given these models the only way to find out what they do will usually be just to run them. But the point is that if the structure of the models is simple enough, and fits in well enough with what can be implemented efficiently on a practical computer, then it will often still be perfectly possible to find out many consequences of the model.

And that, in a sense, is what much of this book has been about.

NOTES FOR CHAPTER 12 OF *A NEW KIND OF SCIENCE*

The Principle of Computational Equivalence

Computational Irreducibility

■ **History.** The notion that there could be fundamental limits to knowledge or predictability has been discussed repeatedly since antiquity. But most often it has been assumed that the origin of this must be inadequacy in models, not difficulty in working out their consequences. And indeed already in the 1500s with the introduction of symbolic algebra and the discovery of formulas for solving cubic and quartic equations the expectation began to develop that with sufficient cleverness it should be possible to derive a formula for the solution to any purely mathematical problem. Infinitesimals were sometimes thought to get in the way of finite understanding—but this was believed to be overcome by calculus. And when mathematical models for natural systems became widespread in the late 1600s it was generally assumed that their basic consequences could always be found in terms of formulas or geometrical theorems, perhaps with fairly straightforward numerical calculations required for connection to practical situations. In discussing gravitational interactions between many planets Isaac Newton did however comment in 1684 that "to define these motions by exact laws admitting of easy calculation exceeds, if I am not mistaken, the force of any human mind". But in the course of the 1700s and 1800s formulas were successfully found for solutions to a great many problems in mathematical physics (see note below)—at least when suitable special functions (see page 1091) were introduced. The three-body problem (see page 972) nevertheless continued to resist efforts at general solution. In the 1820s it was shown that quintic equations cannot in general be solved in terms of radicals (see page 1137), and by the 1890s it was known that degree 7 equations cannot in general be solved even if elliptic functions are allowed. Around 1890 it was then shown that the three-body problem could not be solved in general in terms of ordinary algebraic functions and integrals (see page 972). However, perhaps in part because of a shift towards probabilistic theories such as quantum and statistical mechanics there remained the conviction that for

relevant aspects of behavior formulas should still exist. The difficulty for example of finding more than a few exact solutions to the equations of general relativity was noted— but a steady stream of results (see note below) maintained the belief that with sufficient cleverness a formula could be found for behavior according to any model.

In the 1950s computers began to be used to work out numerical solutions to equations—but this was seen mostly as a convenience for applications, not as a reflection of any basic necessity. A few computer experiments were done on systems with simple underlying rules, but partly because Monte Carlo methods were sometimes used, it was typically assumed that their results were just approximations to what could in principle be represented by exact formulas. And this view was strengthened in the 1960s when solitons given by simple formulas were found in some of these systems.

The difficulty of solving equations for numerical weather prediction was noted even in the 1920s. And by the 1950s and 1960s the question of whether computer calculations would be able to outrun actual weather was often discussed. But it was normally assumed that the issue was just getting a better approximation to the underlying equations—or better initial measurements—not something more fundamental.

Particularly in the context of game theory and cybernetics the idea had developed in the 1940s that it should be possible to make mathematical predictions even about complex human situations. And for example starting in the early 1950s government control of economies based on predictions from linear models became common. By the early 1970s, however, such approaches were generally seen as unsuccessful, but it was usually assumed that the reason was not fundamental, but was just that there were too many disparate elements to handle in practice.

The notions of universality and undecidability that underlie computational irreducibility emerged in the 1930s, but they were not seen as relevant to questions arising in natural science. Starting in the 1940s they were presumably the basis

Note: page numbers in the text refer to the original work *A New Kind of Science* (2002).

for a few arguments made about free will and fundamental unpredictability of human behavior (see page 1135), particularly in the context of economics. And in the late 1950s there was brief interest among philosophers in connecting results like Gödel's Theorem to questions of determinism—though mostly there was just confusion centered around the difficulty of finding countable proofs for statements about the continuous processes assumed to occur in physics.

The development of algorithmic information theory in the 1960s led to discussion of objects whose information content cannot be compressed or derived from anything shorter. But as indicated on page 1067 this is rather different from what I call computational irreducibility. In the 1970s computational complexity theory began to address questions about overall resources needed to perform computations, but concentrated on computations that perform fairly specific known practical tasks. At the beginning of the 1980s, however, it was noted that certain problems about models of spin glasses were NP-complete. But there was no immediate realization that this was connected to any underlying general phenomenon.

Starting in the late 1970s there was increasing interest in issues of predictability in models of physical systems. And it was emphasized that when the equations in such models are nonlinear it often becomes difficult to find their solutions. But usually this was at some level assumed to be associated with sensitive dependence on initial conditions and the chaos phenomenon—even though as we saw on page 1098 this alone does not even prevent there from being formulas.

By the early 1980s it had become popular to use computers to study various models of natural systems. Sometimes the idea was to simulate a large collection of disparate elements, say as involved in a nuclear explosion. Sometimes instead the idea was to get a numerical approximation to some fairly simple partial differential equation, say for fluid flow. Sometimes the idea was to use randomized methods to get a statistical approximation to properties say of spin systems or lattice gauge theories. And sometimes the idea was to work out terms in a symbolic perturbation series approximation, say in quantum field theory or celestial mechanics. With any of these approaches huge amounts of computer time were often used. But it was almost always implicitly assumed that this was necessary in order to overcome the approximations being used, and not for some more fundamental reason.

Particularly in physics, there has been some awareness of examples such as quark confinement in QCD where it seems especially difficult to deduce the consequences of a theory—but no general significance has been attached to this.

When I started studying cellular automata in the early 1980s I was quickly struck by the difficulty of finding formulas for their behavior. In traditional models based for example on continuous numbers or approximations to them there was usually no obvious correspondence between a model and computations that might be done about it. But the evolution of a cellular automaton was immediately reminiscent of other computational processes—leading me by 1984 to formulate explicitly the concept of computational irreducibility.

No doubt an important reason computational irreducibility was not identified before is that for more than two centuries students had been led to think that basic theoretical science could somehow always be done with convenient formulas. For almost all textbooks tend to discuss only those cases that happen to come out this way. Starting in earnest in the 1990s, however, the influence of *Mathematica* has gradually led to broader ranges of examples. But there still remains a very widespread belief that if a theoretical result about the behavior of a system is truly fundamental then it must be possible to state it in terms of a simple mathematical formula.

▪ **Exact solutions.** Some notable cases where closed-form analytical results have been found in terms of standard mathematical functions include: quadratic equations (~2000 BC) (*Sqrt*); cubic, quartic equations (1530s) ($x^{1/n}$); 2-body problem (1687) (*Cos*); catenary (1690) (*Cosh*); brachistochrone (1696) (*Sin*); spinning top (1849; 1888; 1888) (*JacobiSN*; *WeierstrassP*; hyperelliptic functions); quintic equations (1858) (*EllipticTheta*); half-plane diffraction (1896) (*FresnelC*); Mie scattering (1908) (*BesselJ, BesselY, LegendreP*); Einstein equations (Schwarzschild (1916), Reissner-Nordström (1916), Kerr (1963) solutions) (rational and trigonometric functions); quantum hydrogen atom and harmonic oscillator (1927) (*LaguerreL, HermiteH*); 2D Ising model (1944) (*Sinh, EllipticK*); various Feynman diagrams (1960s–1980s) (*PolyLog*); KdV equation (1967) (*Sech* etc.); Toda lattice (1967) (*Sech*); six-vertex spin model (1967) (*Sinh* integrals); Calogero-Moser model (1971) (*Hypergeometric1F1*); Yang-Mills instantons (1975) (rational functions); hard-hexagon spin model (1979) (*EllipticTheta*); additive cellular automata (1984) (*MultiplicativeOrder*); Seiberg-Witten supersymmetric theory (1994) (*Hypergeometric2F1*). When problems are originally stated as differential equations, results in terms of integrals ("quadrature") are sometimes considered exact solutions—as occasionally are convergent series. When one exact solution is found, there often end up being a whole family—with much investigation going into the symmetries that relate them. It is notable that when many of the examples above were discovered they were at first expected to have broad

Note: page numbers in the text refer to the original work *A New Kind of Science* (2002).

significance in their fields. But the fact that few actually did can be seen as further evidence of how narrow the scope of computational reducibility usually is. Notable examples of systems that have been much investigated, but where no exact solutions have been found include the 3D Ising model, quantum anharmonic oscillator and quantum helium atom.

■ **Amount of computation.** Computational irreducibility suggests that it might be possible to define "amount of computation" as an independently meaningful quantity—perhaps vaguely like entropy or amount of information. And such a quantity might satisfy laws vaguely analogous to the laws of thermodynamics that would for example determine what processes are possible and what are not. If one knew the fundamental rules for the universe then one way in principle to define the amount of computation associated with a given process would be to find the minimum number of applications of the rules for the universe that are needed to reproduce the process at some level of description.

■ **Page 743 · More complicated rules.** The standard rule for a cellular automaton specifies how every possible block of cells of a certain size should be updated at every step. One can imagine finding the outcome of evolution more efficiently by adding rules that specify what happens to larger blocks of cells after more steps. And as a practical matter, one can look up different blocks using a method like hashing. But much as one would expect from data compression this will only in the end work more efficiently if there are some large blocks that are sufficiently common. Note that dealing with blocks of different sizes requires going beyond an ordinary cellular automaton rule. But in a sequential substitution system—and especially in a multiway system (see page 776)—this can be done just as part of an ordinary rule.

■ **Page 744 · Reducible systems.** The color of a cell at step t and position x can be found by starting with initial condition

Flatten[With[{w = Max[Ceiling[Log[2, {t, x}]]]},
{2 Reverse[IntegerDigits[t, 2, w]] + 1,
5, 2 IntegerDigits[x, 2, w] + 2}]]

then for rule 188 running the cellular automaton with rule

{{a : (1|3), 1|3, _} → a, {_, 2|4, a : (2|4)} → a,
{3, 5|10, 2} → 6, {1, 5|7, 4} → 0, {3, 5, 4} → 7,
{1, 6, 2} → 10, {1, 6|11, 4} → 8, {3, 6|8|10|11, 4} → 9,
{3, 7|9, 2} → 11, {1, 8|11, 2} → 9, {3, 11, 2} → 8,
{1, 9|10, 4} → 11, {_, a_ /; a > 4, _} → a, {_, _, _} → 0}

and for rule 60 running the cellular automaton with rule

{{a : (1|3), 1|3, _} → a, {_, 2|4, a : (2|4)} → a,
{1, 5, 4} → 0, {_, 5, _} → 5, {_, _, _} → 0}

■ **Speed-up theorems.** That there exist computations that are arbitrarily computationally reducible was noted in work on the theory of computation in the mid-1960s.

■ **Page 745 · Mathematical functions.** The number of bit operations needed to add two n-digit numbers is of order n. The number of operations $m[n]$ needed to multiply them increases just slightly more rapidly than n (see page 1093). (Even if one can do operations on all digits in parallel it still takes of order n steps in a system like a cellular automaton for the effects of different digits to mix together—though see also page 1149.) The number of operations to evaluate $Mod[a, b]$ is of order n if a has n digits and b is small. Many standard continuous mathematical functions just increase or decrease smoothly at large x (see page 917). The main issue in evaluating those that exhibit regular oscillations at large x is to find their oscillation period with sufficient precision. Thus for example if x is an integer with n digits then evaluating $Sin[x]$ or $FractionalPart[x\,c]$ requires respectively finding π or c to n-digit precision. It is known how to evaluate π (see page 912) and all standard elementary functions to n-digit precision using about $Log[n]\,m[n]$ operations. (This can be done by repeatedly making use of functional relations such as $Exp[2\,x] == Exp[x]^2$ which express $f[2\,x]$ as a polynomial in $f[x]$; such an approach is known to work for elementary, elliptic, modular and other functions associated with *ArithmeticGeometricMean* and for example *DedekindEta*.) Known methods for high-precision evaluation of special functions—usually based in the end on series representations—typically require of order $n^{1/s}\,m[n]$ operations, where s is often 2 or 3. (Examples of more difficult cases include *HypergeometricPFQ[a, b, 1]* and *StieltjesGamma[k]*, where logarithmic series can require an exponential number of terms. Evaluation of *BernoulliB[x]* is also difficult.) Any iterative procedure (such as *FindRoot*) that yields a constant multiple more digits at each step will take about $Log[n]$ steps to get n digits. Roots of polynomials can thus almost always be found with *NSolve* in about $Log[n]\,m[n]$ operations. If one evaluates *NIntegrate* or *NDSolve* by effectively fitting functions to order s polynomials the difficulty of getting results with n-digit precision typically increases like $2^{n/s}$. An adaptive algorithm such as Romberg integration reduces this to about $2^{\wedge}\sqrt{n}$. The best-known algorithms for evaluating $Zeta[1/2 + i\,x]$ (see page 918) to fixed precision take roughly $\sqrt{x}$ operations—or $2^{n/2}$ operations if x is an n-digit integer. (The evaluation is based on the Riemann-Siegel formula, which involves sums of about $\sqrt{x}$ cosines.) Unlike for continuous mathematical functions, known algorithms for number theoretical functions such as *FactorInteger[x]* or *MoebiusMu[x]* typically seem to require a number of operations that grows faster with the number of digits n in x than any power of n (see page 1090).

Note: page numbers in the text refer to the original work *A New Kind of Science* (2002).

559

■ **Formulas.** It is always in principle possible to build up some kind of formula for the outcome of any process of evolution, say of a cellular automaton (see page 618). But for there to be computational reducibility this formula needs to be simple and easy to evaluate—as it is if it consists just of a few standard mathematical functions (see note above; page 1098).

■ **Page 747 · Short computations.** Some properties include:

(a) The regions are bounded by the hyperbolas $x\,y == Exp[n/2]$ for successive integers n.

(d) There is approximate repetition associated with rational approximations to π (for example with period 22), but never precise repetition.

(e) The pattern essentially shows which x are divisors of y, just as on pages 132 and 909.

(h) $Mod[Quotient[s, 2^n], 2]$ extracts the digit associated with 2^n in the base 2 digit sequence of s.

(i) Like (e), except that colors at neighboring positions alternate.

(l) See page 613.

(m) The pattern can be generated by a 2D substitution system with rule $\{1 \rightarrow \{\{0, 0\}, \{0, 1\}\}, 0 \rightarrow \{\{1, 1\}, \{1, 0\}\}\}$ (see page 583).

(See also page 870.)

Even though standard mathematical functions are used, few of the pictures can readily be generalized to continuous values of x and y.

■ **Intrinsic limits in science.** Before computational irreducibility other sources of limits to science that have been discussed include: measurement in quantum mechanics, prediction in chaos theory and singularities in gravitation theory. As it happens, in each of these cases I suspect that the supposed limits are actually just associated with a lack of correct analysis of all elements of the relevant systems. In mathematics, however, more valid intrinsic limits—much closer to computational irreducibility—follow for example from Gödel's Theorem.

Note: page numbers in the text refer to the original work *A New Kind of Science* (2002).

560

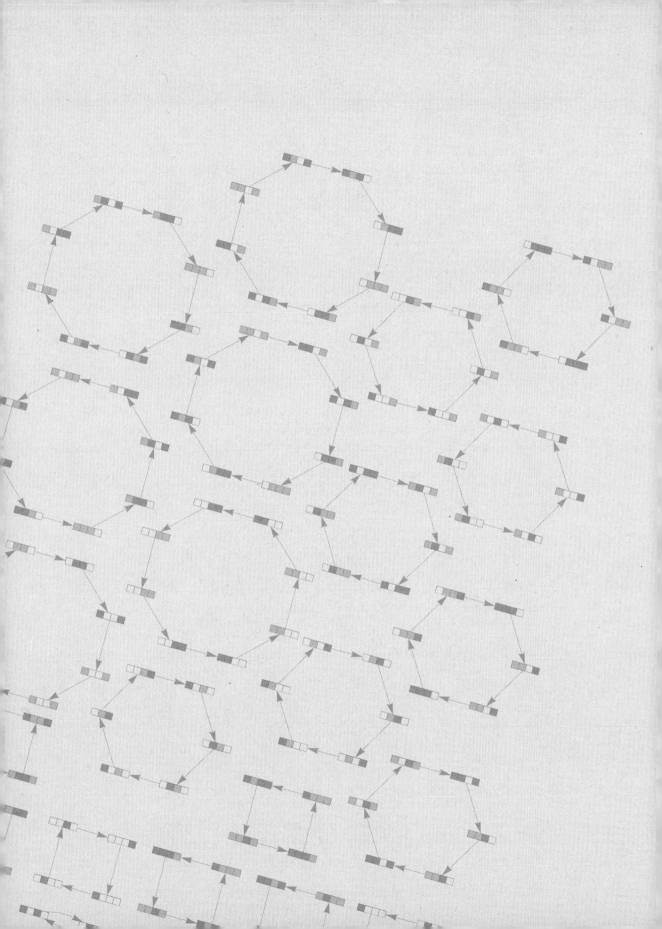

Index

Detailed Sources for Photos and Archival Material

pp. 121–122, 124, 219–221, 396–397: Used with permission of McGraw Hill, from *Statistical Physics*, Frederick Rief, 1967; permission conveyed through Copyright Clearance Center, Inc. **p. 125** © Jim Austin Computer Collection at The Computer Sheds & Neil Barrett Photography. **pp. 154–156:** S. Wolfram (1984), "Computer Software in Science and Mathematics", *Scientific American* September 1984. **p. 157:** S. Wolfram (1985), "Twenty Problems in the Theory of Cellular Automata", *Physica Scripta* 1985, 170. doi: 10.1088/0031-8949/1985/T9/029. **p. 176:** From *The Feynman Lectures on Physics* by Richard P. Feynman, © 2011. Reprinted by permission of Basic Books, an imprint of Hachette Book Group, Inc. **pp. 221–222:** Film created by Berni Alder. **p. 222:** Used with permission of World Scientific Publishing Co., Inc., from *Advances in the Computational Sciences: Proceedings of the Symposium in Honor of Dr Berni Alder's 90th Birthday*, Reic Schwegler & Brenda M. Rubenstein, Eds., 2017; permission conveyed through Copyright Clearance Center, Inc. **p. 224:** Used with permission of the American Association for the Advancement of Science, from "The Liquid State" in *Science*, Volume 80, Issue 2067, 125–133, Joel H. Hindebrand, 1934; permission conveyed through Copyright Clearance Center, Inc. **p. 224:** Used with permission of AIP Publishing, from "Molecular Distribution in Liquids" in *Journal of Chemical Physics*, Volume 7, Issue 10, John G. Kirkwood, 1939; permission conveyed through Copyright Clearance Center, Inc. **p. 225:** Used with permission of AIP Publishing, from "Radial Distribution Functions and the Equation of State of a Fluid Composed of Rigid Spherical Molecules" in *Journal of Chemical Physics*, Volume 18, Issue 8, John G. Kirkwood et al., 1950; permission conveyed through Copyright Clearance Center, Inc. **p. 226:** Used with permission of AIP Publishing, from "Equation of State Calculations by Fast Computing Machines" in *Journal of Chemical Physics*, Volume 21, Issue 6, Nicholas Metropolis et al., 2004; permission conveyed through Copyright Clearance Center, Inc. **p. 226:** Used with permission of AIP Publishing, from "Radial Distribution Function Calculated by the Monte-Carlo Method for a Hard Sphere Fluid" in *Journal of Chemical Physics*, Volume 23, Issue 3, B. J. Alder, S. P. Frankel, V. A. Lewinson, 1955; permission conveyed through Copyright Clearance Center, Inc. **p. 227:** Alder, B. J. and Wainwright, T. E. (1957), "Molecular Dynamics by Electronic Computers", in: Prigogine, I., Ed., *International Symposium on Transport Processes in Statistical Mechanics*, Interscience Publishers (John Wiley & Sons), New York, 97–131. **p. 228:** *Elementary Statistical Physics*, Charles Kittel. © 1958, John Wiley & Sons, Inc. Reproduced with permission of the Licensor through PLSclear. **p. 229:** B. J. Alder and T. E. Wainwright (1959), "Molecular Motions", *Scientific American* October 1959. **p. 230:** Used with permission of AIP Publishing, from "Studies in Molecular Dynamics. I. General Method" in *Journal of Chemical Physics*, Volume 31, Issue 2, B. J. Alder, T. E. Wainwright, 2004; permission conveyed through Copyright Clearance Center, Inc. **p. 231:** AIP Emilio Segrè Visual Archives. **pp. 232–233:** Fermi, E., Pasta, P., Ulam, S., and Tsingou, M. 1955. "Studies of the Nonlinear Problems". United States. doi: 10.2172/4376203. www.osti.gov/servlets/purl/4376203. **p. 310:** Daderot at English Wikipedia; link to license: creativecommons.org/licenses/by-sa/3.0. **p. 312:** Jeremy Norman Collection of Images; link to license: creativecommons.org/licenses/by-sa/4.0. **pp. 393–394:** Used with permission of John Wiley & Sons, from *Thermodynamics*, Herbert Callen, 1959; permission conveyed through Copyright Clearance Center, Inc. **p. 413–416:** S. Wolfram (1985), *Phys. Rev. Lett.*, 54, 735. Copyright (1985) by the American Physical Society. doi: 10.1103/PhysRevLett.54.735. **p. 417–420:** S. Wolfram (1985), *Phys. Rev. Lett.*, 55, 449. Copyright (1985) by the American Physical Society. doi: 10.1103/PhysRevLett.55.449.